T0257900

IET TELECOMMUNICATIONS SERIES 74

Access, Fronthaul and Backhaul Networks for 5G & Beyond

Other volumes in this series:

Access, Fronthaul and Backhaul Networks for 5G & Beyond

Edited by
Muhammad Ali Imran, Syed Ali Raza Zaidi
and Muhammad Zeeshan Shakir

The Institution of Engineering and Technology

Published by The Institution of Engineering and Technology, London, United Kingdom

The Institution of Engineering and Technology is registered as a Charity in England & Wales (no. 211014) and Scotland (no. SC038698).

© The Institution of Engineering and Technology 2017

First published 2017

The Institution of Engineering and Technology
Michael Faraday House
Six Hills Way, Stevenage
Herts SG1 2AY, United Kingdom

www.theiet.org

British Library Cataloguing in Publication Data
A catalogue record for this product is available from the British Library

ISBN 978-1-78561-213-8 (hardback)
ISBN 978-1-78561-214-5 (PDF)

Typeset in India by MPS Limited

Contents

10 NFV and SDN for fronthaul-based systems   231

Maria A. Lema, Massimo Condoluci, Toktam Mahmoodi,
Fragkiskos Sardis, and Mischa Dohler

PART III   Backhaul Network   249

11 Mobile backhaul evolution: from GSM to LTE-Advanced   251

Andy Sutton

18 Fronthaul and backhaul integration (Crosshaul) for 5G mobile transport networks **461**

Xavier Costa-Perez, Antonia Paolicelli, Antonio de la Oliva, Fabio Cavaliere, Thomas Deiß, Xi Li, and Alain Mourad

19 Device-to-device communication for 5G **495**

Rafay Iqbal Ansari, Syed Ali Hassan, and Chrysostomos Chrysostomou

About the editors

Prof. Muhammad Ali Imran is Professor of Communication Systems and Vice Dean at Glasgow College UESTC, University of Glasgow, UK. He was awarded his M.Sc. (Distinction) and Ph.D. degrees from Imperial College London, UK, in 2002 and 2007, respectively. He is an affiliate Professor at the University of Oklahoma, USA, and a visiting Professor at 5G Innovation Centre, University of Surrey, UK. He has over 18 years of combined academic and industry experience, working primarily in the research areas of cellular communication systems. He has been awarded 15 patents, has authored/co-authored over 300 journal and conference publications, and has been principal/co-principal investigator on over £6 million in sponsored research grants and contracts. He has supervised 30+ successful PhD graduates.

He has an award of excellence in recognition of his academic achievements conferred by the President of Pakistan. He was also awarded IEEE Comsoc's Fred Ellersick award 2014, FEPS Learning and Teaching award 2014, Sentinel of Science Award 2016, and he was twice nominated for Tony Jean's Inspirational Teaching award. He is a shortlisted finalist for The Wharton-QS Stars Awards 2014, QS Stars Reimagine Education Award 2016 for innovative teaching and VC's learning and teaching award in University of Surrey. He is a senior member of IEEE and a Senior Fellow of Higher Education Academy (SFHEA), UK.

He has given an invited TEDx talk (2015) and more than 10 plenary talks, several tutorials and seminars in international conferences, events and other institutions. He has taught on international short courses in USA, Pakistan and China. He is the co-founder of IEEE Workshop BackNets 2015 and chaired several tracks/workshops of international conferences. He is an associate Editor for *IEEE Communications Letters, IEEE Open Access* and *IET Communications* journal and has served as guest editor for many prestigious international journals.

Dr Syed Ali Raza Zaidi is a University Academic Fellow (Assistant Professor) at the University of Leeds in the area of wireless communication & sensing systems. From 2013–2015, he was associated with the SPCOM research group working on the United States Army Research Lab funded project in the area of Network Science. From 2011–2013, he was associated with the International University of Rabat working as a Lecturer. He was also a visiting research scientist at Qatar Innovations and Mobility Centre from October–December 2013 working on QNRF funded project QSON. He received his Doctoral Degree at the School of Electronic and Electrical Engineering at Leeds and was awarded the G. W. and F. W. Carter Prize for best thesis and best research paper. He has published more than 90 papers in leading IEEE conferences and journals. From 2014–2015, he served as an editor for *IEEE Communication Letters*. He was also lead guest editor for *IET Signal Processing* journal's special issue on "Signal Processing for Large Scale 5G Wireless Networks." Currently, he is an Associate Technical Editor for *IEEE Communication Magazine*. Dr. Zaidi is EURASIP Local Liaison for United Kingdom and also General Secretary for IEEE Technical Subcommittee on Backhaul and Fronthaul networks. He is also an active member of EPSRC Peer Review College.

Dr Muhammad Zeeshan Shakir is an Assistant Professor in Networks in the School of Engineering and Computing at the University of West Scotland (UWS), UK. His research interests include design, development and deployment of diverse wireless communication systems, including hyper-dense heterogeneous small-cell networks, Green networks and 5G technologies such as D2D communications, Networked-flying platforms (NFPs) and IoT. He has published more than 75 technical journal and conference papers and has contributed to 7 books, all in reputable venues. He is an editor of two research monographs and an author of a book entitled *Green Heterogeneous Wireless Networks* published jointly by Wiley and IEEE Press. He has been/is serving as a Chair/Co-Chair/Member of several workshops/special sessions and technical program committee of different IEEE flagship conferences, including Globecom, ICC, VTC and WCNC. He is an Associate Technical Editor of *IEEE Communications Magazine* and has served as a lead Guest Editor/Guest Editor for *IEEE Communications Magazine*, *IEEE Wireless Communications* and *IEEE Access*. He is serving as a Chair of IEEE ComSoc emerging technical committee on backhaul/fronthaul networking and communications. He is a Senior Member of IEEE, an active member of IEEE ComSoc and IEEE Standard Association.

Preface

The recent widespread use of mobile Internet, complemented by the advent of many smart applications has led to an explosive growth in mobile data traffic over the last few years. This remarkably growing momentum of the mobile traffic will most likely continue on a similar trajectory, mainly due to the emerging need of connecting people, machines, and applications in a ubiquitous manner through the mobile infrastructure. As a result, the current and projected dramatic growth of mobile data traffic necessitates the development of fifth generation (5G) mobile communications technology.

The 5G communications would provide us with the promise of a mobile broadband experience far beyond current 4G systems. The 5G has a broad vision and envisages design targets that include $10–100\times$ peak date rate, $1,000\times$ network capacity, $10\times$ energy efficiency, and $10–30\times$ lower latency. In achieving these expectations, operators and carriers are planning to leverage metro, pico, nano and elastic cells to improve the user experience and, consequently, improve the overall network performance. Efficient and satisfactory operation of all these densely deployed cells hinges on a suitable backhaul and fronthaul provisioning. Hence, there is considerable market interest for the development of innovative/smart backhaul and fronthaul solutions for ultra-dense heterogeneous networks.

It is envisioned that the 5G networks would be mostly deployed for data-centric applications. Therefore, one of the main considerations which the operators are facing today is how to migrate existing backhaul/fronthaul infrastructure towards Internet Protocol (IP)-based solutions for hyper dense small-cell deployment. Moreover, the data rates of the 5G networks will demand the optical backhaul/fronthaul such as fibre. However, it's unlikely that fibre will be economical for all installation sites. Operators will also face deployment restrictions for laying the fibre in many developed areas. Millimetre wave backhaul/front haul is an attractive option, but technological and regulatory challenges are yet to be addressed. Another possible emergent solution is to exploit the interworking and joint design of open-access and backhaul/fronthaul network architecture for hyper dense small cells based on cloud networks. This requires adaptive and smart backhauling/fronthauling solutions that optimise their operations jointly with the access network optimisation. The availability, convergence, and economics of smart backhauling/fronthauling systems are the most important factors in selecting the appropriate backhaul/fronthaul technologies for multiple radio access technologies (including small cells, relays and distributed antennas) and heterogeneous-types of excessive traffic in the future cellular network. Hence, it is imperative to analyse the variety of end-to-end backhaul/fronthaul solutions for

5G networks. Various standard bodies such as Next Generation Mobile Networks (NGMN), Metro Ethernet Forum (MEF) and Broadband Forum (BBF) are also studying the implications on their defined mobile backhaul network architectures. Recently, the European Telecommunications Standards Institute (ETSI) has started a new specification group to study the feasibility of the V-band (57–66 GHz), the E-band (71–76 and 81–86 GHz) and, in the future, higher frequency bands (up to 300 GHz) for large volume backhaul and fronthaul applications to support future mobile network.

Development of 5G technologies including backhaul/fronthaul and its standardisation is a mission-critical task. Since the entire research community is working against an extremely tight timeline to provide 5G innovative technologies with extensive performance evaluation metrics by year 2017/2018 along with required standardisations milestones achieved, 5G hardware and components by 2019 and a fully deployed network by 2020. All these steps are interrelated and there is little or no room for any timeline slippage. There is already an international competition to achieve these goals since early winners will capture the major portions of the market and will be more difficult to unseat.

This book serves as a main platform to further explore the goals and targets set by the research community working in this challenging field. In particular, the key objective is to provide a comprehensive research driven proceedings for both academic and industrial stakeholders to further innovate the future backhaul/fronthaul solutions. The book intends to cover wide spectrum of underlying themes ranging from the recent thrust in edge caching for backhaul relaxation to mmWave based fronthauling for virtualised radio access networks. The book is organised in four main parts: The first part deals with the technological advances in the Access Network and their implications on backhaul and fronthaul technologies. The second part focuses on the Fronthaul Network technologies and the third part on the Backhaul Network technologies. The final part, System Integration and Case Studies, presents the system level and integration perspective for different conventional and futuristic scenarios and also sheds light on some insights gained from a deployed network.

Unlike existing wireless generations for which video has proven to be a killer application, it is envisioned that augmented reality, virtual reality, haptic communication and tele-robotics will be the key killer applications for 5G networks. This new breed of applications require ultra-high throughput at sub 10 ms latency. From Access networks perspective this would mean that: (i) the cell size must shrink to enhance received signal-to-interference-plus-noise ratio (SINR) and (ii) density of deployment must increase by several folds to increase the spatial reuse of limited available spectrum. Both of these design factors need to be attained without significant increase in the aggregate interference. The first three chapters are geared towards the solutions which are emerging as key enablers for addressing these design challenges.

In Chapter 1, Van Nguyen presents comprehensive overview of network densification techniques. The author highlights approaches for modelling ultra-dense network deployments considering spatio-temporal dynamics of propagation channel. Scaling laws for link and network level performance are derived. A key finding is that under strongest cell association, the performance of ultra-dense wireless networks exhibits three distinct regimes of the user performance, namely growth, saturation and deficit regime. The tail behaviour of the channel power and near-field path loss exponent are the key parameters that determine the performance limits.

In Chapter 2, Chen presents an overview of Massive and Network Multiple-Input-Multiple-Output (MIMO) systems. While network densification relies on smaller cell size to physically bring the base station closer to the user, Massive MIMO exploits diversity gain to realise the same effect. In addition, due to increased array size on the base station the interference resilience of Massive MIMO systems is significantly higher than single antenna small-cell base stations. Nevertheless, it is well known that the throughput of Massive MIMO systems saturates beyond certain operational point. Small-cell networks can then complement Massive MIMO systems. Massive MIMO system can also be formed by networking distributed MIMO base station. Under the split plane architecture (discussed in Chapter 17) it is then possible to construct user-centric Networked Massive MIMO systems. In Chapter 2, the author highlights the fundamental aspects of Massive and Network MIMO systems. Uplink Channel estimation, Uplink signal detection and decoding and Downlink precoding are explored in detail. Impairments such as pilot contamination and their impact on time division duplex (TDD) Massive MIMO system have also been investigated.

In Chapter 3, Ksairi *et al.* provide industrial perspective on Massive MIMO systems. The chapter expands on previous discussion by highlighting the issues mentioned in Chapter 2. The authors also provide detailed insight on design of recent advances in the area of frequency division duplex (FDD) Massive MIMO where acquisition of channel state information is more challenging than the traditional the counterpart TDD system.

In Chapter 4, Shrivastava *et al.* reflect on the evolution of time division duplex mobile networks towards 5G. In particular authors provide detailed account of state-of-the-art techniques which 3GPP has proposed for dynamic uplink–downlink reconfiguration in TDD networks by selecting one of the seven proposed frame structures. This can potentially enable network to respond to the changes in traffic requirements on the fly. Also, it enables capability to serve heterogeneous traffic resulting from the mix of traditional and new machine type communication. The authors provide a detailed account of software defined architecture for TDD-LTE network, resource management logic and algorithms for virtual cell creation and reconfiguration of uplink–downlink parameters.

In Chapter 5, Safdar explores the issue of cross-tier interference which is experienced by mobile users due to deployment of small cells. While small-cell deployments improve link margin they are also susceptible to increased cross-tier and co-tier interference. The aggregate co-channel interference needs to be engineered so that the gains rendered via aggressive spatial reuse are not offset by the interference. While co-channel transmitters can employ power control to transmit at lower transmit powers, the cross-tier interference can be significantly higher. In this chapter, authors propose utilising the spectrum interweave technique to mitigate cross-tier interference. The technique relies upon the whitespace of incumbent macro-cells which can be exploited as transmission opportunities for secondary small cellular network deployment.

In Chapter 6, Simsek *et al.* outline the architectural aspects and technical requirements of the Tactile Internet. Tactile Internet or the internet-of-robots require ultra-low latency to enable applications like tele-surgery. In this chapter, authors identify the core design challenges and sketch directions for addressing these design challenges. Joint coding method to the simultaneous transmission of data over multiple independent wireless links is presented to improve latency and throughput. Moreover, radio resource slicing and application-specific customisation techniques are introduced to enable the co-existence of different vertical applications along with Tactile Internet applications.

The second part of this book is dedicated to the recent advancements in front-hauling solutions. Specifically 5G wireless networks will be geared to dynamically allocate resources in on-demand manner. This has led to proposals for decoupling of control signalling from the user data plane. Such architecture not only treats network resources as soft entities but indeed also relies upon virtualisation of network functionality itself. More specifically, the remote radio heads (RRHs) can be separated from the base band units (BBUs) enabling resource pooling, interference and transmission coordination in a central manner. RRHs are then connected to BBUs using wired or wireless front-haul. Chapters 7–10, presented in this section, provide a comprehensive discussion on various front-hauling solutions which are promising candidates for 5G wireless networks.

In Chapter 7, Lema Rosas *et al.* present a detailed discussion on practical aspects of front-hauling in next generation wireless networks. In particular authors present how centralised radio access networks are evolving and how network function virtualisation (NFV) can enable various splits across air interface protocols. It is shown that how existing backhaul solutions can be adapted for next generation fronthauling and what are key requirements for adapting these interfaces.

In Chapter 8, Zhang *et al.* tackle the topic of interference management and resource allocation in backhaul and access networks using the concept of emerging Cloud RAN. They present an approach to optimise the cluster size of cooperating cells in a Cloud RAN cluster assuming linear precoding algorithms (ZF and MRT) under the assumption of CSI latency. They also present a novel framework for joint path selection and scheduling algorithms for a cluster of small cells connected via mmWave backhaul. The proposed algorithms show significant performance improvement in conjunction with the efficient interference coordination strategies like dynamic clustering.

In Chapter 9, Jaber *et al.* argue why self-organised fronthaul is needed for the 5G and beyond systems. The consideration of the fronthaul requirements (e.g., capacity, latency, jitter and synchronisation) are important for the suitable technologies that can meet the expectations of the next generation network. The chapter further delves into the C-RAN/fronthaul versus D-RAN/backhaul dilemma, offering a joint backhaul/RAN perspective and tangible trade-off analysis of the available options. The chapter argues that the joint RAN/fronthaul optimisation has a pivotal role in adjusting the level of centralisation according to the fronthaul capabilities and vice-versa, potentially using SON and software-defined networks (SDN).

In Chapter 10, Condoluci *et al.* present the role of NFV and SDN for the fronthaul-based systems. They present the case for a flexible and dynamic topology in order to ensure the correct usage of the promising C-RAN technology. They discuss how virtualisation of the network functions allows to satisfy the high-level heterogeneity in the network while allowing the use of several "functional splits" ensuring that the cell cooperation is configured at the desired level.

Front-haul and access network relies upon low latency backhaul for providing ultra-low latency. To this end, the third part of this book is dedicated to back-haul networks (BackNets).

In Chapter 11, Sutton presents a detailed account of the evolution of the backhaul technologies from the early days of GSM network to the latest state-of-the-art LTE-Advanced systems with an outlook of the future 5G and beyond systems. This chapter presents a detailed tutorial account of the pros and cons of different generations of the backhaul and how this architectural evolution played a key role in meeting the generational requirements of the networks.

In Chapter 12, Alves and Souza present the case for the wired versus the wireless backhaul options. They support the idea that there is no "one-size-fits-all" solution for this and the emerging network will be a hybrid of the two options. One of the key challenges for future wireless networks design is to choose the solution that offers the best compromise in terms of performance, cost and interoperability for each particular case.

In Chapter 13, Sharma *et al.* explore the topic of spectral coexistence for the next generation of wireless backhaul networks. They provide a detailed review of terrestrial and the satellite backhaul technologies and then present various spectrum sharing techniques. They analyse their challenges and potential for application in practical scenarios through their presentation of a series of case studies. They conclude with a vast range of potential future directions for the hybrid satellite-terrestrial backhauling networks.

In Chapter 14, Mohamed *et al.* discuss the implications on the backhaul networking in order to support the futuristic architecture of a control-data functionality separation in cellular networks. In contrast to the promised gains of the new architecture, there are several challenges and constraints for the future backhaul which are highlighted and analysed in this chapter. The main conclusion is that the new architecture requires a careful deployment of both control and data base stations which is aware of the constraints and potentials of the available backhaul network technologies.

In Chapter 15, Bahmani *et al.* present the significance of content centric network architecture to manage the ever-increasing mobile traffic problem faced by mobile operators mainly due to the video content delivery. An overview of the several content placement approaches is presented to effectively cache the content in mobile networks. Later, the chapter introduces the content placement problem in 5G heterogeneous networks to achieve QoS, backhaul relaxation and energy efficiency. Several simulation scenarios have been presented to discuss the performance of the system in terms of QoS, backhaul relaxation and energy efficiency.

The fourth and the final part of this book provides comprehensive discussion on overall system integration and presents case studies from practical deployments.

In Chapter 16, Hussein *et al.* present a detailed overview of the stringent requirements of the time-critical applications in future heterogeneous networks such as low latency, higher data rate and improved QoS. A new layered architecture is proposed by integrating the centralised and distributed paradigms such as SDN, NFV and edge computing and addressing some of the key requirements of the backhauling/fronthauling in 5G networks. This chapter also briefly presents some challenges to implement such unified architecture in future networks such as scalability, security and privacy.

In Chapter 17, Kourtessis *et al.* discuss a novel SDN architecture featuring SAT>IP and Cached video services, an SDN controllable PHY layer and a SDN enabled access infrastructure that allows cellular mobile, PON and fixed wireless networks to operate independently using their own network controller via network slicing. Extensive discussions and experimental illustrations are provided to stress on the SDN-enabled network optimisation and video content delivery while maintaining the strict latency and demands requirements. Fronthauling approaches are also proposed to allow streaming of SAT>IP content over SDN enabled Xhaul.

In Chapter 18, Costa *et al.* discuss the integration of fronthaul and backhaul for 5G mobile networks by integrating number of interfaces in the data, control and application planes. Followed by an extensive overview of the emerging technologies to design the fronthaul/backhaul for 5G networks, this chapter presents 5G-Crosshaul architecture with features such as decoupled data and control plane; centralised control plane and exposure of state and abstract resources to applications. Detailed interactions between the architecture components are presented and discussed. Enabling applications are discussed in detail followed by some standardisation perspectives and future roadmap of the proposed SDN driven integrated Crosshaul 5G architecture.

In Chapter 19, Ansari *et al.* present a detailed overview of the D2D communication in the context of deployment and its integration with the 5G networks. Several deployment topologies and traffic offloading techniques are presented. Detailed overview of the resource management, interference cancellation and management are also presented. Later, the chapter presents several smart features to be integrated with the future D2D communications such as energy harvesting, pricing/incentives mechanisms and social trust enabled pairing of devices.

In Chapter 20, Bassoy *et al.* present a deployment use case of CoMP from operator perspectives and overview different schemes and their challenges to integrate CoMP with 3GPP standardisation. Later, this chapter presents CoMP trials and discuss operational requirements and implementation challenges of various CoMP schemes operating over LTE-A in the UK. Outcome of the trials are discussed in terms of SINR improvement, throughput increment and reduction in user power, particularly for cell edge users. The chapter concluded by highlighting some limitations of presented CoMP schemes such as dynamic clustering, network load balancing and energy efficiency.

Part I

Access Network

Chapter 1

Network densification

Van Minh Nguyen[1]

Abstract

Network densification is a promising cellular deployment technique that leverages spatial reuse to enhance coverage and throughput. Recent work has identified that at some point ultra-densification will no longer be able to deliver significant throughput gains. Throughout this chapter, we provide a unified treatment of the performance limits and from which shed light on how to leverage the potential of network densification. We firstly show that there are three scaling regimes for the downlink signal-to-interference-plus-noise ratio (SINR), *coverage probability*, and *average rate*. Specifically, depending on the near-field pathloss and the fading distribution, the user performance of ultra dense networks would either monotonically increase, saturate, or decay with increasing network density. Secondly, we show that network performance in terms of *coverage density* and *area spectral efficiency* can benefit from increased network infrastructure better than the user performance does. Furthermore, we analytically prove that enhancing the tail distribution of channel power is a fundamental way to leverage the benefit of network densification.

1.1 Introduction

Mobile traffic has significantly increased over the last decade mainly due to the stunning expansion of smart wireless devices and bandwidth-demanding applications. This trend is forecast to be maintained, especially with the deployment of the Fifth Generation (5G) and beyond networks and machine-type communications. In this context, industry and research community have looked for potential solutions to greatly increase the network capacity. Among those, network densification is a network deployment method that ultimately exploits spatial frequency reuse of the cellular network concept. It basically consists in deploying ultra dense network of small cells either in a heterogeneous architecture with conventional macro cells, or as a standalone hot-spot network.

---

[1]Mathematical and Algorithmic Sciences Laboratory, Huawei Technologies France SASU, France

Back to the past, dense network deployment using the so-called picocells was known as early as the beginning stage of the Second Generation. However, network densification has never received as high attention and expectation as we know today. It is firstly driven by the exponential growth of data traffic while a major part of the mobile throughput growth during the past few years has been enabled by the network densification. It is advocated that network densification will be the main technology enabler for achieving the 5G requirement of 1000× increase in mobile network data throughput compared to 3GPP Long-Term Evolution (LTE). Moreover, recent advances on transport networks, such as high-capacity optical, millimeter wave (mmWave) communication and directional beamforming, may provide reliable and high-capacity links between the core network and small cells. Their development and deployment has been therefore gaining momentum in both wireless industry and research community during the last few years [1,6,12,13,19]. In particular, a Google search with 'network densification' keyword quickly provides a massive amount of general information and discussions. It has also attracted the attention of standardization bodies such as 3GPP LTE-Advanced.

The underlying foundation of the above outlook is the presumed linear capacity scaling with the number of small cells deployed in the network. Nevertheless, there has been noticeable divergence between this outlook and conclusion of various network studies according to which densification is not always beneficial to the network performance. Recent and often conflicting findings based on various modeling assumptions [2,3,9,14–16,24] have identified that densification may eventually stop at a certain point delivering significant throughput gains.

In this chapter, we aim at providing an answer to whether there are any fundamental limits to network densification due to physical limits arising from electromagnetic propagation. In particular, we aim at answering the following questions:

- *How does network performance scale with the network infrastructure?*
- *What should be done to leverage the potential of network densification?*

In what follows, we investigate these questions from a technical point of view. We will firstly develop a comprehensive model of the network in such a way that it is generic to provide a complete characterization of the system, and is precise to correctly capture physical properties. Based on that, we develop fundamental limits of user performance, and then of network performance under the effect of network densification. After that, we investigate the above second question by providing performance ordering results. Finally, we summarize the main figures of network densification. Some related mathematical tools are provided in the chapter appendix.

1.2   Modeling methodology

In wireless communications, radio links are susceptible to time-varying channel impediments, interference, and noise. It includes long-term attenuation due to pathloss, medium-term variation due to shadowing, and short-term fluctuations due to multi-path fading. In principle, these three main variations are usually taken into account by different network processing levels. Network planning usually depends on the effect of pathloss, while radio link-level procedures such as power control and handover aim at compensating for the combining effect of pathloss and shadowing, and fast fading is tackled by symbol-level processing at the physical layer.

However, incorporating meaningfully the aforementioned propagation phenomena in network modeling often leads to complex and cumbersome mathematical derivations. As a result, most existing network analyses, especially the ones using stochastic geometry, has nearly always assumed power law pathloss and Rayleigh fading due to their tractability. Although some works investigated the effect of pathloss singularity, the effect of shadowing is usually ignored and other simplifications are employed for ease of analysis. This has led to unexpected observations in specific scenarios, as well as to divergent or even contrasting conclusions in prior work on the fundamental limits of network densification. Furthermore, the standard assumption of exponential distributed channel power will not apply in a scenario with directional transmissions and multi-antenna processing, as are envisioned for massive MIMO (multiple-input-multiple-output), coordinated multipoint (CoMP), and mmWave systems. Advanced communication and signal processing techniques are expected to enhance the channel gain, which in some cases may have a diffuse power component [10] or regularly varying tail (its definition is given in the sequel) [22]. Therefore, it necessitates to develop a more precise and generic model for path loss, shadowing, as well as general fading.

1.2.1 Pathloss modeling

Let $l : \mathbb{R}_+ \to \mathbb{R}_+$ represent the path loss function where $\mathbb{R}^+ = \{x \in \mathbb{R} \mid x \geq 0\}$, the receive power P_{rx} is related to the transmit power P_{tx} by $P_{\text{rx}} = P_{\text{tx}}/l(r)$ with r being the transmitter–receiver distance. Physics laws impose that $1/l(r) \leq 1, \forall r$. However, in the literature, $l(\cdot)$ has been usually assumed to admit a singular power-law model, i.e., $l(r) \sim r^\beta$ where β is the pathloss exponent satisfying $\beta \geq d$. This far-field propagation model has been widely used due to its tractability. However, for short ranges, especially when $r \to 0$, this model is no longer relevant and becomes singular at the origin. This is indeed more likely to happen in the context of network densification where the inter-site distance becomes smaller. In addition, the dependence of the pathloss exponent on the distance in urban environments and in mmWave communications advocates the use of a more generic pathloss function. These requirements can be satisfied by modeling the pathloss function as follows (see Figure 1.1):

$$l(r) = \sum_{k=0}^{K-1} A_k r^{\beta_k} \, \mathbf{1} \, (R_k \leq r < R_{k+1}), \tag{1.1}$$

where $\mathbf{1}(\cdot)$ is the indicator function, $K \geq 1$ is a given constant characterizing the number of pathloss slopes, and R_k are constants satisfying:

$$0 = R_0 < R_1 < \cdots < R_{K-1} < R_K = R_\infty, \tag{1.2}$$

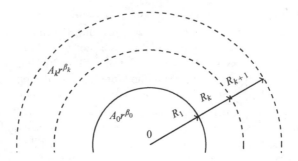

Figure 1.1   Multi-slope pathloss model

where R_∞ represents the maximum distance from a user to its farthest base station in the network (more details about R_∞ are discussed in the network modeling part). Constant β_k denotes the pathloss exponent satisfying:

$$\beta_0 \geq 0, \tag{1.3a}$$

$$\beta_k \geq d - 1, \quad \text{for } k = 1, \ldots, K - 1, \tag{1.3b}$$

$$\beta_k < \beta_{k+1} < \infty, \quad \text{for } k = 0, \ldots, K - 2, \tag{1.3c}$$

and A_k are constants to maintain continuity of $l(\cdot)$, i.e.,

$$A_k > 0 \quad \text{and} \quad A_k R_{k+1}^{\beta_k} = A_{k+1} R_{k+1}^{\beta_{k+1}}, \tag{1.4}$$

for $k = 0, \ldots, K - 2$. For notational simplicity, we also define $\delta_k = d/\beta_k$, for $k = 0, \ldots, K - 1$.

The above general multi-slope model (see (1.1)) captures the fact that the pathloss exponent varies with distance, while remaining unchanged within a certain range. In principle, free-space propagation in $\mathbb{R}^3$ has pathloss exponent equal to 2 (i.e., $\beta = d - 1$), whereas in realistic scenarios, pathloss models often include antenna imperfections and empirical models usually result in the general condition (1.3b) for far-field propagation. Condition (1.3c) models the physical property that the pathloss increases faster as the distance increases. Notice, however, that this condition is not important in the subsequent analytical development. Finally, condition (1.3a) is related to the near field (i.e., it is applied to the distance range $[0, R_1]$).

The pathloss function as defined above has the following widely used special cases:

- $K = 1$, $\beta_0 \geq d$: $l(r) = A_0 r^{\beta_0}$, which is the standard singular (unbounded) pathloss model;
- $K = 2$, $\beta_0 = 0$: $l(r) = \max(A_0, A_1 r^{\beta_1})$, which is the bounded pathloss recommended by 3GPP, in which A_0 is referred to as the minimum coupling loss.

Due to the particular importance of pathloss boundedness for having a realistic and practically relevant model, we have the following definition.

Definition 1.1. *A pathloss function $l : \mathbb{R}^+ \to \mathbb{R}^+$ is said to be* bounded *if and only if $1/l(r) < \infty, \forall r \in \mathbb{R}^+$, and* unbounded *otherwise. Furthermore, the pathloss function $l(\cdot)$ is said to be* physical *if and only if $1/l(r) \leq 1, \forall r \in \mathbb{R}^+$.*

It is clear that the pathloss function (1.1) is bounded if and only if (iff) $\beta_0 = 0$, and is physical iff $\beta_0 = 0$ and $A_0 \geq 1$.

1.2.2 Channel power modeling

Besides pathloss attenuation, shadowing, and small-scale fading – commonly referred to as *fading* in the sequel – are additional sources of wireless link variation. It has been usually adopted that the shadowing effect is modeled by a lognormal distribution of the channel power while fast fading is modeled by a Rayleigh distribution of the channel amplitude. Not only such abstraction is not always verified in the field, but more importantly, mmWave propagation exhibits higher directional and blockage property as compared to conventional cellular frequency bands. Fading thus may follow a different distribution. Furthermore, emerging utilization of advanced communication and signal processing techniques is expected to enhance the channel gain, which in some cases may have a diffuse power component or regularly varying tail. Therefore, it is desirable to consider a more general class of fading in the analysis of emerging dense networks, and we will do so.

Precisely, let m_i be a variable containing all other propagation phenomena and transmission gains except pathloss, including transmit power, small-scale fading, shadowing, and gains due to antenna pattern, beamforming, etc. from ith node to the user, and let us refer to m_i as *channel power*.

The variables $\{m_i\}$ are assumed not identical to zero and independently distributed according to some distribution F_m. Doing so, we provide the ability to capture either the separate effect of small scale fading and of shadowing, or the combining effect of the composite fast fading-shadowing. Our general framework enables us to model any fading distribution and channel gain that can be observed in current wireless communications and networking.

For subsequent analysis, we will classify the space of fading, or any non-degenerate distribution in general, basing on its tail behavior. Precisely, we distinguish between the class of distributions that have *regularly varying tails* and the other class that contains all the remaining distributions, which have lighter tails.

Definition 1.2 (Regular varying function in Karamata's sense). *A positive, Lebesgue measurable function h on $(0, \infty)$ is called* regularly varying *with index $\alpha \in \mathbb{R}$ at ∞ if $\lim_{x \to \infty} h(tx)/h(x) = t^\alpha$ for $0 < t < \infty$. In particular, h is called* slowly varying *(resp.* rapidly varying*) (at ∞) if $\alpha = 0$ (resp. if $\alpha = -\infty$). We denote by $\mathcal{R}_\alpha$ the class of regularly varying functions with index α.*

Note that if h is a regularly varying function with index α, it can be represented as $h(x) = x^\alpha L(x)$ as $x \to \infty$ for some $L \in \mathcal{R}_0$. Moreover, if a function h is regularly

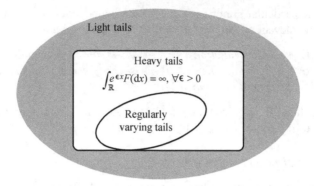

Figure 1.2   Classification of tail behavior

varying with index α, $\alpha \in (-\infty, 0]$, then it is heavy-tailed. In Figure 1.2, we provide an illustration of the tail behavior classes; the interested reader is referred to [7,11,17].

For upcoming numerical examples, the channel power in the case of well-known Rayleigh fast fading and lognormal shadowing is defined as follows.

Definition 1.3 (Composite fading). Composite *fading, say* $X_{\text{Composite}}$, *is defined as the superposition of Rayleigh fast fading, say* X_{Rayleigh}, *and lognormal shadowing, say* $X_{\text{Lognormal}}$, *i.e.,*

$$X_{\text{Composite}} \stackrel{\text{def}}{=} X_{\text{Rayleigh}}^2 X_{\text{Lognormal}}. \tag{1.5}$$

Note that lognormal shadowing, Rayleigh fast fading, and composite fading are all in the class of the rapidly varying tails, $\mathcal{R}_{-\infty}$. Finally, the following definition is useful for tail classification.

Definition 1.4 (Tail-equivalence). *Two distributions F and H are called* tail-equivalent *if they have the same right endpoint, say* x_∞, *and*

$$\lim_{x \uparrow x_\infty} \bar{F}(x)/\bar{H}(x) = c \quad \text{for } 0 < c < \infty.$$

1.2.3 Network modeling

Unlike path loss which has been differently modeled in the literature, cellular networks and dense networks in particular have been commonly modeled by using tools from stochastic geometry in which base stations are located according to a Poisson point process [4]. Physically, this means that base station locations are uniformly random and mutually independent of one another. Although it is not exactly the case in reality, it has been shown by different network analyses that this assumption does not introduce important discrepancy to a perfectly regular and deterministic model. Moreover, suitability of this mathematical abstraction can be reinforced particularly for dense networks due to the fact that network planification of a large number of small base stations is complex, making the resulting network closer to a random distribution model.

To this end, consider a typical downlink user located at the origin and that the network is composed of cell sites located at positions $\{\mathbf{x}_i, i = 0, 1, \ldots\}$. For convenience, cell sites are referred to as nodes, whereas the *typical user* is simply referred to as *user*. As previously discussed, $\{\mathbf{x}_i\}$ are assumed to be random variables independently distributed on the *network domain* according to a homogeneous Poisson point process (PPP) of intensity λ, denoted by Φ. In most existing works, the entire d-dimensional Euclidean space $\mathbb{R}^d$, where $d = 2$ is usually assumed as network domain. However, network domain is in reality limited and far-away nodes are less relevant to the typical user due to path loss, it turns out to be more suitable to consider that the distance from the user to any node is upper bounded by some constant $0 < R_\infty < \infty$. Each node transmits with some power that is independent of the others but is not necessarily constant.

As a matter of seeking the fundamental limits of network densification, the network is considered *fully loaded* in the sequel. Note, however, that the case of partially loaded network can be straightforwardly applied by considering that the network intensity is $\rho\lambda$ in which $0 \leq \rho \leq 1$ is the network load ratio.

The signal power, say P_i, that the user receiving from ith node is expressed as

$$P_i \overset{\text{def}}{=} m_i/l(||\mathbf{x}_i||),$$

where $||\cdot||$ is the Euclidean distance. Using the above notation, the quality of the signal received from ith node, denoted by Q_i, is expressed in terms of its signal-to-interference-plus-noise ratio (SINR) as

$$Q_i \overset{\text{def}}{=} P_i/(I_i + W),$$

where $I_i = \sum_{j \neq i} P_j$ is the aggregate interference with respect to ith node's signal, and thermal noise at the user's receive antenna is assumed Gaussian with average power W. In addition, we define

$$I \overset{\text{def}}{=} \sum_j P_j$$

to be the total interference, and

$$M \overset{\text{def}}{=} \max_i P_i,$$

which is physically the signal power that is the strongest between all the nodes received by the user.

1.2.4 User association modeling

SINR of a user is defined with respect to its serving cell, which in turn depends on the underlying user association scheme. Nearest base station association has been widely employed in the literature of stochastic geometry based analyses mainly due to its mathematical convenience. By contrast, *strongest cell association* is the realistic and practically relevant scheme in which the user is connected to the cell site that provides the best signal quality or the strongest signal strength. Note that if $SINR_{max}$ and $SINR_{near}$ are the user's SINR under strongest cell association and nearest base station association, respectively, then by construction

$$\mathbb{P}(SINR_{near} \geq t) \leq \mathbb{P}(SINR_{max} \geq t), \quad \forall t \in \mathbb{R}.$$

It can be easily shown that equality holds only if pathloss is not decreasing with respect to the distance and the channel power is a deterministic constant (implying that fading is absent or is not considered). In other words, nearest base station association is an under-approximation of the strongest cell association unless channel power is constant. Therefore, strongest cell association is the most appropriate scheme—both in a practical and theoretical sense—to assess the maximum achievable performance.

In light of that, strongest cell association is considered in the sequel, and assuming each base station gives orthogonal resources (e.g., OFDMA) to users associated with it, the SINR of the typical user is given by

$$SINR = \max_{i \in \Phi} Q_i, \tag{1.6}$$

and can be expressed as [17,21]:

$$SINR = \frac{M}{I + W - M}. \tag{1.7}$$

1.2.5 Notation

Quantities whose dependence on the density λ is analyzed are denoted by $\cdot(\lambda)$, e.g., $SINR(\lambda)$, $I(\lambda)$, and $M(\lambda)$. We also denote by r, m, and P, the distance, associated channel power, and received power from a random node, respectively. The distribution function of P is denoted by F_P, and $\bar{F}_P = 1 - F_P$.

We also use notation $\xrightarrow{d}$, $\xrightarrow{p}$, and $\xrightarrow{a.s.}$ to denote the convergence in distribution, convergence in probability, and almost sure (a.s.) convergence, respectively. Notations $\mathbb{P}(\cdot)$ and $\mathbb{E}(\cdot)$ are, respectively, the probability and the expectation operators.

In addition, for real functions f and g, we say $f \sim g$ if $\lim_{x \to \infty} (f(x)/g(x)) = 1$, and $f = o(g)$ if $\lim_{x \to \infty} (f(x)/g(x)) = 0$.

Finally, real intervals formed by numbers $a, b \in \mathbb{R}$ are denoted as follows: $(a, b) = \{x \in \mathbb{R} \mid a < x < b\}$, $[a, b) = \{x \in \mathbb{R} \mid a \leq x < b\}$, $(a, b] = \{x \in \mathbb{R} \mid a < x \leq b\}$, and $[a, b] = \{x \in \mathbb{R} \mid a \leq x \leq b\}$.

1.3 User performance scaling laws

The performance metrics used to characterize user performance are the coverage probability and the average rate that the user experiences from its *serving cell*. Precisely, The *SINR coverage probability*, denoted by $\mathcal{P}_y$, is defined as the probability that SINR is larger than a given target value y, and the *average rate*, denoted by $\mathcal{C}$, is defined as the Shannon rate (in nats/s/Hz) assuming Gaussian codebooks, i.e.,

$$\mathcal{P}_y = \mathbb{P}(SINR \geq y) \quad \text{and} \quad \mathcal{C} = \mathbb{E}(\log(1 + SINR)). \tag{1.8}$$

User performance is determined by its SINR, which in turns jointly depends on the useful signal M and the total interference I. By densifying the network, we increase M but also I (in probabilistic sense). Therefore, the way how M and I scale with the network infrastructure decides the final effect of network densification. The following theorem makes a primary step for understanding the relationship between M and I.

Theorem 1.1 ([20]). *The tail distribution of the received signal power $\bar{F}_P$ depends on the tail distribution of the channel power $\bar{F}_m$ and the pathloss function $l(r)$ as follows:*

- *If $\bar{F}_m \in \mathcal{R}_{-\alpha}$ with $\alpha \in [0, \infty]$, then $\bar{F}_P \in \mathcal{R}_{-\rho}$ where $\rho = \min(\delta_0, \alpha)$ with the convention that $\delta_0 = +\infty$ for $\beta_0 = 0$, and $\min(+\infty, +\infty) = +\infty$;*
- *If $\bar{F}_m(x) = o(\bar{H}(x))$ as $x \to \infty$ with $\bar{H} \in \mathcal{R}_{-\infty}$, then $\bar{F}_P(t)$ and $\bar{F}_m(A_0 t)$ are tail-equivalent for $\beta_0 = 0$, and $\bar{F}_P \in \mathcal{R}_{-\delta_0}$ for $\beta_0 > 0$.*

Theorem 1.1 shows that the tail behavior of the wireless link depends not only on whether the pathloss function is bounded or not, but also on the tail behavior of the channel power. More precisely, a key implication of Theorem 1.1 is that pathloss and channel power have interchangeable effects on the tail behavior of the wireless link. This can also be shown using Breiman's Theorem [8] and results from large deviation of product distributions. Specifically, if the channel power is regularly varying with index $-\rho$ and the ρ-moment of the pathloss is finite, then the tail behavior of the received signal is governed by the regularly varying tail, i.e., the wireless link is also regularly varying regardless of the pathloss singularity. For lighter-tailed channel power, the regular variation property of the wireless link is solely imposed by the pathloss singularity.

More importantly, Theorem 1.1 is a general result and covers all tail behaviors of the channel power. The first case covers the heaviest tails (i.e., $\mathcal{R}_{-\alpha}$ with $0 \leq \alpha < \infty$), such as Pareto distribution, as well as the moderately heavy tails (i.e., the class $\mathcal{R}_{-\infty}$), such as exponential, normal, lognormal, gamma distributions. The second case covers all remaining tails (e.g., truncated distributions). Therefore, for any statistical distribution of the channel power, and in particular of fast fading and shadowing, Theorem 1.1 allows us to characterize the tail behavior of the wireless link, which is essential to understand the behavior of the interference, the maximum received power, and their asymptotic relationship.

In current wireless networks, the signal distribution $\bar{F}_m$ is governed by lognormal or gamma shadowing and Rayleigh fast fading. Since these fading distributions belong to the class $\mathcal{R}_{-\infty}$ and the pathloss is bounded, the tail distribution of the received signal follows $\bar{F}_P \in \mathcal{R}_{-\infty}$. Note also that in most relevant cases, it can be shown that $\bar{F}_P$ belongs to the maximum domain of attraction of a Gumbel distribution, resulting in asymptotic independence of M and I [21].

Based on the above result, now we want to better understand the signal power scaling and the interplay between pathloss function and channel power distribution. The following result can be derived.

Corollary 1.1 ([20]). *The tail distribution $\bar{F}_P$ is classified as follows*:

- $\bar{F}_P \in \mathcal{R}_0$ *if and only if* $\bar{F}_m \in \mathcal{R}_0$;
- $\bar{F}_P \in \mathcal{R}_{-\alpha}$ *with* $\alpha \in (0, 1)$ *if* $\beta_0 > d$ *or* $\bar{F}_m \in \mathcal{R}_{-\alpha}$;
- $\bar{F}_P \in \mathcal{R}_{-\alpha}$ *with* $\alpha > 1$ *if* $0 < \beta_0 < d$ *and* $\bar{F}_m \notin \mathcal{R}_{-\rho}$ *with* $\rho \in [0, 1]$;
- $\bar{F}_P = o(\bar{H})$ *with* $\bar{H} \in \mathcal{R}_{-\infty}$ *if* $\beta_0 = 0$ *and* $\bar{F}_m = o(\bar{H})$.

Using the above characterization, we are able to understand the scaling property of the received SINR under the effect of network densification.

Theorem 1.2 ([20]). *In dense networks (i.e., $\lambda \to \infty$), the received SINR behaves as*

1. SINR $\xrightarrow{\text{p}} \infty$ *if* $\bar{F}_P \in \mathcal{R}_0$;
2. SINR $\xrightarrow{\text{d}} D$ *if* $\bar{F}_P \in \mathcal{R}_{-\alpha}$ *with* $0 < \alpha < 1$, *where D has a non-degenerate distribution*;
3. SINR $\xrightarrow{\text{a.s.}} 0$ *if* $\bar{F}_P \notin \mathcal{R}_{-\alpha}$ *with* $\alpha \in [0, 1]$.

According to Corollary 1.1, $\bar{F}_P \in \mathcal{R}_0$ is due to the fact that $\bar{F}_m \in \mathcal{R}_0$. Since m is the channel power containing the transmit power and all potential gains and propagation phenomena (including fading, array gain, etc.), $\bar{F}_m \in \mathcal{R}_0$ means that the channel power is more probable to take large values. As a result, it compensates the pathloss and makes the desired signal power grow at the same rate as the aggregate interference (i.e., $M/I \xrightarrow{\text{p}} 1$). This provides a theoretical justification to the fact that network densification always enhances the signal quality SINR.

When $\bar{F}_P \in \mathcal{R}_{-\alpha}$ with $0 < \alpha < 1$, SINR $\xrightarrow{\text{d}} D$ implies that the SINR distribution converges to a non-degenerate distribution. Moreover, from Corollary 1.1, this is due to either large near-field exponent or heavy-tailed channel power. In that case, for any SINR target y, the coverage probability $\mathbb{P}(\text{SINR} > y)$ flattens out starting from some network density λ (ceiling effect). This means that further increasing the network density by installing more BSs does not improve SINR. This saturation effect is confirmed by simulation experiments shown in Figure 1.3, where the tail distribution of SINR converges to a steady distribution for both cases: either $\beta_0 > d$ (left plot) or $\bar{F}_m \in \mathcal{R}_{-\alpha}$ with $\alpha \in (0, 1)$ (right plot). In Figure 1.3, by $F_m \sim$ Composite we mean that channel power corresponds to the case with constant transmit power and with commonly known composite Rayleigh-lognormal fading (see Definition 1.3), which belongs to the rapidly varying class $\mathcal{R}_{-\infty}$. Pareto(α) stands for channel power following a Pareto distribution of shape $1/\alpha$ and some scale $\sigma > 0$, i.e.,

$$\text{Pareto}(\alpha): \bar{F}_m(x) = (1 + x/\sigma)^{-\alpha}.$$

Note that Pareto(α) $\in \mathcal{R}_{-\alpha}$.

In practically relevant network configurations, the pathloss attenuation is bounded (i.e., $\beta_0 = 0$) and channel power is normally less heavy-tailed or even

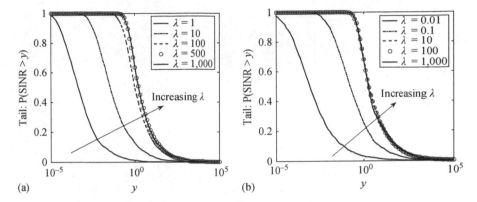

Figure 1.3 *Convergence of SINR to a steady distribution in case of $\bar{F}_P \in \mathcal{R}_{-\alpha}$*
with $\alpha \in (0, 1)$. Simulated parameters include two-slope pathloss
(i.e., $K = 2$) with $A_0 = 1$, $\beta_1 = 4$, $R_1 = 10\,m$, $R_\infty = 40\,km$,
two-dimensional network domain (i.e., $d = 2$), and network density λ
in BSs/km². (a) $\beta_0 = 3$, $F_m \sim$ Composite and (b) $\beta_0 = 0$,
$F_m \sim$ Pareto(0.5)

truncated (i.e., $\bar{F}_m = o(\bar{H})$ for some $\bar{H} \in \mathcal{R}_{-\infty}$). In particular, the conventional case of lognormal shadowing and Rayleigh fading results in channel power belonging to the class $\mathcal{R}_{-\infty}$. As a result, based on Theorem 1.1, we have that $\bar{F}_P \in \mathcal{R}_{-\infty}$, hence SINR $\overset{a.s.}{\to} 0$. Therefore, the SINR is proven to be asymptotically decreasing with the infrastructure density. This means that there is a fundamental limit on network densification and the network should not operate in the ultra dense regime since deploying excessively many BSs would decrease the network performance due to the fact that signal power boosting cannot compensate for the faster growing aggregate interference (i.e., $M/I \overset{a.s.}{\to} 0$). In other words, there exists an optimal density value until which the SINR monotonically increases, and after which the SINR monotonically decreases.

In Figure 1.4, we provide simulation results with $\bar{F}_P \notin \mathcal{R}_{-\alpha}$ with $\alpha \in [0, 1]$, with $\bar{F}_P \in \mathcal{R}_{-2}$ in Figure 1.4(a) and $\bar{F}_P \in \mathcal{R}_{-4}$ in Figure 1.4(b). We observe that the claim that the tail of SINR distribution vanishes and converges to zero when λ increases is confirmed. The convergence of SINR to zero in the high density regime further emphasizes the cardinal importance of performing local scheduling among BSs, as well as signal processing mechanisms for interference mitigation.

To provide a complete characterization of the network performance under densification effect, we investigate now the fundamental limits to the amount of densification, which depend not only on the pathloss but also on the channel power distribution. Based on previous analytical results, we obtain the following theorem on the scaling laws of user performance.

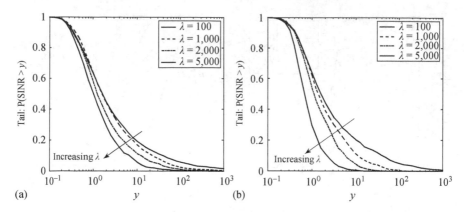

Figure 1.4   *Convergence of SINR to zero when $\bar{F}_P \notin \mathcal{R}_{-\alpha}$ with $\alpha \in [0, 1]$. Simulated parameters include two-slope pathloss (i.e., $K = 2$) with $A_0 = 1$, $\beta_1 = 4$, $R_1 = 10\,m$, $R_\infty = 40\,km$, two-dimensional network domain (i.e., $d = 2$), and network density λ in BSs/km². (a) $\beta_0 = 1$, $F_m \sim$ Composite and (b) $\beta_0 = 0$, $F_m \sim$ Pareto(4)*

Theorem 1.3 ([20]).   *The coverage $\mathcal{P}_y(\lambda)$ for fixed y and the rate $\mathcal{C}(\lambda)$ scale as follows:*

1. $\mathcal{P}_y(\lambda) \rightarrow 1$ *and* $\mathcal{C}(\lambda) \rightarrow \infty$ *as* $\lambda \rightarrow \infty$ *if* $\bar{F}_P \in \mathcal{R}_0$;
2. $\frac{\mathcal{P}_y(u\lambda)}{\mathcal{P}_y(\lambda)} \rightarrow 1$ *and* $\frac{\mathcal{C}(u\lambda)}{\mathcal{C}(\lambda)} \rightarrow 1$ *for* $0 < u < \infty$ *as* $\lambda \rightarrow \infty$ *if* $\bar{F}_P \in \mathcal{R}_{-\alpha}$ *with* $0 < \alpha < 1$;
3. $\mathcal{P}_y(\lambda) \rightarrow 0$ *and* $\mathcal{C}(\lambda) \rightarrow 0$ *as* $\lambda \rightarrow \infty$ *if* $\bar{F}_P \notin \mathcal{R}_{-\alpha}$ *with* $\alpha \in [0, 1]$; *moreover there exist finite densities* λ_p, λ_c *such that* $\mathcal{P}_y(\lambda_p) > \lim_{\lambda \rightarrow \infty} \mathcal{P}_y(\lambda)$ *and* $\mathcal{C}(\lambda_c) > \lim_{\lambda \rightarrow \infty} \mathcal{C}(\lambda)$.

Based on Theorem 1.3, we see that there exists a phase transition when the network density goes to infinity (ultra-densification). Specifically, depending on the pathloss attenuation (singularity and multi-slope) and the channel power distribution, there are three distinct regimes for the coverage and the average rate: monotonically increasing, saturation, and deficit:

- *Growth regime*: When $\bar{F}_P \in \mathcal{R}_0$, meaning that the channel power m is slowly varying $\bar{F}_m \in \mathcal{R}_0$ (see Corollary 1.1), both coverage and rate are monotonically increasing with λ. In particular, the average rate asymptotically grows with the network density. In Figure 1.5, we show simulations with $\bar{F}_m \sim$ Pareto(0.03).[1]

---

[1]Pareto distribution with relatively small varying index α was used as an approximation of $\bar{F}_m \in \mathcal{R}_0$ since standard software packages (e.g., MATLAB®) do not have built-in tools for slowly varying distributions. Note that the smaller the varying index is, the more likely the realizations of the channel power are to have large values, resulting in numerical overflow.

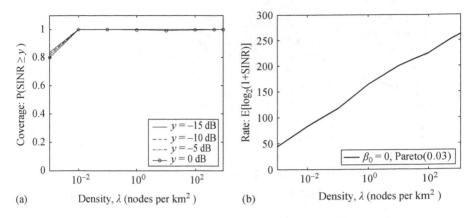

Figure 1.5   *Experiment of growth regime. Approximation of $\bar{F}_P \in \mathcal{R}_0$ by*
$\bar{F}_m \sim Pareto(0.03)$ in MATLAB simulation. Simulated parameters
include two-slope pathloss (i.e., $K = 2$) with $A_0 = 1$, $\beta_1 = 4$,
$R_1 = 10$ m, $R_\infty = 40$ km, two-dimensional network domain (i.e.,
$d = 2$), and network density λ in BSs/km^2. (a) Coverage probability
with $\beta_0 = 0$ and (b) average rate with $\beta_0 = 0$

We can see that, even though the pathloss is bounded, the coverage probability is almost one for all SINR thresholds, while throughput increases almost linearly with the logarithm of the network density in the evaluated range. This growth regime, revealed from Theorem 1.3, shows that the great expectations on the potential of network densification are theoretically possible. However, since slowly varying distributions are rather theoretical extremes and would be rarely observable in practice, the growth regime would be highly unlikely in real-world networks.

- *Saturation regime*: When the channel power is regularly varying with index within -1 and 0 or the near-field pathloss exponent β_0 is larger than the network dimension d, the tail distribution of wireless link behaves as $\bar{F}_P \in \mathcal{R}_{-\alpha}$ with $\alpha \in (0, 1)$. Consequently, both coverage probability and rate saturate past a certain network density. This saturation behavior is also confirmed by simulation experiments as shown in Figures 1.6 and 1.7. In Figure 1.6, it is the pathloss function's singularity with $\beta_0 > d$ that creates performance saturation for any type of channel power distribution not belonging to the class $\mathcal{R}_0$. In Figure 1.7, the saturation is completely due to regularly varying channel power $\bar{F}_m \in \mathcal{R}_{-\alpha}$, $\alpha \in (0, 1)$, regardless of pathloss boundedness. Prior studies have shown that the network performance is invariant of the network density for unbounded pathloss and negligible background noise. Our results (Theorem 1.3) show that this performance saturation may happen in a much larger setting, including (i) with non-negligible thermal noise and (ii) even with bounded pathloss if channel power is in the class $\mathcal{R}_{-\alpha}$ with $\alpha \in (0, 1)$. More importantly, the *unbounded* property

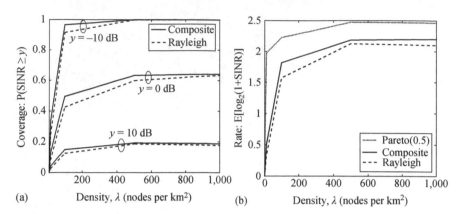

(a)

(b)

Figure 1.6   *Saturation regime due to $\beta_0 > d$. Here, saturation of both coverage probability (left plot) and rate (right plot) is due to large near-field pathloss exponent $\beta_0 = 3$, which leads to $\bar{F}_P \in \mathcal{R}_{-\alpha}$ with $\alpha \in (0, 1)$ for all considered types of the channel power distribution including composite Rayleigh-lognormal, Rayleigh, and Pareto(0.5). Simulated parameters include two-slope pathloss (i.e., $K = 2$) with $A_0 = 1$, $\beta_1 = 4$, $R_1 = 10$ m, $R_\infty = 40$ km, two-dimensional network domain (i.e., $d = 2$), and network density λ in BSs/km^2. (a) Coverage probability with $\beta_0 = 3$ and (b) Average rate with $\beta_0 = 3$*

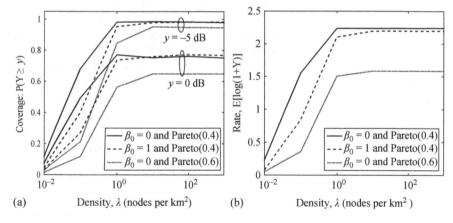

(a)

(b)

Figure 1.7   *Saturation regime due to regularly varying channel power with index in $(-1, 0)$. Here, saturation of both coverage probability (left plot) and rate (right plot) is due to regularly varying channel power $\bar{F}_m \sim Pareto(\alpha)$ with $\alpha = 0.4$ and $\alpha = 0.6$. Simulated parameters include two-slope pathloss (i.e., $K = 2$) with $A_0 = 1$, $\beta_1 = 4$, $R_1 = 10$ m, $R_\infty = 40$ km, two-dimensional network domain (i.e., $d = 2$), and network density λ in BSs/km^2. (a) Coverage probability and (b) average rate*

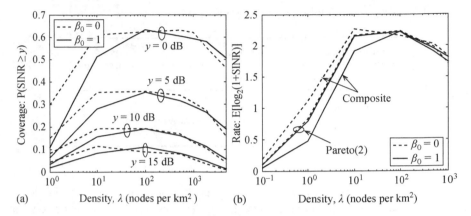

Figure 1.8   Deficit regime when $\bar{F}_P \notin \mathcal{R}_{-\alpha}$ with $\alpha \in [0, 1]$. Simulated parameters include two-slope pathloss (i.e., $K = 2$) with $A_0 = 1$, $\beta_1 = 4$, $R_1 = 10$ m, $R_\infty = 40$ km, two-dimensional network domain (i.e., $d = 2$), and network density λ in BSs/km^2. (a) Coverage probability, $\beta_0 < d$ and $\bar{F}_m \sim$ Composite and (b) average rate, $\beta_0 < d$ and different $\bar{F}_m$

of the pathloss—widely used in the literature—is just a necessary condition to have saturated performance. A sufficient condition is that the near-field pathloss exponent has to be greater than the network dimension, i.e., $\beta_0 > d$. As we will see shortly, when channel power is less heavy-tailed than the class $\mathcal{R}_{-\alpha}$ with $\alpha \in [0, 1]$, then unbounded pathloss with $0 < \beta_0 < d$ results in the same scaling regime as bounded pathloss does.

- *Deficit regime*: The third regime of network densification is determined first by a channel power distribution that is less heavy-tailed, precisely all remaining distributions not belonging to $\mathcal{R}_{-\alpha}$ with $\alpha \in [0, 1]$, and second by a near-field pathloss exponent smaller than the network dimension (i.e., $\beta_0 < d$). In this regime, both coverage probability and rate initially increase in the low density regime, then achieve a maximum at a finite network density, after which they start decaying and go to zero in the ultra dense regime. This behavior is also confirmed by simulations shown in Figure 1.8. More precisely, both coverage probability (left plot) and rate (right plot) exhibit '*inverse U*' curves with respect to the network density λ for different SINR thresholds y and different types of channel power distribution. This suggests that there is an optimal point of network density to aim for. Particularly, this deficit regime can happen even with unbounded pathloss given that the near-field exponent is still smaller than the network dimension (i.e., $\beta_0 < d$); the bounded pathloss used in prior works ($\beta_0 = 0$) to obtain this deficit regime is a special case of this class.

Table 1.1 summarizes the behavior of user performance according to Theorem 1.3 and shows the three different regimes. First, we see that the optimistic expectation of

Table 1.1   *Scaling regimes of user performance*

Scaling regime	$\bar{F}_m \in \mathcal{R}_{-\alpha}$			$\bar{F}_m$ is lighter-tailed
	$\alpha = 0$	$0 < \alpha < 1$	$\alpha > 1$	
$\beta_0 < d$	SINR $\xrightarrow{\text{P}}$ ∞,	saturation	inverse U	
$\beta_0 > d$	$\mathcal{P} \to 1$,		saturation	
	$\mathcal{C} \to \infty$			

ever-growing user's rate and full coverage in an ultra-dense network is theoretically possible, though unlikely in reality. Second, we shed light on the divergence between previous results on the fundamental limits of network densification. It is due to two different assumptions on the pathloss model, one with $\beta_0 > d$ that results in saturation regime, and the other with $\beta_0 = 0$ that results in deficit regime.

1.4   Network performance scaling laws

We now investigate the network-level performance under the effect of densification. For that, we consider two system performance metrics, namely the *coverage density* $\mathcal{D}_y$ (in BSs/m^2) and the area spectral efficiency (ASE) $\mathcal{A}$ (in nats/s/Hz/m^2), which are, respectively, defined as

$$\mathcal{D}_y = \lambda \mathbb{P}(\text{SINR} \geq y) \quad \text{and} \quad \mathcal{A} = \lambda \mathbb{E}(\log(1 + \text{SINR})). \tag{1.9}$$

The coverage density gives an indication of the cell splitting gain [24], i.e., the achievable data rate growth from adding more BSs due to the fact that each user shares its BS with a smaller number of users, as it provides the potential throughput if multiplied by the spectral efficiency $\log_2 (1 + y)$.

Using Theorem 1.3, we can obtain the following result.

Corollary 1.2 ([20]). *The coverage density $\mathcal{D}_y$ for fixed y and the area spectral efficiency $\mathcal{A}$ scale as follows:*

1. $\frac{\mathcal{D}_y(\lambda)}{\lambda} \to 1$ *and* $\frac{\mathcal{A}(\lambda)}{\lambda} \to \infty$ *as* $\lambda \to \infty$ *if* $\bar{F}_P \in \mathcal{R}_0$;
2. $\frac{\mathcal{D}_y(\lambda)}{\lambda} \to c_{y,\alpha}$ *and* $\frac{\mathcal{A}(\lambda)}{\lambda} \to c_\alpha$ *as* $\lambda \to \infty$ *if* $\bar{F}_P \in \mathcal{R}_{-\alpha}$ *with* $0 < \alpha < 1$, *where constant* $c_{y,\alpha} \in [0, 1]$ *depending on* y *and* α, *and constant* $c_\alpha > 0$ *depending on* α;
3. $\frac{\mathcal{D}_y(\lambda)}{\lambda} \to 0$ *and* $\frac{\mathcal{A}(\lambda)}{\lambda} \to 0$ *as* $\lambda \to \infty$ *if* $\bar{F}_P \notin \mathcal{R}_{-\alpha}$ *with* $\alpha \in [0, 1]$.

We can easily see from Corollary 1.2 that the network-level performance scales with the network infrastructure more optimistically than the user performance does. In particular, both the *growth* and *saturation* regimes of user performance result in *growth* regime of network performance.

In the *deficit* regime of user performance, the network performance scales sub-linearly with the network density. Nevertheless, this does not necessarily mean that $\mathcal{D}_y(\lambda)$ and $\mathcal{A}(\lambda)$ always vanish in the same manner as $\mathcal{P}_y$ and $\mathcal{C}$ do when $\lambda \to \infty$. It depends on the precise tail behavior of $\bar{F}_P$; and in the case that the resulting $\mathbb{P}(\text{SINR} > y)$ does not vanish faster than $\lambda^{-\varepsilon}$ for some $0 < \varepsilon < 1$ as $\lambda \to \infty$, then $\mathcal{D}_y$ and $\mathcal{A}$ will be in the same order of $\lambda^{1-\varepsilon}$ as $\lambda \to \infty$. In such cases, $\mathcal{D}_y$ and $\mathcal{A}$ may still increase with the network infrastructure although they will increase with a much lower speed, i.e., $\mathcal{D}_y(\lambda) = o(\lambda)$ and $\mathcal{A}(\lambda) = o(\lambda)$ as $\lambda \to \infty$.

1.5 Network ordering

The mathematical framework developed above characterizes the asymptotic behavior of coverage and throughput in dense spatial networks. In short, knowing the tail behavior of fading and pathloss characteristics, we can see in which regime (growth, saturation, or deficit) the performance falls in. We now want to compare two networks with the same scaling regime but different characteristics (e.g., different number of transmit antennas, different shadowing, etc.) and see where we will achieve better network performance.

One approach to answer this question would be to derive the performance metrics in closed form. Although the exact distribution of SINR is known for some simplified network model [2,18], it is still analytically cumbersome for realistic and practically relevant system models. In the absence of handy analytical expressions, it is difficult to compare different transmission techniques in general network settings.

For that, in addition to the scaling laws, we develop ordering results for both coverage and average rate in order to facilitate more fine comparison between dense networks with different parameters. The ordering approach provides crisp insights and useful design guidelines into the relative performance of different transmission techniques, while circumventing the need to evaluate complicated coverage and rate expressions. As such, the scaling results provide an answer on the asymptotic perfor-mance behavior (increase, saturation, or inverse-U), while the ordering results aim at identifying when we have superior performance in terms of coverage and rate among networks with the same asymptotic performance.

Firstly, from the construction previously developed for the network model, $\{(\mathbf{x}_i, m_i)\}$ forms an independently marked (i.m.) Poisson point process, denoted by $\tilde{\Phi}$. This allows to have the following definition.

Definition 1.5. *With the previously defined notation, a wireless network, denoted by Ξ, is defined as the shot noise [4] of $\tilde{\Phi}$ associated with the response func-tion defined as $v(\mathbf{y}, \mathbf{x}_i) := m_i/l(\|\mathbf{y} - \mathbf{x}_i\|)$, where $\mathbf{y} \in \mathbb{R}^d$ denotes the location of the receiver.*

Now, we can derive the asymptotic Laplace transform of the inverse of SINR, denoted by $Z = 1/\text{SINR}$.

Lemma 1.1 ([20]). *If $\bar{F}_P \in \mathcal{R}_{-\alpha}$ with $\alpha \in [0, \infty)$, then for $s \in \mathbb{R}^+$,*

$$\mathcal{L}_Z(s) = \left(1 + \alpha \int_0^1 (1 - e^{-st}) \frac{\mathrm{d}t}{t^{\alpha+1}}\right)^{-1}, \quad as \ \lambda \to \infty.$$

Lemma 1.1 allows us to obtain the following ordering between two wireless networks that have the same network intensity.

Theorem 1.4 ([20]). *For two networks Ξ_1 and Ξ_2 with the same density λ and with distribution of wireless link F_1 and F_2, resp., if $\bar{F}_1 \in \mathcal{R}_{-\alpha_1}$, $\bar{F}_2 \in \mathcal{R}_{-\alpha_2}$, and $0 \le \alpha_1 \le \alpha_2 \le \infty$, then*

$$\mathbb{E}(\mathrm{SINR}_1) \ge \mathbb{E}(\mathrm{SINR}_2),$$

$$\mathbb{E}(\log(1 + \mathrm{SINR}_1)) \ge \mathbb{E}(\log(1 + \mathrm{SINR}_2)),$$

as $\lambda \to \infty$, where SINR_1 and SINR_2 are the received SINR of Ξ_1 and Ξ_2, respectively.

Theorem 1.4 states that the heavier the tail of wireless link distribution, the better the performance under ultra-densification. As shown by Theorem 1.1, a heavier-tailed wireless link distribution can be obtained through either a heavier-tailed channel power distribution m or greater near-field pathloss exponent β_0. Physically, heavier-tailed channel power m means greater probability of high channel power, taking into account all effects, such as transmit power, array gains, beamforming gain. Since higher channel power may also increase interference, a natural question is whether it is beneficial to have higher beamforming gain (or higher channel power in general) in the high density regime. Note that large beamforming gains will increase the interference toward the users who are located in the beam-direction of the intended user, however since large beamforming gains are achieved with narrow beams, this probability may be low. Here, Theorem 1.4 states that achieving higher beamforming gains or using techniques that render the tail of wireless link distribution heavier are beneficial in terms of network performance. In short, directional transmissions can be beneficial for the network performance, as are envisioned for massive MIMO and millimeter wave systems.

1.6   Summary

Throughout this chapter, we have developed a unified framework for analyzing the performance of wireless network densification and for identifying its potential and challenges. In particular, it includes a practically relevant channel model that captures multi-slope pathloss and general channel power distributions, including transmit power, shadowing, fast fading, as well as associated gains such as antenna pattern and beamforming gain.

We have firstly discovered the interchangeable effect between pathloss and channel power on the network performance with the following conclusions:

- Under the Poisson field assumption, the most affecting component of the pathloss is its near-field exponent β_0. Bounded pathloss (obtained for $\beta_0 = 0$ and widely used in the literature) is just a special case of $\beta_0 < d$ where d is an integer denoting the network dimension.
- The effect of channel power on the performance scaling is as significant as that of pathloss, and channel power following a regularly varying tail distribution has the same effect as the pathloss function's singularity.

On the performance limits, a key finding is that under strongest cell association, the performance of ultra dense wireless networks exhibits three distinct regimes of the user performance, namely growth, saturation, and deficit regime. The tail behavior of the channel power and near-field pathloss exponent are the key parameters that determine the performance limits and the asymptotic scaling. Some particular implications include:

- Monotonically increasing per-user performance (coverage probability and average rate) by means of ultra-densification is theoretically possible, though highly unlikely in reality since it requires slowly varying tail of the channel power distribution, which is a theoretical extreme.
- In practice, installing more access points is beneficial to the user performance up to a density point, after which further densification can become harmful user performance due to faster growth of interference compared to useful signal. This highlights the cardinal importance of interference mitigation, coordination among neighboring cells, and local spatial scheduling.
- The network performance in terms of coverage density and area spectral efficiency benefit from the network densification more than user performance does. In particular, it scales linearly with the network infrastructure when the user performance is in growth or saturation regime.

We have also derived ordering results for both coverage probability and average rate in order to compare different transmission techniques and provide system design guidelines in general ultra dense network settings, which may exhibit the same asymptotic performance, with the following observations:

- Increasing the tail distribution of the channel power using advanced transmission techniques, such as massive MIMO, CoMP, and directional beamforming, is beneficial as it improves the performance scaling regime. Moreover, the effect of emerging technologies (e.g., D2D, mmWave) on near-field pathloss and channel power distribution need to be studied.
- It is meaningful to determine the optimal network density beyond which further densification becomes destructive or cost-ineffective. This operating point will depend on properties of the channel power distribution, noise level, and pathloss in the near-field region, and is of cardinal importance for the successful network densification.

Appendix A

Our characterization of the coverage and average rate scaling under general pathloss and fading models relies upon results from three related fields of study: regular variation, extreme value theory, and stochastic ordering. In the following, we provide a short summary of the most relevant results to our development.

A.1 Regular variation

In the first part of the chapter, we have defined *regularly varying functions*. The following theorem due to Karamata is often useful to deal with it.

Lemma 1.2 (Karamata's theorem). *Let $L \in \mathcal{R}_0$ be locally bounded in $[x_0, \infty)$ for some $x_0 \geq 0$:*

- *For $\alpha > -1$, we have $\displaystyle\int_{x_0}^{x} t^{\alpha} L(t) \mathrm{d}t \sim (\alpha + 1)^{-1} x^{\alpha+1} L(x), \quad x \to \infty$;*

- *For $\alpha < -1$, we have $\displaystyle\int_{x}^{\infty} t^{\alpha} L(t) \mathrm{d}t \sim -(\alpha + 1)^{-1} x^{\alpha+1} L(x), \quad x \to \infty$.*

For rapidly varying functions, we can have similar result as follows.

Lemma 1.3 (Theorem A3.12, [11]). *Suppose $h \in \mathcal{R}_{-\infty}$ is non-increasing, then for some $z > 0$ and all $\alpha \in \mathbb{R}$ we have: $\int_z^{\infty} t^{\alpha} h(t) \mathrm{d}t < \infty$ and*

$$\lim_{x \to \infty} \frac{x^{\alpha+1} h(x)}{\int_x^{\infty} t^{\alpha} h(t) \mathrm{d}t} = \infty. \tag{A.1}$$

Conversely, if for some $\alpha \in \mathbb{R}$, $\int_1^{\infty} t^{\alpha} h(t) \mathrm{d}t < \infty$ and (A.1) holds, then $h \in \mathcal{R}_{-\infty}$.

Another important theorem of Karamata theory is presented in the following.

Lemma 1.4 (Monotone density theorem). *Let $U(x) = \int_0^x u(y) \mathrm{d}y$ (or $\int_x^{\infty} u(y) \mathrm{d}y$) where u is ultimately monotone (i.e., u is monotone on (z, ∞) for some $z > 0$).*

- *If $U(x) \sim cx^{\alpha} L(x)$ as $x \to \infty$ with $c \geq 0$, $\alpha \in \mathbb{R}$ and $L \in \mathcal{R}_0$ (slowly varying), then $u(x) \sim c\alpha x^{\alpha-1} L(x)$ as $x \to \infty$.*
- *For $c = 0$, the above relations are interpreted as $U(x) = o(x^{\alpha} L(x))$ and $u(x) = o(x^{\alpha-1} L(x))$.*

A.2 Stochastic ordering

Let X and Y be two random variables (RVs) defined on the same probability space such that $\mathbb{P}(X > t) \leq \mathbb{P}(Y > t), \forall t \in \mathbb{R}$. Then X is said to be smaller than Y in the *usual stochastic order*, denoted by $X \leq_{\mathrm{st}} Y$, [23]. The interpretation is that X is less likely than Y to take on large values.

The above definition can be generalized for a set of real valued functions $g : (0, \infty) \to \mathbb{R}$ (denoted by $\mathcal{G}$), and X and Y be two non-negative random variables.

The integral stochastic order with respect to $\mathcal{G}$ is defined as $X \leq_{\mathcal{G}} Y \iff \mathbb{E}[g(X)] \leq \mathbb{E}[g(Y)], \forall g \in \mathcal{G}$.

The following ordering using the Laplace transform is relevant to our paper. For the class $\mathcal{G} = \{g(x) : g(x) = e^{-sx}, s > 0\}$, we have that

$$X \leq_{\mathrm{Lt}} Y \iff \mathcal{L}_Y(s) = \mathbb{E}[e^{-sY}] \leq \mathbb{E}[e^{-sX}] = \mathcal{L}_X(s), \quad \forall s > 0. \tag{A.2}$$

For all *completely monotonic* (c.m.) functions $g(\cdot)$ (see the definition in the sequel), we have that $X \leq_{\mathrm{Lt}} Y \iff \mathbb{E}[g(X)] \geq \mathbb{E}[g(Y)]$, whereas for all $g(\cdot)$ that have a completely monotonic derivative (c.m.d.), $X \leq_{\mathrm{Lt}} Y \iff \mathbb{E}[g(X)] \leq \mathbb{E}[g(Y)]$.

A function $g : (0, \infty) \to \mathbb{R}$ is said to be completely monotone (c.m.), if it possesses derivatives of all orders which satisfy $(-1)^n g^{(n)}(x) \geq 0$, $\forall x \geq 0$, and $n \in \mathbb{N} \cup \{0\}$, where the derivative of order $n = 0$ is defined as $g(x)$ itself. From Bernstein's theorem [5], a function is c.m. iff it can be written as a mixture of decaying exponentials. A function $g : (0, \infty) \to \mathbb{R}$ with $g(x) \geq 0, \forall x > 0$, and $\mathrm{d}g(x)/\mathrm{d}x$ being c.m. is called a Bernstein function. Note that a c.m. function is positive, decreasing and convex, while a Bernstein function is positive, increasing and concave.

References

[1] J. G. Andrews, X. Zhang, G. D. Durgin, and A. K. Gupta, "Are we approaching the fundamental limits of wireless network densification?," *IEEE Communications Magazine*, vol. 54, no. 10, pp. 184–190, Oct. 2016.

[2] J. Andrews, F. Baccelli, and R. Ganti, "A tractable approach to coverage and rate in cellular networks," *IEEE Transactions on Communications*, vol. 59, no. 11, pp. 3122–3134, Nov. 2011.

[3] F. Baccelli and A. Biswas, "On scaling limits of power law shot-noise fields," *Stochastic Models*, vol. 31, no. 2, pp. 187–207, 2015.

[4] F. Baccelli and B. Blaszczyszyn, *Stochastic Geometry and Wireless Networks: Theory, Volume 1*. Norwell, MA: NoW Publishers, 2009.

[5] S. Bernstein, "Sur les fonctions absolument monotones," *Acta Mathematica*, vol. 52, no. 1, pp. 1–66, 1929.

[6] N. Bhushan, J. Li, D. Malladi, *et al.* "Network densification: the dominant theme for wireless evolution into 5G," *IEEE Communications Magazine*, vol. 52, no. 2, pp. 82–89, Feb. 2014.

[7] N. H. Bingham, C. M. Goldie, and J. L. Teugels, *Regular Variation*. Cambridge: Cambridge University Press, 1989.

[8] L. Breiman, "On some limit theorems similar to the arc-sin law," *Theory of Probability and Applications*, vol. 10, no. 2, pp. 323–331, 1965.

[9] C. S. Chen, V. M. Nguyen, and L. Thomas, "On small cell network deployment: a comparative study of random and grid topologies," in *Proceedings of the IEEE Vehicular Technology Conference (VTC Fall) 2012*, Sep. 2012.

[10] G. Durgin, T. Rappaport, and D. A. de Wolf, "New analytical models and probability density functions for fading in wireless communications," *IEEE Transactions on Communications*, vol. 50, no. 6, pp. 1005–1015, Jun. 2002.

[11]   P. Embrechts, C. Klüppelberg, and T. Mikosch, *Modelling Extremal Events*. Berlin: Springer-Verlag, 1997.

[12]   X. Ge, S. Tu, G. Mao, C. X. Wang, and T. Han, "5G ultra-dense cellular networks," *IEEE Wireless Communications Magazine*, vol. 23, no. 1, pp. 72–79, Feb. 2016.

[13]   M. Kamel, W. Hamouda, and A. Youssef, "Ultra-dense networks: A survey," *IEEE Communications Surveys Tutorials*, vol. 18, no. 4, pp. 2522–2545, 2016.

[14]   N. Lee, F. Baccelli, and R. W. Heath, "Spectral efficiency scaling laws in dense random wireless networks with multiple receive antennas," *IEEE Transactions on Information Theory*, vol. 62, no. 3, pp. 1344–1359, Mar. 2016.

[15]   J. Liu, M. Sheng, L. Liu, and J. Li, "How dense is ultra-dense for wireless networks: From far-to near-field communications," *arXiv preprint arXiv:1606.04749*, 2016. [Online]. Available: https://arxiv.org/abs/1606.04749

[16]   D. López-Pérez, M. Ding, H. Claussen, and A. H. Jafari, "Towards 1 Gbps/UE in cellular systems: Understanding ultra-dense small cell deployments," *IEEE Communications on Surveys Tutorials*, vol. 17, no. 4, pp. 2078–2101, 2015.

[17]   V. M. Nguyen, "Wireless link modelling and mobility management for cellular networks," Ph.D. dissertation, Communications & Electronics (COMELEC) Dept., Telecom ParisTech, 2011.

[18]   V. M. Nguyen and F. Baccelli, "A stochastic geometry model for the best signal quality in a wireless network," in *Proceedings of the 8th International Symposium on Modelling and Optimisation Mobile, Ad Hoc Wireless Network (WiOpt)*, pp. 465–471, May 2010.

[19]   V. M. Nguyen and M. Kountouris, "Coverage and capacity scaling laws in downlink ultra-dense cellular networks," in *2016 IEEE International Conference on Communications (ICC)*, pp. 1–7, May 2016.

[20]   V. M. Nguyen and M. Kountouris, "Performance limits of network densification," *IEEE Journal on Selected Areas in Communications*, vol. 35, no. 6, pp. 1294–1308, Jun. 2017.

[21]   V. M. Nguyen, F. Baccelli, L. Thomas, and C. S. Chen, "Best signal quality in cellular networks: asymptotic properties and applications to mobility management in small cell networks," *EURASIP Journal of Wireless Communication Network*, vol. 2010, 2010. doi:10.1155/2010/690161.

[22]   A. Rajan, C. Tepedelenlioglu, and R. Zeng, "A unified fading model using infinitely divisible distributions," *arXiv:1508.04804*, 2015.

[23]   M. Shaked and J. G. Shanthikumar, *Stochastic Orders and Their Applications*. San Diego, CA: Academic Press, 1994.

[24]   X. Zhang and J. Andrews, "Downlink cellular network analysis with multi-slope path loss models," *IEEE Transactions on Communications*, vol. 63, no. 5, pp. 1881–1894, May 2015.

Chapter 2
Massive and network MIMO

Yunfei Chen[1]

In this chapter, the techniques of massive multiple-input-multiple-output (MIMO) and network MIMO are studied. First, massive MIMO is compared with the conventional point-to-point MIMO used in current cellular systems in terms of capacity and energy efficiency. The comparison shows that massive MIMO has huge gains even under the modest conditions using simple linear processing. Then, the implementation of massive MIMO is discussed by focusing on the channel estimation method, the signal detection method and the precoding method. Following this, several main issues in massive MIMO are examined and relevant measures are discussed. Finally, network MIMO is studied by introducing cooperation between base stations to completely remove the inter-cell interference, at the cost of even higher complexity that places limits on the backhaul and fronthaul links.

2.1 Introduction

Two facts about mobile communications have never changed so far: the demand for throughput or date rate always increases, while the availability of the radio spectrum never increases. Consequently, we have reached a point where we cannot keep expanding the bandwidth that a mobile system occupies to increase the data rate or throughput but innovative technologies are required to meet future demands. After the 4G systems have been deployed worldwide, 5G is in the agenda now. It is expected that 5G will have an aggregate data rate about 1,000 times that of current 4G with an energy efficiency of 100 times that of current 4G, along with other requirements [1]. To meet these requirements, a major paradigm shift of enabling technologies is necessary. Disruptive technologies that bring in unprecedentedly component, architectural and radical changes to the mobile systems are inevitable [2]. Among these disruptive technologies, millimeter wave communications, small cell and massive MIMO technologies hold the most promise. This chapter focuses on the massive MIMO and the related network MIMO technologies.

[1] School of Engineering, University of Warwick, UK

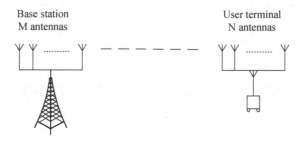

Figure 2.1   Diagram of a point-to-point MIMO with one BS and one terminal

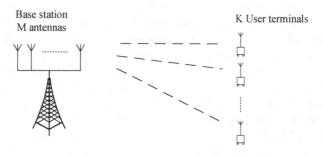

Figure 2.2   Diagram of a massive MIMO with one BS and K user terminals

Figures 2.1 and 2.2 compare massive MIMO with the conventional point-to-point MIMO. Massive MIMO can be considered as an extension of the point-to-point MIMO where the antennas at the terminal are distributed in the cell instead of collocated at one user. Also, the number of antennas at the base station is often very large, more than a hundred or even larger and thus, comes the word of "massive." It will be shown in the following sections that massive MIMO does not have the issues that point-to-point MIMO has and thus, it is scalable. The idea of network MIMO is related to massive MIMO in that it aims to reduce or eliminate the inter-cell interference in massive MIMO by introducing network-wide cooperation to maximize the performance at the cost of extra overhead and complexity.

The rest of the chapter is organized as follows. In Section 2.2, the benefits of massive MIMO will be outlined by comparing it with point-to-point MIMO from an information theoretical point of view. The main techniques used in the implementation of massive MIMO, including channel estimation, signal detection and precoding, will be presented in Section 2.3. Section 2.4 will discuss several important issues in massive MIMO. Section 2.5 focuses on network MIMO. Finally, the chapter is summarized in Section 2.6.

2.2 Benefits of massive MIMO

2.2.1 Point-to-point MIMO

Consider the conventional point-to-point MIMO. In this case, assume that the transmitter has an array of N_t antennas and the receiver has an array of N_r antennas. The received signal is thus given by

$$\mathbf{y} = \sqrt{\gamma}\mathbf{Gx} + \mathbf{n} \tag{2.1}$$

where $\mathbf{y}$ is a $N_r \times 1$ vector representing the received signals at different receiving antennas, γ is the average signal-to-noise ratio (SNR) of the link, $\mathbf{G}$ is the $N_r \times N_t$ matrix representing the channel between transmitter and receiver, $\mathbf{x}$ is the $N_t \times 1$ vector representing the transmitted signals from different transmitting antennas, and $\mathbf{n}$ represents the $N_r \times 1$ noise samples. In the following, we assume $E\{||\mathbf{x}||^2\} = 1$ so that the transmit signal has a unit average power. Also, the noise samples are independent and identically distributed and are circularly symmetric complex Gaussian random variables with mean zero and an identity covariance matrix $\mathbf{I}_{N_r}$. The channel matrix $\mathbf{G}$ is complex and constant within the considered time interval.

From (2.1), when the input is Gaussian-distributed and perfect channel state information (CSI) is available at the receiver, the instantaneous achievable rate can be given by

$$C = \log_2 \det\left(\mathbf{I}_{N_r} + \frac{\gamma}{M}\mathbf{GG}^H\right) \tag{2.2}$$

where $\det(\cdot)$ denotes the determinant operation of a matrix, $\mathbf{I}_{N_r}$ is a $N_r \times N_r$ identity matrix, and $(\cdot)^H$ denotes the Hermitian transpose operation. This rate is bounded as

$$\log_2(1 + \gamma N_r) \leq C \leq \min\{N_t, N_r\} \log_2(1 + \gamma \max\{N_t, N_r\}/N_t). \tag{2.3}$$

The lower bound is achieved in the worst case (rank-1) when there is a strong line-of-sight in a compact array, while the upper bound is achieved in the best case when the channel matrix has independent and identically distributed entries. Hence, the performance of point-to-point MIMO relies heavily on the propagation environment. Also, even under the most favorable channel propagation environment, the SNR at the edge of a cell is normally very small and in this case, $\lim_{\gamma \to 0} C \to \gamma N_r/\ln 2$, where the multiplexing gain is lost and therefore, the multiple transmitting antennas are of no use. Thus, the point-to-point MIMO is not scalable when there is line of sight or when the SNR is too low at the edge of a cell.

Now consider two extreme or limiting cases when the values of N_t or N_r go to infinity. The random matrix theory states that, when the row vectors or the column vectors of the channel matrix $\mathbf{G}$ are asymptotically orthogonal, one has [3]:

$$\lim_{N_t \to \infty} \frac{\mathbf{GG}^H}{N_t} \to \mathbf{I}_{N_r} \tag{2.4}$$

for $N_t \gg N_r$ and

$$\lim_{N_r \to \infty} \frac{GG^H}{N_r} \to I_{N_t} \tag{2.5}$$

for $N_r \gg N_t$. Using these results, when the number of transmitting antennas goes to infinity, one has from (2.2) $\lim_{N_t \to \infty} C \to N_r \log_2(1 + \gamma)$, and when the number of receiving antennas goes to infinity, $\lim_{N_r \to \infty} C \to N_t \log_2(1 + \gamma N_r/N_t)$. From these equations, one sees that the achievable rate approaches the upper bound of the actual achievable rate, when the number of antennas goes to infinity, under the condition that the propagation vectors are asymptotically orthogonal. This observation suggests that one could use an excessively large number of antennas in MIMO to achieve the upper limit of the achievable rate with scalable multiplexing gains. This motivates the extension to massive MIMO from point-to-point MIMO.

2.2.2 Massive MIMO

As seen from the previous section, it is important to assume the favorable propagation environment where the column or row vectors of the channel matrix are asymptotically orthogonal. Thus, we start with the channel matrix for massive MIMO, where $M \gg K$.

Specifically, define the channel matrix of the uplink from the users to the BS as

$$G = AB^{1/2} \tag{2.6}$$

where G is the $M \times K$ channel matrix from the users to the BS, A is an $M \times K$ matrix representing small-scale fading, and B is the $K \times K$ matrix representing large-scale fading. Thus, we have explicitly separated the effect of small-scale fading from the effect of large-scale fading in (2.6). The antenna array at the BS is sufficiently compact such that the propagation paths from different antennas to a particular user have the same large-scale fading. In this case, B is a diagonal matrix and the kth element on the diagonal line represents the large-scale fading as $b_k = P_k \beta_k / d_k^v$, where P_k is the power factor to the kth user, β_k is the shadowing coefficient on the kth path, d_k is the distance from the BS to the kth user and v is the path loss exponent. Before we proceed, we apply the random matrix theory in (2.4) here to obtain:

$$\frac{G^H G}{M} = B^{1/2} \frac{A^H A}{M} B^{1/2} \approx B^{1/2} I_K B^{1/2} = B \tag{2.7}$$

where the approximation comes from the asymptotic orthogonality or favorable propagation when M is very large.

For the uplink, the users transmit data to the BS. Assume that all the users transmit their own data simultaneously, which is the worst case scenario, the received signal at the BS is given by

$$z_u = \sqrt{\gamma_u} G p + n_u \tag{2.8}$$

where z_u is the $M \times 1$ vector representing the received signals at different antennas in the BS, γ_u is the average SNR in the uplink, $p = [p_1, p_2, \ldots, p_K]^T$ is the transmitted signal with p_k being the transmitted signal of the kth user, n_u is the $M \times 1$ receiver noise vector in the uplink. Further, each element in $\mathbf{n}_u$ is a complex Gaussian random variable with mean zero and variance 1, is circularly symmetric, and is independent of each other. The transmitted signal satisfies $E\{||\mathbf{p}||^2\} = 1$ to have a unit average power. From (2.8) and similar to before, the sum achievable rate of the uplink is $C_u = \log_2 \det(\mathbf{I}_K + \gamma_u \mathbf{G}^H \mathbf{G})$. Using (2.7) when the number of antennas at the BS is large, one has the asymptotic sum rate as

$$C_u \approx \sum_{k=1}^{K} \log_2(1 + M\gamma_u b_k) \tag{2.9}$$

which is combined by adding all achievable rates in links with the kth link having a SNR of $M\gamma_u b_k$ together.

Unlike conventional point-to-point MIMO, the optimal capacity of the uplink can be achieved using simple linear processing. For example, using maximum ratio combining (MRC) (or matched-filtering (MF) called in some works), the received signal at the BS becomes:

$$\mathbf{G}^H \mathbf{z}_u = \sqrt{\gamma_u} \mathbf{G}^H \mathbf{G} \mathbf{p} + \mathbf{G}^H \mathbf{n}_u \approx M\sqrt{\gamma_u} \mathbf{B} \mathbf{p} + \mathbf{G}^H \mathbf{n}_u. \tag{2.10}$$

Since $\mathbf{B}$ is a diagonal matrix, the kth element of $\mathbf{G}^H \mathbf{z}_u$ is given by $M\sqrt{\gamma_u} b_k p_k + n_k$, which can then be used to decode p_k. This is a very simple method compared with the sphere decoding commonly used in point-to-point MIMO systems [4]. There is no interference between users and each element can be considered as a received signal in a single-input-single-output (SISO) channel. Thus, the simple MRC or MF operation can separate different users from the received signal to convert massive MIMO into SISO channels. Also, one sees that the kth element of the filtered signal $M\sqrt{\gamma_u} b_k p_k + n_k$ has a SNR of $M\gamma_u b_k$, which gives an achievable rate of $\log_2 (1 + M\gamma_u b_k)$, the same as the kth term in the sum of (2.9). Thus, MRC or MF can achieve the channel capacity.

For the downlink, the BS transmits data to the users, and the received signal at the users can be described as

$$\mathbf{z}_d = \sqrt{\gamma_d} \mathbf{G}^T \mathbf{q} + \mathbf{n}_d \tag{2.11}$$

where $\mathbf{z}_d$ is the $K \times 1$ vector representing the received signals at different users, γ_d is the average SNR in the downlink, $\mathbf{q} = [q_1, q_2, \ldots, q_M]^T$ is the transmitted signal with q_m being the transmitted signal of the mth antenna, $\mathbf{n}_d$ is the $K \times 1$ receiver noise vector in the downlink, and the channel reciprocity in time-division duplex (TDD) is used such that the channel matrix from the BS to the users is a transpose of $\mathbf{G}$. Also, assume that each element in $\mathbf{n}_d$ is a complex Gaussian random variable with mean zero and unit variance, is circularly symmetric, and is independent of each other. The transmitted signal satisfies $E\{||\mathbf{q}||^2\} = 1$ to have a unit average power too. As mentioned before, the users are normally low-cost and low-complexity terminals in the cell. Thus, precoding can be performed to shift the complexity to the BS. In this case, both the BS and the users have the channel CSI and the BS can optimally allocate power. According to previous results [5], for such a link, the sum rate is given by $C_d = \max_{\mathbf{C}}\{\log_2 \det(\mathbf{I}_M + \gamma_d \mathbf{G} \mathbf{C} \mathbf{G}^H)\}$, where $\mathbf{C}$ is a $K \times K$ diagonal matrix with the kth diagonal element c_k being the allocated optimal power for the kth user and $\sum_{k=1}^{K} c_k = 1$. Again, under favorable propagation, one can use (2.7) to obtain:

$$C_d \approx \sum_{k=1}^{K} \max_{c_k} \{\log_2(1 + M\gamma_d c_k b_k)\} \tag{2.12}$$

which can be considered as a sum of achievable rates in links with the kth link having a SNR of $M\gamma_d c_k b_k$. Thus, again, the capacity of the downlink can be achieved when the channel matrix has asymptotically orthogonal columns.

Similar to the uplink, the capacity of the downlink can be achieved by using simple linear precoding. For example, using MF or MRC linear precoder, the transmitted signal at the BS can be expressed as $\mathbf{q} = (1/\sqrt{M})\mathbf{G}^*\mathbf{B}^{-1/2}\mathbf{C}^{1/2}\mathbf{x}$, where $\mathbf{x} = [x_1, x_2, \ldots, x_K]^T$ is the actual modulated symbol to be transmitted to different users and $\frac{1}{\sqrt{M}}\mathbf{G}^*\mathbf{B}^{-1/2}\mathbf{C}^{1/2}$ is the precoding matrix. Using this in (2.11) and invoking the favorable propagation condition, one immediately has

$$\mathbf{z}_d \approx \sqrt{\gamma_d M}\mathbf{B}^{1/2}\mathbf{C}^{1/2}\mathbf{x} + \mathbf{n}_d. \tag{2.13}$$

Since both $\mathbf{B}$ and $\mathbf{C}$ are diagonal, the received signal $\mathbf{z}_d$ can be easily converted into separate users without any multi-user interference and the massive MIMO channel can be converted into K SISO channels. The kth element of $\mathbf{z}_d$ is used to decode the signal for the kth user. Also, the SNR of the kth user is given by $\gamma_d M b_k c_k$, which gives an achievable rate of $\log_2(1 + \gamma_d M c_k b_k)$, identical to the kth term in (2.12). Such capacity-achieving precoding is much simpler than the conventional point-to-point MIMO that uses complicated algorithms like dirty-paper coding [6]. Figure 2.3

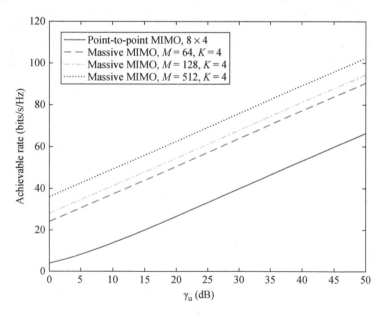

Figure 2.3   Sum rate vs. γ_u for different values of M when $K = 4$

compares the sum rates of point-to-point MIMO and massive MIMO for different system parameters.

In general, the capacity of massive MIMO can be 10 times or more than that of a point-to-point MIMO system and at the same time, its energy efficiency can be 100 times or more than that of a point-to-point MIMO system [7,8]. This can be seen by comparing the achievable rate formulas presented before. For example, from (2.3) and (2.9), if we use 8×8 antennas for point-to-point MIMO in the best case, to achieve 10 times this capacity, we can use 80 users in massive MIMO, which is a modest number of users. Also, for the same capacity, point-to-point MIMO in (2.3) uses a power of γ. To achieve the same capacity, from (2.9), massive MIMO only needs to use a power of $Mb_k\gamma_u = \gamma$ so that $\gamma_u = \gamma/(Mb_k)$ is only a fraction of γ. Since M is usually large in massive MIMO, if $b_k = 1$, we may achieve 100 times energy efficiency by choosing $M = 100$.

2.3   Techniques in massive MIMO

Before we discuss the techniques in massive MIMO, it is necessary to identify the basic modes of operation for the massive MIMO system. One of such modes is the slot structure. Traditionally, we have two slot structures: TDD where the uplink and the downlink operate at the same frequency band in different time slots and frequency-division duplex (FDD) where the uplink and the downlink operate in different frequency bands.

Assume that there are M antennas at the BS and K users in the cell served by this BS. In the FDD system, the downlink and the uplink have different CSI. Thus, they have to be estimated separately [9]. First, the terminals transmit unknown data to the BS. Then, the terminals transmit known pilots to the BS to acquire the uplink CSI. For orthogonal pilots, this requires a training period of $T_u \geq K$. The BS then uses the received pilots from the terminals to acquire the CSI of the uplink for the detection of the unknown data. Second, the BS transmits known pilots to all the terminals in the downlink. Again, since optimal pilots are achieved when they are orthogonal, this pilot transmission requires a time interval of at least $T_d \geq M$. Using the received pilots, the terminals perform channel estimation to acquire the CSI of the downlink. Then, the acquired CSI at the terminal is sent back to the BS for downlink precoding. This transmission requires at least another time period of T_d. Finally, the BS uses the estimated downlink CSI to precode and transmit the unknown data in the downlink. Thus, the whole process of CSI acquisition in FDD can be completed using $2T_d + T_u \geq 2M + K$ pilots. Even when the terminals transmit the estimated downlink CSI through dedicated control channels, that is, the acquired CSI is fed back to the BS through a different frequency band, a minimum time of $T_d + T_u \geq M + K$ is still required.

In the TDD system, since uplink and downlink operate at the same frequency band, we often assume channel reciprocity such that the uplink and the downlink have the same CSI [10]. Thus, only the CSI of the uplink is needed to perform signal detection and transmission precoding at the BS. In this case, all the K users send orthogonal pilots to the BS for channel estimation. Thus, the whole process of CSI acquisition in TDD can be completed using $T_u \geq K$ pilots.

Comparing FDD with TDD, one sees that FDD requires a significantly larger number of pilots than TDD in order to acquire CSI. Also, the number of pilots required depends on the number of antennas at the BS M. However, this acquisition does not assume any channel reciprocity. On the other hand, TDD uses much fewer pilots than FDD, and the number of pilots required does not depend on M. To achieve this, channel reciprocity has to be assumed.

In a practical wireless channel, the CSI is a random process that changes with time. Thus, to achieve accurate channel estimation and to make sure that the acquired CSI is not outdated for signal detection and precoding, the channel estimation operation must be performed before the CSI changes again. Mathematically, this means that the frame length or the total number of symbols in one packet, including the training period or the pilots, must be smaller than the channel coherence time. For TDD, as long as the number of users is moderate, this requirement can be easily satisfied. However, for FDD, since the number of pilots is much larger and it increases with M as well, such requirement is difficult to satisfy in massive MIMO where the value of M is normally hundreds of times larger than that in the conventional point-to-point MIMO. Thus, in massive MIMO systems, although FDD may be used for some special cases [11–13], it is TDD that is more efficient and therefore widely used. In the following, we assume TDD in the discussion.

2.3.1 Channel estimation in uplink

For TDD, due to channel reciprocity, only $T_u > K$ pilots need to be transmitted at each terminal in the uplink. Using similar notations as before, since we have K terminals, the received pilots at the BS can be expressed as

$$\mathbf{Z}_p = \sqrt{\gamma_p T_u} \mathbf{GS} + \mathbf{N}_p \tag{2.14}$$

where $\mathbf{Z}_p$ is a $M \times T_u$ matrix representing the received pilot symbols at different antennas during different symbol periods in the BS, γ_p is the average transmission SNR in the uplink, $\mathbf{S}$ is the $K \times T_u$ matrix representing the transmitted pilot symbols from different terminals during different symbol periods, $\mathbf{N}_p$ is the $M \times T_u$ receiver noise matrix during the pilot transmission in the uplink, and $\mathbf{G}$ is the channel matrix defined as before. Again, each element in $\mathbf{N}_p$ is a complex Gaussian random variable with mean zero and variance 1, is circularly symmetric, and is independent of each other. The transmitted pilot satisfying $\mathbf{SS}^H = \mathbf{I}_K$ is a unitary matrix so that each terminal has a unit average power. For Rayleigh fading channels, $\mathbf{G}$ is a complex Gaussian random matrix with each element having mean zero and variance 1. Note that γ_p could be the same as or different from γ_u in (2.8), depending on whether power control is implemented at the terminals. Note that another important assumption here is that the channel matrix $\mathbf{G}$ does not change during these operations. Thus, this is a block fading channel enabling us to estimate $\mathbf{G}$ for signal detection.

Using the received pilots in (2.14), the most widely used method of channel estimation is minimum-mean-squared-error (MMSE) estimation, where the channel estimate is calculated as [10]:

$$\hat{\mathbf{G}} = \frac{1}{\sqrt{\gamma_p T_u}} \mathbf{Z}_p \mathbf{S}^H = \mathbf{G} + \frac{1}{\sqrt{\gamma_p T_u}} \mathbf{N}_p \mathbf{S}^H. \tag{2.15}$$

One sees from (2.15) that the first term is the true channel matrix and the second term is the channel estimation error caused by noise during the pilot transmission. Also, when $\gamma_p T_u$ is asymptotically large, which can be achieved by increasing the pilot transmission power γ_p or increasing the training period T_u, one has that the variance of the estimation error goes to zero at a rate of $1/\gamma_p T_u$. Thus, (2.15) is an efficient estimator.

In the literature, it is also convenient to write the channel estimate in another form of $\hat{\mathbf{G}} = \chi \mathbf{G} + \sqrt{1 - \chi^2}\mathbf{E}$, where χ is the correlation coefficient between $\mathbf{G}$ and $\hat{\mathbf{G}}$ and also indicates the accuracy of the channel estimate, and $\mathbf{E}$ is the estimation error independent of the true channel matrix. This form captures the essence of channel estimation. However, it does not necessarily include (2.15) as a special case. There are also other channel estimators developed for massive MIMO. For example, in [14], compressive sensing was used to explore the fact that the degrees of freedom in massive MIMO is much larger than the number of independent channel parameters. These estimators are not discussed in detail here, because the MMSE estimators have excellent performances with simple structures and therefore, are desirable in a massive MIMO system. Next, we discuss the signal detection techniques in the uplink.

2.3.2   Detection in uplink

As discussed in the previous section, simple linear techniques can be used to detect the received data symbols at the BS. These simple linear techniques also achieve the optimal capacity asymptotically. In particular, it was shown in (2.10) that one can use MRC to reach the optimal capacity. In this section, we focus on three simple linear detectors: MRC, zero-forcing (ZF), and MMSE. Figure 2.4 gives the diagram of the signal detector in the uplink.

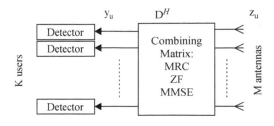

Figure 2.4   Diagram of signal detection at the BS

In general, these techniques multiply the received signal with a combining matrix $\mathbf{D}$ for decoding. The combining matrix is given as follows:

$$MRC : \mathbf{D} = \mathbf{G}$$

$$ZF : \mathbf{D} = \mathbf{G}(\mathbf{G}^H \mathbf{G})^{-1}$$

$$MMSE : \mathbf{D} = \mathbf{G}\left(\mathbf{G}^H \mathbf{G} + \frac{1}{\gamma_u}\mathbf{I}_K\right)^{-1}. \tag{2.16}$$

Thus, if the received signal is given by (2.8), one has the combined signal as $\mathbf{y}_u = \mathbf{D}^H \mathbf{z}_u = \sqrt{\gamma_u}\mathbf{D}^H \mathbf{G}\mathbf{p} + \mathbf{D}^H \mathbf{n}_u$. The kth element of $\mathbf{y}_u$ is used to detect the data symbol sent by the kth user as

$$y_{u,k} = \sqrt{\gamma_u}\mathbf{d}_k^H \mathbf{g}_k p_k + \sqrt{\gamma_u} \sum_{i=1,i\neq k}^{K} \mathbf{d}_k^H \mathbf{g}_i p_i + \mathbf{d}_k^H \mathbf{n}_u \tag{2.17}$$

where $\mathbf{d}_k$ is the kth column of $\mathbf{D}$, $\mathbf{g}_k$ is the kth column of $\mathbf{G}$, and p_k is the data symbol sent by the kth terminal. In (2.17), the first term is the desired signal term, the second term is the inter-user interference and the third term is the noise term. From (2.17), the signal-to-interference-plus-noise ratio (SINR) of the kth user can be derived as

$$\rho_k = \frac{\gamma_u |\mathbf{d}_k^H \mathbf{g}_k|^2}{\gamma_u \sum_{i=1,i\neq k}^{K} |\mathbf{d}_k^H \mathbf{g}_i|^2 + ||\mathbf{d}_k||^2}. \tag{2.18}$$

Using (2.18), the detection performance can be analyzed. For example, if the interference can be assumed Gaussian following the central limit theorem, the instantaneous achievable rate of the uplink is given by $R_{u,k} = \log_2(1 + \rho_k)$ and the instantaneous bit error rate (BER) of the uplink using binary phase shift keying (BPSK) is $P_{u,k} = \frac{1}{2}\text{erfc}(\sqrt{\rho_k})$, where $\text{erfc}(\cdot)$ is the complementary error function. In fading channels, ergodic capacity and average BER will be more useful such that they can be obtained by averaging them over the channel matrix $\mathbf{G}$, respectively.

Using (2.16) in (2.18), specific results for MRC, ZF and MMSE can also be obtained. For example, the instantaneous rates for MRC, ZF, and MMSE can be derived as

$$R_{u,k}^{MRC} = \log_2\left(1 + \frac{\gamma_u ||\mathbf{g}_k||^4}{\gamma_u \sum_{i=1,i\neq k}^{K} |\mathbf{g}_k^H \mathbf{g}_i|^2 + ||\mathbf{g}_k||^2}\right)$$

$$R_{u,k}^{ZF} = \log_2\left(1 + \frac{\gamma_u}{u_k}\right)$$

$$R_{u,k}^{MMSE} = \log_2\left(\frac{1}{v_k}\right) \tag{2.19}$$

where u_k is the kth element on the diagonal line of $(\mathbf{G}^H \mathbf{G})^{-1}$ and v_k is the kth element on the diagonal line of $(\mathbf{I}_K + \gamma_u \mathbf{G}^H \mathbf{G})^{-1}$. It is noted that the MMSE detector is optimum in the sense that it maximizes the instantaneous achievable rate.

The above results assumed perfect CSI, or no estimation errors in the signal detection. In practice, the combining matrix $\mathbf{D}$ is obtained by using the estimated CSI in the previous subsection. In this case, the combining matrix becomes:

$$MRC : \hat{\mathbf{D}} = \hat{\mathbf{G}}$$

$$ZF : \hat{\mathbf{D}} = \hat{\mathbf{G}}(\hat{\mathbf{G}}^H \hat{\mathbf{G}})^{-1}$$

$$MMSE : \hat{\mathbf{D}} = \hat{\mathbf{G}} \left(\hat{\mathbf{G}}^H \hat{\mathbf{G}} + \frac{1}{\gamma_u} \mathbf{I}_K \right)^{-1}. \tag{2.20}$$

where $\hat{\mathbf{G}}$ is the channel estimate given before. Similarly, the SINR of the kth user can be obtained as

$$\rho_k = \frac{\gamma_u |\hat{\mathbf{d}}_k^H \mathbf{g}_k|^2}{\gamma_u \sum_{i=1,i\neq k}^{K} |\hat{\mathbf{d}}_k^H \mathbf{g}_i|^2 + ||\hat{\mathbf{d}}_k||^2} \tag{2.21}$$

where $\hat{\mathbf{d}}_k$ is the kth column of $\hat{\mathbf{D}}$. The effect of channel estimation error on the detection performance is less obvious for ZF and MMSE. However, for MRC, it can be examined as follows. Using $\hat{\mathbf{D}} = \chi\mathbf{G} + \sqrt{1 - \chi^2}\mathbf{E}$:

$$\mathbf{y}_u = \left(\chi\mathbf{G} + \sqrt{1 - \chi^2}\mathbf{E} \right)^H \mathbf{z}_u$$

$$= \sqrt{\gamma_u}\chi\mathbf{G}^H \mathbf{G}\mathbf{p} + \chi\mathbf{G}^H \mathbf{n}_u + \sqrt{\gamma_u}\sqrt{1 - \chi^2}\mathbf{E}^H \mathbf{G}\mathbf{p} + \sqrt{1 - \chi^2}\mathbf{E}^H \mathbf{n}_u. \tag{2.22}$$

Thus, the SINR of the kth user can be obtained for MRC as

$$\rho_k = \frac{\gamma_u \chi^2 ||\mathbf{g}_k||^4}{\gamma_u \chi^2 \sum_{i=1,i\neq k}^{K} |\mathbf{g}_k^H \mathbf{g}_i|^2 + \chi^2 ||\mathbf{g}_k||^2 + (1 - \chi^2)M \left[1 + \gamma_u \left| \sum_{i=1}^{K} \mathbf{g}_i p_i \right|^2 \right]} \tag{2.23}$$

where we assume that the elements in $\mathbf{E}$ have zero mean and unit variance. One sees that the estimation error reduces the effective SINR significantly. This happens to the ZF and MMSE detectors too, at different levels. Similar results and conclusions can also be made if we use other channel estimators. Due to the presence of the estimation error, in the ideal case when M goes to infinity, the SINR will be reduced too. In this case, inter-user interference is inevitable even with asymptotic orthogonality. Also, unlike the case of perfect CSI, when estimation errors occur, MRC, ZF, and MMSE do not give the same combining matrix even when M goes to infinity.

Next, we discuss the important energy efficiency problem in massive MIMO, as this depends on the detectors used. When M is finite and perfect CSI is available, using the convexity of $\log_2(1 + \frac{1}{x})$ and Jensen's inequality, it can be shown that the ergodic capacity of MRC or the average of (2.19) over **G** is lower bounded by [15]:

$$
\begin{aligned}
E\{R_{u,k}^{MRC}\} \quad &\geq \quad \log_2\left(1 + \left(E\left\{\frac{\gamma_u \sum_{i=1,i\neq k}^{K} |\mathbf{g}_k^H \mathbf{g}_i|^2 + ||\mathbf{g}_k||^2}{\gamma_u ||\mathbf{g}_k||^4}\right\}\right)^{-1}\right) \\
&= \quad \log_2\left(1 + \frac{\gamma_u(M-1)b_k}{\gamma_u \sum_{i=1,i\neq k}^{M} b_k + 1}\right) \\
&\to_{M\to\infty} \log_2(1 + M\gamma_u b_k).
\end{aligned}
\tag{2.24}
$$

For ZF, using the Jensen's inequality, the ergodic capacity is shown to be lower bounded by [15]:

$$
\begin{aligned}
E\{R_{u,k}^{ZF}\} \quad &\geq \quad \log_2\left(1 + \frac{\gamma_u}{E\{u_k\}}\right) \\
&= \quad \log_2(1 + \gamma_u(M-K)b_k) \\
&\to_{M\to\infty} \log_2(1 + M\gamma_u b_k).
\end{aligned}
\tag{2.25}
$$

For MMSE, using the Jensen's inequality, the ergodic capacity is lower bounded as [15]:

$$
\begin{aligned}
E\{R_{u,k}^{MMSE}\} \quad &\geq \quad \log_2\left(1 + \frac{1}{E\left\{\frac{u_k}{1-u_k}\right\}}\right) \\
&\approx \quad \log_2\left(1 + \frac{\Gamma(a)}{\Gamma(a-1)}\theta\right) \\
&\to_{M\to\infty} \log_2(1 + M\gamma_u b_k)
\end{aligned}
\tag{2.26}
$$

where a and b are the parameters from a Gamma probability density function to approximate the distribution of $1/u_k - 1$. Thus, one can see that, for a finite M with perfect CSI, the scaling law of $1/M$ applies such that the massive MIMO system still has an energy saving of $10 \log_{10} M$ dB over the SISO system. In the above, Rayleigh fading is assumed for the channel matrix **G**. When M is finite and only imperfect CSI is assumed, similar results have been derived in [15]. In this case, the energy efficiency improves at a rate of $\frac{1}{\sqrt{M}}$ and thus, the energy saving becomes $5 \log_{10} M$ dB in this case, compared with a SISO system using perfect CSI. A detailed discussion of the above results can be found in [15].

2.3.3 Precoding in downlink

The precoding scheme at the BS aims to simplify the signal detection at the terminals. Figure 2.5 shows a diagram of precoding in the downlink. For conventional point-to-point MIMO systems, nonlinear precoding techniques are known to provide near-optimal performances, such as dirty paper [6] and vector perturbation [16]. However, these methods are also known for their high complexity. On the other hand, for massive

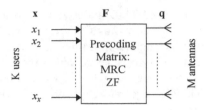

Figure 2.5   Diagram of precoding at the BS

MIMO systems, linear techniques also provide near-optimal performances when the number of antennas is large such that asymptotic orthogonality can be achieved. Thus, we focus on simple linear precoding schemes, such as MRC and ZF.

For MRC, the transmitted signal $\mathbf{x} = [x_1, x_2, \ldots, x_K]^T$ is precoded as $\mathbf{q} = \mathbf{Fx}$, where $\mathbf{F} = \sqrt{\alpha}\mathbf{G}^*$ is the precoding matrix and α is a power normalization factor. Thus, the signal received at the terminals becomes $\mathbf{z}_d = \sqrt{\gamma_d}\mathbf{G}^T\mathbf{q} + \mathbf{n}_d = \sqrt{\gamma_d\alpha}\mathbf{G}^T\mathbf{G}^*\mathbf{x} + \mathbf{n}_d$. For the kth terminal, its received signal is given by $z_{d,k} = \sqrt{\gamma_d\alpha} \sum_{i=1}^{K} \mathbf{g}_k^T \mathbf{g}_i^* x_k + n_{d,k}$. The SNR is given by

$$\rho_{d,k} = \gamma_d\alpha \left| \sum_{i=1}^{K} \mathbf{g}_k^T \mathbf{g}_i^* \right|^2. \tag{2.27}$$

For ZF, the transmitted signal is precoded using $\mathbf{F} = \sqrt{\alpha}\mathbf{G}^*(\mathbf{G}^T\mathbf{G}^*)^{-1}$. Thus, the signal received at the terminals becomes $\mathbf{z}_d = \sqrt{\gamma_d}\mathbf{G}^T\mathbf{q} + \mathbf{n}_d = \sqrt{\gamma_d\alpha}\mathbf{x} + \mathbf{n}_d$. The received signal of the kth terminal is $z_{d,k} = \sqrt{\gamma_d\alpha}x_k + n_{d,k}$. The SNR is given by

$$\rho_{d,k} = \gamma_d\alpha. \tag{2.28}$$

Several observations can be made from the above equations. First, both MRC and ZF decompose the received signals into decoupled components for different terminals. This has greatly simplified the reception at the terminals. The only difference between them is their SNR. Second, ZF needs to perform a matrix inversion for precoding while MRC does not need to. In this sense, MRC is simpler. It is also noted that the SNR of MRC increases with K and M in this case, while the SNR of ZF does not depend on K or M, except in the cases when the power normalization factor is set as a function of K or M. This means one can increase the SNR of MRC by scaling up the massive MIMO system. Finally, for ZF, sometimes a regularized algorithm can also be used, where $\mathbf{q} = \sqrt{\alpha}\mathbf{G}^*(\mathbf{G}^T\mathbf{G}^* + \delta\mathbf{I}_M)^{-1}\mathbf{x}$, where δ is the regularization factor that gives ZF when $\delta = 0$, MMSE when $\delta = 1$, and MRC when $\delta \to \infty$.

2.4   Issues in massive MIMO

In the previous section, we have discussed three important techniques of massive MIMO in different phases. In the discussion, a number of key assumptions have been made. For example, we have used TDD and assumed channel reciprocity in order to save the pilot transmission for downlink CSI acquisition. Also, we have assumed asymptotic orthogonality of the channel matrix or the favorable propagation in order to achieve the capacity and energy efficiency gains. Another important assumption is that we have only considered one cell with one BS and several terminals. We are interested in knowing what will happen if these assumptions are not satisfied or how realistic these assumptions are.

In the following, we will examine these three issues one by one. In particular, since most realistic systems have multiple cells, inter-cell interference is not negligible, which will cause pilot contamination. Since pilot contamination significantly reduces the accuracy of channel estimation, which then affects the performances of signal detection in the uplink and precoding in the downlink, we will examine this issue first, followed by some discussions on channel reciprocity and favorable propagation.

2.4.1 Pilot contamination

Pilot contamination occurs when multiple cells are used and share the same frequency band in the system. The main source of pilot contamination comes from the non-orthogonal pilots in different cells. For simplicity, we assume that there are now L cells in the system. Each cell has one BS serving K terminals, and the BS has M antennas. Denote $\mathbf{S}_l$ as the $K \times T_u$ pilot matrix for the lth cell, where $\mathbf{S}_l \mathbf{S}_l^H = \mathbf{I}_K$ and this pilot matrix is transmitted by K terminals to the lth BS during T_u symbol periods. Also, assume that all the cells transmit the pilots at the same time. This is the worst-case scenario. Also, denote $\mathbf{G}_{l,j} = \mathbf{A}_{l,j}\mathbf{B}_{l,j}^{1/2}$ as the channel matrix from terminals in the jth cell to the BS in the lth cell. Thus, $\mathbf{G}_{l,l}$ gives the channel matrix $\mathbf{G}$ in the single cell case discussed in the previous sections.

Based on the above assumptions, the received pilots for channel estimation at the BS in the uplink become:

$$\mathbf{Z}_{p,l} = \sqrt{\gamma_p T_u}\mathbf{G}_{l,l}\mathbf{S}_l + \sqrt{\gamma_p T_u} \sum_{j=1,j\neq l}^{L} \mathbf{G}_{l,j}\mathbf{S}_j + \mathbf{N}_{p,l}. \tag{2.29}$$

One sees that there is an extra term of $\sqrt{\gamma_p T_u} \sum_{j=1,j\neq l}^{L} \mathbf{G}_{l,j}\mathbf{S}_j$ coming from other cells, where the jth cell is transmitting a pilot matrix of $\mathbf{S}_j$. This is the inter-cell interference. Using the MMSE channel estimator, the CSI of the uplink can be acquired as

$$\hat{\mathbf{G}}_{l,l} = \frac{1}{\sqrt{\gamma_p T_u}}\mathbf{Z}_{p,l}\mathbf{S}_l^H = \mathbf{G}_{l,l} + \sum_{j=1,j\neq l}^{L} \mathbf{G}_{l,j}\mathbf{S}_j\mathbf{S}_l^H + \frac{1}{\gamma_p T_u}\mathbf{N}_{p,l}\mathbf{S}_l^H. \tag{2.30}$$

One sees that, in addition to the channel estimation error caused by the noise as in previous sections, there is an additional channel estimation error caused by the inter-cell interference during the pilot transmission. This term is called pilot contamination.

If the same pilot is used for all cells such that $\mathbf{S}_l = \mathbf{S}_j$, one has $\mathbf{S}_j\mathbf{S}_l^H = \mathbf{I}_K$ such that the MMSE channel estimate becomes $\hat{\mathbf{G}}_{l,l} = \mathbf{G}_{l,l} + \sum_{j=1,j\neq l}^{L} \mathbf{G}_{l,j} + \frac{1}{\gamma_p T_u}\mathbf{N}_{p,l}\mathbf{S}_l^H$. This incurs the largest amount of pilot contamination. If the pilots from different cells are orthogonal such that $\mathbf{S}_j\mathbf{S}_l^H = \mathbf{0}_K$ for $j \neq l$, the MMSE channel estimate becomes $\hat{\mathbf{G}}_{l,l} = \mathbf{G}_{l,l} + (1/\gamma_p T_u)\mathbf{N}_{p,l}\mathbf{S}_l^H$. This completely removes pilot contamination. In practice, the cells with pilots the same as $\mathbf{S}_l$ will maximize the pilot contamination while the cells with pilots orthogonal to $\mathbf{S}_l$ will not cause any pilot contamination. Due to the limited channel coherence time and bandwidth, the training period is limited such that one may have to reuse pilots over different cells [7]. Thus, in practice, non-orthogonal pilots often exist to induce pilot contamination. The pilot contamination will lead to an error floor or capacity upper limit, as most inter-user or inter-cell interferences do.

Next, we examine the effect of pilot contamination. For convenience, we only consider the case when the same pilot is reused in all cells. Also, we only examine MRC. Other schemes can be discussed similarly. When multiple cells exist, the received signal at the BS in the lth cell can be expressed as $\mathbf{z}_{u,l} = \sqrt{\gamma_u} \sum_{j=1}^{L} \mathbf{G}_{l,j}\mathbf{p}_j + \mathbf{n}_{u,l}$, where $\mathbf{p}_j$ is the transmitted data from the jth cell that causes inter-cell interference when $j \neq l$. Then, using the channel estimate for signal detection, one has

$$\mathbf{y}_{u,l} = \left(\sum_{j_1=1}^{L} \mathbf{G}_{l,j_1} + \frac{1}{\gamma_p T_u}\mathbf{N}_{p,l}\mathbf{S}_l^H \right)^H \left(\sqrt{\gamma_u} \sum_{j_2=1}^{L} \mathbf{G}_{l,j_2}\mathbf{p}_{j_2} + \mathbf{n}_{u,l} \right). \tag{2.31}$$

One sees that transmission of other cells causes pilot contamination in the channel estimate as well as inter-cell interference in the data reception of the terminals. When the number of antennas M goes to infinity, using the asymptotic orthogonality, it can be shown that the SINR of the kth user in the lth cell becomes [7]:

$$\rho_{u,k} \approx \frac{b_{lkl}^2}{\sum_{j=1,j\neq l}^{L} b_{lkj}^2} \tag{2.32}$$

where b_{lkj} is the kth element of $\mathbf{B}_{l,j}$ and $\sum_{j=1,j\neq l}^{L} b_{lkj}^2$ is the sum of squares of $L-1$ coefficients from different terminals to the same BS. Thus, pilot contamination causes an error floor and it will not disappear with an increase in the number of antennas. This is a major deteriorating factor of massive MIMO. However, the SINR only depends on large-scale fading, and it is not related to small-scale fading or noise. It can be shown that the same result applies to ZF and MMSE detectors.

Pilot contamination affects the downlink too. Again, use MRC as an example. The precoded signal for transmission in the downlink can be expressed as $\mathbf{z}_{d,l} = \sqrt{\gamma_d} \sum_{j=1}^{L} G_{l,j}^T \mathbf{q}_j + \mathbf{n}_{d,l}$, where $\mathbf{q}_j$ is the precoded data from the BS in the jth cell. Using MRC precoding, we have

$$\mathbf{z}_{d,l} = \sqrt{\gamma_d}\alpha \sum_{j=1}^{L} G_{l,j}^T \hat{G}_{j,j}^* \mathbf{x}_j + \mathbf{n}_{d,l}$$

$$= \sqrt{\gamma_d}\alpha \sum_{j=1}^{L} G_{l,j}^T \left[\sum_{j_1=1}^{L} \mathbf{G}_{l,j_1} + \frac{1}{\gamma_p T_u}\mathbf{N}_{p,l}\mathbf{S}_l^H \right]^* \mathbf{x}_j + \mathbf{n}_{d,l}. \tag{2.33}$$

Using the asymptotic orthogonality, when M is large, the SINR is approximately:

$$\rho_{d,k} \approx \frac{b_{lkl}^2}{\sum_{l=1,l\neq j}^{L} b_{lkj}^2} \tag{2.34}$$

where $\sum_{l=1,l\neq j}^{L} b_{lkj}^2$ is the sum of squares of $L - 1$ coefficients from different BSs in different cells to the same terminal. Again, pilot contamination causes error floor that is independent of the number of antennas.

2.4.2   Channel reciprocity

Channel reciprocity is an important assumption that we have made when we go for the TDD mode. This is normally the case when the uplink and the downlink operate at the same frequency band. However, there may still be some deviation from this assumption due to hardware impairment.

Assume that T_1 and R_1 are the power amplifier and the low-noise amplifier of the transmitter and the receiver at the BS, respectively, and T_2 and R_2 are the power amplifier and the low-noise amplifier of the transmitter and the receiver at the terminal, respectively. Thus, the uplink point-to-point channel from the terminal to the BS can be expressed as $G(t, \tau) = T_2(\tau) * C(t, \tau) * R_1(\tau)$ and the downlink point-to-point channel from the BS to the terminal can be expressed as $H(t, \tau) = T_1(\tau) * C(t, \tau) * R_2(\tau)$, where τ is the delay variable, $C(t, \tau)$ is the impulse response of the electromagnetic channel, and $*$ represents convolution. Since the BS and the terminal have different requirements, it is likely that their amplifiers may not have the same characteristics. Consequently, $G(t, \tau)$ may not be the same as $H(t, \tau)$, which violates the channel reciprocity assumption. Also, in practical transceivers, frequency offset is often seen. This offset may accumulate over different operations and can make the downlink channel and uplink channel non-reciprocal in a few seconds [17]. Thus, robust calibration must be considered to estimate the difference between $G(t, \tau)$ and $H(t, \tau)$ using real channel measurements and then compensate the channel estimator with this difference in signal detection or precoding. Without this calibration, effects similar to pilot contamination may occur to reduce the SINR further.

2.4.3   Favorable propagation

The favorable propagation assumption has been used in almost all the discussions in the previous sections. It is important to check if this assumption holds in reality, as this will determine the usefulness of the previous results. Fortunately, it has been shown in several experiments that this assumption holds in many cases of massive MIMO [8,18,19].

Physically, the favorable propagation assumption requires that the channel responses from the BS to different terminals are sufficiently different so that different terminals can be separated. Mathematically, as we used before, this requires that the small-scale fading matrix satisfies $(1/M)\mathbf{AA}^H \to \mathbf{I}$ when M is very large. To verify this, reference [19] used a 128-element circular array at the BS serving 3 indoor terminals and 3 outdoor terminals at 2.6 GHz with a bandwidth of 50 MHz.

Using 100 snapshots at 161 frequency points, channel measurements are taken and then normalized to examine small-scale fading. The cumulative distribution functions of the ordered eigenvalues of $\mathbf{GG}^H$ were plotted and compared with the ideal case of independent and identically distributed complex Gaussian channels. It was shown that the large antenna array provides eigenvalues with very stable behaviors and the spread between the largest and the smallest eigenvalues is less than 7 dB, much smaller than a conventional 6×6 MIMO system that has a spread of about 23 dB. Reference [8] also examined a linear array of 128 elements at different positions. Using similar setting but for 4 terminals, the linear array was shown to have a spread less than 3 dB between the largest and smallest eigenvalues, very close to the ideal case of an identity matrix. These experimental results support the favorable propagation assumption used in the discussions.

In summary, the favorable propagation is largely valid in practice. The actual convergence rate of $(1/M)\mathbf{AA}^H$ toward the identity matrix depends on the exact configuration of the antenna array, the precoding method used and other factors. Next, we discuss network MIMO before we finish this chapter.

2.5 Network MIMO

Network MIMO is also called multi-cell MIMO. It aims to reduce or completely remove the inter-cell interference by introducing cooperation between base stations. A similar idea was proposed as macro diversity decades ago to fight large-scale fading, such as shadowing, by using multiple base stations [20]. Recently, it has been re-considered due to the high spectral efficiency requirements of 5G, while the inter-cell interference is the main limiting factor of spectral efficiency in current cellular systems. For comparison, massive MIMO is a type of multi-user MIMO to combat the multi-user interference, when different users have asymptotically orthogonal channels for large numbers of antennas, while network MIMO is a type of multi-cell MIMO to combat the inter-cell interference by coordinating base stations in the system.

The idea of network MIMO is to exploit the inter-cell interference such that all the base stations in the network become a whole virtual array of antennas serving all users in the network, as opposed to conventional systems that treat inter-cell interference as harmful disturbance and combat it. This virtual MIMO can be achieved by connecting all the base stations to a central processor via high-speed backhaul links for information exchange. The price that one has to pay is the increased complexity of multi-cell cooperation as well as increased hardware for backhaul connections. In fact, the equivalence of multi-cell and virtual MIMO is only valid when the backhaul condition is ideal. In the following, we discuss the results of network MIMO.

There are different levels of multi-cell cooperation, ranging from interference coordination or coordinated beamforming, where the BSs only share the CSI of different links via feedback controls to jointly mitigate the inter-cell interference, to rate-limited MIMO cooperation, where the BSs share the CSI of different links as well as limited amount of data from different users, to full MIMO cooperation, where the BSs share the CSI of all links and the data of all users to maximize the benefit of

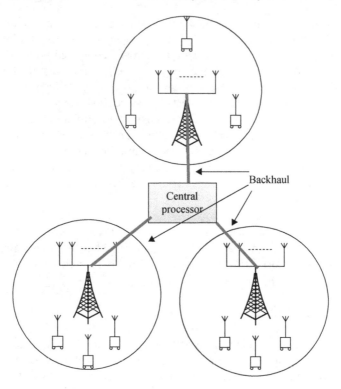

Figure 2.6   Diagram of network MIMO

multi-cell processing. Network MIMO is only equivalent to multi-user MIMO when full MIMO cooperation is explored, which is the case considered in this chapter. As a result, many results for massive MIMO and network MIMO are similar, except that one at the cell level and the other at the network level.

Assume that there are L cells each with one BS in the system and K terminals in each cell. Each BS in the cell has M antennas, while the terminals in the cell have a single antenna. The BSs cooperate with each other and operate at the same frequency band. Figure 2.6 shows a diagram of network MIMO.

Next, consider the achievable rates first. Most of these results in the literature assume a Wyner-type model, where all BSs have single antennas ($M = 1$) and interference is only present between adjacent cells. In particular, the following discussion focuses on the linear Wyner model, where the BSs are on a line, such as a highway or a long corridor. Essentially, this model implies that both the terminals and the BSs form distributed MIMO, as each terminal and each BS can be considered as a distributed antenna in the network. In this case, the received signal in the uplink can be expressed as $z_{u,l} = \sum_{j=-N_l}^{N_u} g_{l,l-j} p_{l-j} + n_{u,l}$, where $z_{u,l}$ is the received signal at the lth BS and is a scalar as the BS has a single antenna now, N_l and N_u are the numbers of interfering cells at each side of the lth cell, $g_{l,l-j}$ is the channel vector from the $(l-j)$th cell to the lth cell, $\mathbf{p}_{l-j}$ is the transmitted signal from the $(l-j)$-th cell, and n_l is the receiver noise at the BS of the lth cell. This model includes two special cases: Gaussian Wyner channel, where $N_u = N_l = N$ and $\mathbf{g}_{l,l-j} = a_j \mathbf{1}_K$ and Gaussian soft-handoff channel, where $N_l = 0, N_u = N$, and $\mathbf{g}_{l,l-j} = a_j \mathbf{1}_K$, with $\mathbf{1}_K$ being an all one vector. In fading channels, random variables can also be assumed where the channel vectors become independent and its elements follow a joint distribution. Using the received signal, the achievable rate per cell can be derived as [21]:

$$R_u = \frac{1}{L} \log_2 \det\left(\mathbf{I}_L + \frac{P}{K}\mathbf{GG}^H\right) \tag{2.35}$$

where P is the total transmission of each cell and $\mathbf{G}$ is a $L \times LK$ channel matrix whose lth row contains all channel coefficients to the lth BS from other cells and terminals in the same cell. This assumes unlimited backhaul capacity in the ideal case. When the backhaul is limited so that rate-limited MIMO cooperation has to be performed, the achievable rate will be the minimum of R_u and the backhaul capacity. Similarly, for the downlink, the achievable rate is given by [22]:

$$R_d = \frac{1}{L} \min_{\mathbf{\Lambda}} \max_{\mathbf{C}} \log_2 \frac{\det(\mathbf{\Lambda} + P/K\mathbf{GCG}^H)}{\det(\mathbf{\Lambda})} \tag{2.36}$$

where $\mathbf{\Lambda}$ and $\mathbf{C}$ are $LK \times LK$ diagonal matrices with $\text{trace}(\mathbf{\Lambda}) \leq L$ and $\text{trace}(\mathbf{C}) \leq L$. Again, unlimited backhaul capacity is assumed.

Next, we consider signal processing in the multi-cell or network MIMO setting. In particular, we focus on precoding in the downlink and signal detection in the uplink, as before. Assume that $g_{mlkj} = a_{mlkj}\sqrt{b_{lkj}}$ as the channel gain between the mth antenna in the lth cell and the kth terminal in the jth cell, a_{mlkj} represents small-scale fading and b_{lkj} represents large-scale fading. The elements of g_{mlkj}, a_{mlkj}, and b_{lkj} can be stacked to give $\mathbf{G}_{l,j}$, $\mathbf{A}_{l,j}$, and $\mathbf{B}_{l,j}$ in the previous sections.

For precoding in the downlink, consider the simple MRC scheme. Instead of performing beamforming on the transmitted signal $\mathbf{x}_j$ directly, as in single-cell processing in (2.33), network MIMO performs an additional transform on $\mathbf{x}_j$ as $x'_{k,l} = \sum_{j=1}^{L} \psi_{l,k,j} x_{k,j}$, where $x_{k,j}$ is the transmitted signal for the kth terminal in the jth cell and $\psi_{l,k,j}$ is determined by the large-scale fading coefficients b_{lkj} to cancel the effect of inter-cell interference. For detection in the uplink, a simple linear MRC scheme can also be applied. In this case, after the normal combining as performed in (2.31) for single-cell processing, network MIMO performs an additional combination by multiplying the combined signals for the kth terminals in all L base stations as $y_{u,l,k} = \sum_{j=1}^{L} \phi_{l,k,j} y_{u,j,k}$, where $\phi_{l,k,j}$ is also determined by the large-scale fading coefficients b_{lkj}. It was shown in [23] that this extra transformation can completely remove the inter-cell interference in the asymptotic case such that the SINR does not have an upper limit with an increase of transmission power. One sees that the principle here for network MIMO is to perform extra transformation or combination to cancel the inter-cell interference at the network level. One sees that in doing this, CSI and data from all cells, such as $x_{k,j}$ and $y_{u,k,j}$, need to be shared by all cells to be sent to the central processor. More complicated cases can be found in [24].

2.6   Conclusion

In this chapter, we have discussed massive MIMO and network MIMO. The massive MIMO is compared with point-to-point MIMO in terms of capacity and energy efficiency. The analysis has shown that massive MIMO is able to achieve a capacity gain of at least 10 and an energy efficiency gain of at least 100 over the conventional point-to-point MIMO used in current cellular systems, even under the modest conditions using simple linear processing. Thus, massive MIMO is a very promising solution to meet the requirements of 5G. However, these gains are achieved at some cost. In particular, inter-cell interference will cause pilot contamination, which could seriously degrade the performance of massive MIMO. To tackle this problem, several measures can be taken. Among them, network MIMO introduces cooperation between base stations to completely remove the inter-cell interference, at the cost of even higher complexity. These techniques mainly focus on the radio access of the 5G network. They can have implications on the other parts of the system. For example, in network MIMO, the amount of overhead required for coordination places a great pressure on the backhaul to base stations as well as on the fronthaul to the mobile terminals. This may seriously reduce the achievable rate of the whole network. This coordination also necessitates sophisticated system level control to make it work.

References

[1]   Andrews J.G., Buzzi B., Choi W., *et al.*, "What will 5G be?," *IEEE Journal of Selected Areas of Communication*, vol. 32, pp. 1065–1082, Jun. 2014.

[2]   Boccardi F., Heath R.W., Lozano A., Marzetta T.L., and Popovski P., "Five disruptive technology directions for 5G," *IEEE Communications Magazine*, vol. 52, pp. 74–80, Feb. 2014.

[3]   Matthaiou M., McKay M.R., Smith P.J., and Nossek J.A., "On the condition number distribution of complex Wishart matrices," *IEEE Transactions on Communications*, vol. 58, no. 6, pp. 1705–1717, Jun. 2010.

[4]   Jalden J. and Ottersten B., "On the complexity of sphere decoding in digital communications," *IEEE Transactions on Signal Processing*, vol. 53, no. 4, pp. 1474–1484, Apr. 2005.

[5]   Vishwanath S., Jindal N., and Goldsmith A., "Duality, achievable rates, and sum-rate capacity of Gaussian MIMO broadcast channels," *IEEE Transactions on Information Theory*, vol. 49, no. 2, pp. 2658–2668, Oct. 2003.

[6]   Costa M., "Writing on dirty paper," *IEEE Transactions on Information Theory*, vol. IT-29, no. 3, pp. 439–441, May 1983.

[7]   Marzetta T.L., "Noncooperative cellular wireless with unlimited numbers of base station antennas," *IEEE Transactions on Wireless Communications*, vol. 9, no. 11, pp. 3590–3600, Nov. 2010.

[8]   Larsson E.G., Edfors O., Tufvesson F., and Marzetta T., "Massive MIMO for next generation wireless systems," *IEEE Communications Magazine*, vol. 52, no. 2, pp. 186–195, Feb. 2014.

[9]   Marzetta T.L. and Hochwald B.M., "Fast transfer of channel state information in wireless systems," *IEEE Transactions on Signal Processing*, vol. 54, pp. 1268–1278, Apr. 2006.

[10] Marzetta T.L., "How much training is required for multiuser MIMO?," *Proceedings of the 40th Asilomar Conference on Signals, Systems, Computing*, Pacific Grove, USA, pp. 359–363, Oct. 2006.

[11] Adhikary A., Nam J., Ahn J.-Y., and Caire G., "Joint spatial division and multiplexing – The large-scale array regime," *IEEE Transactions on Information Theory*, vol. 59, no. 10, pp. 6441–6463, Oct. 2013.

[12] Nam J., Ahn J.-Y., Adhikary A., and Caire G., "Joint spatial division and multiplexing: realizing massive MIMO gains with limited channel state information," in *Proceedings of the 46th Annual Conference on Information Science Systems (CISS)*, 2012, pp. 1–6.

[13] Choi J., Love D., and Kim T., "Trellis-extended codebooks and successive phase adjustment: a path from LTE-advanced to FDD massive MIMO systems," *IEEE Transactions on Wireless Communications*, vol. 14, no. 4, pp. 2007–2016, Apr. 2014.

[14] Nguyen S. and Ghrayeb A., "Compressive sensing-based channel estimation for massive multiuser MIMO systems," *Proceedings of the IEEE Wireless Communications and Networking Conference (WCNC)*, Shanghai, China, pp. 2890–2895, Apr. 2013.

[15] Ngo H.Q., Larsson E.G., and Marzetta T.L. "Energy and spectral efficiency of very large multiuser MIMO systems," *IEEE Transactions on Communications*, vol. 61, no. 4, pp. 1436–1449, Apr. 2013.

[16] Hochwald B.M., Peel C.B., and Swindleust A.L., "A vector-perturbation technique for near-capacity multiantenna communications – Part II: perturbation," *IEEE Transactions on Communications*, vol. 53, no. 5, pp. 537–544, May 2005.

[17] Guillaud M. and Kaltenberger F., "Towards practical channel reciprocity exploitation: relative calibration in the presence of frequency offset," *Proceedings of the IEEE Wireless Communications Network Conference (WCNC)*, 2013, pp. 2525–2530.

[18] Gao X., Edfors O., Tufvesson F., and Larsson E.K., "Massive MIMO in real propagation environments: do all antennas contribute equally?," *IEEE Transactions on Communications*, vol. 63, no. 11, pp. 3917–3928, Nov. 2015.

[19] Rusek F., Persson D., Lau B.K., and Larsson E.G., "Scaling up MIMO: opportunities and challenges with very large arrays," *IEEE Signal Processing Magazine*, pp. 40–60, Jan. 2013.

[20] Stuber G.L., *Principles of Mobile Communication*, 2nd Ed. London: Kluwer, 2001.

[21] Wyner A.D., "Shannon-theoretic approach to a Gaussian cellular multiple-access channel," *IEEE Transactions on Information Theory*, vol. 40, no. 6, pp. 1713–1727, Nov. 1994.

[22] Somekh O., Zaidel B.M., and Shamai S., "Sum rate characterization of joint multiple cell-site processing," *IEEE Transactions on Information Theory*, vol. 53, no. 12, pp. 4473–4497, Dec. 2007.

[23]   Ashikhmin A. and Marzetta T.L., "Pilot contamination precoding in multi-cell large scale antenna systems," *Proceedings of the IEEE ISIT 2012*, pp. 1137–1141, 2012.

[24]   Zakhour R. and Hanly S.V., "Base station cooperation on the downlink: large system analysis," *IEEE Transactions on Information Theory*, vol. 58, no. 4, pp. 2079–2106, Apr. 2012.

Chapter 3

The role of massive MIMO in 5G access networks: potentials, challenges, and solutions

Nassar Ksairi[1], Marco Maso[1], and Beatrice Tomasi[1]

Abstract

The massive multiple–input/multiple–output (MIMO) technique has been gaining momentum lately as a potential key enabler of the network spectral efficiency (and hence throughput) increase expected from the fifth generation (5G) of cellular network technology. This technique consists in equipping the base stations (BSs) with arrays having a large number, typically from tens to few hundreds, of antenna elements. This, in theory, guarantees the possibility of serving larger numbers of concurrent transmissions to or from the BSs than the multiplexing techniques used in previous generations of cellular technology, without compromising the throughput of each transmission. By exploiting the high spatial multiplexing gain that can be realized through massive MIMO systems, 5G networks could provide the expected boost to the overall throughput with respect to previous generations. On the one hand, having more antenna elements in the BS array increases the spatial resolution of both outgoing and incoming signals. On the other hand, it diminishes the effect of the additive white gaussian noise (AWGN) at the receiver. These features lead to several advantages both in terms of access mechanisms and associated signal processing: little intra-cell and multicell interference leakage, optimality of simple linear precoding/detection schemes, capability of simultaneously serving a multitude of competing wireless connections in each cell area without compromising their individual throughput, just to name a few. However, this comes at the cost of the acquisition of accurate uplink/downlink channel state information (CSI) at the BS. This operation is challenging in many respects, and has a non-negligible impact in terms of both algorithms implemented at the physical/access/network layer and practical hardware design. It is worth noting that the extent of this impact strongly depends on the frequency bands over which the transmissions are performed. In this regard, it is a common belief that signals transmitted in future 5G networks will likely span a wide spectrum of frequencies.

[1]Mathematical and Algorithmic Sciences Laboratory, Huawei Technologies France SASU, France

In other words, their radio carrier wavelengths will range from decimeters, as in legacy cellular networks, to millimeters, as in millimeter wave (mmWave) communications. Evident peculiarities will characterize such signals, and future massive MIMO systems will have to leverage them in effective ways in order to achieve the promised network throughput increase. This chapter starts from these observations to present a discussion on the potential and the challenges associated to the deployment of massive MIMO systems for 5G networks.

3.1   Introduction

According to recent reports, global mobile data traffic grew 74% in 2015, reaching 3.7 exabytes per month at the end of 2015, up from 2.1 exabytes per month at the end of 2014. In this context, mobile data traffic will reach the following milestones by 2020 [1]:

- Monthly global mobile data traffic will be 30.6 exabytes;
- The number of mobile-connected devices per capita will reach 1.5;
- Seventy-five percent of the world's mobile data traffic will be video by 2020;
- New services are expected to emerge in the telecommunications ecosystem, e.g., virtual reality and autonomous vehicles, in addition to the existing mobile services.

As a matter of fact, long-term evolution (LTE) and long-term evolution advanced (LTE-A) may not offer suitable tools to address the expected data traffic and mobile-connected devices increases. This inadequacy could make issues such as spectrum scarcity, capacity shortfalls and poor coverage reoccur in the cellular network, despite all the efforts made in the last decade to address them. A clear response to the traffic explosion generated by new-generation wireless devices and novel mobile services is then necessary. The identification of effective technologies to handle the constant and relentless increase of mobile data traffic is one of the main priorities at the dawn of the definition of the so-called 5G cellular networks. This goal is certainly ambitious, however the aforementioned figures do not leave much room for doubts about the picture at large. Based on the so-far available results of theoretical analysis, extensive simulations and early prototypes and test deployments, massive MIMO can be rightfully considered one of the most promising access technologies toward achieving the above goal by merit of its spectrum efficiency [2–4]. This technology simply consists in using large arrays of antennas at the BSs to simultaneously serve a large number of user terminals (UTs) over the same time/frequency resources by means of adapted signal processing and data detection techniques. In this regard, it should be noted that equipping the BSs with multiple antennas, or using them to simultaneously serve many UTs, is not a new idea in itself. However, the number of these antennas and the way they are intended to be used in massive MIMO, i.e., to spatially multiplex a multitude of concurrent transmissions between UTs and their serving BS, has no precedent in legacy wireless networks.

3.2   The role of MIMO techniques in access networks

The maximum throughput offered by a wireless link can be increased in several ways. The most efficient approach to achieve this goal is arguably to act on the system parameters that affect the throughput linearly*, e.g., the transmission bandwidth and the number of streams multiplexed over the link. The feasibility of these two solutions strongly depends on the considered system. For instance, if we focus on the bands suitable for cellular coverage, we observe that available resources in this portion of the spectrum are indeed very scarce. Enlarging the bandwidth of the transmissions performed to and from the UTs is often impossible. In this context, throughput enhancements must necessarily come from the spatial multiplexing of several data streams on the same bandwidth, a technique that increases the spectral efficiency of a wireless link provided that both the transmitter and the receiver nodes on that link are equipped with multiple antennas. A straightforward way to increase the total throughput on the network level is thus to resort to MIMO point-to-point technologies, and adopt suitable signal processing at the physical layer to be able to concurrently serve several UTs. This particular MIMO setting is referred typically to as multiuser (MU) MIMO.

3.2.1   Multiuser MIMO

The performance of MU MIMO systems has been vastly analyzed in the last two decades [5]. MU-MIMO technologies are nowadays considered mature both theoretically and practically. For this reason, recent communication standards, e.g., LTE-A and 802.11n, already include several operating modes in which BSs are equipped with more than one antenna. It is a well-accepted belief that this trend will not only continue in future communications standards, but also intensify [4]. The advantages of a MU-MIMO system over its single user (SU) counterpart are evident: (i) several UTs can be served simultaneously by the MIMO BS, (ii) the spatial multiplexing gain can be achieved by the BS even in presence of single-antenna UTs, and (iii) MU diversity is more robust than per-user diversity to rank-deficient and/or line-of-sight (LOS) channels. However, these benefits come at a price. In particular, full CSI at the BS with respect to the downlink (resp. uplink) is necessary to be able to perform linear precoding in the downlink (resp. linear combining in the uplink) and to observe the multiplexing gain. Acquiring the CSI in a MU setting may be extremely complex and place a non-negligible burden on uplink capacity in most systems [5]. A more detailed discussion on this aspect is deferred to Section 3.3. A perfect CSI assumption will be made in the remainder of this section for the sake of simplicity.

Let us now switch our focus to possible approaches that can help attain the aforementioned spatial multiplexing gain. Consider the downlink of a single cell with a pool $\mathcal{K}$ of $K > 1$ single-antenna UTs associated to a BS with $M > 1$ antennas. Assume that a subset $\mathcal{M} \subset \mathcal{K}$ of these UTs is selected by the scheduler for downlink transmission and denoted by $\mathbf{h}_k$ the $M \times 1$ channel vector between the BS and any

---

*As opposed to system parameters that affect the throughput algorithmically, e.g., the transmit power.

user $k \in \mathcal{M}$. One possible downlink MU-MIMO implementation is *linear precoding*. In this scheme, the per-antenna baseband signals are multiplied with complex coefficients so that they add up in a constructive manner at the intended UT after the wireless propagation, while optionally reducing or nulling the interference experienced by the co-scheduled UTs. More precisely, if we denote by $\mathbf{w}_{k,n}$ the precoding vector for the signal of user k at the position of the resource element (RE) n, then for all $k \in \mathcal{M}$ the baseband signal sample y_k received by the terminal of user k is given by

$$y_{k,n} = \sqrt{\alpha} \sum_{j \in \mathcal{M}} \mathbf{w}_{j,n}^{\mathrm{H}} \mathbf{h}_{k,n} x_{j,n} + z_{k,n}, \tag{3.1}$$

where $x_{k,n}$ is the zero-mean unit-variance data symbol destined for user j on n and $z_{k,n}$ is an additive noise random variable with variance σ^2. Here, $\mathbf{h}_{k,n}$ is the vector of baseband-equivalent discrete-time channel coefficients from the BS to the receiver of user k at RE n and α is a normalization factor chosen to constrain the average per-user transmit power $\mathbb{E}[\|\mathbf{w}_{k,n}^{\mathrm{H}} x_{k,n}\|^2]$ to be equal to P^{DL}. In practice, each linear precoder can be designed according to several criteria. For instance, the so-called downlink zero forcing beamforming (ZFBF) consists in choosing $\mathbf{w}_{k,n}$ such that inter-user interference is eliminated, i.e., $\mathbf{w}_{j,n}^{\mathrm{H}} \mathbf{h}_k = 0$, $\forall j \neq k$. Let $\mathbf{W}_n \stackrel{\text{def}}{=} [\mathbf{w}_{k_1,n} \cdots \mathbf{w}_{k_M,n}]$ and $\mathbf{H}_n \stackrel{\text{def}}{=} [\mathbf{h}_{k_1,n} \cdots \mathbf{h}_{k_M,n}]$, where $\{k_1, k_2, \ldots, k_M\} = \mathcal{M} \subset \mathcal{K}$ are the indexes of the scheduled users. The ZFBF precoding vectors can be obtained from the channel matrix as $\mathbf{W}_n = \mathbf{W}_n^{\mathrm{ZF}}$, where

$$\mathbf{W}_n^{\mathrm{ZF}} \stackrel{\text{def}}{=} \left(\mathbf{H}_n \mathbf{H}_n^{\mathrm{H}} \right)^{-1} \mathbf{H}_n. \tag{3.2}$$

In general, ZFBF requires a matrix inversion, which can be a challenging operation in large MU-MIMO systems. A more feasible alternative to compute the precoding vectors is the so-called maximum-ratio transmission (MRT) beamforming. This approach, whose goal is to maximize the amount of power conveyed to each UT, i.e., its experienced signal to noise ratio (SNR), simply consists in setting $\mathbf{W}_n = \mathbf{W}_n^{\mathrm{MRT}}$, where

$$\mathbf{W}_n^{\mathrm{MRT}} \stackrel{\text{def}}{=} \mathbf{H}_n. \tag{3.3}$$

Now, denote by $R_n(\mathcal{M})$ the achievable instantaneous spectral efficiency on the considered RE when users $\mathcal{M}$ are scheduled for downlink transmission on that element. Whether using ZFBF or MRT, $R_n(\mathcal{M})$ can be written as

$$R_n(\mathcal{M}) = \sum_{k \in \mathcal{M}} \log_2(1 + \mathsf{sinr}_{k,n}) \tag{3.4}$$

where $\mathsf{sinr}_{k,n}$ is the signal to interference plus noise ratio (SINR) at user k over RE n given by

$$\mathsf{sinr}_{k,n} = \frac{\alpha |\mathbf{w}_{k,n}^{\mathrm{H}} \mathbf{h}_{k,n}|^2}{\alpha \sum_{j \neq k} |\mathbf{w}_{j,n}^{\mathrm{H}} \mathbf{h}_{k,n}|^2 + \sigma^2}. \tag{3.5}$$

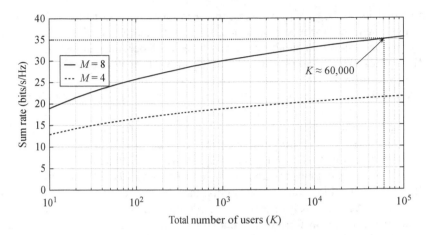

Figure 3.1 *Sum rate performance of ZFBF in a single-cell scenario with optimal user scheduling (not accounting for CSI overhead)*

where $\mathbf{w}_{k,n}$ are the column vectors of (3.2) if ZFBF is applied and of (3.3) if MRT is used instead. The maximum instantaneous spectral efficiency, denoted as R_n^*, is achieved when the selection of $\mathcal{M}$ out of $\mathcal{K}$ can be optimized as follows:[1]

$$R_n^* \stackrel{\text{def}}{=} \max_{\mathcal{M} \subset \mathcal{K}} R_n(\mathcal{M}). \tag{3.6}$$

When the size of the users pool, K, is much larger than M, asymptotic analysis tools can be used to obtain the scaling of R_n^* with respect to K [6]. For instance, consider the downlink of a single-cell MU-MIMO system with respectively $M = 4$ and $M = 8$ BS antennas and in which ZFBF is adopted. Now assume that the scheduler can choose, respectively, four and eight users from a much larger pool of K users with diverse channel conditions, i.e., $\mathbf{h}_{k,n}$ and $\mathbf{h}_{j,n}$ are independent random vectors if $k \neq j$ with independent and identically distributed (i.i.d.) entries, each having a zero-mean complex Gaussian distribution with a variance η_k and η_j, respectively. Then, the ergodic spectral efficiency achievable in this case, i.e., $\mathbb{E}[R_n^*]$, grows with K, as illustrated in Figure 3.1. In this regard, we note that the figure has been obtained assuming that the SNR experienced by user k, i.e., $\mathsf{snr}_k \stackrel{\text{def}}{=} P^{\text{DL}} \eta_k / \sigma^2$, is equal to $10 - 10 \log_{10}(M)$ dB. This SNR value is obtained, for instance, if $K = 8$ users are located on a circle of radius 300 m around a BS that transmits at a power level equal to $KP^{\text{DL}} = 46$ dBm and if the distance-dependent path-loss follows a COST-231 Hata model [7] with a carrier frequency 2.4 GHz. The fact that the achievable (sum) spectral efficiency increases with K is referred to as MU diversity. Indeed,

---

[1]Ideally, the maximum spectral efficiency is achieved when the optimization of user scheduling is done jointly on all the available resource elements, not separately as in (3.6), so that the gain from time and frequency diversity can be attained.

with a sufficiently large pool of users, the scheduler can almost surely find a group of M users with homogeneous path-loss conditions and whose channel vectors are nearly orthogonal to one another.[2] In this case, MRT becomes almost equivalent to ZFBF [6]. However, relying on this desirable property to achieve orders-of-magnitude gains in spectral efficiency, while having a moderate number of antennas at the BS turns out to be impractical for two main reasons. First, such gains are only possible if $K \gg M$, a condition that is rarely satisfied in real-world systems. For instance, achieving a tenfold increase with respect to the spectral efficiency of current cellular communications systems[3] requires that the cell area contain $K \approx 60,000$ service-seeking users, out of which only four to eight users will be simultaneously served! Second, MU diversity gains are conditioned on having timely CSI about the channels of all the K users asking for service, so that the scheduler can choose the fit M users among them. As previously mentioned, collecting this CSI could quickly become an excessive burden on the system capacity in this case, as further discussed in Section 3.3. In this regard, it is worth noting that the CSI overhead, needed both for taking scheduling decisions in (3.6) and for setting the precoding coefficients in (3.2) and (3.3), is not accounted for in Figure 3.1.

As for the uplink, one practical way to implement MU MIMO is to apply *linear combining*. In this scheme, the baseband signals at the output of the radio frequency (RF) chains of the BS antennas are multiplied with complex coefficients before being added to boost the signal of the intended UT at the expense of the co-scheduled UTs. We cite zero forcing combining (ZFC) and maximum-ratio combining (MRC) as two uplink schemes that can be used to select the combining coefficients for the co-scheduled users. These schemes are the counterparts of the downlink precoding schemes given, respectively, by (3.2) and (3.3). Accordingly, the MU diversity gain achievable with ZFC and MRC, and the uplink spectral efficiency, can be analyzed as we did with for the downlink, and similar trends as in Figure 3.1 can be observed.

3.2.2 From MU MIMO to massive (MU) MIMO

As we showed in the previous section, the key property behind MU diversity is that when the users pool $\mathscr{K}$ is large enough, there is almost surely a subset of the pool with nearly mutually orthogonal channel vectors. The basic idea behind massive MIMO is the ability to retain all the benefits of MU diversity regardless of the size of the users pool in each cell. It turns out that this is achievable by equipping the BS with a large number $M \gg 1$ of antennas, at least under certain statistical propagation conditions that are referred to as *favorable propagation* [4].

Indeed, when the distribution of the $M \times K$ random matrix $\mathbf{H}_{\text{tot},n} \stackrel{\text{def}}{=} [\mathbf{h}_{1,n} \cdots \mathbf{h}_{K,n}]$ satisfies some favorable-propagation conditions, letting M grow to infinity induces an

---

[2]At least under the assumption of independent fading across different users.

[3]In a transmission without spatial multiplexing, the achievable spectral efficiency when $\mathsf{snr} = 10$ dB is upper-bounded by $\log(1 + \mathsf{snr}) \approx 3.5$ bits/s/Hz. This value is in the middle of the range of spectral efficiency values (0.2–5.6 bits/s/Hz) achievable in LTE.

asymptotic property of users' channel random vectors, known as the *massive MIMO effect*, expressed by

$$\frac{1}{M}\|\mathbf{h}_{k,n}\|^2 \to 1, \text{ and } \frac{1}{M}\mathbf{h}_{k,n}^{\mathrm{H}}\mathbf{h}_{j,n} \to 0, \forall k,j \in \mathcal{K}, k \neq j, \tag{3.7}$$

where convergence in (3.7) should be in some relevant sense for random variables, e.g., almost-sure convergence. One example of favorable propagation that matches a wide range of experimental measurements in rich-scattering environments on sub-6 GHz carriers is the Rayleigh i.i.d. channel model. As its name indicates, the entries of $\mathbf{H}_{\mathrm{tot},n}$ in this model are independent zero-mean Gaussian random variables. This channel model guarantees favorable propagation since with it (3.7) holds with almost-sure convergence due to the law of large numbers.

Another example of favorable propagation is the uniform LOS channel model [8] for uniform linear arrays (ULAs). In this model, channel vectors $\mathbf{h}_{k,n}$ satisfy:

$$\mathbf{h}_{k,n} = e^{\iota \phi_{k,n}} \begin{bmatrix} 1 & e^{-\iota 2\pi \frac{d}{\lambda} \sin(\theta_{k,n})} & \cdots & e^{-\iota 2\pi (M-1)\frac{d}{\lambda} \sin(\theta_{k,n})} \end{bmatrix}, \tag{3.8}$$

$\forall k \in \mathcal{K}$, where d is the antenna spacing of the ULA, $\phi_{k,n}$ is a random variable that is uniformly distributed on $[0, 2\pi]$, $\{\theta_{k,n}\}_{k\in\mathcal{K}}$ are independently distributed random arrival angles with respect to the array boresight, and λ is the frequency carrier wavelength. With this channel model, the first condition of favorable propagation in (3.7) is satisfied with equality, while the second is satisfied provided that $|\theta_{k,n} - \theta_{j,n}| > 1/M, \forall k,j \in \mathcal{K}, k \neq j$ [8]. Note that this model can be used to approximate mmWave channels in LOS environments which are characterized by a LOS component that is typically much stronger than the reflected paths. However, if we consider a more general mmWave channel model with both LOS and non-line-of-sight (NLOS) components, and where the NLOS paths could be due to reflectors/scatterers that are shared by several users in the cell area, then favorable propagation might no longer strictly hold. In this case, an appropriate user scheduling might be necessary to restore the massive MIMO effect [9].

3.2.3 Benefits and potentials of massive MIMO

When the massive MIMO effect holds, an orders-of-magnitude spectral efficiency increase can be achieved without the high computational cost and the heavy protocol overhead involved in optimal (or even near-optimal) user scheduling. This can be shown by analyzing, in the limit of an infinite number of antennas, the asymptotic behavior of the instantaneous spectral efficiency $R_n(\mathcal{K})$, given by (3.4), when all the $\mathcal{K}$ users asking for service are spatially multiplexed, i.e., $\mathcal{M} = \mathcal{K}$. To make the analysis more realistic, we should also make the number of co-scheduled users K tend to infinity, but in such a way that K/M tend to a constant value smaller than one. As a matter of fact, this asymptotic regime is better suited to match the practical setting where M could be large, but not extremely large as compared to K.

At this stage, it is important to observe that $R_n(\mathcal{K})$ is a random variable that depends on the (random) realizations of the users' channel vectors. Nevertheless, a deterministic approximation of $R_n(\mathcal{K})$, called a *deterministic equivalent*, which becomes increasingly accurate as the system setting approaches the asymptotic regime described above, can be found by using tools of random matrix theory [10]. For instance, the spectral efficiency achievable under Rayleigh i.i.d. favorable propagation in the downlink of a single cell with MRT can be shown to satisfy:

$$R_n(\mathcal{K}) - \sum_{k \in \mathcal{K}} \log\left(1 + \overline{\mathsf{sinr}}_k\right) \overset{a.s.}{\to} 0, \qquad \text{(without pilot overhead)} \qquad (3.9)$$

where $\overline{\mathsf{sinr}}_k$ is the deterministic equivalent of sinr_k from (3.5). In the case of perfect channel state information at the transmitter (CSIT), $\overline{\mathsf{sinr}}_k$ writes as

$$\overline{\mathsf{sinr}}_k \overset{\text{def}}{=} \frac{\eta_k P^{\mathrm{DL}}}{(K/M)\eta_k P^{\mathrm{DL}} + \sigma^2/M}, \qquad (3.10)$$

where η_k is the large-scale fading coefficient of the channel of user k. The deterministic equivalent of $R_n(\mathcal{K})$ is plotted in Figure 3.2 for four and eight spatially multiplexed users, respectively, when the SNR given by $\mathsf{snr}_k = \eta_k P^{\mathrm{DL}}/\sigma^2$ is equal to $10 - 10 \log_{10}(K)$ dB for all scheduled users. For the sake of comparison with Figure 3.1, the overhead due to CSI acquisition needed to perform MRT is not accounted for in Figure 3.2 as well. From the figure it can be seen that when the array size is smaller than 200 antenna elements, which nowadays is a practically implementable setup [11], the system can theoretically achieve a spectral efficiency approximately ten times larger than what can be achieved with fourth generation (4G) technology. This theoretical gain has been confirmed with both numerical simulations [10] and experimental results [11]. Moreover, the figure shows that the spectral efficiency is higher when transmitting to 8 co-scheduled UTs rather than 4. This can be explained by the fact that the linear increase in the total spectral efficiency due to the summation in (3.9) dominates the per-user logarithmic spectral efficiency decrease,

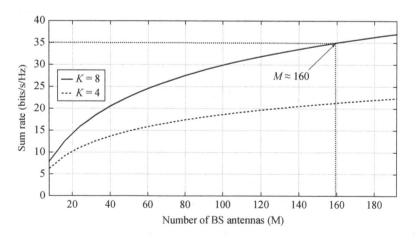

Figure 3.2   Sum rate performance of massive MIMO with maximum-ratio beamforming in a single cell without any user scheduling (not accounting for CSI overhead)

caused by the higher MU interference induced by the presence of a larger number of multiplexed users. We later show that this trend continues to hold up to a certain threshold beyond which the gain from user multiplexing is outweighed by the loss due to the increase in the overhead needed to gather CSI from a larger number of users.

It is worth pointing out at this stage that increasing the system spectral efficiency is not the only advantage brought by the massive MIMO effect. Indeed, another attractive feature of massive MIMO systems is the possibility to reduce both the UT and the BS transmit powers while achieving at least the same individual (per link) data rates [3]. Moreover, massive MIMO can also offer an alternative to cell shrinking as a way of increasing the network capacity [10].

3.2.4 System level implications

From a system level perspective, massive MIMO can be deployed according to two major paradigms. In practice, the large number of antennas necessary to observe the massive MIMO effect can be either geographically *colocated* or *distributed*. These two approaches provide fundamentally different sets of advantages and are certainly characterized by diverse deployment costs. For instance, colocated massive MIMO arrays offer robustness against small-scale distortions, such as the fast fading or the thermal noise at the receiver. Conversely, distributed massive MIMO arrays effectively combat large-scale effects, such distance-dependent path-loss and shadow fading. Regardless of the adopted approach, a key condition needs to be satisfied in order to observe these advantages, i.e., timely and accurate CSI should be made available at the transmitter. This constraint is not easy to fulfill, especially for distributed antenna array implementations. In this case in fact, the presence of a backhaul that can guarantee very low latency and high data rate would be necessary to ensure the feasibility of effective downlink transmissions toward the scheduled users, in particular if coordinated multipoint (CoMP) is performed. In this regard, it is worth observing that similar constraints would apply to the fronthaul in the case of distributed massive MIMO systems making use of remote radio heads (RRH) [12]. As a consequence, if factors like mobility, Doppler shifts, and hardware impairments at both transmitter and receiver are considered, then colocated massive MIMO arrays seem to simplify the acquisition of accurate and timely CSI as compared to their distributed counterparts [4]. However, the potential higher spectral efficiency of the latter makes them an extremely promising candidate for future 5G networks as well. To summarize, it is rather evident that the superiority of one approach as compared to the other cannot be easily assessed, and is certainly the result of both theoretical and practical considerations. In particular, it will be strongly affected by the context and considered application.

3.3 CSI acquisition for massive MIMO

As mentioned in Sections 3.2.1 and 3.2.3, the spectral efficiency analysis presented so far did not take the CSI acquisition overhead into account, for the sake of simplicity.

However, the impact of such overhead may actually represent a stringent bottleneck for the massive MIMO system performance, if appropriate countermeasures are not taken. A simple consideration to highlight the extent of the aforementioned impact is the following. Consider a BS adopting ZFBF or MRT to serve a set of co-scheduled UEs. As a matter of fact, this process requires that the BS knows all the co-scheduled users' downlink channel vectors as can be seen from (3.2) and (3.3). It is straightforward to see that the amount of information necessary to the BS to serve the user grows with the size of the user pool in the system (and the number of antennas per UE). In general, the same observation holds for uplink MU-MIMO operations such as ZFC and MRC as well.

In the remainder of this chapter we will specifically focus on the CSI acquisition overhead that is arguably the most peculiar to massive MIMO, i.e., the one experienced to acquire the necessary CSI to serve UEs in the downlink efficiently. In fact, this is the setting where the throughput gains from massive MIMO are the most needed in practical systems. As a first step, we will show that the way this acquisition is done, and the way it affects the achievable spectral efficiency, are notably different depending on whether a wireless propagation property called *channel reciprocity* holds or not.

3.3.1   CSI acquisition with channel reciprocity: massive MIMO for TDD cellular systems

Channel reciprocity is a property of the wireless link thanks to which the channel from the BS to each UT matches, or can be derived, from the reverse link, i.e., from the UT to the BS. This property holds, for instance, in wireless systems that use time-division duplexing (TDD) to coordinate uplink and downlink transmissions. As a result, when channel reciprocity is applicable, uplink CSI acquired at the multi-antenna BS can be used to spatially multiplexed UTs in the downlink by means of precoding techniques such as ZFBF and MRT. In other words, the BS can obtain CSIT simply by estimating the uplink channel, without any explicit CSI feedback from the UTs. The main advantage from such a possibility is that CSI acquisition overhead does no longer need to grow with the number of BS antennas, as the channels from the UT to all of these antennas can be learned all at once. Another advantage is that existing uplink reference signals in current wireless systems can be re-purposed so that they can help with downlink MU-MIMO transmissions by providing the necessary CSIT. These uplink reference signals are special signals known by the BS and normally referred to as *user-specific reference signals* or *user-specific pilots*.

One kind of user-specific pilots is the reference signal sent by the UTs along with their uplink data in wireless multiuser access scenarios. The CSI obtained from this kind of user-specific pilots can be used to coherently demodulate the uplink data symbols sent by the terminals.[4] Furthermore, if the BS has multiple antennas, the additional degrees of freedom provided by them make it possible to schedule multiple user transmissions on the same time-frequency resources. Such spatially multiplexed

---

[4]Coherent demodulation is known to provide higher spectral efficiencies than non-coherent demodulation.

transmissions interfere with each other at the BS side, thus making the CSI obtained from user-specific pilots crucial for the detection of the interfering signals, e.g., by using linear combining.

Another kind of user-specific pilots are the uplink reference signals that are sent periodically by all active UTs even when they have no uplink data to transmit. Their original purpose was to help the BS in keeping updated its collected CSI so that it can take scheduling decisions for subsequent transmissions to/from the user terminals to determine which of them will transmit/receive on which time-frequency radio resources. One example of such reference signals is the sounding reference signal (SRS) pilots sent during some transmit time interval (TTI) in LTE systems.

In practice, the CSI learned at the BS from any of these existing user-specific pilots can be used in TDD systems for downlink MU-MIMO transmissions performed, if the latter takes place shortly after the reception of the uplink pilots. In other words, the validity of such CSI is limited by the coherence time of the wireless channel, that is the maximum interval between two highly correlated channel realizations or, alternatively, the time duration over which the channel impulse response is considered to be not varying. Conversely, if the BS needs to transmit data to a number of spatially multiplexed UTs after a longer period of time from the last uplink pilot transmission from them, then the CSIT at the BS will be outdated. As a matter of fact, using outdated CSIT in MU-MIMO precoding schemes could lead to a severe reduction of their achievable throughput [13]. Remarkably, an interesting solution to this problem is offered by the pilot transmission scheme illustrated in Figure 3.3, and recently proposed for future 5G systems in which the so-called flexible TDD frame structure is adopted [14]. In this structure, every downlink transmission to one or multiple co-scheduled UTs is preceded by a variable-length uplink transmission that contains both user-specific pilots, needed to acquire the CSIT, and possibly data symbols if at least one of these UTs happens to have data to send. In addition to providing the CSIT needed for downlink MU-MIMO precoding, the CSI obtained from the uplink pilots can also be used for spatial combining and coherent demodulation of users' signals at the BS if the frame contains uplink data. Whether from in-data or periodic pilots, getting the most precise CSI about concurrent uplink transmissions requires that the reference signals of the co-scheduled UTs are mutually orthogonal in some sense. This orthogonality can be achieved using time division multiplexing (TDM), frequency division multiplexing (FDM), code division multiplexing (CDM) (including orthogonal cover codes (OCCs)[5] and cyclic shift (CS)[6] sequences) or any combination of these multiplexing methods. What all these orthogonal multiplexing schemes have in common is that the minimum number of pilot symbols they require is proportional to the number of concurrent user transmissions.

---

[5] OCC sequences are columns (or rows) of orthogonal code matrices such as Walsh–Hadamard matrices.
[6] Different CS sequences can be derived from one sequence, called the base sequence, by multiplying it with different phase shifts. One example is given by Zadoff-Chu sequences used as SRS in the uplink of LTE systems [15].

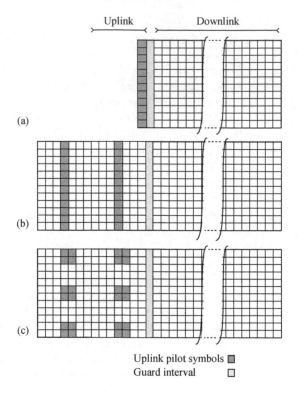

Figure 3.3   Three examples of uplink pilot transmission in a flexible TDD frame structure: (a) no uplink data, (b) with uplink data, (c) with uplink data

From a practical perspective, two challenging limitations to the sought spectral efficiency gain should be addressed in TDD massive MIMO systems where CSIT at the BS is obtained with any of the aforementioned orthogonal uplink pilot schemes. The first is the excessive overhead needed when collecting CSIT from a large number of co-scheduled UTs. The second is the special kind of interference, dubbed *pilot contamination*, user pilot transmissions in neighboring cells cause to each other when some/all of these pilots are reused across the network.

3.3.1.1   CSI overhead for massive MIMO in TDD cellular systems

Whether used for coherent demodulation, scheduling decisions, spatial combining or spatial precoding, the resources (in time and/or frequency) devoted to any of the aforementioned user-specific pilots are subtracted from those used for data transmission, and thus represent an overhead for the wireless system. If we denote by N_d^{DL} the number of resource elements available for downlink data transmission, by N_p^{UL} the number of pilot symbols used for uplink CSI estimation and by $R^{\mathrm{TDD}}(\mathscr{K})$

the effective spectral efficiency over the block of $N_d^{\text{DL}} + N_p^{\text{UL}}$ resource elements, then

$$R^{\text{TDD}}(\mathscr{K}) = \sum_{k \in \mathscr{K}} \frac{1}{N_d^{\text{DL}} + N_p^{\text{UL}}} \sum_{n=1}^{N_d^{\text{DL}}} R_n(\mathscr{K}). \qquad \text{(with pilot overhead)} \qquad (3.11)$$

In particular, $R^{\text{TDD}}(\mathscr{K})$ can be approximated using (3.9) as follows:

$$R^{\text{TDD}}(\mathscr{K}) \approx \sum_{k \in \mathscr{K}} \left(1 - \frac{N_p^{\text{UL}}}{N_d^{\text{DL}} + N_p^{\text{UL}}} \right) \log\left(1 + \overline{\text{sinr}}_k \right), \qquad (3.12)$$

where $\overline{\text{sinr}}_k$ is given by (3.10). For instance, if the pilot design of part (b) of Figure 3.3 is used to transmit LTE-like CS uplink pilots, then at least $N_p^{\text{UL}} = 24$ symbols per each physical resource block (PRB) of size $N_d^{\text{DL}} = 168$ are needed for 8 co-scheduled UTs. The effect of this pilot overhead on the spectral efficiency is plotted in Figure 3.4. In this regard, it is worth noting that if the uplink CSI overhead is accounted for, then the BS needs to be equipped with 275 antennas to experience the tenfold gain in spectral efficiency with respect to 4G systems illustrated in Figure 3.2, instead of 170 if this overhead is ignored.

When the number of UTs to be served is larger, the reduction in spectral efficiency due to pilot overhead will be even more severe, especially if the adopted pilot sequences, like the ones used in all the schemes covered so far, have orthogonal designs. Indeed, when user-specific pilots are defined using orthogonal sequences, an ideally zero inter-pilot interference can be achieved. This property certainly yields the best channel estimation quality. However, it also makes the pilot overhead increases quickly with the number of co-scheduled UTs. In this context, if we want to serve 24 users instead of 8 with the same pilot scheme that was used to get Figure 3.4, then the total number of pilot symbols in each PRB should be increased from $N_p^{\text{UL}} = 24$ to 72 if their mutual orthogonality is to be conserved. This amounts to a pilot overhead that is approximately equal to 42.9%. We now show that once this pilot overhead, or the overhead of any other orthogonal pilot scheme, reaches the value of 50%, increasing the number of co-scheduled users can no longer increase the system throughput. To that end, we refer to (3.12) and write:

$$N_p^{\text{UL}} = pK, \text{ (orthogonal pilots)} \qquad (3.13)$$

where p is an integer factor which reflects the fact that the number of pilot symbols in orthogonal designs should be proportional to K. Assume now that $\text{sinr}_k = \text{sinr}, \forall k$. If we optimize this quantity with respect to K while M is kept fixed, we obtain that the maximum $R^{\text{TDD}}(\mathscr{K})$ is achieved by setting $pK/N_d^{\text{DL}} = 0.5$. Beyond this value, $R^{\text{TDD}}(\mathscr{K})$ starts to drop.[7] From the perspective of sum-rate maximization, the number of co-scheduled users in TDD massive MIMO should thus be chosen such that the

---

[7]Note that, since in practical scenario sinr is not constant but decreases with K, due to the growing multiuser interference, the peak of $R^{\text{TDD}}(\mathscr{K})$ will be actually reached at a value of the overhead that is smaller than a half.

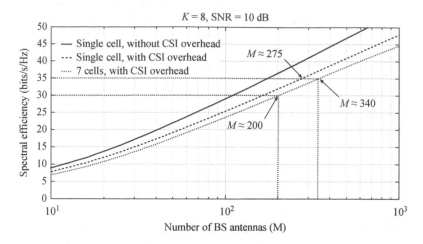

Figure 3.4   Sum rate performance of massive MIMO with maximum-ratio beamforming (with and without accounting for uplink CSI overhead and pilot contamination)

overhead required to acquire their CSI does not exceed 50% of the time-frequency resources available for the transmission to these users.

When the uplink part of the frame structure contains data sent by the UTs, e.g., as in sub-figures (b) and (c) of Figure 3.3, CSI overhead can be reduced to some extent by resorting to *pilot pattern adaptation*. This scheme consists in grouping users based both on the Doppler frequency shift and the maximum delay spread of their respective channels, and to constrain with this grouping step the scheduling of users to the available PRBs of the system. Now, since the required density of pilot repetitions in a pilot pattern is related to the level of correlation of the channel coefficients along the time and the frequency axes, pilot patterns with a reduced density can be used on a PRB in which all scheduled users have milder requirements on pilots. For instance, the density of pilot repetitions in the resource grid can be reduced from its original value of two OFDM pilot symbols (respectively, four in the case of higher number of multiplexed UTs) shown in sub-figure (b) of Figure 3.3 and which assumes a worst-case Doppler frequency shift of 300 Hz to one (respectively, two) for UTs with smaller Doppler frequency shifts as shown in Figure 3.5. While significant pilot overhead reduction can be achieved with uplink pilot pattern adaptation [16], further measures might be needed when the resulting overhead is still prohibitive, e.g., if the number of co-scheduled users is very large. In such cases, the adoption of non-orthogonal (or hybrid orthogonal/non-orthogonal) sequences as uplink pilots is advisable, as discussed in the following subsection.

Non-orthogonal pilot sequences for TDD massive MIMO systems

So far we have seen that when CSI is acquired through orthogonal pilots, there is a maximum number of spatially multiplexed users beyond which the system spectral

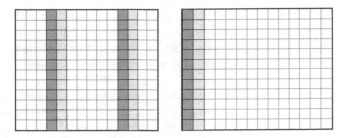

Figure 3.5   An example of adaptive pilot patterns in the time domain

efficiency starts to drop. However, scenarios in which the BS needs to operate beyond this boundary exist in practical relevant settings. This is the case, for instance, in multi-service scenarios in which the BS is required to serve a large number of machine-to-machine (M2M) connections while continuing to support the more conventional broadband links to/from handheld devices. Another example is the crowd scenario in which the BS needs to operate in an area with a very dense deployment of devices, e.g., sports venues. In these scenarios, non-orthogonal pilot sequences would be clearly more suitable than their orthogonal counterparts.

Non-orthogonal pilot designs are typically characterized with a sequence length N_p^{UL} that satisfies:

$$N_p^{\text{UL}} = pL, \quad \text{(non-orthogonal pilots)} \tag{3.14}$$

where L is an integer smaller than the number of pilot sequences K, and where p is the factor from (3.13). Generally speaking, non-orthogonal pilots might yield larger channel estimation errors than what can be achieved through orthogonal pilots, due to the ensuing interference among the pilot signals of co-scheduled users. This degradation occurs even if the non-orthogonal pilot sequences are selected in an optimized manner by finding K points on a N_p^{UL}-dimensional complex sphere of a radius that is dictated by the pilot transmit power constraint [17].

An alternative approach to keep the impact of the aforementioned degradation on the system sum-rate performance within acceptable levels is to use hybrid orthogonal/non-orthogonal pilot designs instead of purely non-orthogonal pilots. In such designs, two subsets of pilot sequences are generated, one composed of mutually orthogonal sequences and the other of mutually non-orthogonal sequences. These subsets are next assigned to two disjoint groups of UTs. In this context, if the two subsets are also made orthogonal to one another, then the intra-cell pilot contamination due to the non-orthogonality of the sequences of the second subset can be maintained within the corresponding group of users. Hybrid constructions offer the possibility of fine-tuning the respective proportions of UTs with mutually orthogonal or non-orthogonal pilots. As a direct consequence of this flexibility, they are guaranteed to outperform both their orthogonal and non-orthogonal counterparts in terms of sum throughput and/or the number of simultaneous connections they can support. One way to build such subsets of pilot sequences is to use a code domain spreading technique called hierarchical class dependent spreading [18] that was originally conceived for data transmission but which can be re-purposed for transmitting hybrid pilot sequences.

Another way to limit performance degradation due to non-orthogonal pilot sequences is to make use of the side information the BS might have about the users' channels in the form of second-order statistics. This is described in the following subsection. In this regard, we note that in the remainder of the channel we will implicitly assume that the channel vector $\mathbf{h}_{k,n}$ for each user k is constant during the channel learning phase and during data transmission, for the sake of simplicity of the notations. This allows us to remove the subscript n from our equations, e.g., notation $\mathbf{h}_{k,n}$ simply becomes $\mathbf{h}_k$. It is worth observing that this rather conventional assumption generally holds for low-mobility UTs and for systems with small enough PRBs.

Pilot design with known per-user spatial covariance matrices

When channel coefficients are correlated, the orthogonality condition can be relaxed, thus making pilot sequences shorter than the number of transmit antennas. Massive MIMO channels typically exhibit a large degree of correlation to the point of resulting in a rank deficient spatial covariance matrix. This property has been validated (for example [19, Lemma 2]) using the one-ring model, in which the scattering objects form rings around each UT. If the BS is installed on a relatively high site compared to the UTs, the channels angle of arrivals (AoAs) have finite supports. Interestingly, this correlation can potentially help reduce the pilot lengths, since the requirement to maintain pilot orthogonality across many users can be relaxed, as follows.

Denote by $\mathbf{R}_k$ the spatial covariance matrix between the kth single-antenna UT and the M antennas at the BS, i.e., $\mathbf{R}_k = \mathbb{E}[\mathbf{h}_k \mathbf{h}_k^H] \in \mathbb{C}^{M \times M}$. As previously discussed, such covariance matrices have rank $r_k \leq M$, with strict inequality typically holding in massive MIMO systems, i.e., $r_k < M$, due to the aforementioned rank deficiency [19]. We assume that matrices $\mathbf{R}_k$ are known at the BS and constant, i.e., the random process representing the channel coefficients across the antenna array is stationary. We also assume that $\mathbf{h}_k$ is Gaussian, i.e., $\mathbf{h}_k \sim \mathscr{CN}(0, \mathbf{R}_k)$. Now, each covariance matrix can be decomposed as $\mathbf{R}_k^{\frac{1}{2}} = \mathbf{U}_k \boldsymbol{\lambda}_k^{\frac{1}{2}}$, where $\boldsymbol{\lambda}_k$ and $\mathbf{U}_k$ are the diagonal matrix containing the r_k non-zero eigenvalues of $\mathbf{R}_k$ and the matrix of the associated eigenvectors, respectively. Then, the column vector of channel coefficients, $\mathbf{h}_k \in \mathbb{C}^M$ can be expressed as the product between the square root of the covariance matrix and a vector of complex Gaussian i.i.d. random variables, $\boldsymbol{\eta}_k \sim \mathscr{CN}(0, \mathbf{I}_{r_k})$, that represents the fast fading process, i.e.,

$$\mathbf{h}_k = \mathbf{R}_k^{\frac{1}{2}} \boldsymbol{\eta}_k, \qquad \forall k \in \mathscr{K}. \tag{3.15}$$

If the users are separated by distances of the order of at least few wavelengths, it is reasonable to assume that vectors $\boldsymbol{\eta}_k$ are statistically independent.

In non-orthogonal uplink pilot design, K non-orthogonal sequences $\mathbf{p}_k \in \mathbb{C}^{N_p^{\mathrm{UL}}}$ for all $k \in \mathscr{K}$ are sent simultaneously from all co-scheduled terminals to the BS in order to estimate the channel coefficients vectors $\mathbf{h}_k$. Here, N_p^{UL} is the length of the pilot sequence given by (3.14). Now, let $n_1, \ldots, n_{N_p^{\mathrm{UL}}}$ be the positions of the REs bearing the pilot symbols. The signal received at the BS, $\mathbf{Y} = [\mathbf{y}_{n_1}, \dots,$ $\mathbf{y}_{n_{N_p^{\text{UL}}}}] \in \mathbb{C}^{M \times N_p^{\text{UL}}}$, is obtained as

$$\mathbf{Y} = \mathbf{H}\mathbf{P}^T + \mathbf{N}, \tag{3.16}$$

where $\mathbf{H} = [\mathbf{h}_1, \dots, \mathbf{h}_K]$ is the column concatenation of the channel vectors from the K terminals to the M antennas at the BS, $\mathbf{P} = [\mathbf{p}_1, \dots, \mathbf{p}_K] \in \mathbb{C}^{N_p^{\text{UL}} \times K}$ is the matrix containing the training sequences sent by the UTs, and $\mathbf{N} \in \mathbb{C}^{M \times N_p^{\text{UL}}}$ is a vector of additive noise samples.

By vectorizing the received signal in (3.16), it can be expressed as

$$\text{vec}(\mathbf{Y}) = \tilde{\mathbf{P}}\text{vec}(\mathbf{H}) + \text{vec}(\mathbf{N}), \tag{3.17}$$

where $\tilde{\mathbf{P}} = (\mathbf{P} \otimes \mathbf{I}_M)$. Combining (3.15) with (3.17) yields:

$$\mathbf{y} = \text{vec}(\mathbf{Y}) = \tilde{\mathbf{P}}\tilde{\mathbf{R}}^{\frac{1}{2}}\boldsymbol{\eta} + \text{vec}(\mathbf{N}), \tag{3.18}$$

where

$$\tilde{\mathbf{R}}^{\frac{1}{2}} = \begin{pmatrix} \mathbf{R}_1^{\frac{1}{2}} & \cdots & 0 \\ 0 & \mathbf{R}_k^{\frac{1}{2}} & 0 \\ 0 & \cdots & \mathbf{R}_K^{\frac{1}{2}} \end{pmatrix},$$

and $\boldsymbol{\eta} \in \mathbb{C}^{\sum_{i \in \mathcal{K}} r_i}$ is the vector concatenation of the fast fading process coefficients, i.e., $\boldsymbol{\eta}^T = [\boldsymbol{\eta}_1^T, \dots, \boldsymbol{\eta}_K^T]$. The objective of the receiver in this case is to jointly estimate $\{\mathbf{h}_k\}_{k \in \mathcal{K}}$ from the received signal $\mathbf{Y}$.

Roughly speaking, the case in which all the covariance matrices are mutually orthogonal, i.e., $\text{Tr}(\mathbf{R}_i \mathbf{R}_j) = 0, \forall i, j \in \mathcal{K}$ such that $i \neq j$, is the best-case scenario from joint channel estimation perspective. Indeed, in this case, the matrix $\tilde{\mathbf{P}}\tilde{\mathbf{R}}$ in (3.18) is full column rank and $\boldsymbol{\eta}$ is *identifiable*[8] based on $\mathbf{y}$ even when $N_p^{\text{UL}} = 1$ [20]. In practice, this means that the mutual orthogonality of the covariance matrices of the co-scheduled users guarantees "good" channel estimates even with short non-orthogonal pilot sequences. It is worth mentioning here that this mutual orthogonality holds for instance when the supports of the AoA of all the scheduled users' channels are non-overlapping [21].

However, channel subspaces across users are in general neither orthogonal nor identical. Therefore, it becomes a combinatorial problem to find the set partition with the smallest number of subsets, each containing users with mutually orthogonal subspaces. The complexity of finding this set partition scales with the number of users and is thus impractical. An alternative way of designing short non-orthogonal pilot sequences in this case consists in formulating an optimization problem with an estimation error constraint. Let us focus, for instance, on the linear minimum

---

[8]In the estimation theory sense.

mean square error (LMMSE) estimator. The LMMSE estimator of the fast fading coefficients between all users and the BS array, $\hat{\boldsymbol{\eta}}$, can be written as

$$\hat{\boldsymbol{\eta}} = \mathbf{C}_{\eta y} \mathbf{C}_y^{-1} \mathbf{y}, \tag{3.19}$$

where $\mathbf{C}_{\eta y} = \tilde{\mathbf{R}}^{\frac{H}{2}} \tilde{\mathbf{P}}^H$, and $\mathbf{C}_y = \tilde{\mathbf{P}} \tilde{\mathbf{R}} \tilde{\mathbf{P}}^H + \sigma^2 \mathbf{I}_{LM}$. The covariance matrix of the estimation error for $\boldsymbol{\eta}$ is then given by [22]:

$$\mathbf{C}_{e,\eta} = \mathbb{E}\left[(\hat{\boldsymbol{\eta}} - \boldsymbol{\eta})(\hat{\boldsymbol{\eta}} - \boldsymbol{\eta})^H\right] \tag{3.20}$$

$$= \mathbf{I}_r - \tilde{\mathbf{R}}^{\frac{H}{2}} \tilde{\mathbf{P}}^H (\tilde{\mathbf{P}} \tilde{\mathbf{R}} \tilde{\mathbf{P}}^H + \sigma^2 \mathbf{I}_{LM})^{-1} \tilde{\mathbf{P}} \tilde{\mathbf{R}}^{\frac{1}{2}}. \tag{3.21}$$

Considering (3.15), we define $\hat{\mathbf{h}} = \tilde{\mathbf{R}}^{\frac{1}{2}} \hat{\boldsymbol{\eta}}$. The covariance matrix of the estimation error on $\mathbf{h}$ is therefore:

$$\mathbf{C}_e = \mathbb{E}\left[(\hat{\mathbf{h}} - \mathbf{h})(\hat{\mathbf{h}} - \mathbf{h})^H\right] \tag{3.22}$$

$$= \tilde{\mathbf{R}}^{\frac{1}{2}} \mathbf{C}_{e,\eta} \tilde{\mathbf{R}}^{\frac{H}{2}} \tag{3.23}$$

$$= \tilde{\mathbf{R}} - \tilde{\mathbf{R}} \tilde{\mathbf{P}}^H (\tilde{\mathbf{P}} \tilde{\mathbf{R}} \tilde{\mathbf{P}}^H + \sigma^2 \mathbf{I}_{LM})^{-1} \tilde{\mathbf{P}} \tilde{\mathbf{R}}. \tag{3.24}$$

In order to control the accuracy of the channel estimation process across all the users, one approach is to uniformly bound the estimation error on all dimensions of $\mathbf{h}$ by a given constant $\varepsilon > 0$. This can be done by requiring that all the eigenvalues of $\mathbf{C}_e$ are lower or equal to ε, which we denote[9] $\mathbf{C}_e \preceq \varepsilon \mathbf{I}$.

Now, the pilot length minimization is a rank minimization problem over a convex set [20]. Thus, we can seek the minimum N_p^{UL} for which there exists a $N_p^{\text{UL}} \times K$ matrix $\mathbf{P}$ that satisfies $\mathbf{C}_e \preceq \varepsilon \mathbf{I}$. This can be achieved by solving the following optimization problem:

$$\min_{\mathbf{P} \in \mathbb{C}^{N_p^{\text{UL}} \times K}} N_p^{\text{UL}} \tag{3.25}$$

$$\text{s.t.} \quad \mathbf{C}_e \preceq \varepsilon \mathbf{I}.$$

If we let $\mathbf{X} = \mathbf{P}^H \mathbf{P} \in \mathbb{C}^{K \times K}$, then minimizing N_p^{UL} is equivalent to minimizing rank($\mathbf{X}$), i.e., (3.25) becomes:

$$\min_{\mathbf{X} \succeq 0} \text{rank}(\mathbf{X}) \tag{3.26}$$

$$\text{s.t.} \quad \mathbf{C}_e \preceq \varepsilon \mathbf{I}.$$

We give here an example of the achievable performance of a pilot length reduction method based on this formulation. Consider a set of covariance matrices generated by a ray-tracing procedure based on the one-ring channel model [23], with a central frequency of 1.8 GHz, and noise variance at each BS antenna element equal to 10^{-4}. The average length pilot sequence length N_p^{UL} obtained as a result of (3.26) is illustrated in Figure 3.6, as a function of the error threshold ε, and for different values of

---

[9]For two positive semidefinite matrices $\mathbf{A}$ and $\mathbf{B}$, $\mathbf{A} \preceq \mathbf{B}$ is a shorthand notation for the condition that $\mathbf{B} - \mathbf{A}$ is positive semidefinite.

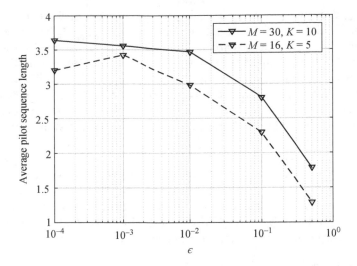

Figure 3.6 Average length N_p^{UL} of the pilot sequences as a function of the error threshold ε for $(M, K) = (30, 10)$ (solid line) and $(16, 5)$ (dashed)

M and K. The reader interested in further details on the algorithm that can be used to solve (3.26) may refer to [20].

3.3.1.2 Pilot contamination and its effects on TDD massive MIMO systems

All the spectral efficiency results presented so far have been obtained assuming that all UTs belong to one cell that is isolated from any other wireless transmission. However, practical network implementations are typically characterized by the presence of multiple cells. Thus, to guarantee the practical relevance of the results of any analysis of massive MIMO schemes, the latter should be analyzed, and possibly redesigned, accounting for the effect of multicell interference on both the CSI acquisition step and the actual data transmission. This is the subject of the current subsection. Now, let $\mathscr{C}$ designate the set of indexes of cells interfering with each other due to their geographical proximity, and denote by $\mathscr{K}_c$ the set of UTs in cell $c \in \mathscr{C}$. In this context, index $c = 0$ will be used to designate the cell under investigation.

Denote by $y_{c,n}^{MC}$ the baseband signal sample (the superscript MC stands for "multicell") received by the BS of cell c on a pilot resource element n. In cell $c = 0$, we have

$$y_{0,n}^{MC} = \sum_{j \in \mathscr{K}_0} \mathbf{h}_{k,0} p_{j,0,n} + \sum_{c \in \mathscr{C} \backslash \{0\}} \sum_{j \in \mathscr{K}_c} \mathbf{h}_{k,0} p_{j,c,n} + z_{0,n}, \quad (n : \text{pilot RE}) \qquad (3.27)$$

where $p_{k,c,n}$ is the element of pilot sequence $\mathbf{p}_k$ sent by user $k \in \mathscr{K}_c$ on the position of resource element n. Furthermore, denote by $\mathcal{O}$ the set of available pilot sequences from which $\mathbf{p}_k$ is taken and note that $\mathcal{O}$ is a finite set in all practical pilot designs.

As for $\mathbf{h}_{k,c}$, it denotes the baseband discrete-time channel of the link between the BS of cell c and the terminal of user k. Let $\mathscr{P}_k \subset \bigcup_{c \in \mathscr{C} \backslash \{0\}} \mathscr{K}_c$ be the set of users from the interfering cells who share the same pilot sequence assigned to user k in the considered cell. In this regard, we observe that the total number of users $| \bigcup_{c \in \mathscr{C}} \mathscr{K}_c|$ in the network is typically larger than the number of available pilot sequences,[10] and the set $\mathscr{P}_k$ is typically non-empty. Hence, the estimate $\hat{\mathbf{h}}_{k,0}$ computed at the BS of the cell under examination is corrupted (or contaminated) by the pilot signals transmitted by the users in $\mathscr{P}_k$. For example, if a least squares (LS) channel estimation is performed, then:

$$\hat{\mathbf{h}}_{k,0} = \mathbf{h}_{k,0} + \sum_{j \in \mathscr{P}_k} \mathbf{h}_{j,0} + \mathbf{z}_{k,0,n}, \tag{3.28}$$

where $\mathbf{z}_{k,0,n}$ is a channel estimation noise vector that is independent of the channel coefficients. Similarly, for a user j in cell c that belongs to $\mathscr{P}_k$, the channel estimate $\hat{\mathbf{h}}_{j,c}$ has an interference term due to the pilot signal of user k, and reads:

$$\hat{\mathbf{h}}_{j,c} = \mathbf{h}_{j,c} + \mathbf{h}_{k,c} + \sum_{l \in \mathscr{P}_k, l \neq k} \mathbf{h}_{l,c} + \mathbf{z}_{j,c,n}. \tag{3.29}$$

The channel estimation error due to the second term in the right hand side (RHS) of (3.28) and (3.29) can have a significant impact on the spectral efficiency when transmitting to user k. The reason is as follows. We start by defining $x_{j,c,n}$ as the data symbol destined to user j in cell c. Additionally, we let $\{\hat{\mathbf{h}}_{k,0}\}_{k \in \mathscr{K}}$ and $\{\hat{\mathbf{h}}_{j,c}\}_{j \in \mathscr{K}_c}$ be the contaminated uplink CSI. Finally, we let $\hat{\mathbf{w}}_{k,0}$ and $\hat{\mathbf{w}}_{j,c}$ be the linear precoders employed by the BS of the examined cell and of the interfering cell c to serve users $k \in \mathscr{K}_0$ and $j \in \mathscr{K}_c$, respectively. In this regard, we remark that the computation of $\hat{\mathbf{w}}_{k,0}$, $\hat{\mathbf{w}}_{j,c}$ is based on the contaminated CSI. For instance, if MRT is employed then $\hat{\mathbf{w}}_{k,0} = \hat{\mathbf{h}}_{k,0}$ and $\hat{\mathbf{w}}_{j,c} = \hat{\mathbf{h}}_{j,c}$. The received sample on a data resource element n, i.e., $y_{k,0,n}^{\mathrm{MC}}$, can then be written as

$$y_{k,0,n}^{\mathrm{MC}} = \sqrt{\alpha} \sum_{j \in \mathscr{K}_0} \hat{\mathbf{w}}_{j,0}^{\mathrm{H}} \mathbf{h}_{j,0} x_{j,0,n} + \sqrt{\alpha_c} \sum_{j \in \mathscr{K}_c} \hat{\mathbf{w}}_{j,c}^{\mathrm{H}} \mathbf{h}_{k,c} x_{j,c,n} + \mathbf{z}_{k,0,n}, \quad (n : \text{data RE}) \tag{3.30}$$

where α_c is the average transmit power normalization parameter in cell c. In particular, by looking at (3.29), the term $\hat{\mathbf{w}}_{j,c}^{\mathrm{H}} \mathbf{h}_{k,c}$ in (3.30) can be written as

$$\frac{1}{M} \hat{\mathbf{w}}_{j,c}^{\mathrm{H}} \mathbf{h}_{k,c} = \frac{1}{M} \mathbf{h}_{k,c}^{\mathrm{H}} \mathbf{h}_{k,c} + \frac{1}{M} \left(\mathbf{h}_{j,c} + \sum_{l \in \mathscr{P}_k, l \neq k} \mathbf{h}_{l,c} \right)^{\mathrm{H}} \mathbf{h}_{k,c}, \quad (j \in \mathscr{P}_k). \tag{3.31}$$

At this stage, it is straightforward to note that while the second term in (3.31) vanishes as $M \to \infty$, due to the massive MIMO effect, the first one tends to

---

[10]This is at least true in fully loaded scenarios.

a non-zero constant. As a consequence, the effective SINR of the transmission to user k in a multicell setting could suffer a significant detriment due to pilot contamination. To assess the extent of this loss, we apply the results of the analytical study that has been proposed in [24] to a system in which pilot contamination from six neighboring cells in a hexagonal 7-cell network layout occurs. Please refer to Figure 3.4 for a graphical comparison of the per-cell spectral efficiency of such system as compared to its single-cell counterpart when the inter-site distance and BS transmit power are set at 900 m and a 46 dBm, respectively. Examining Figure 3.4 reveals that in order to achieve the same tenfold gain in spectral efficiency with respect to 4G systems that was illustrated in Figure 3.2, a BS in the considered multi-cell setting must be equipped with up to 340 antennas, as opposed to 275 in a single-cell setting. Conversely, If the BS array size is constrained to be no larger than 200 antennas for practical reasons, then the achievable system spectral efficiency drops to 30 bits/s/Hz. This corresponds to a spectral efficiency 8.6 times larger than what is achieved in 4G systems.

Pilot contamination reduction methods

Several methods have been recently proposed in the literature to achieve an effective pilot contamination reduction [19,25,26]. Let us switch our focus back to (3.28) and (3.29), i.e., the contaminated LS channel estimations independently performed by each BS in the considered multicell network. A significant reduction in pilot contamination in this case could be achieved if the channel coefficients of all the users in each one of the sets $\{\mathscr{P}_k\}_{k \in \mathscr{K}_c}$ were jointly estimated by the BSs [19]. Naturally, this would come at the cost of additional delay and backhaul overhead, needed for the joint multicell processing. However, this additional cost might not always be affordable for the system. If this were the case, one of the following advanced per-cell channel estimation methods, that discriminate against pilot interference originating from neighboring cells, could be used [25]:

- *Amplitude based projection*, which uses the empirical instantaneous covariance matrix computed from the received data samples at the BS to discriminate against multicell pilot interference, by assuming that this interference is weaker than the useful pilot signals in terms of received power.
- *Covariance based projection* can be used as either an alternative or a complementary method when covariance matrices of users' channels are known at the BS, as assumed in the previous subsection [25]. This approach yields its best performance when the AoA of the interfering pilot signals overlap the least with the aggregate AoA region of the useful pilot signals.

Another method to reduce pilot contamination consists in coordinating pilot sequence assignment in clusters of neighboring cells. For instance, the BSs of neighboring cells can make sure that the respective users of the sets $\mathscr{P}_k$ are kept as far away from each other as possible. A simple way to achieve this coordination is the so-called *fractional pilot reuse* [26]. In this scheme, the set $\mathscr{O}$ of available orthogonal pilot sequences is divided into a number $\alpha \geq 1$ of subsets, denoted as $\mathscr{I}$ and $\{\mathscr{E}_i\}_{1 \leq i \leq \alpha}$, where α is the *pilot reuse factor*. As shown in Figure 3.7, pilot sequences from $\mathscr{I}$ can be reused

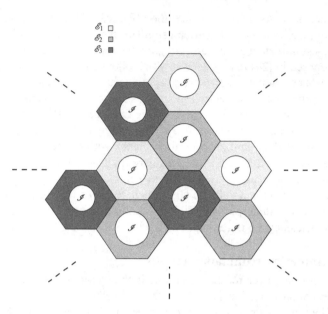

Figure 3.7   Fractional pilot reuse with reuse factor α = 3

in all the cells of the network, but can only be assigned to users each cell center, whereas the sequences taken from $\{\mathscr{E}_i^o\}_{1 \le i \le \alpha}$ are assigned in an orthogonal manner to α neighboring cells and reused throughout the network on the basis of groups of α cells. In this context, if a proper tuning of the factor α depending on the radii of the central cell regions is performed, then fractional pilot reuse can theoretically achieve a 10%–125% gain in per-cell spectral efficiency, depending on the adopted MU-MIMO scheme [26].

3.3.2   *CSI acquisition without channel reciprocity: massive MIMO for frequency-division duplexing (FDD) cellular systems*

When channel reciprocity does not hold, the CSI needed for MU-MIMO transmission in the downlink can no longer be obtained based on uplink pilot transmissions. In this case, the BS needs to send a downlink pilot signal from each of its antennas so that the channels from different BS antennas to each UT can be estimated at the users side and fed back to the BS. The frame structure of Figure 3.3 should thus be modified in FDD systems to allow for a downlink CSI estimation to take place before the uplink transmission. In this regard, it is worth observing that uplink pilots are still necessary for both successful uplink data transmission and successful downlink CSI feedback reporting. The CSI acquisition overhead in this case may rapidly increase. The extent of this phenomenon could be such that some massive MIMO system configurations may deliver extremely poor performance when operating in FDD mode, as shown in

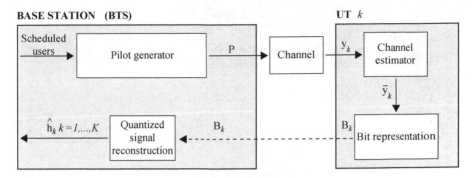

Figure 3.8   Functional representation of a CSI acquisition approach for FDD systems with massive MIMO with channel estimation at the UT. Only one among the K UTs is shown

Figure 3.9, unless appropriate countermeasures are adopted [27]. A simple quantitative example can be used to further assess this extent. Consider the typical procedure adopted in many state-of-the-art FDD systems, illustrated in Figure 3.8 for the sake of simplicity. In practice, first a set of pilot sequences is simultaneously broadcast by the M antennas at the BS. Subsequently, UT k estimates the downlink channels by evaluating the received signal and comparing it to the known pilot sequence(s). After this step, a quantization of this estimation is performed UT k, which typically yields a binary representation of the downlink channel. Such binary representation is then fed back to the BS which finally performs a suitable quantized signal reconstruction to obtain the desired downlink CSI estimation. Now, as for (3.11), we let N_d^{DL} and N_p^{UL} be the number of resource elements available for downlink transmission in an uplink-downlink frame, and the corresponding number of uplink pilot symbols, respectively. Additionally, we define $N_{\mathrm{FB}}^{\mathrm{UL}}$ as the number of uplink symbols needed to feedback the estimated downlink CSI to the BS. The effective spectral efficiency over the frame of $N_d^{\mathrm{DL}} + N_p^{\mathrm{UL}} + N_{\mathrm{FB}}^{\mathrm{UL}}$ resource elements, designated by $R^{\mathrm{TDD}}(\mathscr{K})$, can then be written as

$$R^{\mathrm{FDD}}(\mathscr{K}) = \sum_{k \in \mathscr{K}} \frac{1}{N_d^{\mathrm{DL}} + N_p^{\mathrm{UL}} + N_{\mathrm{FB}}^{\mathrm{UL}}} \sum_{i=1}^{N_d^{\mathrm{DL}} - N_p^{\mathrm{DL}}} R_{n_i}(\mathscr{K}), \quad \text{(with pilot overhead)}$$

(3.32)

where n_i for $i \in \{1, 2, \ldots, N_d^{\mathrm{DL}} - N_p^{\mathrm{DL}}\}$ are the positions of the REs available for data transmission in the downlink frame. Consider now the general setting where no prior information on the users' channels is available at the BS, e.g., the second-order statistics are not known. In this case, orthogonality of the pilot signals sent by the M BS antennas can be guaranteed only if $N_p^{\mathrm{DL}} \geq M$. Similarly, N_p^{UL} should be at least equal to K. Switching our focus to $N_{\mathrm{FB}}^{\mathrm{UL}}$, we observe that if analog feedback is used, then the value of this parameter typically is of the order of M [27]. In total,

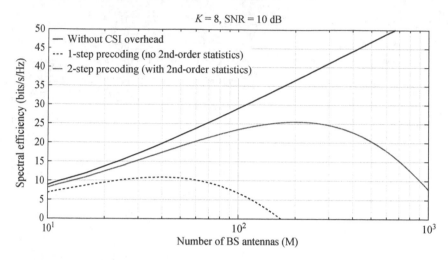

Figure 3.9   Sum rate performance of FDD massive MIMO with maximum-ratio beamforming

this amounts to an overhead of the order of $\frac{2M+K}{N_d^{DL}+K+M}$. The fact that this overhead increases with M leads to the performance shown in Figure 3.9 where the achievable spectral efficiency starts to drop when the number of BS antennas is increased beyond a threshold value.

Non-orthogonal pilot sequences for FDD massive MIMO systems

Similarly to the method described in Section 3.3.1, this section describes a way to reduce CSI overhead in FDD massive MIMO systems by using side information on the downlink channel covariance matrices. We consider the same channel model as in eq.(3.15) and assume to know the downlink spatial covariance matrices at the BS, i.e., $\mathbf{R}_k$, $\forall k$. Similarly to Section 3.3.1, we specifically focus on the downlink CSI estimation at the BS. A pictorial representation of CSI acquisition is given in Figure 3.10, First, a set of pilot sequences is simultaneously broadcast by the M antennas at the BS toward the UTs. Let these sequences have overall length N_p^{DL}, and let $\mathbf{P} \in \mathbb{C}^{N_p^{DL} \times M}$ denote the complete set of pilot sequences used in the system. The signal received at UT k over the corresponding N_p^{UL} REs, and denoted by $\mathbf{y}_k \in \mathbb{C}^{N_p^{DL}}$, can then be expressed as

$$\mathbf{y}_k = \mathbf{P}\mathbf{h}_k + \mathbf{n}_k, \tag{3.33}$$

with $\mathbf{n}_k \sim \mathscr{CN}(0, \sigma_{N,k}^2 \mathbf{I}_{N_p^{DL}})$ defined as the per-user AWGN at UT k. Each UT then quantizes the received signal and transmits it back to the BS, e.g., in the form of a binary representation $\mathbf{B}_k$. Let us assume for simplicity that the feedback link is perfect,

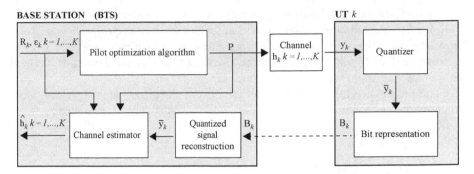

Figure 3.10 *Functional representation of a CSI acquisition approach for FDD systems with massive MIMO without channel estimation at the UT. Only one among the K UTs is shown*

and that the BS can perfectly recover the quantized version of $\mathbf{y}_k$, for all $k = 1, \ldots, K$, that is

$$\bar{\mathbf{y}}_k = \mathbf{y}_k + \mathbf{z}_k = \mathbf{P}\mathbf{h}_k + \mathbf{n}_k + \mathbf{z}_k, \tag{3.34}$$

where $\mathbf{z}_k$ represents the per-user quantization error, which is assumed to be Gaussian with covariance $\sigma_{Q,k}^2 \mathbf{I}_{N_p^{\mathrm{DL}}}$ and independent from the other variables, i.e., the worst case scenario for estimation purposes. The BS can finally compute the channel estimate $\hat{\mathbf{h}}_k$ based on the combined knowledge of $\bar{\mathbf{y}}_k$, $\mathbf{R}_k$ and $\mathbf{P}$, using a LMMSE estimator.

The most straightforward way to reduce the CSI overhead in FDD massive MIMO systems is arguably the design of pilot sequences such that $N_p^{\mathrm{DL}} < M$. However, $\mathbf{P}$ would not admit a left inverse in this case, hence an exact recovery of $\mathbf{h}_k$ from $\mathbf{y}_k$ would not be possible even in the noise-free case. For this reason, according to the previously outlined procedure, the UT does not directly try to estimate the channel in this case. Conversely, the BS estimates $\mathbf{h}_k$ thanks to the assumed available second-order statistics of the CSI.

Both the LMMSE channel estimator and the estimation error covariance matrix of each user can be derived at the BS as a function of the common pilot sequences. For any $\mathbf{P}$, the LMMSE estimator for the fast fading coefficients, $\hat{\boldsymbol{\eta}}_k$, can then be written as

$$\hat{\boldsymbol{\eta}}_k = \mathbf{C}_{\eta_k \bar{\mathbf{y}}_k} \mathbf{C}_{\bar{\mathbf{y}}_k}^{-1} \bar{\mathbf{y}}_k, \tag{3.35}$$

with $\mathbf{C}_{\eta_k \bar{\mathbf{y}}_k} = \mathbf{P}\mathbf{R}_k \mathbf{P}^H + \sigma_k^2 \mathbf{I}_{N_p^{\mathrm{DL}}}$, and where $\sigma_k^2 = \sigma_{N,k}^2 + \sigma_{Q,k}^2$ is the variance of the combination of the thermal and quantization noise terms. The per-user channel estimation error covariance matrix becomes:

$$\mathbf{C}_{e,k}(\mathbf{P}) = \mathbf{U}_k \left(\lambda_k^{-1} + \frac{\mathbf{U}_k^H \mathbf{P}^H \mathbf{P} \mathbf{U}_k}{\sigma_k^2} \right)^{-1} \mathbf{U}_k^H. \tag{3.36}$$

In this formulation we have as many estimation error constraints as users. Interestingly, the accuracy of the channel estimation for user k can be controlled by uniformly bounding the estimation error on all dimensions of $\mathbf{h}_k$ by a given constant $\varepsilon_k > 0$. This can be done by requiring that all the eigenvalues of $\mathbf{C}_{e,k}$ are lower than or equal to a per-user constant ε_k, which we denote as $\mathbf{C}_{e,k} \preceq \varepsilon_k \mathbf{I}_M$.

The problem of minimizing the length N_p^{DL} of the pilots, subject an additional per-antenna maximum pilot energy constraint $E_{\max}$, can then be formulated as

$$\min_{\mathbf{P} \in \mathbb{C}^{N_p^{DL} \times M}} N_p^{DL} \tag{3.37}$$

$$\text{s.t.} \quad \mathbf{C}_{e,k} \preceq \varepsilon_k \mathbf{I}_M \quad \forall k \in \mathscr{K}$$

$$X_{mm} \leq E_{\max} \quad \forall m = 1, \ldots, M$$

where $\mathbf{X} = \mathbf{P}^H \mathbf{P} \in \mathbb{C}^{M \times M}$, with X_{mm} as the mth diagonal element of $\mathbf{X}$.

Now, a useful metric that carries information on both the energy and the pilot length is the total energy dedicated by the BS to pilot sequences, i.e., $\text{Tr}(\mathbf{X})$. This naturally yields the following alternative formulation for the pilot sequences length optimization problem:

$$\min_{\mathbf{P} \in \mathbb{C}^{N_p^{DL} \times M}} \text{Tr}(\mathbf{X}) \tag{3.38}$$

$$\text{s.t.} \quad \mathbf{C}_{e,k} \preceq \varepsilon_k \mathbf{I}_M \quad \forall k \in \mathscr{K}$$

$$X_{mm} \leq E_{\max} \quad \forall m = 1, \ldots, M,$$

that is a convex optimization problem in which the total pilot energy is minimized under the same CSI accuracy constraints as in (3.37).

In practice, both (3.37) and (3.38) can be solved numerically by simulating the communication systems under study, e.g., by means of numerical solvers such as CVX [28]. In this regard, the interested reader may refer to [29] for further details on an example of numerical solution which confirms the potential of this method. The obtained results for this scenario are remarkable: when only the second-order statistics for downlink CSI are available at the BS, the aforementioned procedure results in pilots whose length can be up to a fourth of the pilot length required in the absence of statistical CSI.

3.3.2.1   Reducing CSI acquisition overhead for FDD massive MIMO systems through grouping algorithms

Pilot design optimization is not the only possible option to reduce the CSI acquisition overhead in FDD massive MIMO systems. Alternative families of solutions to achieve this goal have been lately proposed by researchers of both academia and industry, e.g., [21,30,31]. The algorithms adopted by these solutions are typically referred to as *grouping algorithms* and are generally based on two properties of propagation in practical massive MIMO scenarios:

- The effective signal space observed by each UT is much smaller than the column space of the covariance matrix of the channel between the BS array and the UT,

due to correlations among antenna elements. In other words, the covariance matrix of a downlink massive MIMO channel is often rank deficient.
- UTs have similar second-order channel statistics, i.e., covariance matrices, and similar signal spaces to what their close neighbors observe.

State-of-the-art solutions exploit these observations to enforce a smart dimensionality reduction of the signal space at the BS, in turn decreasing the CSI overhead due to the downlink CSI feedback performed by the UTs. The key idea underlying these approaches is to achieve the signal space reduction at the BS by creating either groups of UTs [21,31] or groups of antennas [30]:

- **UTs Grouping**: In this case, UTs are grouped according to metrics depending on the location and/or on the second-order channel statistics of the UTs, e.g., [21]. As a result, UTs with similar channel statistics (or close to each other) belong to the same group. In this context, the association between groups and UTs does not change until the covariance matrices of the downlink channels experience a non-negligible variation. However, this variation usually occurs at a slower pace as compared to the variation experienced by many other relevant system parameters, e.g., QoS indicators. This may impact the fairness of resource allocation and scheduling procedures. A possible solution to this issue is offered by the adaptive grouping strategy proposed in [31]. By means of this approach, a massive MIMO BS operating in FDD can dynamically change the structure of each group, depending on how the aforementioned QoS indicators vary over time. In particular, the geographical distance among UTs is neglected by this algorithm. Only considerations on the considered time varying parameter are made, e.g., the QoS indicators, and the Hausdorff distance between the column space of the covariance matrices. The interested reader may refer to [31] for further details. Regardless of the chosen approach, the goal of any grouping algorithm is to obtain a set of $G \geq 1$ groups, indexed in the sequel using $g \in \{1, \ldots, G\}$. Now, let K_g be the number of UTs in the gth group and $\mathbf{R}_g$ be their representative group covariance matrix. In this context, the BS processes the information streams in order to first enforce (semi-)orthogonality between the groups and then among the UTs. This is achieved by means of a two-stage precoder in which the precoding vector $\mathbf{w}_k$ used for each UT k from the group $g_k \in \{1, \ldots, G\}$ is obtained as the product of an outer and an inner component:

$$\mathbf{w}_k = \mathbf{B}_{g_k} \tilde{\mathbf{w}}_k, \tag{3.39}$$

where $\mathbf{B}_{g_k}$ is a per-group *pre-beamforming* matrix and $\tilde{\mathbf{w}}_k$ is a per-UT reduced-dimensionality precoding vector, defined as follows:
1. The interference among scheduled groups is canceled or mitigated by the *outer component*, i.e., $\mathbf{B}_{g_k}$. The G outer precoders for the G groups scheduled for transmission by the BS in the considered slot are computed based on $\mathbf{R}_g$, e.g., using a block diagonalization procedure [32]. In this case, $\mathbf{B}_g$ for $g \in \{1, \ldots, G\}$ is of size $M \times r_g$, where r_g is the number of non-zero eigenvalues of $\mathbf{R}_g$. At the end of this step, the column space of the outer

precoder of each scheduled group lies in the intersection of the null-spaces of the counterpart covariance matrices of all the others groups.

2. The interference experienced by the S_g scheduled UTs belonging to the same group g is canceled or mitigated by the *inner precoder* $\tilde{\mathbf{w}}_k \in \mathbb{C}^{r_g}$, This step can be performed by means of any linear precoding criterion, e.g., ZFBF [32], as long as it is based on the projection of the downlink channel vector $\mathbf{h}_k$ on the signal space generated by the outer precoder, i.e., on the vector η_k defined in (3.18). Note that this projection is the only instantaneous CSI needed in this approach. It is also worth observing that this projection has a dimension r_g which is an order of magnitude smaller than M. Two-stage precoding can thus achieve non-negligible spectral efficiency gains as compared to an FDD system where no second-order channel statistics are available at the BS and UTs grouping algorithms are not adopted. The reader may refer to Figure 3.9 for an example of spectral efficiency gain when UTs are clustered in just two groups, a simple MRT scheme is adopted by the BS, $r_g = M/8$ and $S_g = 4$. In this context, it is evident that S_g is a design parameter that should be selected carefully before the two-stage precoding procedure takes place.

Quantitatively, the performance of these approaches strongly depends on the distribution of the UTs, their behavior and requirements. The main factors that can induce a significant performance detriment are network sparsity, very heterogeneous quality of service (QoS) requirements across UTs/groups and/or high velocity of the latter.

*   **Antenna Grouping**: In this case, multiple correlated and adjacent antenna elements are mapped to a single representative value using pre-designed patterns [30]. Subsequently, each UT estimates such value and performs the feedback according to the specific chosen implementation. The dimensionality reduction achieved by these methods is generally lower than the performance of their counterparts based on UTs grouping. reduction. Additionally, and similar to many UTs grouping algorithms, state-of-the-art antenna grouping solutions do not offer a straightforward way to incorporate very heterogeneous QoS UTs requirements in their design, hence they can perform rather poorly when this heterogeneity occurs.

Grouping algorithms provide fairly good sum-rate performance, even comparable to their TDD counterparts for some BS array sizes (compare for example the relevant curves from Figures 3.4 and 3.9 at point $M = 200$). However, if the number of UTs in the system is large, their complexity may impede their practical implementation, especially if sub-space decomposition methods are required to perform grouping, precoding and UTs' scheduling. As a matter of fact, finding effective solutions to this problem, while guaranteeing both high spectral efficiency and the possibility of incorporating heterogeneous QoS constraints in the algorithm design, is arguably one of the hardest challenges to face for future implementation of FDD massive MIMO systems.

References

[1] Cisco, "Cisco visual networking index: global mobile data traffic forecast update, 2015-2020," White Paper, http://www.cisco.com/c/en/us/solutions/collateral/service-provider/visual-networking-index-vni/mobile-white-paper-c11-520862.pdf, 2016.

[2] F. Rusek, D. Persson, B. K. Lau, *et al.*, "Scaling up MIMO: opportunities and challenges with very large arrays," *IEEE Signal Processing Magazine*, vol. 30, no. 1, pp. 40–60, Jan. 2013.

[3] L. Lu, G. Y. Li, A. L. Swindlehurst, A. Ashikhmin, and R. Zhang, "An overview of massive MIMO: benefits and challenges," *IEEE Journal of Selected Topics in Signal Processing*, vol. 8, no. 5, pp. 742–758, Oct. 2014.

[4] E. G. Larsson, O. Edfors, F. Tufvesson, and T. L. Marzetta, "Massive MIMO for next generation wireless systems," *IEEE Communications Magazine*, vol. 52, no. 2, pp. 186–195, Feb. 2014.

[5] D. Gesbert, M. Kountouris, R. W. Heath, C. B. Chae, and T. Salzer, "Shifting the MIMO paradigm," *IEEE Signal Processing Magazine*, vol. 24, no. 5, pp. 36–46, Sep. 2007.

[6] T. Yoo and A. Goldsmith, "On the optimality of multiantenna broadcast scheduling using zero-forcing beamforming," *IEEE Journal on Selected Areas in Communications*, vol. 24, no. 3, pp. 528–541, Mar. 2006.

[7] V. Abhayawardhana, I. Wassell, D. Crosby, M. Sellars, and M. Brown, "Comparison of empirical propagation path loss models for fixed wireless access systems," in *2005 IEEE 61st Vehicular Technology Conference (VTC Spring)*, pp. 73–77, May 2005.

[8] H. Q. Ngo, E. G. Larsson, and T. L. Marzetta, "Aspects of favorable propagation in massive MIMO," in *22nd European Signal Processing Conference (EUSIPCO)*, pp. 76–80, Sep. 2014.

[9] A. L. Swindlehurst, E. Ayanoglu, P. Heydari, and F. Capolino, "Millimeterwave massive MIMO: the next wireless revolution?" *IEEE Communications Magazine*, vol. 52, no. 9, pp. 56–62, Sep. 2014.

[10] J. Hoydis, S. ten Brink, and M. Debbah, "Massive MIMO in the UL/DL of cellular networks: how many antennas do we need?" *IEEE Journal on Selected Areas in Communications*, vol. 31, no. 2, pp. 160–171, Feb. 2013.

[11] R. University, "Argos, practical many-antenna base stations," http://argos.rice.edu/.

[12] H. Q. Ngo, A. Ashikhmin, H. Yang, E. G. Larsson, and T. L. Marzetta, "Cell-free massive MIMO: uniformly great service for everyone," in *2015 IEEE 16th International Workshop on Signal Processing Advances in Wireless Communications (SPAWC)*, pp. 201–205, Jun. 2015.

[13] K. T. Truong and R. W. Heath, "Effects of channel aging in massive MIMO systems," *Journal of Communications and Networks*, vol. 15, no. 4, pp. 1229–2370, Sep. 2013.

[14] Q. Wang, Z. Zhao, Y. Guo, *et al.*, "Enhancing OFDM by pulse shaping for self-contained TDD transmission in 5G," in *2016 IEEE 83rd Vehicular Technology Conference (VTC Spring)*, pp. 1–5, May 2016.

[15] F. Khan, *LTE for 4G Mobile Broadband: Air Interface Technologies and Performance*. Cambridge University Press, 2009.

[16] N. Ksairi, B. Tomasi, and S. Tomasin, "Pilot pattern adaptation for 5G MU-MIMO wireless communications," in *17th IEEE International Workshop on Signal Processing Advances in Wireless Communications (SPAWC)*, pp. 1–6, Jul. 2016.

[17] H. Wang, W. Zhang, Y. Liu, Q. Xu, and P. Pan, "On design of non-orthogonal pilot signals for a multi-cell massive MIMO system," *IEEE Communications Letters*, vol. 4, no. 2, pp. 129–132, Apr. 2015.

[18] N. Ksairi, S. Tomasin, and M. Debbah, "A multi-service oriented multiple-access scheme for next-generation mobile networks," in *2016 25th European Conference on Networks and Communications (EUCNC)*, pp. 1–5, Jun. 2016.

[19] H. Yin, D. Gesbert, M. Filippou, and Y. Liu, "A coordinated approach to channel estimation in large-scale multiple-antenna systems," *IEEE Journal on Selected Areas in Communication*, vol. 31, no. 2, pp. 264–273, 2013.

[20] B. Tomasi and M. Guillaud, "Pilot length optimization for spatially correlated multi-user MIMO channel estimation," in *Proceedings of the Asilomar Conference on Signals, Systems and Computers*, Pacific Grove, CA, USA, 2015.

[21] J. Nam, A. Adhikary, J.-Y. Ahn, and G. Caire, "Joint spatial division and multiplexing: opportunistic beamforming, user grouping and simplified downlink scheduling," *IEEE Journal on Selected Topics in Signal Processing*, vol. 8, no. 5, pp. 876–890, Oct. 2014.

[22] T. Kailath, A. H. Sayed, and B. Hassibi, *Linear Estimation*. Upper Saddle River, NJ: Prentice-Hall, 2000.

[23] D. Shiu, G. J. Foschini, M. J. Gans, and J. M. Kahn, "Fading correlation and its effect on the capacity of multielement antenna systems," *IEEE Transactions on Communications*, vol. 48, no. 3, pp. 502–513, 2000.

[24] J. Hoydis, S. ten Brink, and M. Debbah, "Massive MIMO in the UL/DL of cellular networks: how many antennas do we need?" *IEEE Journal on Selected Areas in Communications,*, vol. 31, no. 2, pp. 160–171, 2013.

[25] H. Yin, L. Cottatellucci, D. Gesbert, R. R. Müller, and G. He, "Robust pilot decontamination based on joint angle and power domain discrimination," *IEEE Transactions on Signal Processing*, vol. 64, no. 11, pp. 2990–3003, 2016.

[26] I. Atzeni, J. Arnau, and M. Debbah, "Fractional pilot reuse in massive MIMO systems," in *Proceedings of the IEEE International Conference on Communications Workshop (ICCW)*, London, UK, Jun. 2015.

[27] E. Björnson, E. G. Larsson, and T. L. Marzetta, "Massive MIMO: ten myths and one critical question," *IEEE Communications Magazine*, vol. 54, no. 2, pp. 114–123, Feb. 2016.

[28] M. Grant and S. Boyd, "CVX: MATLAB® software for disciplined convex programming, version 2.1," http://cvxr.com/cvx.

[29]   B. Tomasi, A. Decurninge, and M. Guillaud, "SNOPS: short non-orthogonal pilot sequences for downlink channel state estimation in FDD massive MIMO," in *Proceedings of the IEEE Global Telecommunications Conference (GLOBECOM)*, Washington, DC, USA, 2016.

[30]   B. Lee, J. Choi, J.-Y. Seol, D. Love, and B. Shim, "Antenna grouping based feedback reduction for FDD-based massive MIMO systems," in *IEEE International Conference on Communications (ICC)*, pp. 4477–4482, Jun. 2014.

[31]   A. Destounis and M. Maso, "Adaptive clustering and CSI acquisition for FDD massive MIMO systems with two-level precoding," in *2016 IEEE Wireless Communications and Networking Conference*, pp. 1–6, Apr. 2016.

[32]   Q. Spencer, A. Swindlehurst, and M. Haardt, "Zero-forcing methods for downlink spatial multiplexing in multiuser mimo channels," *IEEE Transactions on Signal Processing*, vol. 52, no. 2, pp. 461–471, Feb. 2004.

Chapter 4

Towards a service-oriented dynamic TDD for 5G networks

*Rudraksh Shrivastava[1,2], Konstantinos Samdanis[1],
and David Grace[2]*

The emerging 5G networks should accommodate a plethora of heterogeneous services with diverse Service Level Agreements, such as Internet of Things, location and social applications and multimedia. Dynamic Time Division Duplex (TDD) networks have the potential to support asymmetric services providing resource flexibility especially for small data applications and for social applications with interchanging uplink (UL) and downlink (DL) demands. TDD networks can leverage the benefits of network programmability via the means of the Software-Defined Network (SDN) paradigm that enables a logically centralized control plane capable of delivering efficient, optimized and flexible network resource management matching specific application UL and DL traffic requirements. This chapter describes the main mechanisms and components for evolving TDD networks toward 5G. In particular, it provides an overview of various dynamic TDD proposals, before elaborating the concept of TDD virtual cells, which allows users residing at the cell edge to utilize resources from multiple base stations forming a customized TDD frame. The adoption of SDN for programming the network resources and frame (re)configuration of TDD virtual cells follows, elaborating the SDN architecture, the resource programmability logic and the resource sharing in heterogeneous environments considering Frequency Division Duplex macros and TDD small cells. The adoption of SDN for enabling a TDD-specific network slice framework is described next, allowing a different TDD frame configuration to be employed within a certain amount of isolated resources, enhancing in this way the network utilization while optimizing the application perceived performance. Finally, some further considerations and challenges are analyzed considering the adoption of TDD in future 5G networks.

[1]NEC Europe Ltd, Germany
[2]Department of Electronics, University of York, UK

4.1   Introduction

The rapid proliferation of new type of devices, such as smartphones, tablets and wearable electronics combined with the widespread of high speed mobile networks, have led to an evolution of diverse mobile services, generating huge amounts of data traffic [1]. Some of the emerging 5G applications need faster, higher capacity networks that can deliver content-rich services, whereas other applications have stringent latency, reliability and scalability requirements such as critical communications [2]. Social media and cloud services have changed the way humans communicate and acquire information from the Internet, being also more interactive due to "always on" features, with asymmetric uplink (UL)/downlink (DL) demands [3]. For instance, massive Internet-of-Things (IoT) applications may need to upload or exchange a high data volume at different times, creating more diverse UL traffic patterns with increased Quality-of-Service (QoS) demands.

Such heterogeneous services may create performance conflicts while coexisting on the top of a common network infrastructure. Hence, it is evident that the efficient support of emerging 5G applications requires an increased degree of network resource flexibility for assuring the respective QoS. Alternatively, mobile network operators would need to upgrade their network infrastructure increasing capital expenditure and operational expenditure costs. To achieve the desired flexibility, mobile network operators are exploring network virtualization, resource programmability and softwarization solutions via the means of Software-Defined Networks (SDNs) and network functions virtualization (NFV) [4]. Network virtualization through SDN and NFV can introduce innovation, that is by rapidly testing and deploying new services, into the mobile operator's network and can create new revenue means by opening the network to different tenants, for example vertical segments, allowing a customized network functionality and resource provision [5].

The evolution of time division duplex (TDD) toward 5G will leverage the benefits of SDN for enabling application and service providers as well as Mobile Virtual Network Operators to program the TDD UL and DL resource configuration and resource allocation. In TDD, such programmability opportunities enable further innovation by allowing users residing at the edge of overlapping cells to utilize resources from different base stations creating in this way a customized TDD frame or a virtual cell. In addition, it offers the opportunity for enabling service-oriented network slices by isolating network resources for exclusive use by particular tenants considering their service requirements.

The remaining of this chapter is organized as follows. Section 4.2 describes the fundamental technologies that enable programmability for the emerging 5G TDD systems, while Section 4.3 overviews the TD-LTE state of the art considering also initial enhancements for dynamic TDD deployments. Section 4.4 elaborates TD-LTE virtual cell concept considering also SDN-enabled resource programmability. Section 4.5 considers multitenancy and resource sharing among Frequency Division Duplex (FDD) and TDD systems as well as TDD network slicing via the means of SDN. Finally, Section 4.6 provides the conclusions.

4.2 Enabling technologies for the emerging 5G TDD systems

5G systems should provide a flexibility resource allocation leveraging the benefits of network programmability through network virtualization via the means of SDN and NFV. The notion of virtualization has slightly distinct meaning in SDN and NFV communities. SDN focuses on abstracting a particular set of underlying network resources allowing open access and programmability for multiple tenants that correspond to particular clients, applications or services. In the context of NFV, virtualization refers to a software entity in a container or virtual machine over a hypervisor on a commercial-of-the-shelf server or dedicated hardware, that is a switch or base station. The term hypervisor refers to a hardware virtualization technique that allows logical functions or otherwise software instances to share a single hardware platform while appearing as individual elements with their own hardware resources such as processor, memory and hard disk.

At present, mobile networks are rigid in nature formed by monolithic network functions that enable firm services, which are difficult to be adjusted and customized. The application of virtualization in the evolving 5G networks aims to transform the way networks are managed and operated, enabling rapid deployments and testing of innovative services. The SDN architecture as defined in [6,7] consists of three different sets of functional layers illustrated in Figure 4.1. The SDN application layer allows SDN applications such as analytics, optimization algorithms, management mechanisms and network control etc. or third parties to communicate service and resource requirements to the SDN controller via the Application-Controller Plane Interface (A-CPI). The control layer contains the SDN controller, a logical entity with

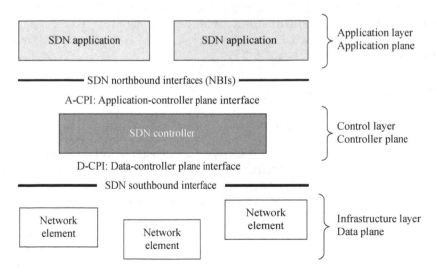

Figure 4.1 Overview of the SDN architecture [6]

a global network view, which allows SDN applications and third parties to control and program the network via the Data-Controller Plane Interface (D-CPI) based on an abstracted network resource view that hides insight information. The SDN controller separates the control from the data plane, that is decoupling the control from traffic forwarding and processing and collecting network statistics. The infrastructure layer involves the network devices that control the forwarding and data processing, that is the data plane.

The use of SDN allows resource-flexibility-supporting customized services and logical network instances on the top of a common-network infrastructure, meeting the service constraints of application providers, vertical segments and virtual operators. In the context of mobile networks, SDN can assure a unify control across heterogeneous radio infrastructures and dense RAN deployments [8], while providing a joint control of mobile and transport network layers [9]. It is anticipated that the adoption of the NFV paradigm for both access and core networks have the potential to enhance further the service flexibility by scaling appropriately Virtual Network Functions (VNFs) for supporting efficiently the 5G services [10]. For an SDN controller, VNFs appear as another type of resource [11], which forms a network graph that the SDN controller can chain together forming a particular service [12]. Hence, SDN and NFV are complementary technologies that can be used together forming tailored services considering evolving traffic conditions, mobility patterns and target QoS demands.

The use of SDN/NFV can also enable network slicing, defining logical self-contained networks that can accommodate different business requirements, on the top of a common physical network infrastructure [13]. The SDN controller can play the role of mediator facilitating admission control and network slice allocation offering programmability to third parties [14]. Network slicing can assure isolation between different services and tenants, allowing an efficient coexistence of heterogeneous applications with often conflicting requirements. At present, 3GPP is performing a study named NexGen [15] within the Architecture Group SA 2, considering various slice operations in the data and control plane for the next generation of mobile networks.

4.3   TD-LTE state of the art

4.3.1   Overview of TD-LTE

3GPP LTE is designed to support both paired spectrum, that is Frequency Division Duplex (FDD) and unpaired spectrum via TDD, using the same radio access scheme, subframe structure and configuration protocols. In contrast with FDD, the main difference lies in the fact that TDD supports unpaired spectrum, where transmissions in UL and DL are separated in the time domain. The fact that unpaired spectrum is easier to acquire makes the deployment of TD-LTE less complex in many occasions. As per [16], the International Telecommunication Union (ITU) is considering a spectrum allocation for IMT-Advanced in the 698–803, 2,300–2,400, 2,500–2,600 and 3,400–3,600 MHz bands, where a large chunk of unpaired spectrum is expected to be allocated for TDD.

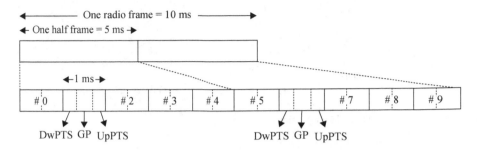

Figure 4.2 Overview of the TDD frame

Table 4.1 Uplink–downlink allocation

				Subframe number					
UL/DL config.	**0**	**1**	**2**	**3**	**4**	**5**	**6**	**7**	**8**	**9**
0	DL	S	UL	UL	UL	DL	S	UL	UL	UL
1	DL	S	UL	UL	DL	DL	S	UL	UL	DL
2	DL	S	UL	DL	DL	DL	S	UL	DL	DL
3	DL	S	UL	UL	UL	DL	DL	DL	DL	DL
4	DL	S	UL	UL	DL	DL	DL	DL	DL	DL
5	DL	S	UL	DL	DL	DL	DL	DL	DL	DL
6	DL	S	UL	UL	UL	DL	S	UL	UL	DL

Each TDD frame is 10 ms long and consists of DL, UL and special (S) subframes with 1 ms duration each as illustrated in Figure 4.2. The S subframe is used for switching from DL to UL transmission direction and contains a conventional DL part named the DL Pilot Time Slot, a Guard Period (GP) of a blank gap that assists the User Equipment (UE) to switch from the DL to UL and an UL Pilot Time Slot part that carries the synchronization information that assists the US to establish UL connectivity. The S subframe is included at least once within each 10 ms frame or in certain UL/DL configuration cases this is included twice, that is once every 5 ms.

3GPP has defined seven different TDD frame configurations as shown in Table 4.1 [17], with the UL/DL portion of each frame configured accordingly, reflecting estimated traffic demands. Such a TDD feature allows resource configuration flexibility that helps in supporting various asymmetric applications such as sensor measurements and IoT, social applications with high UL demands and others, besides the conventional video streaming and Voice over IP.

Besides its inherent flexibility, the first generation of TD-LTE networks supported a static synchronous configuration where a set of base stations serving a greater geographical area follow a TDD configuration with a common UL/DL ratio that suits the overall average long-term traffic demands. Such synchronous operation

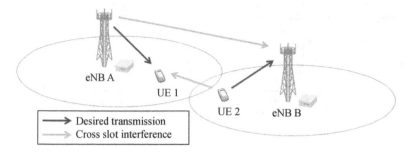

Figure 4.3   Cross slot interference

simplifies interference control assuring no cross slot interference among UE served by neighboring base stations or direct base station interference as illustrated in Figure 4.3. As interference could severely degrade the overall system performance, [18] specifies strict synchronization requirements for TD-LTE between base stations with overlapping coverage areas.

It is evident that with static TD-LTE deployments, it is difficult to satisfy instantaneous cell-specific traffic demands, because traffic alternations at particular cell locations cannot influence significantly the entire system limiting in this way the inherent flexibility of TDD. It has been anticipated that the variation of traffic in UL and DL directions may rise with the emerging applications, hence more dynamic TDD solutions should be considered.

4.3.2   Dynamic TDD

To alleviate the rigid attributes of the static TD-LTE deployments, 3GPP initiated a study in [19], to investigate dynamic TDD focusing on small cells considering scenarios of isolated small cells, various small-cell deployments with limited interference or by using almost blank frames. Such a study leads to the specification of the TDD-enhanced Interference Management and Traffic Adaptation [17], based on power control.

Other efforts for achieving dynamic TDD also took place around the same time, with the Cell-Specific Dynamic Reconfiguration (CSDR) [20] being one of the most significant. The CSDR enhances the resource flexibility by allowing each cell to select independently a UL/DL frame that satisfies best the UL/DL traffic demands of the residing users. In other words, each base station or evolved Node B (eNB) in the 3GPP terminology can dynamically change the UL/DL configuration at specific timescales, in order to match best the instantaneous traffic demands. The issue of interference caused by neighboring cells can be handled via deployment and cell-clustering means or by the use of almost blank frames and power control. Almost blank subframes allows the base-station DL power to be completely blanked or muted in specific subframes where cross slot interference is likely to occur, but this may impact the spectral efficiency.

The notion of interference management and control among neighbor TDD cells, by regulating the cell-edge regions is introduced in [21], where a hybrid approach is introduced combining static and dynamic resource allocation for the inner and outer cell regions, respectively. The dynamic allocation area is negotiated among the neighbor eNBs by selecting and indicating UL/DL subframes, whereas the cross slot interference is avoided by introducing power control. In [22,23], the authors considered the buffer status to flexibly switch between different TD-LTE frame configurations, introducing extra UL/DL resource diversity for the eNB's inner cell regions. The asymmetric UL/DL subframe allocation is also investigated in [24], which exploits the backhaul resources in an efficient way with the aim of reshaping the interference on the channels.

The UL/DL subframes can be reconfigured via updating the broadcast system information. This process depends on the general system information update procedures, in which a paging indication is used to notify users about an upcoming configuration change. As elaborated in [20], the system information update procedure has some drawbacks when the UL/DL frames are modified to support the instantaneous traffic conditions. These limitations include the following:

- According to the standards, the modification time to successfully notify all the users about the upcoming changes is restricted. As a consequence, the reconfiguration timescale is restricted and cannot be done more frequently than every 640 ms [25].
- An ambiguity period from the system information modification boundary until the time when a particular UE has successfully received and decoded the new system information. Such ambiguity period causes UL/DL frame reconfiguration knowledge to become unsynchronized between the network node and the UE. Consequently, this leads to a discontinuity of the Hybrid Automatic Repeat Request (HARQ) processes and hence a degradation in throughput.

To perform the UL/DL frame reconfiguration more frequently, alternative methods need to be explored. For example, a UE-specific higher layer signaling can be used instead of a system information update to indicate a new UL/DL frame configuration, which can be initiated by the scheduler. This involves sending a DL control message, that is, either a radio resource configuration (RRC) or medium access control (MAC) message, toward the target UE to update its UL/DL configuration. RRC and MAC messages can be transmitted more frequently depending on the system requirements, as frequently as a few tens of milliseconds, allowing faster UL/DL reconfiguration timescales reflecting instantaneous traffic demands. The HARQ discontinuity may still operate in the same way as with the system broadcast reconfiguration mechanism. However, here, since the RRC and MAC messages are acknowledged, the ambiguity period is shorter compared to the system broadcast reconfiguration mechanism. The ambiguity period in this case is mainly caused by the feedback delay to transmit the acknowledgement. Some additional ideas for performing an UL/DL frame reconfiguration have been presented in [20].

4.4   TD-LTE virtual cells

Despite the plethora of different UL/DL configurations and the use of dynamic TDD schemes, typical TD-LTE deployments introduce limitations that prevent operators from fully exploiting TDD resource flexibility. Such limitations are mainly centered around the problem of pseudocongestion and on lack of an appropriate TDD multiconnectivity schemes that can help enhancing the resource flexibility. Pseudocongestion is a type of a congestion phenomenon, wherein adequate resources exist at a particular base station, but in the opposite transmission direction than the one that the user requests, thereby causing congestion.

To resolve pseudocongestion and enhance TDD flexibility, the concept of virtual cells is introduced in [26] that enables UEs residing at the cell edge between overlapping cell regions to access radio resources utilizing subframes from multiple eNBs. Virtual cells appear as one logical eNB with its own UL/DL configuration, offering an additional degree of flexibility by facilitating a customized virtual frame for users residing in overlapping cell regions. Such virtual frames are composed by UL/DL subframes derived from multiple eNBs, which are tailored to accommodate particular application demands via the means of TDD specific multiconnectivity. In this way, eNBs and UEs exploit both the spatial and time domain load balancing. Virtual cells can also benefit the inner cell users, that is the users that are not served by the virtual frame, by freeing up additional resources. A simple example of a UE residing within a virtual cell, utilizing DL resources from eNB B and UL resources from eNB A is illustrated in Figure 4.4, resolving in this way pseudocongestion. Note that virtual cells do not only separate the UL from DL, but can flexibly select any subframes irrespective of the transmission direction from overlapping eNBs.

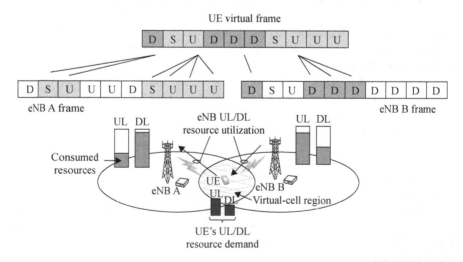

Figure 4.4   A simple example of the virtual cell concept [27]

It is worth noting that due to the device hardware limitations, UEs within the virtual cell region do not access the UL and DL resources at the same time. This process of accessing UL and DL resources from multiple eNBs requires UEs to align the corresponding transmit and receive modes accordingly. For control purposes, additional signaling is required to facilitate virtual cells. Synchronization for transmission and reception via multiple eNBs is necessary to ensure that the data appears as a single stream to and from the UEs. This may be achieved by the following:

- Reception of multiple copies of data at all eNBs and accordingly transmitting the data in selected subframes
- Reception of all the data at a single eNB and use of X2 interface to transfer the data to other eNB participating in the transmission process
- Splitting of data sessions in the PDN/S-GW (Packet Data Network/Serving Gateway) before it reaches the appropriate eNBs

It is worth noting that the transmission and reception sessions via multiple eNBs should be transparent to the end user. After the configuration process of the virtual cells is finalized, mechanisms to perform management and maintenance are crucial in order to adapt to the evolving traffic demand. The SDN paradigm can address such management and maintenance requirements allowing also network resource programmability based on third parties demands and offer flexible TDD reconfiguration and virtual cell allocation [27].

SDN can take advantage of the global network view considering the network state in terms of traffic load and interference as elaborated in [28]. To realize such global network view, the SDN controller can rely on the conventional 3GPP Operations Administration and Management (OAM) system, which collects network monitoring information regularly or upon request. Such network monitoring information may include UL and DL load, RAN topology and Key Performance Indicators, for example, handover (HO) failures, latency, throughput, etc. as specified in [29]. The SDN controller may also obtain certain additional information related to specific rules for counting the occurrence of an event, for example getting statistics on the number of UEs that consumes a certain UL/DL ratio, or the TDD frame reconfiguration changes per eNB via the D-CPI. Such SDN-based architecture provides the pillars for 5G TDD systems enabling flexibility in resource allocation and service customization at particular areas via the use of virtual cells as illustrated in Figure 4.5.

The global network view allows the SDN controller to provide efficient resource allocation enabling virtual cells at particular geographical areas by adaptively modifying the power and subcarrier allocation profile independently for each eNB across the RAN via the D-CPI. Consequently, if the traffic conditions vary or the desired application performance cannot be achieved, the SDN controller can enforce a selective UL/DL ratio reconfiguration even for the sake of facilitating virtual cells, an activity that is not feasible otherwise via the use of dynamic TDD performed locally at the eNB level. Such a resource allocation process via virtual cells should be performed while introducing no negative effects on the performance of other users, that is users residing in the inner cell areas. This is assured by employing the capacity gain and

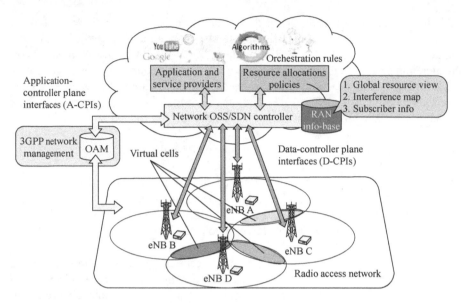

Figure 4.5   SDN-based 5G TDD network architecture enabling virtual cells [27]

capacity loss concept considering the effective bandwidth model presented in [30]. Hence, the size of virtual cells is not fixed but is user centric depending on the traffic load, the number of users and the interference levels.

Algorithms related with the establishment of virtual cells are elaborated in [27] considering network parameters as well as user and application performance. Typically, these algorithms are executed at the SDN controller to take advantage of the scalability for performing intensive data analysis, that is considering several parameters from the entire RAN. Additional distributed mechanisms to control interference and manage the UL/DL frame reconfiguration should be considered as the network interference conditions may sometimes vary at short time interval. This may occur frequently within virtual cell regions due to their small area and user mobility. Therefore, by employing such mechanisms, the workload on the SDN controller may be reduced, which in turn allows for a long-time tolerance of the 5G TDD frame configuration decisions suggested for the RAN.

4.5   Network virtualization and multitenancy in 5G TDD networks

This section analyzes the notion of network virtualization and multitenancy in 5G TDD network considering two different scenarios focusing on: (i) flexible FDD/TDD coexistence assuming a common infrastructure provider and (ii) network slicing, wherein different services are allocated an isolated amount of resources with a different UL/DL configuration ratio.

4.5.1 A flexible FDD/TDD coexistence in a multitenant environment

The deployment of small cells to anticipate the growing traffic demands can enhance the Quality of Experience (QoE) by reducing latency and enhancing throughput by bringing the cellular network close to the end user, which gave rise to heterogeneous networks (HetNets). The adoption of small cells can provide flexibility allowing rapid and unplanned deployments, accommodating bursty and asymmetric traffic such as social media applications and IoT. For these types of applications, TDD is a suitable technology for high frequency bands, that is 3.5 GHz and above, which can also take advantage of the dynamic UL/DL ratio configuration.

A range of different FDD/TDD coexistence deployment scenarios is elaborated in [31] for augmenting the capacity of FDD deployments considering colocated cells with the same and different coverage footprint, TDD hotspots for data offloading and independent TDD RAN. An elastic-infrastructure-sharing framework for FDD/TDD HetNets considering FDD macros and TDD pico cells is elaborated in [32]. Such resource sharing framework relies on the SDN paradigm assuming a common infrastructure provider. The SDN controller, which acts as a resource broker with a global network view, can allocate dynamically spectrum resources between the FDD macros and TDD picos reflecting evolving traffic demands, while programming the UL/DL ratio of the TDD picos considering the residing users. Hence, the FDD/TDD framework can assure an efficient network utilization.

An example of the SDN architecture considering two operators is illustrated in Figure 4.6, assuming that the RAN can accommodate network and spectrum sharing capabilities with the help of SDN-based network management architecture.

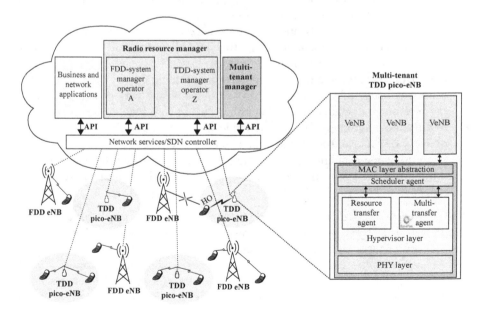

Figure 4.6   SDN-based FDD/TDD network resource management architecture [32]

A scenario with LTE-A HetNet is considered where, FDD macro eNB operator is called operator A and the TDD pico cell operator is called operator Z. All types of eNBs are assumed capable of communicating with the central SDN controller (NY-aware eNBs). The scenario considered focuses on leasing of resources from the TDD system to the FDD system and vice-versa. For the sake of simplicity, a two tenant scenario with operator A and operator Z is considered. However, different scenarios with multiple tenants may also be applied. To facilitate a dynamic network and spectrum sharing, the different tenants, that is the FDD and TDD operators, need to establish a Service Level Agreement (SLA) with the infrastructure provider. The SDN architecture is based on the Open eNB paradigm [33] and can also enable users to connect to the closest eNB in the vicinity provided that it can support multitenant capabilities.

The SDN architecture consists of a number of logical entities and functions. SDN controller enables multitenant connectivity, considering different operators and application providers, and provides the intelligence for resource management; it contains the following:

- **Multitenant manager**, which is responsible for managing multitenant policies, perform HO decisions and provide instructions to the multitenant agent within TDD pico for carrying out efficient resource allocation and sharing procedures.
- **Radio resource manager**, which contains resource availability and utilization, which is used to implement resource-transfer-related decisions between FDD and TDD systems.

Considering the RAN, the macro-eNB-related architecture is described in the OpeneNB [33], while for TDD pico eNBs, the main functions contain the following:

- **VeNB** is a software application emulating the upper layers of the protocol stack, which are managed by different tenants.
- **Hypervisor** that logically abstracts physical resources to enable sharing among multiple operators, while at the same time providing isolation to avoid any type of interference or performance degradation.
- **Resource-transfer agent** is responsible for executing the resource transfer processes, collects information from different tenants and periodically updates the radio resource manager module regarding the network state.
- **Multitenant agent** performs a dynamic slice reconfiguration and enables the delivery of packets from users to the appropriate VeNB.
- **Scheduler agent** facilitates cooperation between the resource transfer agent and the multitenant agent, providing MAC layer abstractions to the corresponding tenant.

It should be noted that both resource transfer agent and the multitenant agent are remotely assisted by two distinct SDN applications at the SDN Controller named the multitenant manager and the radio resource manager.

Therefore, when an FFD tenant experiences a high traffic load and needs additional resources, the TDD pico network can transfer selected resources to the FDD macro (or vice versa). Details regarding the resource management logic and mechanisms that carry out such operations are elaborated in [32].

4.5.2   5G TDD network slicing

The adoption of 5G network slicing for TDD networks allows the usage of a separate UL/DL configuration for each isolated amount of resources that form a logical network, enabling the concurrent deployment of different UL/DL ratios on the top of the same network infrastructure.

The reason for employing such a slicing scheme for TDD networks is to allow different services to optimize the user performance, while at the same time enhance the network utilization. Such 5G network slicing concept, referred to as service-oriented slice explicit (SE) dynamic TDD [34], allows third party services and applications to program the network infrastructure. Such programmability feature allows a different UL/DL ratio per slice, enabling an independent dynamic UL/DL frame reconfiguration to accommodate best the QoS demands of the respective service.

A simple example that provides an overview of the proposed TDD network slicing concept highlighting the difference with the baseline dynamic TDD named CSDR described in [20] is illustrated in Figure 4.7. In the CSDR, eNBs across the RAN may adopt a different UL/DL frame configuration, which can dynamically be adapted to match the aggregated traffic demands considering all residing cell users. Unlike CSDR, in which each eNB employs the same UL/DL frame configuration across all the available subcarriers, the service-oriented SE dynamic TDD or SE-CSDR divides

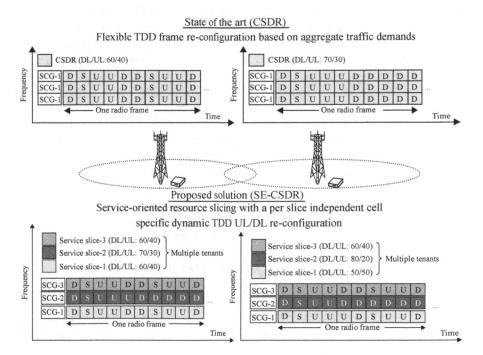

Figure 4.7   A simple example comparing the dynamic TDD with the proposed service-oriented slice explicit CSDR [34]

the available bandwidth and assigns to each subcarrier group (SCG) a particular UL/DL configuration.

For the matter of simplicity, we represent in this example a single SCG being isolated for supporting a single service or tenant, establishing a network slice, but such slicing concept can be applied to any size group of SCGs. Each TDD network slice can independently support CSDR operations at the same or different reconfiguration timescales providing another degree of flexibility compared with CSDR where UL/DL frame reconfiguration updates are performed periodically at predefined timescales.

In TDD network slicing, an additional cross slot interference that may occur between the users served by different network slices, that is interslice users, since different slices may adopt a different UL/DL configuration. To avoid such an interslice interference, each network slice is configured to have one size across the applied RAN area with different slices having separate spectrum bands, that is distinct SCGs. This implies that for a particular tenant, the size of the allocated network slice and respective SCGs across the RAN remains the same for a fixed period of time. The isolation property provided by network slicing offers resource customization by allowing service programmability and operation of independent schedulers, thereby improving the resource utilization efficiency. Isolation also ensures that traffic variations within a particular network slice do not have a negative impact on the performance of the users being served by other network slices. However, the isolation property of network slicing introduces losses of multiplexing gains, as resource scheduling is performed considering the resources per slice instead of the entire spectrum band.

To establish a network slice, third parties communicate service requests with specified SLAs, user characteristics and time duration toward the SDN controller via the standard interface of A-CPI. Once the request is received, the SDN controller analyzes the requirements and in order to be able to perform admission control and resource negotiation in case of not adequate resource availability. To achieve this, the SDN controller not only should rely on a global network view from the 3GPP legacy OAM, but also needs forecasting information regarding the network load and user mobility to be able to assure the desired slice perforce for the complete duration of the request. For this purpose, different forecasting methods can be adopted such as exponential smoothing, machine learning, and artificial intelligence considering both UL and DL directions.

Once a network slice is instantiated, service customization is assured by converting the SLA requirements into a policy, which is communicated via D-CPI to guide the behavior of network equipment. Such a policy may include various rules governing for instance the slice and traffic prioritization, the UL/DL ratio and frame reconfiguration timescales, etc.. The policy installed at eNBs can resolve short term alternating traffic patterns in an autonomous manner, for each particular network slice instance independently. This can relax the need for adjusting regularly the resources allocated per slice to reflect evolving traffic demands via the means of the SDN controller. Instead, the SDN controller needs to allocate besides the corresponding slice resources the appropriate policy.

The SDN controller provides also a QoE assessment control service through the D-CPI for particular applications using corresponding performance analysis and evaluation models. In particular, application-specific information is collected from base stations through a function referred to as QoE acquisition function, which periodically or on-demand collects application specific measurements such as latency, loss, throughput and others and feeds the QoE assessment service at the SDN controller. Such QoE assessment control function makes the application performance visible to the SDN controller and hence to external third parties.

4.6 Conclusions

The next generation of TDD networks toward 5G will push the barrier of flexibility and customization to the limits through the means of resource virtualization. This chapter introduced three different means of resource virtualization including the following:

1. Virtualized cells by combing the resources of neighboring base stations within particular geographical regions considering the application UL/DL ratio demands
2. Sharing spectrum among FDD/TDD HetNets allowing an infrastructure provider to support efficiently multitenancy, maximizing the network utilization
3. Network slicing on the top of a common network infrastructure, allowing the allocation of a different UL/DL configuration and reconfiguration time scales per slice taking into account the service characteristics and the degree of traffic asymmetry

Network virtualization allows advanced network resource management by means of dual or multiconnectivity in virtual cells allowing service customization. The notion of service customization and flexibility for enhancing the QoE considering particular applications is elaborated in [35], wherein the perceived performance is monitored and analyzed before identifying and programing virtual cells at certain geographical locations. The concept of virtual cells can also give rise to TDD-based cell-less architectures as introduced in [36] within a cloud-RAN arrangement. In such cell-less architectures the Baseband Unit can form virtual cells between selected Remote Radio Heads.

The adoption of network virtualization in resource sharing FDD/TDD HetNets and TDD network slicing exploits load balancing in spatial and time domain considering future traffic demands and user mobility as well as service-oriented characteristics, enhancing the network resource utilization.

The SDN paradigm consists of a key technology enabler for TDD network virtualization allowing resource programmability and multitenancy, that is openness to third parties, for supporting diverse UL/DL traffic demands. The use of SDN can also support innovation by introducing rapidly new services, while at the same time it can offer a QoE-aware resource management, providing also feedback to the service/application layer. The use of SDN to provide flexibility in resource allocation can minimize the loss in multiplexing gain that may occur as a consequence of network slicing.

References

[1]   Cisco networks, Cisco Visual Networking Index: Global Mobile Data Traffic Forecast Update, 2016–2021, Feb. 2017.

[2]   3GPP TR 22.891, Study on New Services and Markets Technology Enablers (SMARTER), Technical Report, 3GPP Standards Organization, Rel.14, Sep. 2016.

[3]   GSMA Alliance, The Mobile Economy, 2017.

[4]   5G Vision—The 5G infrastructure public private partnership: the next generation of communication network and services, White paper, 5G Infrastructure Association, Feb. 2015.

[5]   P. Rost, A. Banchs, I. Berberana, *et al.*, "Mobile network architecture evolution towards 5G," *IEEE Communications Magazine*, vol. 54, no. 5, pp. 84–91, May 2016.

[6]   ONF TR-502, SDN Architecture (1), June 2014.

[7]   ONF TR-521, SDN Architecture (1.1), 2016.

[8]   C.J. Bernardos, A. De La Oliva, P. Serrano, *et al.*, "An architecture for software defined wireless networking," *IEEE Wireless Communications*, vol. 21, no. 3, Jun. 2014.

[9]   K. Pentikousis, Y. Wang, and W. Hu, "Mobile flow: towards software defined mobile networks," *IEEE Communications Magazine*, vol. 51, no. 7, Jul. 2013.

[10] ETSI GS NFV-MAN, v1.1.1, Network function virtualization (NFV); Management and orchestration, Dec. 2014.

[11] Open Networking Foundation (ONF) TR-518, Relationship of SDN and NFV, Oct. 2015.

[12] P. Quinn, and T. Nadeau, Problem statement for service function chaining, Internet Engineering Task Force (IETF) RFC 7498, Apr. 2015.

[13] Next Generation Mobile Networks (NGMN) Alliance, 5G White Paper, Feb. 2015.

[14] Open Networking Foundation (ONF) TR-526, Applying SDN Architecture to 5G Slicing, Apr. 2016.

[15] 3GPP TR 23.799, Study on Architecture for Next Generation System, Technical Report, 3GPP Standards Organization, Rel.14, Dec. 2016.

[16] S. Chen, S. Sun, Y. Wang, G. Xiao, and R. Tamrakar, "A comprehensive survey of TDD-based mobile communication systems from TD-SCDMA 3G to TD-LTE(A) 4G and 5G directions," *China Communications*, vol. 12, no. 2, pp. 40–60, Feb. 2015.

[17] 3GPP TS 36.300, Evolved Universal Terrestrial Radio Access (E-UTRA) and Evolved Universal Terrestrial Radio Access Network (E-UTRAN); Overall description; Stage 2, 3GPP Standards Organization, Rel. 14, Dec. 2017.

[18] 3GPP TS 36.133, Technical Specification Group Radio Access Network; Evolved Universal Terrestrial Radio Access (EUTRA); Requirements for Support of Radio Resource Management, 3GPP Standards Organization, Rel.14, Jan. 2017.

[19] 3GPP TR 36.828, Further Enhancements to LTE TDD for DL-UL Interference Management and Traffic Adaptation, Rel.11, Jun. 2012.

[20] Z. Shen, A. Khoryaev, E. Eriksson, and X. Pan, "Dynamic uplink–downlink configuration and interference management in TD-LTE," *IEEE Communications Magazine*, vol. 50, no. 11, Nov. 2012.

[21] H. Lee, and D.-H. Cho, "Combination of dynamic-TDD and static-TDD based on adaptive power control," *IEEE VTC-Fall*, Calgary, Sep. 2008.

[22] Y. Wang, K. Valkealahti, K. Shu, R. Sankar, and S. Morgera, "Performance evaluation of flexible TDD switching in 3GPP LTE system," *2012 35th IEEE Sarnoff Symposium*, Newark, NJ, pp. 1–4, 2012. 2012.

[23] A. Khoryaev, A. Chervyakov, M. Shilov, S. Panteleev, and A. Lomayev, "Performance analysis of dynamic adjustment of TDD uplink–downlink configurations in outdoor picocell LTE networks," *2012 IV International Congress on Ultra Modern Telecommunications and Control Systems*, St. Petersburg, pp. 914–921, 2012

[24] P. Rost, and K. Samdanis, "The two-way interference channel: robust cooperation to exploit limited backhaul efficiency," *Ninth ITG SCC*, Munich, Jan. 2013.

[25] 3GPP TS 36.331, v13.0.0, LTE; Evolved Universal Terrestrial Radio Access (E-UTRA); Resource Control (RRC); Protocol Specification, 3GPP Standards Organization, 2016.

[26] K. Samdanis, R. Shrivastava, A. Prasad, P. Rost, and D. Grace, "Virtual cells: enhancing the resource allocation efficiency for TD-LTE," *IEEE VTC-Fall*, May 2014.

[27] K. Samdanis, R. Shrivastava, A. Prasad, D. Grace, and X. Costa-Perez, "TD-LTE virtual cells: an SDN architecture for user-centric multi-eNB elastic resource management," *Computer Communications*, vol. 83, pp. 1–15, Jun. 2016.

[28] A. Gudipati, D. Perry, L.E. Li, and S. Katti, "SoftRAN: software defined radio access network," *ACM HotSDN*, Hong Kong, Aug. 2013.

[29] 3GPP TS 32.425, Performance measurements Evolved Universal Terrestrial Radio Access Network, 3GPP Standards Organization, Rel.13, Sep. 2015.

[30] P. Mogensen, W. Na, I. Z. Kovacs, *et al.*, "LTE capacity compared to the Shannon bound," *2007 IEEE 65th Vehicular Technology Conference – VTC2007-Spring*, Dublin, pp. 1234–1238, 2007.

[31] TDD and FDD Co-existence in LTE: Synergies and Enhancements, Nokia White Paper, June 2014.

[32] R. Shrivastava, S. Costanzo, K. Samdanis, D. Xenakis, D. Grace, and L. Merakos, "An SDN-based framework for elastic resource sharing in integrated FDD/TDD LTE-A HetNets," *IEEE CloudNet*, Luxemburg, Oct. 2014.

[33] S. Costanzo, D. Xenakis, N. Passas, and L. Merakos, "OpeNB: a framework for virtualizing base stations in LTE networks," *IEEE ICC*, Sydney, Jun. 2014.

[34] S. Costanzo, R. Shrivastava, K. Samdanis, D. Xenakis, X. Costa-Pérez, and D. Grace, "Service-oriented resource virtualization for evolving TDD networks towards 5G," *IEEE WCNC*, Doha, May 2016.

[35]   S. Costanzo, R. Shrivastava, D. Xenakis, K. Samdanis, D. Grace, and L. Merakos, "An SDN-based virtual cell framework for enhancing the QoE in TD-LTE pico cells," *IEEE QoMEX*, Costa-Navarino, May 2015.

[36]   V. Sciancalepore, K. Samdanis, R. Shrivastava, A. Ksentini, and X. Costa-Perez, "A service-tailored TDD cell-less architecture," *IEEE PIMRC*, Sep. 2016.

Chapter 5

Traffic aware scheduling for interference mitigation in cognitive femtocells

Ghazanfar Ali Safdar[1]

Femtocells are designed to co-exist alongside macrocells, providing spatial frequency re-use, higher spectrum efficiency and data rates. However, interference between two networks is imminent; therefore, ways to manage it must be employed to efficiently avoid problems such as coverage holes. In this chapter, we employ cognitive radio enabled femtocells (CFs) and propose a novel scheduling algorithm to address the problem of cross and co-tier interference in a two-tier network system. Macrocell user equipments (MUEs) usually transmit with varying traffic loads and at times it is highly likely for their assigned resource blocks (RBs) being empty. Based on the interweave concept of spectrum assignment, CFs in our proposed scheme assign the RBs of MUEs with a low-data traffic load and low-interference temperature to its FUEs, thereby mitigating cross-tier interference, whereas co-tier interference is mitigated by resolving the contention for the same RBs by employment of matching policy among the coordinating CFs. System-level simulations are performed to investigate the performance of proposed scheme and results obtained are compared with best channel quality indicator and proportional fair schemes both in no fading and fading conditions. It is found that our proposed scheduling scheme outperforms compared schemes in no fading conditions; however, it provides competitive results when fading conditions are considered. The concept of matching policy successfully mitigates the effect of co-tier interference in collocated femtocells by providing improved signal-to-interference-plus-noise ratio, throughput and spectral efficiency results, thereby proving the effectiveness of scheme.

5.1 Introduction

Emerging wireless networks support the simultaneous mix of traffic models which demand more flexible and efficient use of the scarce spectral resource. The usage of these traffic models differs per user at different times. Simulations with a mix of

[1]School of Computer Science and Technology, University of Bedfordshire, UK

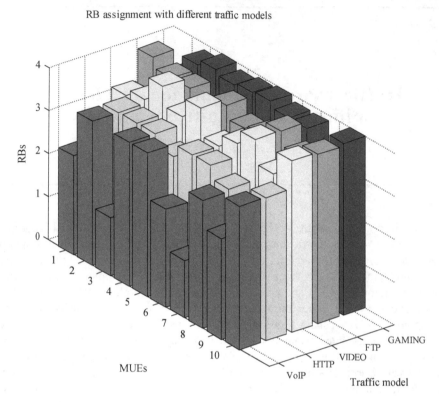

Figure 5.1   Macrocell scenario with varying RB usage/traffic model

realistic traffic models such as VoIP, HTTP, video, FTP and gaming can be mimicked with specifications as defined in [1] to analyse spectral use in a wireless network (Figure 5.1). In cellular networks, data-traffic models have excelled voice services with exponential growth projections expected within the next few years [2]. The limited spectrum cannot accommodate this continuous increase in traffic; therefore, measures have to be put in place to increase spectrum availability while mitigating any inherent interference. 3GPP's Long Term Evolution (LTE) has been developed to meet traffic and high-data rates; however, with the advent of heterogeneous networks such as the deployment of femtocells to complement the deficiencies of the larger macrocell base station (MBS) introduces mutual interference.

The emergence of cognitive radio (CR) which is the ability of radio frequency (RF) to sense its environment and automatically alters its characteristics such as frequency, modulation, power and other operating parameters to dynamically re-use whatever spectrum is available has been widely adopted [3–5]. The recent integration of CR with femtocells [6,7] introduces new and different opportunities of mitigating interference. One of such ways is the ability of a cognitive femtocell (CF) defined as a secondary user to simultaneously use the resources of a primary user (MBS) in an opportunistic manner to avoid interference. There have been many studies on interference mitigation from the perspective of resource/spectrum allocation using CFs [8–11]. Efficient scheduling schemes seek to exploit the time-varying channel conditions to not only mitigate interference but also improve spectrum efficiency, thereby achieving multi-user diversity gain. A scheme based on Gale-Shapely spectrum sharing (GSOIA) is presented in [12] where CFs calculate the utility of each channel and orthogonally assign resources based on a policy to avoid interference. To limit the downlink cross-tier interference, a novel spectrum access scheme is proposed in [13] where an MUE joins a nearby open access FAP while freeing up its allocated sub-channel. The FAP adds the MUE on the list of its UEs and subsequently takes control of its available sub-channel.

Distributed power control schemes to reduce the effect of interference are proposed in [14,15]. In [16], the efficacy of a self updating power control algorithm based on pilot power of an FAP was found to be less significant for FUEs located close to neighbouring FAPs in densely deployed urban femtocells due to low co-tier signal to interference plus noise ratio (SINR). Also, the performance analysis of a coverage radius-based power control scheme to circumvent the problems caused by blind placement of FAPs was laid bare in [17]. Antenna techniques which employ beam directivity of the antennas in FAP and FUEs have also been exploited. In [18], an adaptive pattern switching and adaptive beam-forming antennas are designed and independently employed to exploit the available white spaces in TV White Space (TVWS). Joint or hybrid schemes introduce the combination of two more schemes with the aim of complementing and improving the overall performance of a single scheme. In [19], Fractional Frequency Donation (FFD) is employed which require each FAP with a better throughput to donate bands to poor ones. To find a suitable donor, an FAP measures a FUE's noise power plus interference power and estimates its average channel gain. A survey of interference mitigating schemes specific to CFs are extensively highlighted in [20].

In this chapter, a novel traffic aware scheduling algorithm (SA) is proposed to address the interference in a two-tier network system. This algorithm is based on the premise that MUEs transmit with different traffic loads and it is highly likely for their assigned resource units or resource blocks (RBs) being either empty or with a low interference (Figure 5.1). In the proposed SA, CFs assign the RBs of macrocell user equipments (MUEs) with a low-data traffic load which experiences a low interference temperature to its femtocell user equipments (FUEs). Specifically, it focuses on mitigating the downlink femtocell access point (FAP) interference to nearby macrocell MUE as well as mitigating the uplink interference from MUE towards FAP. As the proposed SA requires non-existence of macro-femto backhaul coordination and no modifications, it is promising for applications in the LTE-advanced (LTE-A) cellular systems that employ heterogeneous networks. The performance of the SA is investigated and the results show an improved SINR, throughput and spectral efficiency as compared to existing state-of-the-art SAs.

5.2   Traffic aware scheduling algorithm (SA)

The scheme could be best described with the help of Figure 5.2. The MBS periodically broadcasts the traffic load $n(t)$ of all its MUEs via signalling interface. CF is able to retrieve traffic load data of MUEs up to a distance of 40 m. The choice of sensing up to 40 m is to enable a FAP sense and retrieve as many empty RBs from MUEs as possible at a distance slightly above its coverage radius which is pegged at 20 m. On the other hand, exceeding a 40 m will introduce sensing overhead which will be detrimental to the algorithm as it is a time-dependent opportunistic approach of using resources. Moreover, it is fair to assume this radius in a densely deployed femtocell. The proposed algorithms both for cross-tier and co-tier interference mitigation are explained in the following sections.

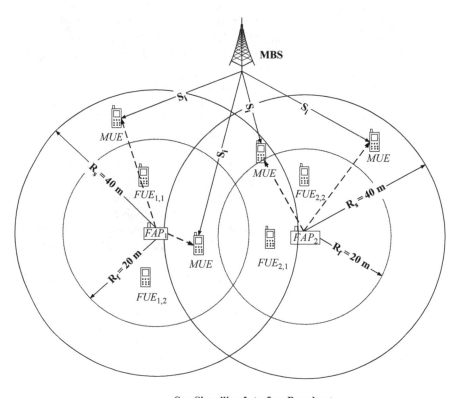

S_i = Signalling Interface Broadcast

R_f = Femtocell Radius

R_s = Sensing Radius

– ➞   Sensing

Figure 5.2   A collocated femtocell scenario with MBS signalling interface

5.2.1 SA cross-tier interference mitigation

RB retrieving strategy: During each scheduling interval, the MUEs are scheduled with RBs from the set:

$$A = m_i: 1 \leq i \leq n(t) \tag{5.1}$$

where m_i denotes ith 180 kHz sub-band over the system bandwidth and $n(t)$ represents the instantaneous traffic load generated from MUEs at time τ.

Let $\tau = \text{TTI} = 1 \, \text{ms}$

In SA algorithm, a CF schedules a FUE on the RB of a MUE with the least traffic load or minimum performance metric for each RB from its throughput analysis and can be expressed as in [21]:

$$k_{i,j}(\tau) = \arg \min \left(\frac{R_{i,j,k}(\tau)}{R_{k,i}(\tau)} \right) \tag{5.2}$$

where $k_{i,j}(\tau)$ is the chosen UE in component carrier (CC)i and in RBj at a specific time τ, $R_{i,j,k}(\tau)$ is the throughput, and $R_{k,i}(\tau)$ is the average throughput of user k [21]. Therefore, the average throughput of each UE is updated for each j according to:

$$R_{k,i}(\tau + 1) = \begin{cases} R_{k,i}(\tau) \left(1 - \frac{1}{t_c} \right) + \frac{1}{t_c} R_{i,j,k}(\tau), & k = k_{i,j}(\tau) \\ R_{k,i}(\tau) \left(1 - \frac{1}{t_c} \right), & k \neq k_{i,j}(\tau) \end{cases} \tag{5.3}$$

where t_c is the time constant [22]. In SA, each CF retrieves this traffic load $R_{k,i}(\tau + 1)$ via the broadcast signalling interface and makes a table known as the traffic load map of all the MUEs within the distance as shown in Table 5.1.

In Table 5.1, at every TTI the traffic load of all the RBs attached to all the MUEs is retrieved and interpreted as:

1 = High-data traffic

0 = Low-data traffic

Table 5.1 Dynamic traffic load map

RB	TTI_1	TTI_2	TTI_3
RB_1	1	1	0
RB_2	0	1	0
RB_3	0	1	1
RB_4	0	1	1
.	.	.	.
.	.	.	.
.	.	.	.
RB_n	1	0	1

Therefore, assume for every femtocell i, ($i \in \{1, 2, \ldots, F\}$), we have an $N \times 1$ vector f_i, where each element $f_i^w \in \{0, 1\}$ denotes whether a FAP will use the RB w when it is 0 or not when it is 1.

5.2.2 SA co-tier interference mitigation

Co-tier interference mitigation focuses on the effective usage of retrieved RBs which is denoted as available RBs among FAPs. In the solution, an RB is assigned to a FUE by its FAP for its sole use at every TTI (Table 5.2).

In SA, each FAP will maintain a table similar to Table 5.2, although it will be an extensive table highlighting all the TTIs (say up to TTl_{10}). Importantly, the RBs that will be available to share will be ones with low-data traffic, such as RB_2, RB_3, RB_4, RB_8 and RB_9. For instance, if we have 50 RBs available and 10 collocated CFs, the RBs are shared equally which will be 5 RBs/CF. The basis for an outright sharing formula is because, through cognition, retrieved RBs of MUEs can be used by FUEs while keeping interference at its barest minimum. Moreover, the number of RBs will always be more than FUEs as a CF (in this scenario a home CF) will only accommodate four FUEs.

Assume for every femtocell, f_n:

$$f_n = \alpha_n + \beta_n \tag{5.4}$$

$$\alpha_n = \{\text{RB}_{n,1}, \text{RB}_{n,2}, \ldots \text{RB}_{n,n}\} \tag{5.5}$$

$$\beta_n = \{\text{RB}_{n,1}, \text{RB}_{n,2}, \ldots \text{RB}_{n,n}\} \tag{5.6}$$

where α_n and β_n are the list of occupied RBs and available RBs, respectively, in the nth FAP.

After sensing, each CF is able to deduce its list of α_n and β_n. As femtocells may be collocated with no coordination between them, they can have similar α_n and β_n. In order to resolve contention among available RBs, a matching policy is introduced to uniquely match retrieved (available) RBs to each CF. The matching policy determines which CF utilises a particular RB from its list of β_n to avoid co-tier interference. To implement the matching policy in every CF_i, we also have a $N \times 1$ vector p_i where each of the elements p_i^w represents the transmit power of a RBw and similarly $N \times 1$ vector q_i where each of the elements q_i^w represents the transmit power in the MBS.

Table 5.2   RBs assignment to FUEs/TTI

RB	RB_1	RB_2	RB_3	RB_4	RB_5	RB_6	RB_7	RB_8	RB_9	RB_{10}
TTI	1	0	0	0	1	1	1	0	0	1

In the proposed SA, a cognitive-enabled scheduling engine to create a matching policy is introduced as follows:

Through sensing, the scheduling engine E, is aware of the locations of the FAPs (FAP$_{LOC}$) and MUEs (MUE$_{LOC}$) of the retrieved RBs in the network. CR enables the femtocell to identify the SINR of the FUE in FAPi on the RBw. To derive the SINR, let the following channel gain variables be introduced as the following:

Channel gain L from: FAPj to FAPi: $L^w_{j,i}$; FAPi to FUEi: $L^w_{i,i}$; MBS to FAPi: $L^w_{b,i}$; FAPi to MBS: $L^w_{i,b}$ and MBS to MUE: $L^w_{b,b}$.

Therefore, the SINR as perceived by a FUE on an available RBw in β_n is given by the following:

$$\text{SINR}^w_i = \frac{f^w_i p^w_i L^w_{i,i}}{f^w_i q^w L^w_{b,i} + \sum_{j=1, j\neq i}^{F} f^w_i f^w_j p^w_j L^w_{j,i} + N} \tag{5.7}$$

where N is the Gaussian noise.

Co-tier interference mitigation is further explained as follows:

Let $E = \{1, 2, \ldots, K\}$

(i) Each FAP in the network submits its list of β_n to E. This list is periodically updated due to the availability of retrieved RBs in β_n.

(ii) E selects a RB from β_n iteratively and maps it against its list of FAP.

(iii) E calculates the SINR^w_i of an FUE on an available RBw in β_n as in Equation (5.7).

(iv) An RB with the highest SINR^w_i is assigned to a FAP and the process continues until all the *RBs* in β_n (which is denoted in the algorithm as $N_{RB}w\_left$) are uniquely assigned to each FAP in the network as shown in Figure 5.3 and the algorithm pseudo code given below in Table 5.3.

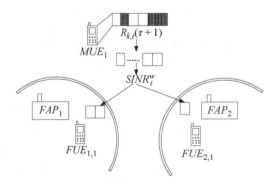

Figure 5.3   Matching policy-based scheduling for co-tier interference mitigation

Table 5.3   Co-tier interference mitigation – RBs matching policy

---

Initialization;
$E = \{1, 2, \ldots, K\}$
1: For RBw1: $N_{\mathrm{RB}}w$
2: Calculate RBw_1,SINR$_i^w$ = max_*value*: min_*value*
3: Set RBw_1 = max_*value*$\{f_i\}$
4: For RBw_2
4: Calculate SINR$_i^w$ = max_*value*: min_*value*
6: Set RBw_2 = max_*value*$\{f_i\}$
7: Continue loop until $N_{\mathrm{RB}}w\_left$ = zero
8: End

5.3   System model

Consider an OFDMA system where the femtocell and macrocell are deployed in a co-channel fashion. The MBS is in the centre of the network and serves the randomly distributed MUEs within its coverage area. As there is more activity of MUEs in the downlink (DL) and less activity in the UL, more focus is on downlink interference in the system model. N_{FAP} femtocells are randomly located within the coverage area of the MBS and the number of MUEs within the coverage area of jth femtocell is denoted as $j = \{1, 2, \ldots, N_{\mathrm{FAP}}\}$. In the system model, MUEs locations are assumed to change per TTI. The CFs in the network operate as a CSG which means MUEs outside its coverage area are not allowed to avail services from it. The simulation parameters are based on 3GPP LTE specifications and given in Table 5.4. The considered network topology consists of a MBS with 30 CFs randomly distributed with four FUEs attached to each CF in a CSG fashion.

5.4   Fading modelling

To highlight the performance of the algorithm, two types of fading, namely Claussen fading and multi-path fading are incorporated in the simulations. The choice of these two fading models is to present the effect of two different levels of destructive interference on the radio channel.

5.4.1   Claussen fading

Claussen fading model is incorporated in the system-level simulation which makes it possible to generate and save the environment maps of path loss and shadowing of all potential locations as introduced in [23]. This significantly reduces the computational complexity of deriving the shadow fading correlations from conventional means such as based on mobile velocity of UEs [24]. In modelling Claussen fading in the simulations, the fading values are derived based only on the correlation with respect to the neighbouring values in the map as illustrated in Figure 5.4. Each of

Table 5.4   System parameters

Parameter		Value
Macrocell		
Cell layout		Dense-urban 3-sector MBS
Carrier frequency		2.14 GHz
Bandwidth		20 MHz
Antenna configuration		SISO (for both BS and UE)
Transmit power		46 dBm (MBS)
Inter-site distance		500 m
Total number of MBS/MUE		1/60
Channel model		Winner+
Traffic model	VoIP, HTTP, video, FTP, gaming	Per cent of users $= [30, 20, 20, 10, 20]$ (as defined in 3GPP RAN R1-070674) [5]
Log-normal shadowing		8 dB
Mobility model		Random-walking model
TTI		1 ms
Path loss: MUE to macro MBS	Outdoor MUE Indoor MUE	PL (dB) $= 15.3 + 37.6 \log_{10} R$ PL (dB) $= 15.3 + 37.6 \log_{10} R + 10$
Femtocell		
Transmit power		20 dBm
Total number of FAP/FUE		30/120
Penetration loss		10 dB indoor
Path loss: FUE to FAP	Dual-stripe model	PL (dB) $= 127 + 30 \log_{10} R/1000$

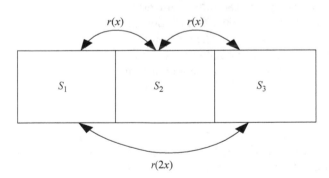

Figure 5.4   Claussen fading

the fading values denoted S_1, S_2 and S_3 is calculated with the correlation matrix R and L from [24]:

$$R = \begin{bmatrix} 1 & r(x) \\ r(x) & 1 \end{bmatrix} \quad \text{and} \quad L = \begin{bmatrix} 1 & 0 \\ r(x) & \sqrt{1 - r^2(x)} \end{bmatrix} \tag{5.8}$$

$$S_1 = a_1 \tag{5.9}$$

$$S_2 = r(x)S_1 + \sqrt{1 - r^2(x)}a_2 \tag{5.10}$$

$$S_3 = r(x)S_2 + \sqrt{1 - r^2(x)}a_3 \tag{5.11}$$

S_3 can also be written as:

$$r^2(x)a_1 + r(x)\sqrt{1 - r^2(x)}a_2 + \sqrt{1 - r^2(x)}a_3$$

where a_1, a_2 and a_3 are generated in dB as distributed random variables with standard deviation.

The mutual information-effective SINR metric (MIESM) [25] is adopted for the simulation. To achieve this, the average mutual information (MI) on all RBs over all UEs is deduced. Subsequently, the effective SINR, $y_{\text{eff(dB)}}$, per RB in every TTI is calculated.

The system level spectral efficiency per RB per TTI can be expressed as follows:

$$R_{\text{sys}} = \propto \log_2\left(1 + 10^{\frac{y_{\text{eff(dB)}}}{10}}\right) \tag{5.12}$$

5.4.2 Multi-path fading

In cellular communication, the RF signal propagates from the transmitter to the receiver via multiple paths due to the obstructions and reflectors existing in the wireless channel. These multi-paths are caused by diffraction, reflection and scattering from buildings, structures and other obstacles existing in the environment. When an MUE is considerably far from the MBS or a FUE is in a different location from its serving FAP, the line of sight signal is not achievable and reception occurs mainly from the indirect signal paths. These multiple paths have various propagation lengths and thus will cause phase and amplitude fluctuations and the received signal time will be delayed. Therefore, the main effect of multi-path propagation can be described in terms of fading and delay spread. The multi-path fading implemented in the simulation is the Rayleigh fading. It is modelled as follows:

The transmitted signal $s(t)$ is assumed to be from an un-modulated carrier and it takes the following form:

$$s(t) = \cos(2\pi f_c t) \tag{5.13}$$

where f_c is the carrier frequency of the radio signal.

The transmitted signal is modelled to be propagated over N reflected and scattered paths. The received signal is calculated as the sum of these N components with random amplitude and phase for each component. The received signal $r(t)$ can be written as follows:

$$r(t) = \sum_{i=1}^{N} a_i \cos(2\pi f_c t + \phi_i) \tag{5.14}$$

where a_i is a random variable equivalent to the amplitude of the ith signal component and ϕ_i is a uniformly distributed random variable equivalent to the phase angle of the ith signal component.

5.5 Performance analysis

In this section, the effectiveness of the SA for interference mitigation in a two-tier CF network is evaluated and compared with best channel quality indicator (CQI) [26] and proportional fair (PF) schedulers [27]. Moreover, the effect on throughput and spectral efficiency are investigated through simulations. The cross-tier analysis is divided into two so as to look at the SAs in varying channel conditions with a fading and no fading cross-tier analysis. Further analysis is conducted with the increment of MUEs in the network to reflect how it affects the algorithms in both channel conditions while keeping the FUEs constant.

5.5.1 Cross-tier interference mitigation (no fading)

Figure 5.5 presents comparative analysis of the average UEs wideband SINR. Clearly, SA results into higher values of SINR (*ca* 22.5 dB) compared to best CQI (*ca* 19 dB) and PF (*ca* 18 dB). The reasons behind this achievement are laid out as follows. In a two-tier co-channel network where FAPs are generally randomly deployed with no coordination, RBs which are used by MUEs are re-used by FUEs which introduces mutual interference on both parties. In best CQI, RBs are assigned to the MUEs with the best radio link conditions; an MUE which is not close to its serving MBS can suffer greatly in the presence of collocated FAPs. These FAPs which are un-coordinated vie to re-use same RBs as of this MUE even though they may have a higher channel gain over the RB. This increases the interference due on that particular MUE resulting into reduced SINR (as shown in Figure 5.5). In SA, scheduling is coordinated to avoid a situation like this by only re-using RBs available and unused by an MUE. On the

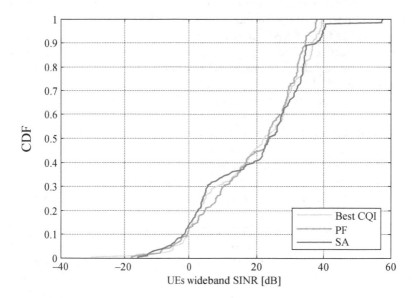

Figure 5.5   Average UEs wideband SINR (no fading)

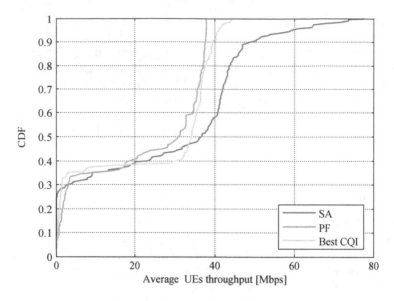

Figure 5.6   Average UEs throughput (no fading)

other hand, best CQI performs slightly better than PF because scheduling is primarily based on fairness in PF, with lesser emphasis on interference mitigation.

Figure 5.6 shows the average throughput gap of the UEs of SA against best CQI and proportional fair (PF) schedulers. The UEs were randomly distributed scattered around from the cell centre to the cell edge to reflect diversity in deployment and channel conditions. In line with the results of SINR, SA scheduler achieves better throughput, around 37 Mbps, compared to 25 and 21 Mbps for best CQI and PF schedulers, respectively. The SA scheduler benefits from the CFs diversity where available and unused RBs are allocated to FUEs while fulfilling a coordinated strategy. Thus, irrespective of the channel conditions, the UEs are able to achieve higher and considerate throughput levels. On the other hand, although the best CQI uses a channel indicator coefficient to allocate RBs; however, higher interference values accordingly result into lesser throughput for CQI as compared to SA. PF scheduler respects fairness among UEs, thereby resulting into reduced average system throughput.

Figure 5.7 presents comparative analysis of the average spectral efficiency. In line with the results of SINR and throughput, SA has a spectral efficiency of 10 bits/cu compared to best CQI (8 bits/cu) and PF (around 7 bits/cu). The reason SA performs better over both other schedulers can be described by looking into parameters α_n and β_n. In SA scheme, it is considered that the RBs in α_n are always fully utilised by MUEs. The RBs in β_n on the other hand are shared between the CFs in the network. Home based CFs are incorporated in the simulation which accommodates up to 4 FUEs. With a total of 120 CFs in tri-sector network, the RBs were shared equally among the 30 CFs in the network resulting into higher spectral efficiency. Using enterprise FAPs which accommodate up to 16 FUEs per FAP, it is assumed

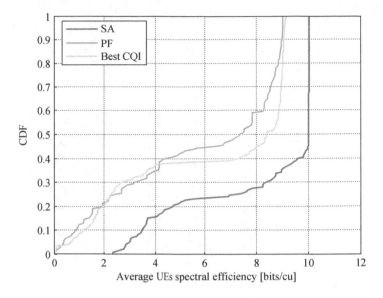

Figure 5.7　Average UEs spectral efficiency (no fading)

that the spectrum efficiency will be even higher. Best CQI has slightly higher spectral efficiency compared to PF, as it increases cell capacity and in turn spectral efficiency at the expense of fairness.

Figures 5.8 and 5.9 conclude the non-fading analysis, where the SINR and throughput values are computed and compared for the three schemes by varying number of MUEs, that is lighter to denser deployment analysis. As shown in Figure 5.8, compared to SINR value of 22 dB (for 20 MUEs), SINR drops to 17 dB (30 MUEs) and 13 dB (50 MUEs) for SA scheme due to the added strain on the network. It only reflects a reduced SINR which is proportional to increased interference from augmented number of MUEs. However, SA mirrors better SINR compared to best CQI and PF schedulers. In line with the SINR results, Figure 5.9 shows how UEs average throughput change for all the compared schemes with varying number of MUEs. Clearly, SA provides encouraging results compared to the best CQI and PF schedulers when more MUEs are formed part of the network.

5.5.2　Cross-tier interference mitigation (with fading)

Simulations were performed with the same system parameters in the presence of Claussen and Rayleigh fading. For clarity, straight line plots with subscript 'c' represent Claussen fading, whereas dotted line plots with subscript 'r' represent Rayleigh fading. Figure 5.10 shows the average UEs wideband SINR. In Claussen fading, 50 per cent of the UEs in SA have an SINR of 19.5 dB. It is evident that fading has somehow affected SA more compared to best CQI (*ca* 19 dB) and PF (*ca* 18 dB). Even though SA is somehow independent of channel conditions, however, compared

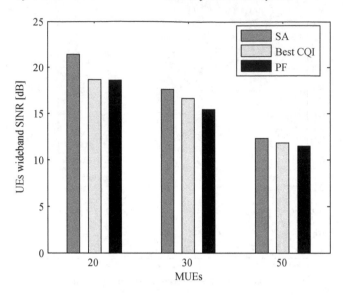

Figure 5.8   UEs wideband SINR with varying number of MUEs (no fading)

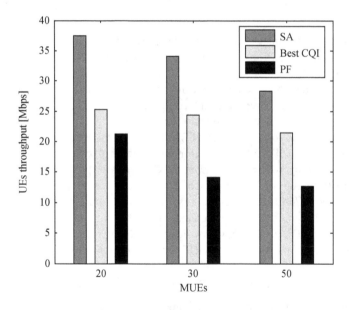

Figure 5.9   UEs throughput with varying number of MUEs (no fading)

to no fading, SA has slightly reduced SINR value in the presence of fading as there are now lesser RBs available to be utilised due to employment of channel conditions.

However, in the presence of Rayleigh fading, the average received power of each UE is lower due to obstacles which increase the outage probability. This outage

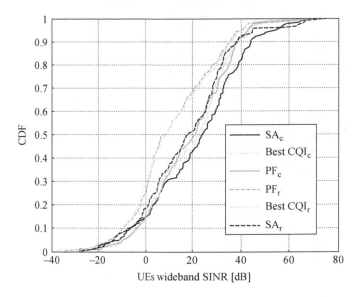

Figure 5.10   Average UEs wideband SINR (with fading)

probability increases further as a result of doppler effect which is directly proportional to UE mobility (speed). The average SINR in best CQI and PF are similar with 50 per cent of the UEs at 9.8 dB. As SA opportunistically utilises the resources of MUEs, it is capable of re-using the resources of MUEs who suffer less degradation, hence a much better average SINR of 18.5 dB.

Figure 5.11 shows the variation of the throughout for both Claussen and Rayleigh fading. SA experiences lesser throughput (33 Mbps, Figure 5.11) in Claussen fading, compared to no fading. The throughput for CQI is not affected much by the channel conditions because of the fact that CQI takes into account channel conditions. PF again performs poorer in fading conditions (17 Mbps) because of its preference to fairness. The combined propagation in Rayleigh fading, on the other hand, yields lower channel capacity, which affect all schemes. The transmitted packets experience fading and the receiver may not detect the faded packets even without collision. With an average throughput of 15 Mbps, PF performs poorly because in the presence of interference from other packets in multi-path fading, it does not offer service differentiation even for traffic with different QoS requirements. Best CQI on the other hand offers service differentiation as it assigns the best links during channel degradation to the users with the best channel quality, thereby experiencing an average throughput of 22 Mbps. SA, with an average throughput of 24 Mbps performs better than both PF and SA because it capitalises on schemes like SA where it is able to utilise resources of the UEs with the best links.

In Figure 5.12, the spectral efficiency was simulated for each scheme. Compared to no fading results, the average spectral efficiency for best CQI and PF under Claussen fading changes significantly with best CQI at 6.779 bits/cu compared to PF at 5.588.

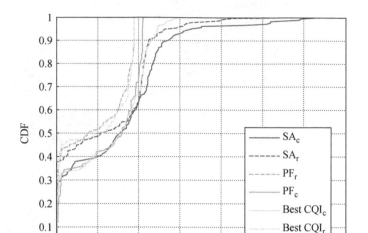

Figure 5.11   Average UEs throughput (with fading)

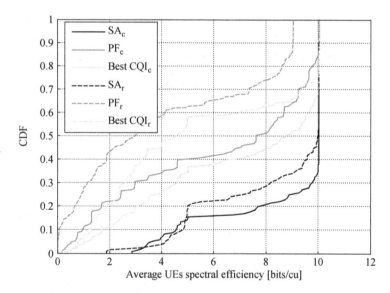

Figure 5.12   Average UEs spectral efficiency (with fading)

In the presence of further degradation from Rayleigh fading, spectral efficiency is reduced significantly for best CQI and PF with 50 per cent of the users utilising 4.6 bits/cu and 3.2 bits/cu, respectively. This is because the multi-path channel has the effect of spreading or broadening the transmitted pulse, thereby reducing channel availability. However, the effect of Claussen and Rayleigh in SA is not noticeable as

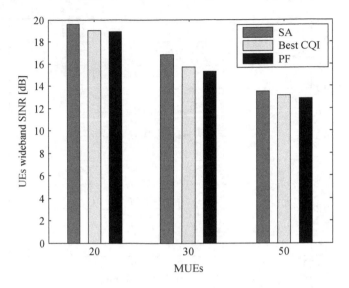

Figure 5.13   UEs wideband SINR with varying MUEs (with fading)

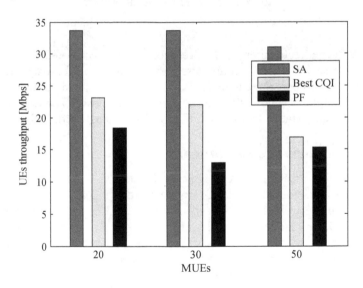

Figure 5.14   UEs throughput with varying MUEs (with fading)

50 per cent of the users still have an efficiency of 10 bits/cu. This is simply because in SA, unlike in PF and best CQI, SA has the capability to utilise as many available and non-degraded resource units.

Figures 5.13 and 5.14 present Claussen fading analysis and provides SINR and throughput values against MUEs. Even though SA still provides competitive and slightly better values over the compared schemes, it is clear that the presented

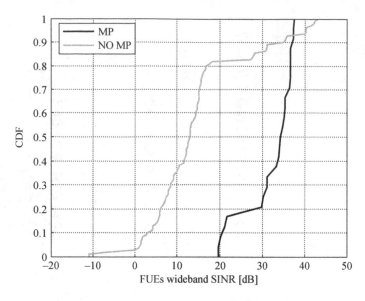

Figure 5.15   FUEs wideband SINR

scheme has been affected more in the presence of fading compared to best CQI and PF. The main reasons are the fact that lesser RBs are available for assignment by SA in fading conditions compared to no fading conditions resulting into slightly reduced SINR and throughput values. The added strain on the network applies equally both for Claussen fading and no fading conditions and accordingly reflected in the results.

5.5.3   Co-tier interference mitigation

The effect of the matching policy algorithm (Section 5.2) for co-tier interference mitigation is evaluated and the results are presented in Figures 5.15–5.17 for SINR, throughput and spectral efficiency, respectively. To analyse the co-tier environment, the simulation is streamlined to include only collocated CFs and FUEs to reflect higher interference temperature. The performance of the presented algorithm in Section 5.2.2, denoted by matching policy 'MP' with a conventional scheduling denoted as 'NO MP', is investigated and compared. Unlike 'MP' which is presented in Section 5.2.2, 'NO MP' represents a scenario where no coordination exists between FAPs and all FAPs vie for same resources simultaneously. Figure 5.15 represents the average SINR values of all the FUEs in the collocated scenario when RBs are shared among the CFs as deduced in Equation 5.7 and represented in Figure 5.3. The surge in SINR values in MP reflects the contention free access to the RBs as opposed to NO MP and this directly affects the throughput value in Figure 5.16 and spectral efficiency in Figure 5.17. As reflected in Figures 5.15–5.17, clearly MP algorithm helps mitigate co-tier interference significantly thereby resulting into hugely improved statistics in a collocated scenario.

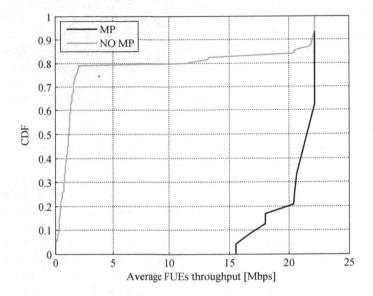

Figure 5.16 Average FUEs throughput

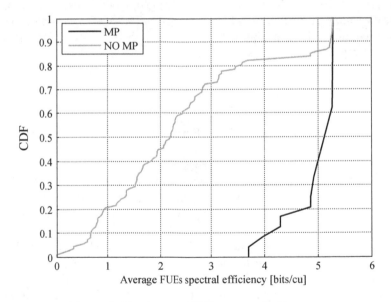

Figure 5.17 Average FUEs spectral efficiency

5.6 Summary

A multi-user channel provides multi-user diversity because of the statistical independence of the individual users' fading processes. Exploiting this diversity by scheduling users on favourable resources is one of the major opportunities to reduce interference and increase the system capacity. Network providers face the problem of delivering services to users with strongly varying channel quality. In two-tier networks, where un-coordinated FAPs are overlaid in the MBS network, interference is eminent. In some cases, system capacity needs to be traded off against user satisfaction, requiring some fairness metric to be considered in the resource allocation process. In this chapter, a SA to mitigate both cross-tier and co-tier interference in two-tier networks was presented. Cross-tier interference is mitigated by employing cognitive FAPs (CFs) which assign the RBs of MUEs with a low-data traffic load to its FUEs based on the interweave concept in spectrum assignment. Whereas the co-tier interference is mitigated by resolving the contention for the same RBs by employment of matching policy among the coordinating CFs. Performance analysis of the proposed scheme is done by system level simulations and results obtained are compared with best CQI and PF schemes. Simulations are performed both for no fading and fading channel conditions represented by Claussen fading. It is found that the proposed scheduling scheme outperforms other schemes in no fading conditions; however, it provides competitive results when fading conditions are considered. The concept of matching policy successfully mitigates the effect of co-tier interference in collocated femtocells by providing improved SINR, throughput and spectral efficiency results, thereby proving the effectiveness of scheme. As the proposed interference mitigation scheme does not require major modifications and existence of macro-femto backhaul coordination, thus it is promising for applications in the LTE-A cellular systems that employ heterogeneous networks.

References

[1]   3GPP RAN WG1 TSGR1-48, "LTE physical layer framework for performance verification," February 2007.

[2]   Ericsson, "Mobile data traffic surpasses voice," March 2010. [Online]. Available: http://www.ericsson.com/thecompany/press/releases/2010/03/1396928.

[3]   J. Mitola III and G. Q. Maguire Jr., "Cognitive radio: making software radios more personal," *IEEE Personal Communications*, vol. 6, no. 4, 1999, pp. 13–18.

[4]   S. Cheng, S. Lien, F. Chu and K. Chen, "On exploiting cognitive radio to mitigate interference in macro/femto heterogeneous networks," *Wireless Communications, IEEE*, vol. 18, no. 3, June 2011, pp. 40–47.

[5]   A. Goldsmith, S.A. Jafar, I. Maric and S. Srinivasa, "Breaking spectrum gridlock with cognitive radios: an information theoretic perspective," *Proceedings of the IEEE*, vol. 97, no. 5, May 2009, pp. 894–914.

[6]   G. Gur, S. Bayhan and F. Alagoz, "Cognitive femtocell networks: an overlay architecture for localized dynamic spectrum access [dynamic spectrum management]," *IEEE Wireless Communications*, vol. 17, no. 4, August 2010, pp. 62–70.

[7] Q. Li, Z. Feng, W. Li, Y. Liu and P. Zhang, "Joint access and power control in cognitive femtocell networks," *International Conference on Wireless Communications and Signal Processing (WCSP)*, November 2011, pp. 1–5.

[8] X. Y. Wang, P.-H. Ho and K.-C. Chen, "Interference analysis and mitigation for cognitive-empowered Femtocells through stochastic dual control," *IEEE Transactions on Wireless Communications*, vol. 11, no. 6, June 2012, pp. 2065–2075.

[9] S. Cheng, W. C. Ao and K. Chen, "Efficiency of a cognitive radio link with opportunistic interference mitigation," *IEEE Transactions on Wireless Communications*, vol. 10, no. 6, 2011, pp. 1715–1720.

[10] A. Attar, V. Krishnamurthy and O. N. Gharehshiran, "Interference management using cognitive base-stations for UMTS LTE," *Communications Magazine, IEEE*, vol. 49, no. 8, August 2011, pp. 152–159.

[11] M. E. Sahin, I. Guvenc, M. Jeong and H. Arslan, "Handling CCI and ICI in OFDMA femtocell networks through frequency scheduling," *IEEE Transactions on Consumer Electronics*, vol. 55, no. 4, November 2009, pp. 1936–1944.

[12] L. Huang, G. Zhu and X. Du, "Cognitive femtocell networks: an opportunistic spectrum access for future indoor wireless coverage," *Wireless Communications, IEEE*, vol. 20, no. 2, April 2013, pp. 44–51.

[13] L. Li, C. Xu and M. Tao, "Resource allocation in open access OFDMA femtocell networks," *Wireless Communications Letters, IEEE*, vol. 1, no. 6, December 2012, pp. 625–628.

[14] F. H. Khan and Y. Choi, "Towards introducing self-configurability in cognitive femtocell networks," *Ninth Annual IEEE Communications Society Conference on Sensor, Mesh and Ad Hoc Communications and Networks (SECON)*, 18–21 June 2012, pp. 37–39.

[15] H. Saad, A. Mohamed and T. ElBatt, "Distributed cooperative Q-learning for power allocation in cognitive femtocell networks," *Vehicular Technology Conference (VTC Fall), 2012 IEEE*, 3–6 September 2012, pp. 1–5.

[16] H. O. Kpojime and G. A. Safdar, "Efficacy of coverage radius-based power control scheme for interference mitigation in femtocells," *Electronics Letters*, vol. 50, no. 8, April 2014, pp. 639–641.

[17] H. O. Kpojime and G. A. Safdar, "Coverage radius bounds and impact on SINR in blindly placed LTE femtocells," *International Journal of Wireless Information Networks*, vol. 22, no. 3, 2015, pp. 262–271, DOI 10.1007/s10776-015-0277-9.

[18] Z. Zhao, M. Schellmann, H. Boulaaba and E. Schulz, "Interference study for cognitive LTE-femtocell in TV white spaces," *Telecom World (ITU WT), 2011 Technical Symposium at ITU*, October 2011, pp. 153–158.

[19] G. Zhao, C. Yang, G. Y. Li and G. Sun, "Fractional frequency donation for cognitive interference management among femtocells," *Global Telecommunications Conference (GLOBECOM 2011), 2011 IEEE*, December 2011, pp. 1–6.

[20]   H. O. Kpojime and G. A. Safdar, "Interference mitigation in cognitive-radio-based femtocells," *IEEE Communications Surveys & Tutorials*, vol. 17, no. 3, pp. 1511–1534, thirdquarter 2015.

[21]   Y. Wang, K. I. Pedersen, T. B. Sorensen and P. E. Mogensen, "Carrier load balancing and packet scheduling for multi-carrier systems," *IEEE Transactions on Wireless Communications*, vol. 9, no. 5, May 2010, pp. 1780–1789.

[22]   Y. Wang, K. I. Pedersen, M. Navarro, P. E. Mogensen and I. B. Srensen, "Uplink overhead analysis and outage protection for multi-carrier LTE-advanced systems," *2009 IEEE 20th International Symposium on Personal, Indoor and Mobile Radio Communications*, Tokyo, 13–16 September 2009, pp. 17–21.

[23]   H. Claussen, "Efficient modelling of channel maps with correlated shadow fading in mobile radio systems," *IEEE 16th International Symposium on Personal, Indoor and Mobile Radio Communications, PIMRC*, vol. 1, 11–14 September 2005, pp. 512–516.

[24]   M. Gudmundson, "Correlation model for shadow fading in mobile radio systems," *Electronic Letter*, vol. 27. no. 2, November 1991, pp. 2145–2146.

[25]   H. J. Bang, T. Ekman and D. Gesbert, "Channel predictive proportional fair scheduling," *IEEE Transactions on Wireless Communications*, vol. 7, no. 2, pp. 482–487, 2008.

[26]   E. Dahlman, S. Parkvall, J. Skold, and P. Beming, *3G Evolution HSPA and LTE for Mobile Broadband*, Elsevier, Amsterdam, 2008.

[27]   F. P. Kelly, A. K. Maulloo and D. K. H. Tan, "Rate control in communication networks: shadow prices, proportional fairness and stability," *Journal of the Operational Research Society*, vol. 49, 1998, pp. 237–252.

Chapter 6
5G radio access for the Tactile Internet
M. Simsek[1], A. Aijaz[2], M. Dohler[3], and G. Fettweis[1]

Abstract

In this chapter, the key features of the emerging fifth generation (5G) wireless networks are highlighted briefly together with 5G architecture and the 5G radio access network (RAN), as specified by the standardisation bodies. The flexible nature of 5G networks and the anticipated 5G RAN design are introduced to be the enabling wireless edge for the Tactile Internet by providing ultra-reliable, ultra-responsive, and intelligent network connectivity. After discussing the architectural aspects and technical requirements of the Tactile Internet, the core design challenges are identified and some directions for addressing these design challenges are also provided. A joint coding method to the simultaneous transmission of data over multiple independent wireless links is presented in this context. Moreover, radio resource slicing and application-specific customisation techniques are introduced to enable the co-existence of different vertical applications along with Tactile Internet applications. Finally, edge-intelligence techniques are investigated to enable the perception of real time and to overcome the physical limitation due to the finite speed of light.

6.1 Introduction

Recently, the notion of the Tactile Internet is emerging, which is envisioned to enable the delivery of real-time control and physical haptic experiences, remotely. Unlike the conventional Internet embodiments, the Tactile Internet provides a medium for remote physical interaction in real time, which requires the exchange of haptic information. Ongoing standardisation efforts define the Tactile Internet as 'A network or network of networks for remotely accessing, perceiving, manipulating or controlling real or virtual objects or processes in perceived real time by humans or machines'. Currently, the conventional Internet embodiments are widely used for delivering content services (voice telephony, text messaging, file sharing, etc.). The Tactile Internet

[1] Faculty of Electrical and Computer Engineering, Technical University Dresden, Germany
[2] Telecommunications Research Laboratory, Toshiba Research Europe Ltd., UK
[3] Department of Informatics, King's College London, UK

will provide a true paradigm shift from content delivery to remote skill-set delivery, thereby revolutionising every segment of the society [1,2].

State-of-the-art fourth generation (4G) mobile communications networks do not largely fulfil the technical requirements of the Tactile Internet. The design requirements for emerging fifth generation (5G) mobile communications networks largely coincide with those of real-time interaction. Therefore, it is expected that 5G networks will underpin the Tactile Internet at the wireless edge.

It is anticipated that the legacy Long-Term Evolution Advanced (LTE-A) will continue to develop in a backward-compatible way and become an integral component of the 5G ecosystem. The overall 5G wireless access solution will consist of evolved LTE-A radio access network (RAN), complemented with novel technological enhancements and architectural designs, co-existing with potentially new radio access technologies (RATs) (in new spectrum) and will support various emerging use cases, such autonomous driving and industrial automation.

6.2 Architecture and requirements

In the new exciting era of 5G, new communication requirements pose challenges on existing networks in terms of technologies and business models. The 5G use cases demand very diverse and sometimes extreme requirements that can be categorised as enhanced Mobile Broad Band (eMBB), massive Machine Type Communication (mMTC), and ultra-Reliable and Low-Latency Communication (URLLC). The current architecture of 4G networks utilises a relatively monolithic network and transport framework to accommodate a variety of services such as mobile traffic from smart phones. The current architecture is not flexible and scalable enough to support efficiently a wider range of use cases, as discussed for 5G networks, when each has its own specific set of performance, scalability, and availability requirements. Furthermore, the introduction of new network services should be made more efficient. Nevertheless, several use cases are expected to be active concurrently in the same operator network, thus requiring a high degree of flexibility and scalability of the 5G network.

The architectural requirements, assumptions, and principles together with the key issues for 5G networks are presented in [1]. The key high-level architectural requirements among multiple other requirements can be summarised as follows:

- The support of new RATs, the Evolved E-UTRA, and non-3rd Generation Partnership Project (3GPP) access types (e.g. WLAN access).
- The support of multiple simultaneous connections of a user equipment via multiple access technologies.
- The support of separation of control plane and user plane functions.
- The support of services with different latency requirements between the user equipment/device and the Data Network.
- The support of different levels of resilience/reliability for the services provided by the network.

- The support of network slicing. Network slicing is a concept for running multiple logical networks as virtually independent business operations on a common physical infrastructure [2]. A Network Slice is, hereby, a network created by the operator customised to provide an optimised solution for a specific market scenario, which demands specific requirements [3].
- The support of architecture enhancements for vertical applications (e.g. agriculture, e-health, etc.).
- The support of critical communications, including mission-critical communications [4].
- To leverage techniques, such as network function virtualisation (NFV) and software defined networking (SDN) to reduce total cost of ownership, improve operational efficiency, energy efficiency, and simplicity and flexibility for offering new services. NFV decouples software implementations of network functions from the compute, storage, and networking resources through a virtualisation layer. This decoupling requires a new set of management and orchestration functions and creates new dependencies between them, thus requiring interoperable standardised interfaces, common information models, and the mapping of such information models to data models. One of the key benefits of NFV is the elasticity provided by the infrastructure for capacity expansion and the roll out of new network functions. Recommendation ITU-T Y.3300 [5] defines SDN as 'a set of techniques that enables to directly program, orchestrate, control and manage network resources, which facilitates the design, delivery and operation of network services in a dynamic and scalable manner'. Although this broad definition translates in many different ways in terms of specifications and implementations, most SDN-labelled solutions relocate the control of network resources to dedicated network elements, namely SDN controllers.

The key architectural assumption is that the functional split between the core and access network in 5G shall be defined with support for new RAT(s), the Evolved E-UTRA, and non-3GPP access types. Further details are going to be specified in future.

With these architectural specifications, 5G is accepted to be the first enabler of the Tactile Internet and will thus be the underlying network of the Tactile Internet architecture. The Tactile Internet requires ultra-reliable and ultra-responsive network connectivity that would enable typical reliabilities and latencies for physical real-time interaction. The underlying 5G-driven communication architecture, composed of the RAN and Core Network (CN), is expected to meet the key requirements in realising the Tactile Internet. As shown in Figure 6.1, the end-to-end architecture for the Tactile Internet can be split into three distinct domains: a master domain, a network domain, and a controlled domain.

The master domain consists of an operator, i.e. a human or machine, and an operator system interface. This interface is actually a master robot/controlling device which converts the operator's input to a 'Tactile Internet input' through various coding techniques. In case, the controlling device is a haptic device; it allows a human to touch, feel, manipulate, or control objects in real or virtual environments. It primarily

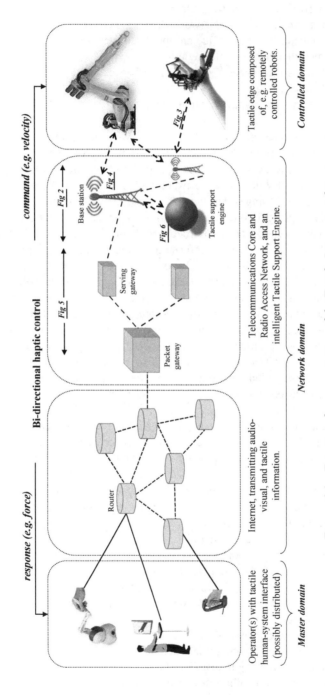

Figure 6.1 *Functional representation of the Tactile Internet architecture*

controls the operation of the controlled domain [6]. In case of a networked control system, the master domain contains a controller, which gives command signals to a sensor/actuator system.

The controlled domain consists of controlled robot and is controlled directly by the master domain through various command signals for interaction in a remote environment. In case of remote operation with haptic feedback, energy is exchanged between the master and controlled domains, thereby closing a global control loop.

The network domain provides the medium for bilateral communication between the master and controlled domains and couples the operator to the remote environment. In case of a human operator, this coupling is kinaesthetically.

6.3   5G Radio access network

Next-generation 5G networks will cater for a wide range of new business opportunities and use cases, some of which have yet to be conceptualised. The service-driven 5G network architecture aims to meet flexibly and efficiently diversified mobile service requirements. Therefore, both the 5G architecture and the 5G RAN require significant changes as compared to former wireless generations. As the RAN remains the key enabler of ubiquitous wireless connectivity, significant rework is targeted for 5G to enable various use cases with stringent requirements. The new 5G radio will integrate existing technologies, such as LTE-A and Wi-Fi, together with revolutionary technologies for emerging eMBB, mMTC, and URLLC applications. The 5G RAN is thus expected to optimally exploit a broad portfolio of enabling technologies and frequency bands towards becoming a service-driven network. Besides the support of heterogeneous RATs, massive multiple input multiple output, and millimetre wave frequency bands, future RAN will leverage virtualisation and cloud computing to adjust execution logic to specific service requirements. With allocation of mobile network functions to the most appropriate location so that access and networking functions, and corresponding states, no longer necessarily reside in different locations, joint performance optimisation is anticipated.

Supporting emerging 5G use cases requires, among other things, a flexible and programmable infrastructure, which enables agile service composition and scaling in a cost- and resource-efficient manner. SDN and NFV are promising technologies with the potential to fulfil the flexibility and programmability requirements of 5G. While SDN separates the network control plane from the data plane, introduces abstractions, and means for network programmability, NFV decouples network functions from hardware appliances allowing the realisation of network functions on general-purpose processors (i.e. decoupling network services from platforms). SDN and NFV are complementary approaches which combined have the potential to foster opportunities for new ways to design, deploy, operate, and manage networks and their services.

The introduction of SDN and NFV in 5G networks will lead to a drastic change to take place in the network architecture, allowing traditional structures to be broken down into customizable elements that can be chained together programmatically to provide just the right level of connectivity. To build a future CN architecture that is highly flexible, modular, and scalable will require a much higher degree of programmability and automation than exists in today's networks. To this end, the 5G network will be in an environment that is cloud based, with a high degree of NFV for scalability and SDN for flexible networking, dynamic orchestration of network resources.

This is the concept of network slicing that will enable CNs to be built in a way that maximises flexibility. The 5G RAN is expected to support a differentiated handling of different network slices, which have been pre-configured by the operator. How RAN supports the slice enabling in terms of RAN functions (i.e. the set of network functions that comprise each slice) is implementation dependent.

6.4   Tactile Internet design challenges

The Tactile Internet, wherein humans or machines will remotely access and control real and virtual objects or systems in real time, will not be realised without overcoming the enormous system design challenges. The information and communications infrastructure enabling the envisioned Tactile Internet has to meet a number of stringent design challenges.

Especially, cases where human beings interact with their environment are crucial because humans' perceptual processes limit the speed of our interaction with their environment. The interaction with a technical system is experienced as intuitive and natural only if the feedback of the system is adapted to the human reaction time. Consequently, the requirements for technical systems enabling real-time interactions depend on the human senses. Realising reaction times for perceived real-time interactivity, all human senses can interact, in principle, with machines. Therefore, transmitting accurately the equivalent of human touch via data networks is aimed to be realised by the Tactile Internet. Key design challenges of the Tactile Internet are as follows:

- This requires *ultra-responsive* network connectivity. This is especially important for technical systems with tactile and haptic interaction or for mission-critical communications, e.g. machine-type communication which enables real-time control and automation of dynamic processes in industrial automation, manufacturing, traffic management, etc.
- Another key challenge is to be able to provide carrier grade access *reliability*. Hence, ultra-reliable network connectivity is an important requirement for the Tactile Internet, whereas specific reliability requirements may differ for various types of applications.
- A potential approach to reduce the impact of latency on haptic control is to deploy predictive and interpolative/extrapolative modules closer to the edge of the network in any advanced cloud infrastructure. Such *edge-intelligence* techniques will play a critical role in making the Tactile Internet a reality.

- *Radio resource allocation*, which is a key component of radio resource management, has a direct impact on throughput, latency, reliability, quality-of-service, and the performance of higher layers. Due to stringent latency requirements, radio resources must be provided on priority for mission-critical Tactile Internet applications. Besides, for the co-existence of haptic and other vertical applications, flexible approaches to radio resource management, capable of providing on-demand functionality, will be needed in 5G networks.
- For enabling remote control of objects and systems some information, e.g. kinaesthetic information, signals are exchanged bi-directionally over the network. As a result, a system supporting Tactile Internet applications requires *control feedback loops*.

6.5 Addressing Tactile Internet RAN design challenges

6.5.1 Reliability

Reliability and related terminology have been used in various areas and different contexts; thus, commonly accepted definitions exist. A general definition of reliability is as follows:

Reliability describes the probability that a system which is in an operative state will perform the required functions for a specified period. Assuming that the time to failure is denoted by t_f with the cumulative distribution function F_{t_f}, the reliability $R(t)$ is given by

$$R(t) = \mathbb{P}[t_f > t] = 1 - F_{t_f}(t) \qquad \text{for } t > 0 \qquad (6.1)$$

Thus, $R(t)$ denotes the probability that the system does not fail until time t. In wireless communications, however, the most common definition of reliability is slightly different. In its common interpretation, reliability in wireless networks characterises the probability that a certain amount of data is successfully transmitted from the source to the destination within a specified period of time. Reliability has been defined by 3GPP as 'the amount of sent network layer packets successfully delivered to a given node within the time constraint required by the targeted service, divided by the total number of sent network' [7,8]. Hence, the end-to-end transmission of data under given (service specific) latency constraints is considered for the determination of the reliability of wireless systems. Therefore, it is important to consider all components of a system in the analysis of reliability. Exemplary causes of failure in the open systems interconnection layer model starting from the physical layer up to the application layer are power loss and hardware/software failures, shadowing, fading and interference, control channel or hardware limitation, limited capacity or scheduling, too late or too early handovers, failures in routing, and backhaul, respectively.

With the advent of discussions about potential use cases and requirements for emerging Tactile Internet applications and their requirements of reliability and latency, the Tactile Internet has gained more and more attention. Several related works have been discussing different techniques to enhance reliability, e.g. [8–12].

The wireless channel is characterised by fading effects, namely the small-scale and large-scale fading which may lead to significant receive power degradation and cause unpredictability. To compensate these effects, diversity is one of the key approaches. In diversity, the basic concept is to transmit the same data via several independent diversity branches to get independent signal replicas via space diversity, frequency diversity, and time diversity. This leads to a high probability and not all transmissions fade simultaneously; the deepest fades can be avoided and, hence, to a protection against fading. The different diversity branches can be detailed as follows:

- Space diversity: In space diversity, the transmission of the same data from or to separate locations leads to diversity. It is separated into micro- and macro-diversity. In micro-diversity, the signals from antennas positioned at separate locations are combined. Typically, these antennas are located at the same base station and their spacing is a few wavelengths. The utilisation of multiple antennas can, hereby, significantly increase the signal quality and the reliability performance due to reduced small scale fading. However, it does not compensate large-scale fading effects. In micro-diversity, a much larger spatial separation of transmit and receive antennas is considered. This leads to uncorrelated large-scale fading and, hence, to improved reliability.
- Time diversity: In time diversity, the same data are transmitted repeatedly at time intervals that exceed the coherence time of the channel so that the re-transmitted data undergo independent fading. One known time diversity technique is the hybrid automatic repeat request, a combination of retransmissions and forward error correction coding which leads to improved reliability as long as the retransmissions happen within the given time constraint of the service, e.g. [13,14]. Hence, a key drawback of time diversity is the increased latency.
- Frequency diversity: In frequency diversity, the small-scale fading effects are compensated if the frequencies of multiple signals are separated by at least the coherence bandwidth [15]. Hence, frequency diversity can be achieved either within one frequency band if the bandwidth is large enough or by combining multiple signals of different carrier frequencies. The latter technique is also known as multi-connectivity.

Besides the described diversity techniques, coding approaches may also lead to improved reliability by achieving low error rates under given channel conditions. One emerging approach is network coding where intermediate network nodes do not simply forward packets but combine and recode them so that multipath diversity is exploited and data rate, security, and reliability are improved [16].

Another approach is the simultaneous transmission of jointly encoded data over multiple wireless channels, which can be realised, for example, by spatially coupled low-density parity-check (SC-LDPC) codes. While reliability is, hereby, achieved by the transmission over multiple independent channels, low latency is kept on the feasible level by using short codewords that can be efficiently decoded. Furthermore, the decrease in the number of retransmissions (via increased reliability) also affects the latency. A SC-LDPC code-based wireless link-combining approach shall be applicable to any number of wireless links/channels. The main idea of such a SC-LDPC code-based link-combining method, proposed as spatially coupled combining (SCC), can be summarised as follows: The SCC method is based on root-LDPC codes, whereby the code structure aims at full diversity over L different frequency links. The number of links L describes the redundancy of the transmitted data, i.e. the code rate R is given by $1/L$. Each codeword contains n information bits and $(L-1)n$ redundant bits, i.e. in each link, there are n/L information bits and $(L-1)n/L$ redundant bits [17]. A combining method comprising three steps has been proposed [18]:

In the first step, a regular code that employs the root-check concept to provide full diversity over block fading channels is generated. Hereby, a parity check matrix H yielding the following code structure is created:

A regular root-LDPC code is generated which provides full diversity over L links. The parity check matrix H is a $(L-1)n \times Ln$ matrix and consists of sub-matrixes: $A^{(f,e)}$, $B^{(f,e)}$, $C^{(f,e)}$, $D^{(f,e)}$, defined as follows:

- Matrix $A^{(f,e)}$ of size $n/L \times n/L$ describes the 'root' connections between check nodes and information bits.
- Matrix $C^{(f,e)}$ of size $n/L \times n/L$ describes the 'redundant' connections between check nodes and information bits.
- Matrix $D^{(f,e)}$ of size $n/L \times (L-1)n/L$ defines the 'redundant' connections between redundant bits and check nodes.
- Matrices $B^{(f,e)}$ of size $(n/L) \times ((L-1)n)/L$ and $F^{(f,e)}$ of size $(n/L) \times (n/L)$ are all zeros matrices and keep the root-LDPC rule, i.e. each root-check provides the connections between two different fading blocks.

Then, the next set of matrixes is defined by the combination of the matrices $A^{(f,e)}, B^{(f,e)}, C^{(f,e)}, D^{(f,e)}, F^{(f,e)}$:

$$
R^{(f,e)} = \begin{bmatrix} A^{(1,1)} & B^{(1,2)} \\ \cdots & \cdots \\ A^{(L-1,1)} & B^{(L-1,2)} \end{bmatrix}, \quad
K^{(f,e)} = \begin{bmatrix} F^{(1,1)} & B^{(1,2)} \\ \cdots & \cdots \\ F^{(e-1,1)} & B^{(e-1,2)} \\ C^{(e,1)} & D^{(e,2)} \\ F^{(e+1,1)} & B^{(e+1,2)} \\ \cdots & \cdots \\ F^{(l-[e+1],1)} & B^{(l-[e+1],2)} \end{bmatrix},
$$

$$
W^{(f,e)} = \begin{bmatrix} F^{(1,1)} & B^{(1,2)} \\ \cdots & \cdots \\ F^{(e-2,1)} & B^{(e-2,2)} \\ C^{(e-1,1)} & D^{(e-1,2)} \\ F^{(e,1)} & B^{(e,2)} \\ \cdots & \cdots \\ F^{(l-e,1)} & B^{(l-e,2)} \end{bmatrix}. \tag{6.2}
$$

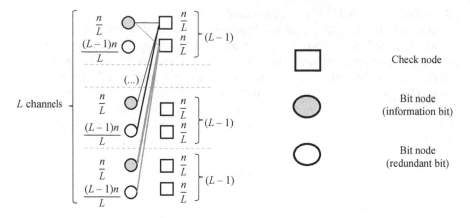

Figure 6.2   Generated LDPC code structure in first step of the proposed SCC method

It is assumed that the row index for matrices $B^{(f,e)}$ and $F^{(f,e)}$ is larger than zero; otherwise, the row is empty. Finally, the parity-check matrix is defined as

$$H = \begin{bmatrix} R^{(1,1)} & W^{(1,2)} & W^{(1,3)} & & W^{(1,l)} \\ K^{(2,1)} & R^{(2,2)} & W^{(2,3)} & \cdots & W^{(2,l)} \\ K^{(3,1)} & K^{(3,2)} & R^{(3,3)} & & W^{(3,l)} \\ & \vdots & & \ddots & \vdots \\ K^{(l,1)} & K^{(l,2)} & K^{(l,3)} & \cdots & R^{(l,l)} \end{bmatrix}. \tag{6.3}$$

Based on the obtained parity-check matrix, the root-LDPC code is generated. The generated root-LDPC code provides full diversity, but, as depicted in Figure 6.2, the code structure contains many cycles, which increase the error floor of, especially, short codewords. In the second step of the SCC method, this error floor for high signal-to-noise-ratio (SNR) values is decreased by modifying the parity-check matrix H in such a way that one connection within each 4-cycle of the structure in the parity-check matrix is erased yielding an irregular root-LDPC code. In a final step, it is checked if the code rate is still kept. If this is not the case, the steps are repeated until the code rate is achieved. The details of the approach are presented in [18].

Figure 6.3 depicts the word error rate (WER) over the SNR as a comparison of different combining methods, namely selection combining, equal gain combining, maximum ratio combining, and the proposed SCC methods for the case of $L = 5$. It can be observed that SCC outperforms the baseline combining methods so that significant gains in terms of WER are achieved. In Figure 6.4, the required number of wireless links to achieve an outage probability of 10^{-7} under given SNR values is depicted. It can be observed that the number of links can be reduced (increased) if the SNR value is improved (decreased).

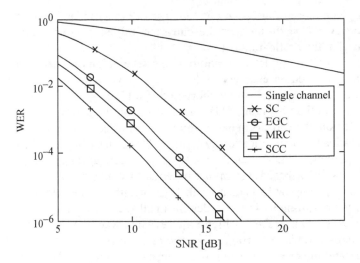

Figure 6.3 *Word error rate (WER) over SNR for different combining methods over five wireless links*

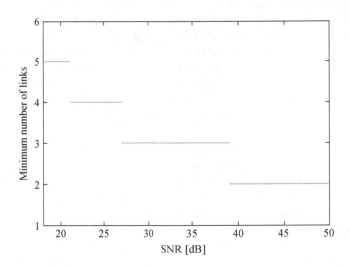

Figure 6.4 *Minimum number of wireless links to achieve an outage probability below 10^{-7}*

6.5.2 *Latency*

In order to enable certain services related to Tactile Internet, a minimal level of reliability and latency is required to guarantee the user experience and/or enable the service initially. The end-to-end latency is defined by 3GPP as the time it takes to transfer a given piece of information from a source to a destination, measured at the application level, from the moment the data is transmitted by the source to the moment it is received at the destination [6].

To meet challenging latency requirements, several improvements and new techniques are discussed as enablers of low-latency communication. First of all, the wireless transmission needs to be shortened significantly. It is proposed to reduce the transmission time interval (TTI), which is 1 ms in LTE to 0.1–0.25 ms or below [19]. In addition, the waveform has to support short packets and low latency. Most of today's broadband communication systems are based on Orthogonal Frequency Division Multiplexing (OFDM) mainly due to its robustness against multi-path channels [20]. Today's OFDM-based systems are based on 14 symbols and 12 sub-carriers with a sub-carrier spacing of 15 kHz at a sub-frame length of 1 ms. However, to achieve low latency transmissions in the order of a millisecond, the physical transmission must have very small packets, which requires a one-way PHY layer transmission of several micro-seconds, e.g. 100 μs. As the packet error correction encoding at the transmitter and the error correction decoding and detection at the receiver limit the packet size to less than the target latency, a packet must be smaller than 100 μs packet duration. In current LTE cellular systems, hence, the numerology together with the reference symbols design, channel estimation, and channel coding requires significant revision of the cellular PHY for the Tactile Internet applications in 5G systems. One way to overcome these limitations for achieving latencies of 1 ms in OFDM-based systems is to change the OFDM numerology, i.e. symbol duration, sub-carrier spacing, etc. together with the sub-frame length, and enable high levels of diversity, and fast channel estimation together with fast channel decoding (e.g. with convolutional codes). In References [21,22], it has been shown that an OFDM-based system with changed OFDM numerology can achieve low latency. Further, a recent investigation [23] practically demonstrates achieving 1 ms latency with modified OFDM numerology.

Another important aspect is the uplink channel access. In LTE, users send scheduling requests to inform the base station on data transmission. Then, the base station schedules the request and sends a scheduling grant to indicate the resource allocation. The delay of this channel access procedure can be reduced by semi-persistent scheduling and pre-reserved time-frequency resources as discussed in [19,24].

To support the co-existence of vertical applications, while achieving latency of 1 ms, it is desirable to have a flexible frame structure. One approach is to have a dynamic adjustment of TTI in accordance with the service requirements [25].

6.5.3   Edge intelligence

Intelligence plays a substantial role in overcoming the milliseconds delay barrier. This is because the finite speed of light is the biggest adversary in facilitating the required real-time experience advocated by the Tactile Internet. Improvements of hardware, protocols, and systems are paramount in diminishing end-to-end delays; however, the ultimate limit is set by the finite speed of light. Very sophisticated techniques thus need to be used to allow for the required paradigm shift. We believe that these will be provided by unprecedented edge artificial intelligence (AI) engines which are cached and then executed in real time close to the tactile experience. Emerging techniques related to this are intelligent caching, edge cloud, and model-mediated teleoperation approaches (MMTA). These are reviewed briefly below before outlining suggested approaches on edge cloud and functioning of the AI engines.

6.5.3.1 State of the Art

The current trend is to locate content closer to the user through adaptive caching technologies and the expectation is that caching will reach to the level of access points and even terminals [24]. Such intelligent distribution of popular content at the edge of the network in a user-centric manner has allowed for significant de-congestion of the CN and for increased performance in terms of latency [25–30]. In terms of hosting the edge intelligence in a scalable and secure way, Cloud-RAN [31,32] and even cloudlet [33] approaches have recently been proposed. Related to the optimisation of caching and use of process prediction, advanced machine learning and other AI tools have been used in the recent past [31,34]. Pertinent to this are the recent developments in MMTA. Hereby, rather than directly sending back the haptic (force) signals, the parameters of the object model, which approximate the remote environment, are estimated and transmitted back to the master in real time [35,36]. An alternative approach is to model the behaviour of the human operator with the estimated model parameters on the master side being transmitted to the slave to guide the slaves motion [37–39].

6.5.3.2 Tactile artificial intelligence engines

As explained above, the most important contents to be stored are AI engines which predict the haptic/tactile experience, i.e. acceleration of movement on one end and the force feedback on the other. That allows to spatially decouple the active and reactive end(s) of the Tactile Internet as the tactile experience is virtually emulated on either end; this, in turn, allows a much wider geographic separation between the tactile ends, beyond the 1 ms-at-speed-of-light-limit.

The algorithmic framework is currently based on a wide gamut of algorithms, ranging from complex MMTA approaches to some simple linear regression algorithms which are able to predict movement and reaction over the next tens of milliseconds (mainly because our skillset-driven actions are fairly repetitive and exhibit strong patterns across the six degrees of freedom). The principle is illustrated in Figure 6.5.

When the predicted action/reaction deviates from the real one by a certain amount ε, then the coefficients are updated and transmitted to the other end allowing for corrections to be put in place before damage is done at, e.g. a deviation of δ.

More sophisticated algorithms have become available, ranging from [40–51]. For instance, [41] employed a prediction method for three-dimensional position and force data by means of an advanced first-order autoregressive model. After an initialisation and training process, the adaptive coefficients of the model are computed for the predicted values to be produced. The algorithm then decides if the training values need to be updated either from the predicted data or from the current real data.

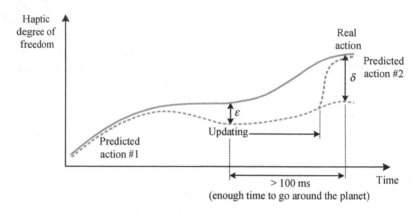

Figure 6.5   Illustration of how predictive edge-AI gives the perception of a
1-ms delay whilst the actual latency due to communications can be
much larger

The AI algorithms, ported through intelligent edge caching and stored on appropriate cloud technologies, aid in enabling the perception of real-time interaction, stabilising the tactile system, and consequently enhancing the Quality of Experience (QoE) of the Tactile Internet user.

6.5.4   Radio resource management

With the introduction of tactile applications into the 5G ecosystem, daunting new requirements arise for radio resource management. Moreover, the available radio resources would be shared between tactile applications and other human-to-human or machine-to-machine applications, having different and often conflicting service requirements. Because of stringent service requirements, radio resources must be provided on priority, without any external competition, for tactile applications. A promising approach to achieve this is to allocate a separate slice of radio resources to tactile applications, which remains dedicated for any ongoing operation. Therefore, slicing of radio resources, in context of 5G, becomes particularly important. Such slicing can be achieved through virtualisation of radio resources. Compared to virtualisation in the wired domain, unique challenges get introduced when virtualising radio resources. The virtualisation-based radio resource slicing must be able to achieve tight *isolation*, provide application-specific *customisation*, and ensure efficient *utilisation* across different radio slices. Isolation means that any change in slice due to traffic load, channel conditions, etc. should not affect other slices. Customisation provides the flexibility of application-specific radio resource management schemes with the allocated slice. Efficient utilisation becomes particularly important due to the scarce nature of radio resources.

One way of achieving radio resource slicing is through independent radio resource slicing at base station level. However, this approach may perform sub-optimally

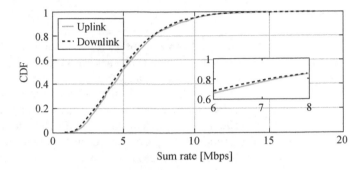

Figure 6.6   CDF of sum rate in the uplink and the downlink for the greedy heuristic algorithm

(from a utilisation perspective), as user distribution, average channel conditions, and user traffic requirements may vary significantly across base stations at fine timescales. Therefore, network-wide radio resource slicing across multiple base stations is much more attractive. The network-wide radio resource slicing not only provides dynamic allocation of resources across a set of base stations but also holds minimum footprint for wide-scale adoption by requiring minimal changes to legacy base stations. Further, any radio resource slicing strategy should not only ensure maximum utilisation of the radio resources but also guarantee the utility requirements of different vertical applications. Recently, a radio resource slicing framework, termed as Hap-SliceR [52], has been proposed for 5G networks that provides a network-wide radio resource slicing strategy while meeting the aforementioned requirements (Figure 6.6).

Finally, it is important to develop radio resource customisation algorithms for emerging applications that would be enabled by the Tactile Internet. One of the primary applications is haptic communications, which creates unique requirements for radio resource allocation. The problem of radio resource allocation for haptic communications has been recently addressed in [53]. The radio resource allocation requirements of haptic communications, along with the specific resource allocation constraints of SC-FDMA and OFDMA, have been translated into a unique radio resource allocation problem. To overcome the analytic and computational complexity, a greedy heuristic algorithm has been proposed for radio resource allocation. Performance evaluation, based on a recently proposed 5G air-interface design, demonstrates that the proposed algorithm achieves symmetric resource allocation in uplink and downlink, while accounting for the delay bound and date requirements.

The results demonstrate that, from a system-design perspective, joint design of uplink and downlink becomes crucial to support emerging 5G Tactile Internet applications. One way to achieve joint design is through centralised scheduling in both uplink and downlink. Moreover, from an operator's perspective, an important consideration is whether to partition the radio resources for uplink and downlink or provide a single pool of radio resources.

6.6   Conclusion

The Tactile Internet is expected to revolutionise almost every segment of the society by enabling unprecedented applications. It is a quantum leap prospect for the global economy. It is expected that the emerging 5G mobile communications networks will underpin the Tactile Internet at the wireless edge through ultra-reliable, ultra-responsive, and intelligent network connectivity. It was the aim of this chapter to investigate 5G RAN design for enabling the Tactile Internet. After discussing the architectural aspects and technical requirements of the Tactile Internet, the core design challenges for RAN design in context of 5G and the Tactile Internet are identified. Some recommendations and directions for addressing these design challenges, such as diversity or joint coding techniques, are also provided. It is expected that diversity techniques will play a key role in achieving ultra-reliable wireless connectivity. To achieve the envisioned latency requirements, optimisations in air-interface protocols and frame structure would be crucial. Moreover, radio resource slicing and application-specific customisation techniques would enable co-existence of different vertical applications along with fulfilling the requirements of critical applications. Finally, edge-intelligence techniques would be indispensable in enabling the perception of real-time and overcoming the physical limitation due to the finite speed of light.

References

[1] A. Aijaz, M. Dohler, A. H. Aghvami, V. Friderikos and M. Frodigh, "Realizing the tactile internet: haptic communications over next generation 5G cellular networks," [Online]. Available: http://arxiv.org/abs/1510.02826, 2015.

[2] M. Simsek, A. Aijaz, M. Dohler, J. Sachs and G. Fettweis, "5G-enabled tactile internet," *IEEE Journal on Selected Areas of Communication (JSAC), SI on Emerging Technologies*, vol. 32, no. 3, pp. 1–14, 2016.

[3] 3GPP, "Study on architecture for next generation system," 3GPP TR 23.799 V0.7.0, 2016.

[4] NGMN, "5G project requirements and architecture – work stream E2E architecture," NGMN Version 1.0, 2016.

[5] 3GPP, "Study on new radio access technology: radio access architecture and interfaces," 3GPP TR 38.801 V0.2.0, 2016.

[6] E. Steinbach, S. Hirche, M. Ernst, *et al.*, "Haptic communications," *Proceedings of the IEEE*, vol. 100, no. 4, pp. 937–956, 2012.

[7] 3GPP, "Study on scenarios and requirements for next generation access technologies," 3GPP TR 38.913, 2016.

[8] ITU-T, "Frameworks of software defined networking," Recommendation ITU-T Y.3300, 2014.

[9] 3GPP, "Feasibility study on new services and markets technology enablers," 3GPP TR 22891, 2014.

[10] 3GPP, "Study on scenarios and requirements for next generation," 3GPP TR 38.913, 2016.

[11] P. Popovski, "Ultra-reliable communication in 5G systems," in *International Conference on 5G for Ubiquitous Connectivity*, 2014.

[12] H. D. Schotten, R. Sattiraju, D. G. Serrano, Z. Ren and P. Fertl, "Availability indication as key enabler for ultra-reliable communication in 5G," in *European Conference on Networks and Communication*, 2014.

[13] D. Öhmann, A. Awada, I. Viering, M. Simsek and G. Fettweis, "SINR model with best server association for high availability studies of wireless networks," *IEEE Wireless Communications Letters*, vol. 5, no. 1, pp. 60–63, 2016.

[14] F. Kirsten, D. Öhmann, M. Simsek and G. Fettweis, "On the utility of macro- and microdiversity for achieving high availability in wireless networks," in *IEEE International Symposium on Personal*, 2015.

[15] D. Öhmann, A. Awada, I. Viering, M. Simsek and G. Fettweis, "Achieving high availability in wireless networks by inter-frequency multi-connectivity," in *IEEE International Conference on Communications*, 2016.

[16] B. Soret, G. Pocovi, K. I. Pedersen and P. Mogensen, "Increasing reliability by means of root cause aware HARQ and interference coordination," in *IEEE Vehicular Technology Conference Fall*, 2015.

[17] H. Shariatmadari, Z. Li, M. A. Uusitalo, S. Iraji and R. Jäntti, "Link adaptation design for ultra-reliable communications," in *IEEE International Conference on Communications*, 2016.

[18] A. Goldsmith, *Wireless Communications*, Cambridge University Press, Cambridge, 2005.

[19] D. Szabo, A. Gulyas, F. H. P. Fitzek and D. E. Lucani, "Towards the tactile internet: decreasing communication latency with network coding and software defined networking," in *Proceedings of European Wireless Conference*, 2015.

[20] N. Hassan, M. Lentmaier, I. Andriyanova and G. Fettweis, "Improving code diversity on block-fading channels by spatial coupling," in *IEEE International Symposium on Information Theory*, 2014.

[21] M. Soszka, M. Simsek and G. Fettweis, "On link combining methods for highly reliable future wireless communication," in *IEEE International Symposium on Wireless Communication Systems*, 2016.

[22] 3GPP, "Study on latency reduction techniques for LTE," 3GPP TR 36.881, 2016.

[23] M. Mirahmadi, A. Al-Dweik and A. Shami, "BER reduction of ODFM-based broadband communication systems over multipath channels with impulsive noise," *IEEE Communications Magazine*, vol. 61, no. 11, pp. 4602–4615, 2013.

[24] O. N. C. Yilmaz, Y.-P. E. Wang, N. A. Johansson, N. Brahmi, S. A. Ashraf and J. Sachs, "Analysis of ultra-reliable and low-latency 5G communication for factory automation use case," in *IEEE International Conference on Communications*, 2015.

[25] N. A. Johansson, Y.-P. E. Wang, E. Eriksson and M. Hessler, "Radio access for ultra-reliable and low-latency 5G communications," in *IEEE International Conference on Communications*, 2015.

[26] J. Pilz, M. Mehlhose, T. Wirth, D. Wieruch, B. Holfeld and T. Haustein, "A tactile Internet demonstration: 1 ms ultra low delay for wireless communications towards 5G," in *IEEE International Conference on Computer Communications (INFOCOM)*, 2016.

[27] S. A. Ashraf, F. Lindqvist, B. Lindoff and R. Baldemair, "Control channel design trade-offs for ultra-reliable and low-latency communication systems," in *IEEE Globecom Workshop on Ultra-Low Latency and Ultra-High Reliability in Wireless Communication*, 2015.

[28] Nokia, "5G for mission critical communication," *White Paper* [Online]. Available: http://info.networks.nokia.com/5GforMissionCriticalCommunication_01. LP.html, 2016.

[29] E. Bastug, M. Bennis and M. Debbah, "Living on the edge: the role of proactive caching in 5G wireless networks," *IEEE Communications Magazine*, vol. 52, no. 8, pp. 82–99, Aug. 2014.

[30] H. Ahlehagh and S. Dey, "Video-aware scheduling and caching in the radio access network," *IEEE/ACM Transactions on Networking*, vol. 22, no. 5, pp. 1444–1462, Oct. 2014.

[31] X. Wang, M. Chen, T. Taleb, A. Ksentini and V. Leung, "Cache in the air: exploiting content caching and delivery techniques for 5G systems," *IEEE Communications Magazine*, vol. 52, no. 2, pp. 131–139, Feb. 2014.

[32] W. Dron, M. Uddin, S. Wang, *et al.*, "Caching for non-independent content: improving information gathering in constrained networks," in *Military Communications Conference, MILCOM 2013 – 2013 IEEE*, pp. 720–1725, 18–20 Nov. 2013.

[33] F. Sardis, G. Mapp, J. Loo and M. Aiash, "Dynamic edge-caching for mobile users: minimising inter-AS traffic by moving cloud services and VMs," in *2014 28th International Conference on Advanced Information Networking and Applications Workshops (WAINA)*, pp. 144–149, 13–16 May 2014.

[34] E. Bastug, M. Bennis and M. Debbah, "Social and spatial proactive caching for mobile data offloading," in *2014 IEEE International Conference on Communications Workshops (ICC)*, 10–14 Jun. 2014.

[35] G. Alfano, M. Garetto and E. Leonardi, "Content-centric wireless networks with limited buffers: when mobility hurts," *IEEE/ACM Transactions on Networking*, 2015.

[36] D. Wubben, P. Rost, J. S. Bartelt, *et al.*, "Benefits and impact of cloud computing on 5G signal processing: flexible centralization through cloud-RAN," *IEEE Signal Processing Magazine*, vol. 31, no. 6, pp. 35–44, Nov. 2014.

[37] A. Maeder, M. Lalam, A. de Domenico, *et al.*, "Towards a flexible functional split for cloud-RAN networks," in *European Conference on Networks and Communications (EuCNC)*, 23–26 Jun. 2014.

[38] Y. Li and W. Wang, "The unheralded power of cloudlet computing in the vicinity of mobile device," in *IEEE Global Communications Conference (GLOBECOM)*, 9–13 Dec. 2013.

[39] Z. Shao, M. J. Er and G.-B. Huang, "Receding horizon cache and extreme learning machine based reinforcement learning," in *12th International Conference on Control Automation Robotics & Vision (ICARCV)*, pp. 1591–1596, 5–7 Dec. 2012.

[40] B. Hannaford, "A design framework for teleoperators with kinesthetic feedback," *IEEE Transactions on Robotics and Automation*, vol. 5, no. 4, pp. 426–434, Aug. 1989.

[41] P. Mitra and G. Niemeyer, "Model-mediated telemanipulation," *The International Journal of Robotics Research*, vol. 27, no. 2, pp. 253–262, 2008.

[42] P. Prekopiou, S. Tzafestas and W. Harwin, "Towards variable-time delays-robust tele manipulation through master state prediction," in *IEEE/ASME International Conference on Advanced Intelligent Mechatronics*, pp. 305–310, 1999.

[43] I. B. France, "Predictive algorithms for distant touching," in *EuroHaptics Conference*, pp. 1–5, Jul. 2002.

[44] C. Weber, V. Nitsch, U. Unterhinninghofen, B. Farber and M. Buss, "Position and force augmentation in a telepresence system and their effects on perceived realism," in *EuroHaptics Conference*, pp. 226–231, Mar. 2009.

[45] N. Sakr, N. D. Georganas and J. Zhao, "Exploring human perception-based data reduction for haptic communication in 6-DOF telepresence systems," in *IEEE International Symposium on Haptic Audio-Visual Environments and Games (HAVE)*, pp. 1–6, Oct. 2010.

[46] N. Sakr, N. Georganas, J. Zhao and X. Shen, "Motion and force prediction in haptic media," in *IEEE International Conference on Multimedia and Expo*, pp. 2242–2245, Jul. 2007.

[47] X. Xu, J. Kammerl, R. Chaudhari and E. Steinbach, "Hybrid signal-based and geometry-based prediction for haptic data reduction," in *IEEE International Workshop on Haptic Audio Visual Environments and Games*, pp. 68–73, Oct. 2011.

[48] X. Xu, B. Cizmeci, A. Al-Nuaimi and E. Steinbach, "Point cloud-based model-mediated teleoperation with dynamic and perception-based model updating," *IEEE Transactions on Instrumentation and Measurement*, vol. 63, no. 11, pp. 2558–2569, Nov. 2014.

[49] F. Brandi and E. Steinbach, "Prediction techniques for haptic communication and their vulnerability to packet losses," in *IEEE International Symposium on Haptic Audio Visual Environments and Games*, pp. 63–68, Oct. 2013.

[50] F. Guo, C. Zhang and Y. He, "Haptic data compression based on a linear prediction model and quadratic curve reconstruction," *Journal of Software*, vol. 9, no. 11, pp. 2796–2803, 2014.

[51] N. Sakr, J. Zhou, N. D. Georganas, J. Zhao and E. Petriu, "Network traffic reduction in six degree-of-freedom haptic, telementoring systems," in *IEEE*

International Conference on Systems, Man and Cybernetics, pp. 2451–2456, Oct. 2009.

[52]   A. Aijaz, "Hap-SliceR: a radio resource slicing framework for 5G networks with haptic communications," *IEEE Systems Journal*, pp. 1–12, Jan. 2017. doi: 10.1109/JSYST.2017.2647970.

[53]   A. Aijaz, "Towards 5G-enabled tactile internet: radio resource allocation for haptic communications," in *IEEE Wireless Communications and Networking Conference (WCNC) Workshops*, 2016.

Chapter 7

Fronthauling for 5G and beyond

Anvar Tukmanov[1], Maria A. Lema[2], Ian Mings[1],
Massimo Condoluci[2], Toktam Mahmoodi[2],
Zaid Al-Daher[1], and Mischa Dohler[2]

The need for highly-efficient spectral techniques to manage and coordinate interference across the footprint of a network, and in particular cell edge regions, goes back to the early days of cellular networks and is largely driven by densification of the radio networks to meet growing traffic demands. The concept was first envisaged in 1947 by D.H. (Doug) Ring in an internal Bell Labs Journal [1] not publicly available at the time. In this paper, the fundamentals of a wide-area coverage system for New York were laid out, and the designs of various coverage and service layers, nowadays referred to as HetNets, were described. The concept of deploying small cells in areas of high traffic, adoption of frequency reuse plans, frequency discrimination in adjacent cells, and interference management are some of the techniques presented at the time to improve coverage and capacity of the New York radio network. The key messages from the paper directly implied the need for operators to manage cellular interference as they, (i) increased the number of cells, and (ii) shrank the reuse distance between sites. Both of these concepts remain applicable in today's mobile networks and in the forthcoming 5G radio networks.

One of the key technologies to improve cellular network performance is the adoption of small cells – though the term "'small cell'" is a relative definition to the typical underlying macro footprint range. When Bell labs introduced the concept of small cells in 1947 they had a radius of about 8 km. In current deployments a small cell radius is typically 500 m or less; a range reduction by a factor of 16 which may increase even more in 5G. Such small cells along with indoor femto basestations necessitate techniques such as Self-Organised Networks (SONs) to manage the interference levels between each other and between higher metro and macro layers. LTE in particular is designed to operate with a frequency reuse factor of 1 (i.e., all cells use the full channel bandwidth available to an operator) to optimise the network spectral efficiency. However, operating in this way causes interference between

[1]Research and Innovation, British Telecommunications plc, UK
[2]Department of Informatics, King's College London, UK

neighbouring cells – particularly at cell edges where users will experience a poor signal to interference and noise ratio (SINR) which, in turn, reduces the spectral efficiency. This drove the need for mechanisms to mitigate interference between cells and layers such as Inter-cell interference coordination (ICIC) and enhanced Inter-cell interference coordination (eICIC). SON has several other advantages including automation and rapid deployment which are highly desirable features from an operator's perspective.

A number of implementation challenges related to deployment of small cells in a way that achieves optimum performance are currently being addressed by Small Cell Forum and NGMN including (i) reliability and interoperability in a multi-vendor, multi-operator environment, (ii) complexity in parametrisation, optimisation and algorithm development, (iii) network performance variations based use case scenarios and (iv) the added complexity of network topologies. The success of SON is also dependent on X2, which is the standard interface between base stations. X2 is currently used for handover and for basic interference coordination techniques. More advanced cooperative and interference mitigation techniques such as CoMP and Further eICIC (FeICIC) together with any further enhanced coordination techniques currently being proposed for 5G and the anticipated cell densification within different HetNeT layers, would require much more stringent latency constraints on X2 and also S1 links.

Another technology which facilitates cell densification and interference mitigation is the concept of CRAN which was first introduced by IBM [2] and further developed by China Mobile Research Institute [3]. Conceptually, base stations in CRAN typically consist of baseband units (BBU) and remote radio heads (RRH) as illustrated in Figure 7.1. In many CRAN systems, the BBU which is usually located at the bottom of the base station contains RLC, PDCP, MAC and PHY layers and the RRH is usually placed near the antennas at the top of the mast and contains mainly the RF functions. The interconnection between the BBU and RRH typically carries a form of In-phase and Quadrature (IQ) digitised signal components and is based on the Common Public Radio Interface (CPRI), a specification developed by a number of vendors including Ericsson, Huawei Technologies, NEC Corporation, Alcatel Lucent and Nokia Solution Networks.

In some CRAN based deployments, co-located BBUs support a cluster of base stations from a central location or in a data centre, allowing the complex RAN signal processing for a number of cells to be centralised in one place. This centralisation can enable more efficient implementation of SON features as the inter-basestation X2 connections are contained within one location. It can also provide the ability to adapt capacity resources to different traffic trends more efficiently and can support someLTE Advanced features and transmission schemes such as MU-MIMO and CoMP. Figure 7.2 highlights other benefits of CRAN, such as scalability, reduced operational and energy costs and virtualisation of the RAN, which are out of the scope of this chapter. In order to understand the tradeoffs associated with the design of CRAN systems, the next section describes options for RAN functional splits between the functions performed centrally, and the functions retained in the RRH or RU.

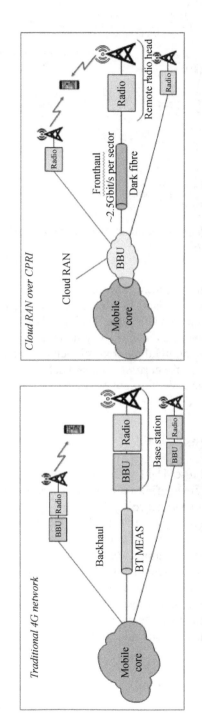

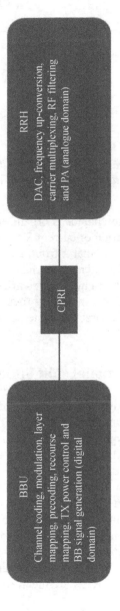

Figure 7.1   Base stations typically consist of a baseband unit (BBU) and a remote radio head (RRH)

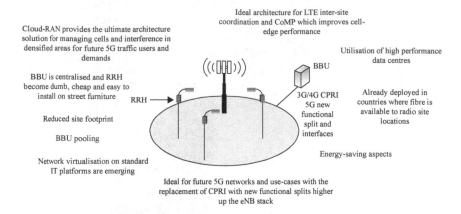

Figure 7.2   CRAN benefits include scalability, reduced operational and energy costs and virtualisation of the RAN

The rest of the chapter is organised as follows. Section 7.1 provides a deeper look into the operation of current RAN protocols in a base station, also highlighting the typical challenges that a fronthaul link serving a split 5G base station is likely to inherit from 4G systems. We then provide a review of key future network architecture elements and radio features affecting fronthaul transmissions in Section 7.2. Sections 7.3 and 7.4 respectively focus on the current fronthaul solutions that will be available for the first 5G systems, and on the industry view on aspects of 5G fronthauling.

7.1 RAN functional split options

Large channel bandwidths and lower end-to-end latency are some of the likely characteristics of 5G air interface, entering an active definition stage in standards at the time of writing. The requirement for the channels exceeding the current LTE maximum bandwidth of 20 MHz originates from ever growing demand for broadband speeds, and may be realised through the use of microwave and millimetre wave (mmW) spectrum together with spectrum aggregation techniques. Reduced latency, in the order of a few milliseconds, in turn is required for some of 5G use cases featuring various system control applications, such as robotics and virtual reality, and may be achieved using shorter and more flexible radio frame structures.

Functional separation of RAN with these features is explored in this section, focusing on the coupling between functions and on the potential impact on fronthauling for split RAN architectures.

7.1.1 Splitting RAN air interface protocols

Efficient multiplexing of traffic and control data for multiple users within a single air interface frame is likely to remain the core function of RAN in 5G. In LTE this is

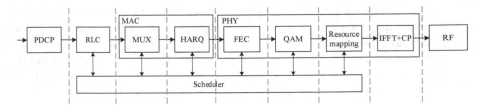

Figure 7.3   Candidate functional splits

realised through a scheduling function, present in every base station and working to achieve different objectives, such as consistent service performance for all connected users or minimisation of effects from interference. Scheduling algorithms defining the eventual mapping of information bits to waveform states are not standardised, however a scheduler is typically coupled with a number of base station's radio functions, affecting fronthaul requirements for different base station split options.

Figure 7.3 illustrates the sequence of LTE RAN protocols utilised in downlink transmission, their potential relationship to scheduling, and options for splits between the functions performed in a central unit (CU) and a remote unit (RU) connected with a fronthaul link. Operation of individual protocols in the radio stack of a base station is well described in the literature, e.g., [4,5], hence we provide a brief description of RAN protocols, focusing more on their interaction through scheduling and potential effects on fronthaul.

User plane radio protocol stack in LTE base stations consists of four main functional blocks. Packet Data Convergence Protocol (PDCP) is responsible for encryption, header compression and in-sequence delivery of user and control data. Radio-Link Control (RLC) layer performs segmentation of PDCP protocol data units (PDUs) and slower-rate retransmissions through ARQ mechanisms. Segmented data is then passed to the Medium Access Control (MAC) layer for multiplexing and scheduling, resulting in Transport Blocks (TBs). Physical layer (PHY) applies forward error correction coding and maps encoded TBs onto resource elements to be transmitted over the air.

In order to understand the role of scheduling in the RAN protocol stack and its possible interaction with four protocol layers above, consider an example of a base station mapping information bits entering the PDCP layer onto resource elements (REs) within the LTE time–frequency resource grid at the PHY layer. Each RE represents a subcarrier carrying one OFDM symbol represented by a state of the carrier waveform. Depending on the type of transmitted information and channel conditions, between 2 and 8 coded bits per symbol can be carried in one RE.

However note that the bits mapped onto REs are *coded* bits, comprising codewords that are derived from the TBs at the PHY layer. Therefore codeword bit grouping information based on the modulation scheme selected at MAC layer is required at the PHY modulation stage in an explicit form. While this information can be derived from control messaging such as Transport Format, the scheduler may need to be aware of granular channel quality in order to benefit from adaptive modulation and coding. Depending on implementation, these selection decisions may have dependencies spanning as far as RLC layer. For example, the base station may implement a scheduler adapting to near real-time channel conditions, or optimise device's memory requirements by reducing the need for data buffering. In this case user traffic multiplexing decisions within MAC and segmentation in RLC may depend on the distribution of supported modulation orders across REs in the PHY.

7.1.2   PDCP-RLC split

Forms of RAN split between PDCP and RLC layers already have applications in such features as Dual Connectivity and LTE-WiFi Aggregation (LWA). In particular, both technologies are based on the idea of providing additional transmission capacity by means of a connection point to the device. In case of Dual Connectivity, some of PDCP PDUs are transmitted to another cell's RLC layer, similar to CU-RU interaction. In the case of LWA, PDCP PDUs are sent to an RU operating a different air interface technology. Main advantages of such RAN partitioning are modest bandwidth and latency fronthaul requirements. In fact, Figure 7.4 illustrates potential reduction in the required fronthaul bandwidth between RU and CU assuming a reduction in overhead per PDU due to header compression utilised in PDCP layer.

Advantages include centralised over-the-air encryption, greater potential for coordination of mobility and handover procedures, e.g., data forwarding from the

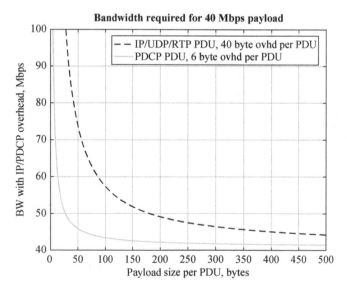

*Figure 7.4   Illustration of the required bandwidth to transfer 40 Mbit/s of user
payload over PDCP PDUs with header compression and over
datagrams with IP/UDP/RTP headers*

old serving cell to a new serving cell can be simplified in a CU hosting base station functions above the RLC layer. One of the drawbacks of PDCP-RLC split is limited potential for coordinated scheduling between multiple RUs. Due to the nature of the split, main scheduling operations are performed at the RU without an explicit exposure to PDCP processing.

7.1.3 RLC-MAC split

RLC produces segments containing PDCP layer PDUs on demand based on the notifications from the MAC layer about the total size of RLC PDUs that MAC layer can transport. A functional split between RLC and MAC layers therefore can be impractical especially in the case of shorter subframe sizes in 5G air interface compared to LTE. In particular, shorter subframe sizes allow for more frequent decisions by the scheduler, adapting better to traffic demands or channel conditions, however this results in more frequent notifications to RLC from MAC specifying the size of the next batch of RLC PDUs. Flow control mechanisms and pre-emptive transmission of RLC PDUs of fixed size ahead of the MAC requests, as described in [6], can alleviate the resulting fronthaul latency and jitter requirements, however implementation of additional flow control mechanisms reduces the gains from centralising RLC processing.

7.1.4 Split MAC

MAC layer is responsible for multiplexing RLC data from logical channels onto transport channels provided by PHY layer, retransmissions via HARQ and nominally for scheduling, although as mentioned before, scheduling in practice may have dependencies across the whole base station radio stack. The case for separation of MAC layer is typically motivated by the desire to achieve performance gains and reductions in equipment complexity through offloading most MAC functions, such as multiplexing and some of the scheduling decisions to CU, leaving in the RU only those functions that require real-time communication with PHY, such as HARQ and fine scheduling decisions.

Fronthaul links in this case would transport some form of pre-multiplexed higher-layer protocol datagrams along with scheduling commands. While the bandwidth requirements for this split should be comparable to the requirements for RAN splits at higher layers, the latency requirements are highly dependent on the realisation and interaction of scheduling functions in the CU and RU. One possible realisation of this split could involve CU delivering data to the RU in advance in the order decided in the higher-MAC scheduler, leaving the tactical decisions to the lower-MAC scheduler based on the HARQ and fine channel measurement reports [6].

While the final protocol architecture of 5G RAN is not known at the time of writing, implementation specifics and inter-dependency of lower and higher MAC layer scheduling functions may pose an interoperability challenge for base station equipment vendors and operators.

7.1.5 MAC-PHY split

The main motivation for this type of split is in enhanced capabilities from joint scheduling and coordination among RUs connected to a common CU, in addition to the benefits realised by splits at higher layers. The output from the CU to RU in this case would consist of TBs for further FEC encoding in PHY layer, and Transport Format (TF) carrying necessary information for correct processing of a TB. There are two main challenges associated with realisation of this split.

First, channel measurements utilised in the scheduling decisions need to be transported from RUs to the CUs in a timely manner in order to benefit from coordinated multi-cell scheduling and adaptive modulation schemes. For schedulers in 5G eNBs, potentially making decisions more frequently, fronthaul delays may reduce the ability fully benefit from shorter subframes and wider channel bandwidth, however exact performance trade-offs would need to be studied for specific 5G RAN technologies.

Second, data retransmissions handled through HARQ processes in LTE were designed based on compromises between performance and complexity of implementation. In particular, uplink HARQ responses from the eNB to the UE were designed to occur at pre-determined moments, hard-limiting time budget for baseband processing at the UEs and eNB. Based on LTE cell size of 100 km, the UE and eNB baseband processing needs to complete, respectively, within 2.3 and 3 ms in order for the UE uplink to be correctly acknowledged [5], with any transport latency reducing these budgets. Fronthaul requirements for next generation RAN would therefore depend on such factors as the type and number of HARQ processes expected to be supported by the base stations and UEs, as well as the typical cell sizes and use cases.

7.1.6 PHY split: FEC performed at CU

Assuming the latency, possibly a jitter, requirements on the fronthaul can be satisfied through the design of RAN protocols, appropriate choice of transport technology, or both, the PHY layer splits become possible offering opportunities for tighter coordination of transmissions and RU hardware simplification.

Among all PHY functions, the first candidate for a move to CU is FEC functionality. In LTE, FEC operates closely with MAC, providing multiple levels of CRC procedures, turbo or convolutional encoding on TBs that is very tightly coupled with MAC HARQ processes through redundancy version indications necessary for soft combining at the receiver side. The result of FEC procedures is further scrambled to provide an additional level of protection against interference, and resulting codewords are mapped onto the symbols of the modulation scheme chosen by the scheduler based channel conditions and other system indications.

Fronthaul transport for this split would therefore carry codewords together with additional scheduling information required to perform further PHY functions in the RU such as MIMO processing, precoding, antenna mapping and power allocation. As in the case of split MAC, bandwidth requirements would be comparable to higher-layer splits, while latency requirements would depend on the implementation of the scheduling functions in the RU and CU. However the latency and jitter requirements are likely to be tighter compared to split MAC or MAC-PHY split since the RU now has much less flexibility in the scheduling decisions and has to execute CU's commands on symbol-by-symbol basis.

7.1.7 PHY split: modulation performed at CU

This is the first split where fronthaul would transport quantised in-phase and quadrature (IQ) components of the symbols from the modulation scheme chosen within CU to carry user and control data that was eventually encoded into codewords. The RU responsibilities would now be limited to conversion of the sampled frequency-domain modulation symbols onto time-domain through IFFT operation, followed by CP insertion, parallel-to-serial and digital-to-analogue conversions.

Fronthaul bandwidth requirements for this split are dependent on the number of symbols transmitted and on the number of bits used to quantise each symbol, plus any control information necessary for further PHY processing in the RU. Calculations for LTE 20 MHz channel bandwidth suggest approximately 900 Mbit/s throughput would be required to transport 150 Mbit of user data [6]. Channel bandwidths expected for 5G are likely significantly exceed current conventions, especially in mmWave frequencies, resulting in multi-gigabit throughput requirements to support this split.

7.2 Radio access network technologies, architecture and backhaul options

This section outlines some of 5G radio features and components of the emerging 5G architectures with potential impact on fronthaul/midhaul.

7.2.1 Modern network architecture

New architectural advancements in 5G need to be designed in accordance to the requirements of the majority of services and applications that will be carried out in the network. To create one network that is able to simultaneously satisfy the needs of multiple devices of different nature, it is necessary to introduce flexibility, programmability and virtualisation.

This section goes through the main features considered to be key in the design of the 5G architecture, with a specific focus on how a C-RAN with fronthauled access will impact the applicability or the final performance of such features.

7.2.1.1 Backhaul, fronthaul and midhaul

Cloud or C-RAN is a logically centralised set of eNB baseband and higher layer functionalities. The baseline architecture of a fully centralised Cloud RAN is shown in Figure 7.5. Based on the NGMN definitions [7], the Fronthaul spans distances between the RRH and the BBU. The classical form of fronthaul is a point-to-point link that transports baseband radio samples, also known as common public radio interface (CPRI), which is a synchronous interface transporting digitised base band signals over a symmetric high speed physical or radio link.

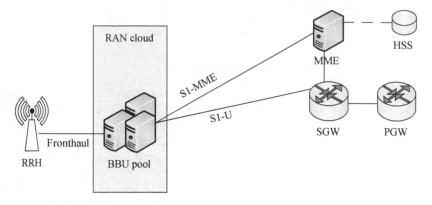

Figure 7.5   Cloud RAN general architecture overview

Based on the Metro-Ethernet Forum definition [8], Midhaul is the interconnection of a small-cell and a macro-cell via Ethernet links, with the assumption that the small cell is covered by an eNB. Without loss of generality, in this chapter the term midhaul is used to refer to a point-to-point link or network that transports signals beyond the physical layer; the term fronthaul is used to refer to the point-to-point link or network that transports physical layer signals.

As part of the core network, the Backhaul spans the section between the baseband and the evolved packet core (EPC) elements: MME, SGW, PGW, etc.

7.2.1.2   Network function virtualisation

Network Function Virtualisation (NFV) has already shown great potential in the virtualisation of core network functionalities, since it increases the flexibility of the core network implementation. In this way, control or user plane functions related to mobility management or gateways can be virtualised and placed in data centres anywhere in the network. The possibility of decoupling network functions allows configuring dedicated core networks, so that the system can better meet the service requirements [9,10].

Thus, some of the NFV general advantages are that the virtualised version of the functionality can be placed in standard IT servers and switches, reducing the cost of proprietary platforms, and allowing for more flexible implementations.

Virtualisation is very challenging in RAN low layers due to its real-time nature; synchronisation requirements that ensure a good wireless signal processing performance are in the order of microsecond and nanosecond. However, the base band radio processing may be virtualised with the use of software defined radio (SDR) techniques, and allow the mobile network to introduce the advantages of NFV in the RAN.

7.2.1.3   Network slicing

5G networks should have a high degree of programmability, configurability and flexibility to support heterogeneous deployment solution, scenarios and applications: to this aim, a key aspect of 5G systems is considered the network slicing [11]. According to NGMN, a network slice (a.k.a. 5G slice) is composed of a collection of 5G network functions (i.e., specific features enabled on Control/User-planes) and specific radio access technologies (RAT) settings that are combined together for the specific application or business model [12]. This means that all slices will not contain the same functions, i.e., a 5G slice will provide only the traffic treatment that is necessary for the service, and avoid all other unnecessary functionalities [13].

A 5G slice can span all domains of the network:

- specific configurations of the transport network;
- dedicated radio configuration or even a specific RAT;
- dedicated configuration on the fronthaul and the backhaul.

For any of the mentioned domains, network slicing determines reservation of resources to handle the enabled slices in order to fulfil the QoS constraints/features of the slice. This might introduce issues in the case of fronthaul solutions which do not offer mechanisms for resource reservation/prioritisation.

Slicing based on fronthaul or midhaul availability

A slice is commonly considered as a set of QoS features to be guaranteed in the whole network domain, where QoS from the slicing point of view is considered as the whole set of features which involves not only the typical QoS parameters (such as data rate, latency, jitter, reliability) but also other service-related functionalities (such as mobility, security, authentication, etc.) characterising the slice.

From one perspective, network slicing poses challenging requirements on the fronthaul/midhaul interface(s) which should be open and flexible, offering multivendor operation and good forward and backward compatibility, while providing options for transport bandwidth reduction. From another perspective, the characteristics of a given fronthaul/midhaul interface (latency, supported data rate, reliability, etc.) poses constraints on the possible slices to be enabled on that segment of the network. Under this point of view, the availability of different fronthaul, midhaul and backhaul solutions drastically influences the slices to be enabled: this might introduce "slice holes" in the provisioning of network slices only in those areas where the fronthaul/backhaul characteristics fulfil the requirements of the slice with consequent implications on the business models of the provider.

Slicing for multiple RAT aggregation

5G wireless systems are likely to jointly utilise 3GPP and non-3GPP RATs [14]. In this field, the fronthaul or midhaul technology available for each different RAT influences the network slicing in terms of selection of the best RAT as well as in terms of inter-RAT management. Usually the "best" RAT is the one which is able to guarantee the best performance on the radio interface (e.g., lowest interference, highest data rate or lowest congestion); nevertheless slicing takes into consideration the whole network domain and this might introduce issues when selecting a specific RAT to handle a specific slice (e.g., a RAT able to offer high data rates but with limited capabilities in terms of latency on the underlying infrastructure is not suitable for slices that need to guarantee low-latency handover mechanisms). From another point of view, the aggregation of multiple RATs might introduce issues for inter-RAT communications: in this case, the latency experienced over the fronthaul or midhaul (as well as on the backhaul) could avoid achieving the goal of a seamless handover procedure because of the need to perform additional authentication/security mechanisms when moving from one RAT to another one.

7.2.1.4 Mobile edge computing

Mobile edge computing (MEC) can be considered as a service provided through slicing. The idea behind this approach is that content, services and applications might be available at the edge of the network, i.e., as close as possible to the end-user. This will help the operator to generate additional revenue by saving the utilisation of network resources.

Although there is not an obvious connection at first glance, mobile edge computing and Cloud RAN are synergistic technologies. Cloud RAN with virtualisation of the eNB functions uses general purpose processors (GPPs) to run baseband functions, at a data centre, or locally at the venue itself. The same GPP platform can be used for MEC applications which run at the edge and not the core, even though the edge is centralised.

From a capacity point of view, mobile edge computing does not introduce any additional requirements compared to legacy data provisioning procedures on the fronthaul. The main benefits are in terms of backhaul offload (as once data is available on the edge, no transmission in backhaul will be further necessary). Nevertheless, the backhaul might limit the provisioning of mobile edge computing services because of its capacity. Indeed, when a content, service or application needs to be moved to the edge, the backhaul is expected to handle the traffic needed to move the service to the edge plus all the related signalling traffic necessary to manage this shift. As a consequence, low-capacity backhaul links might limit the provisioning slices involving mobile edge computing.

From a C-RAN (and thus, fronthaul and midhaul) perspective, it is worth to underline the following aspect. When a data/service/application is available on the edge it is expected to be provide quicker to the end user, as the traffic will not reach the core network. In this case, the delay that may limit the service latency is the Layer 2 reliability procedure (hybrid automatic repeat request HARQ), which poses a high requirement to the fronthaul not to further increase that delay, given that is the most susceptible part of the network when increasing latency.

7.2.1.5 Service virtualisation

Service virtualisation allows network functionalities to become a software package that might be moved in the network according to the specific requirements/conditions of the network in a given instant/period. Service virtualisation puts constraints in terms of reliability of the fronthaul/backhaul, to avoid retransmissions when a service is moving from one location to another one. In addition, fronthaul/backhaul latency needs to be considered as it will affect the time to run of the virtualised service.

A specific scenario for service virtualisation is the inter-connection with non-3GPP RANs, where solutions such as moving authentication functionalities close to the edge (i.e., AAA services run in a C-RAN instead of being provided by the PGW) can be exploited to offload the backhaul links and to cut delays. Nevertheless, this is only applicable in case the fronthaul is suitable to offer high capacity and low latency.

7.2.2 5G technologies and use cases

The new generation of RAN that will enable 5G is going to consider most of the current innovative features that are being discussed today by both standardisation bodies and research community. 5G is going to co-exist with the evolution of the LTE-A standard, and new radio access technologies (RATs) are also going to be integrated, to form a unique heterogeneous network, capable of providing multi-connectivity from different points of view: multiple scenarios, multiple radios and multiple cells.

In general, 5G enablers are all those technologies that can contribute to a large scale cooperative network, and at the same time enable the main features of 5G, such as enhanced mobile broadband experience or mission-critical machine-type communications with ultra-reliable and ultra-low latency communications. It is particularly interesting to identify the applicability of such 5G enablers in a C-RAN architecture and how the underlying fronthaul network impacts the overall network performance.

7.2.2.1 Integration of multiple air interfaces

Besides the enhancement of the already mature LTE-A, a set of new radio access technologies is required to satisfy future requirements in terms of spectral efficiency and availability and throughput. In particular, these new RATs need to exploit efficiently higher frequency bands and bandwidths, which require the use of new time and frequency numerology:

- Support for low latency: Shorten transmission time interval (TTI) to 0.2–0.25 ms [15]
- Low overhead to cope with time dispersion: Cyclic Prefix of 1 μs [16,17]
- Common clock with LTE to support different RATs
- Robustness against phase noise and frequency offset: use of large subcarrier spacing [15].

Since the new waveform requirements are set to satisfy the 5G ambitions, the challenges to be imposed in a C-RAN architecture are then transferred: very low latency to satisfy the TTI reduction and high capacity to satisfy the increase in spectral efficiency.

In a 5G network, the RAN architecture needs to enable the aggregation of multiple RATs, including new air interfaces and legacy ones, i.e., 4G/3G and possibly fixed services. In this sense, system convergence and integration are two "musts" for 5G.

Hence, high frequency bands, in the range of millimetre wave (mmW), need to be integrated with the use of low frequency bands, such as LTE and 4G communications. Work in [18] discusses the integration of LTE and new 5G air interfaces, where UEs are capable of simultaneous transmit and receive in both radio technologies.

In a centralised RAN context the architecture that supports the integration needs to be evaluated, since sharing resources is more than likely. This work proposes an architecture relying on common protocols, called integration layers. Due to difficult synchronisation at lower layers, the integration point is recommended to be located at least at the PDCP and RRC layers. It is assumed that the LTE and the 5G air interface are in a co-located RAN. Indeed, similar ideas as in the Dual Connectivity architecture discussed in [19] are presented for the user plane aggregation; single data flows can be aggregated over multiple air interfaces, or different flows may be mapped to different air interfaces. In a scenario with RAN cloudification, where PDCP and RRC layers of integration are centralised, there is a strong requirement on the midhaul network to support multiple RAT data integration, mainly in terms of capacity and latency.

Recent work by the 3GPP involve two options of data offloading, one is to the wireless LAN via WiFi/LTE aggregation and another one is using LTE in unlicensed spectrum. 3GPP has defined several WLAN offloading mechanisms which rely on the connection between the LTE core network and WLAN. The recent work on data aggregation at the LTE base station allows for better control of offloading with improved system and user performance while leveraging the existing LTE features of carrier aggregation (CA) and Dual Connectivity [20,21].

7.2.2.2   Support for massive MIMO

With massive MIMO, the system uses antenna arrays, with a few hundred antennas, that simultaneously serve many tens of terminals in the same time–frequency resource. The basic premise behind massive MIMO is to capture all the benefits of conventional MIMO, but on a much larger scale [22].

In a centralised RAN scenario, the use of Massive MIMO would dramatically increase the data rate requirement in the fronthaul, proportionally to the number of antennas. In fact, if digitised radio signals are transmitted through the fronthaul links capacity requirements would increase to the order of 2 Tbps for 500 MHz bandwidth with the use of mmW, as shown in [23]. As a matter of fact, in [24] authors estimate the line rate to transmit sampled radio signals over the fronthaul for a 20 MHz bandwidth using 64 antennas: nearly 80 Gbps are required.

Therefore, to allow for efficient C-RAN implementation, other architectures need to be evaluated. Work in [23] suggests that beamforming operations can be shifted close to the RRH to alleviate these data rate requirements. In the same line, [25] suggests that MIMO precoding, detection and modulation/demodulation functions should be located at the RRH; thus the information to be transported are modulation information bits. In this case the required bandwidth for a C-RAN is one order of magnitude lower that transporting sampled base band signals.

7.2.2.3   Massive cooperation: multi-connectivity networks

Multi-connectivity is a disruptive technology where devices will simultaneously transmit and receive to and from different access points. One of the key issues when moving towards a user or service centric network is to provide the mobile network with sufficient flexibility to select the serving cell(s) that better suits the device or service requirements.

In fact, the whole concept of cellular association is believed to change with 5G, and one device has no singular connection, but a set of antennas that provide service. In such a context, cooperative and decoupling techniques are massively exploited. Typical cooperative features include, joint transmission and reception, coordinated scheduling and beamforming, enhanced inter-cell interference coordination, among others. As well, decoupling techniques being considered by the literature are the Control and User plane split and the Uplink and Downlink split.

Both, the 3GPP and the research community have contributed actively with new architectural alternatives that enable easy cooperation among base stations. Coordinated multi-point (CoMP), Dual Connectivity or inter-site Carrier Aggregation are some examples in the available literature. It is true that to enable efficient cooperative networks the information exchange among the different access points is essential. Current LTE-A standards consider this information exchange through the X2 low latency interface, which at some extent can limit the performance of cooperative applications. In this sense, centralised processing of control information and management facilitates the implementation of such features by localising the information exchange. In this sense, centralised cooperative processing requires a fronthaul or midhaul network to aggregate traffic from multiple access points.

Inter-cell interference is also a crucial aspect in massive cooperation. Each cell autonomously restricts the resource allocation with the objective of limiting interference between adjacent cells. To this end, cells require data exchange which can tolerate significant latency. Nowadays inter-cell interference coordination (ICIC) techniques may be applied with any level of centralisation, since either fully centralised or distributed architectures allow for either static or dynamic ICIC algorithms. However, the fast coordination among medium access control (MAC) layers in a C-RAN environment can result in the integration of smarter enhanced ICIC (eICIC) techniques that allow to efficiently carry out high-speed and ultra-reliable communications in the cell edge (i.e., where the UE suffers from strong interference).

Coordinated multi-point transmission and reception
CoMP refers to the wide range of techniques that enable dynamic coordination among multiple cells that belong to the same "cluster". Two major forms of CoMP can be recognised in 4G LTE-A: Joint Processing (JP) transmission or reception, and Coordinated Scheduling or Beamforming (CS or CB, respectively).

In particular, CS/CB's main goal is to identify the worst interferer and avoid collisions by preventing the use of the most destructive precoding matrices (precoding is the process in which the incoming layered data is distributed to each antenna port). To this end, cells need to negotiate their beams and exchange MAC layer information; if this is done through the X2 link then bandwidth and latency requirements may not be sufficient to satisfy a high number of users. In [26] it is concluded that a MAC level centralisation can strongly limit the performance of CoMP due to additional latencies for data paths and channel state information (CSI) feedback over the midhaul network. On the other hand, PHY layer centralisation can allow all kinds of CoMP schemes. Along the same lines, work in [27] discusses that most of the CoMP schemes (Downlink joint transmission and CS/CB) can be achieved with a MAC layer centralisation, provided that information on the CSI is forwarded to the central scheduling entity. The CSI acquisition process for multiple RRHs is quite demanding in terms of overhead and reporting delays and both have a direct impact on the system performance.

The particular cases of interference rejection combining (IRC) or successive interference cancellation (SIC) the PHY layer should be centralised at some extent to avoid additional communication among the central and remote unit for CoMP purposes. Similarly, work in [28] proposes an architecture with physical layer centralisation to support UL JP techniques.

Dual connectivity

Dual connectivity is one of the 3GPP potential solutions to improve user performance by combining the benefits of the Macro cell coverage and the Small cell capacity. This new technology introduced in Release 12 [19] is defined as the simultaneous use of radio resources from two eNBs connected via non-ideal backhaul link over the X2 interface. One of the new advances is the introduction of the bearer split concept, which allows a UE to receive simultaneously from two different eNBs, known as Master eNB and Secondary eNB, MeNB and SeNB, respectively. The 3GPP has proposed several architectural alternatives for downlink dual connectivity in [19], an architecture with a centralised PDCP layer can effectively support the dual connectivity with user plane bearer split.

Downlink and uplink decoupling

The UL and DL split, or DUDe, has been covered by the literature recently as a means to reduce the UL and DL imbalance that occur in heterogeneous networks, due to the transmit power disparities between small and macro cells. The DUDe technology is the most device-centric enabler feature being investigated so far; it allows to have two different serving cells, one for the DL and another for the UL. As well, the UL feasibility of adopting the bearer split (i.e., dual connectivity) has been argued by the literature in terms of power consumption, and UL data should be transmitted directly to the best cell in terms of received power [29].

Architecture solutions that allow to support DUDe while maximising the capacity must include at least a shared MAC layer among both serving cells, since Layer 2 control information needs to be forwarded from one serving cell to another (i.e., HARQ protocol acknowledgements). As well, Layer 3 RRC ought to be centralised and shared among both serving cells, since parallel RRC connections would add too much complexity in the UE side [30].

Device-to-device communications integration

Devices themselves can as well collaborate in the RAN, by allowing direct transmission between devices controlled by the serving eNB. One of the key aspects in Device-to-Device (D2D) is the control plane information, managed by the eNB. Control information sharing in the network can improve the spectral and energy efficiency of the devices having direct communications. Since cognitive and instantaneous decisions can be made in the RAN, to allow for improved management of resources and the contention of the interference levels, a centralised control and resource management can potentially improve the outcomes of D2D.

7.2.3   Practical backhaul technologies

The fronthaul interface (as the transmission of base band sampled signals) distance is limited by the implementation of the HARQ protocol in the uplink in LTE (i.e., 8 ms) since it is the lowest round trip time (RTT) timer imposed by the MAC layer. Due to the synchronous nature of the HARQ in the UL and to its explicit dependency with the sub-frame number, is the MAC procedure that poses the most stringent requirement on latency. Relevant fronthaul solutions must fulfil the requirements outlined for CPRI transmissions, and assure a correct performance of all procedures. Fronthaul options available for native CPRI transport discussed in [4] are classified into technologies that can either multiplex or perform addressing to the native CPRI signal:

Dark fibre
Upon availability, it is a very straightforward solution, but requires high CAPEX. Point-to-point distance is limited by HARQ timers. Only one link can be transmitted since no multiplex is carried out.

WDM type
- Passive WDM: allows for transmission rates up to 100 Gbps, and the distance is limited by latency requirements (i.e., HARQ). Performance is similar to dark fibre but better reuse of facilities due to the multiplexing capabilities.
- CWDM: A single fibre with bidirectional transmission can be used to reduce costs.
- DWDM: Good for large aggregate transport requirements.
- WDM-PON: Alternative to DWDM, WDM PON with injection-locked SFPs.

Microwave
In Millimetre radio solutions the distance is capped due to processing and modulation (few hundred metres to 7 km). Capacity is typically between 1.25 and 2.5Gbps. However, Ahmadi [5] highlights that considering channel sizes, physical modulations and coding rates that are supported by recent implementations, a capacity in the range of 10 Gbps can be achieved. Microwave solutions require high bandwidth availability and high spectrum.

Optical transport networks
Good network solution that meets the jitter requirements of CPRI. One good advantage is forward error correction (FEC) which makes links less sensitive to bit errors, however the FEC added latency would further reduce the achieved distance.

XGPON and GPON
GPON is used in connection with FTTH, which is available in many urban areas. Distances are limited to 20 km, and it is impractical for fronthaul applications because it is asymmetric and bandwidth limited.

Network solutions that are able to transmit native CPRI are those that can cope with the increased capacity demands and very low jitter; Table 7.1 summarised the most typical fronthaul network options.

Table 7.1   Transport network capabilities

Transport	Throughput	Latency	Multiplexing capabilities
Dark fibre	10 Gbps+	5 μs/km	None
DWDM/CWDM	100 Gbps+	5 μs/km	High
TDM – PON	10 Gbps	Dynamic BW allocation >1 ms	High
GPON (FTTx)	DL: 2.5 Gbps UL: 1.25 Gbps	<1 ms	High
EPON (FTTx)	1–10 Gbps	<1 ms	High
OTN (FTTx)	OTU4 – 112 Gbps	FEC latency	High
Millimetre wave	2.5–10 Gbps	0.5 ms–100 μs	High
xDSL (G.fast)	10–100 Mbps (1 Gbps)	5–35 ms (1 ms)	High

7.3   Current fronthaul solutions

7.3.1   CPRI in C-RAN

CPRI is the most common transmission mode between the BBU and the RRH, and it carries sample base band signals. Thus, capacity demands for native CPRI transmission are based on several factors. The fronthaul bandwidth is proportional to the system's available bandwidth, the number of antennas and the quantisation resolution (the number of bits per I or Q sample are 8-20 bits for LTE [31]) and in any case it is dependent on the cell load and the user data rates [32]. For example, macro sites generally have three-to-six sectors combining different mobile RATs (i.e., 2G, 3G and 4G in multiple frequencies). According to the NGMN alliance report, one MCell generates approximately 15 Gbps of uncompressed sampled base band signals [26].

Hence, the basic fronthaul requirements to be considered for the transmission of native CPRI are [7]:

- Capacity: from CPRI option 3–9, i.e., 2.457 Gbps to 12.165 Gbps [32]. Note that this capacity has been calculated considering LTE bandwidth configurations and different number of antennas: CPRI option 3 capacity link is for an LTE bandwidth of 10 MHz with 4 antennas, or a 20 MHz bandwidth with 2 antennas. As remarked in Section 1.2 the use of multiple antennas or mmW can increase capacity demands to the order of terabytes.
- Jitter: in the range of nanoseconds, according to the physical layer time alignment error (TAE) (i.e., 65 ns)
- Latency: maximum round trip delay excluding cable length 5 μs, to assure the efficient implementation of frequency division duplex (FDD) inner loop power control [32]
- Scalability: Support for multiple RATs and RAN sharing
- Distance: 1–10 km for most deployments, 20–50 km for large clouds

Also, the fronthaul must deliver synchronisation information from the BBUs to the RRHs, which is natively supported by CPRI through the control and management.

Given these stringent requirements, new scalable and efficient solutions need to be explored in the context of CPRI. One option is the CPRI compression that allows to significantly reduce the required bitrate while assuring that it meets the transparency requirements of the CPRI. Also the literature is exploring the so-called functional split, where some of the eNB functionalities remain in the remote unit, allowing to relax the bandwidth and delay constraints.

Also, the fronthaul must deliver synchronisation information from the BBUs to the RRHs, which is natively supported by CPRI through the control and management.

Given these stringent requirements, new scalable and efficient solutions need to be explored in the context of CPRI. One option is the CPRI compression that allows to significantly reduce the required bitrate while assuring that it meets the transparency requirements of the CPRI. Also the literature is exploring the so-called functional split, where some of the eNB functionalities remain in the remote unit, allowing to relax the bandwidth and delay constraints.

7.3.2 CPRI compression

Compressed CPRI may be used to reduce capacity requirements in places where fronthaul bit stream transport is limited, CPRI compressed and decompressed function may be used, which can provide 2–3 times more utilisation.

Point-to-point fronthaul compression methods, where the central unit independently compresses each remote unit baseband signal, includes techniques such as filtering, block scaling and non-linear quantisation [33]. These solutions allow to remove redundancies in the spectral domain by down sampling the input signal, mitigate peak variations. Compression reduces the data rate by a factor of three with respect to uncompressed CPRI signals. For example, the work in [34] applies the same techniques and obtained good system performance with 1/3 compression rates in a practical propagation environment. In particular, results show that compression can effectively reduce the amount of data rate transmission on the CPRI without comprising the actual baseband data at low compression ratio.

Moreover, Park *et al.* [35,36] present another approach for compression in the downlink, named multivariate fronthaul compression. The proposed solution is based on joint design of precoding and compression of the baseband signals across all base stations; results show that this CPRI compression scheme outperforms the conventional approaches of point-to-point compression. As a rule of thumb, compressed CPRI techniques are seen to reduce the fronthaul rate by a factor around 3 [27].

7.3.3 Fronthaul or midhaul over ethernet

When evaluating CPRI as the transport service for the C-RAN fronthaul some issues arise:

- provides no statistical multiplexing gain (due to the continuous data transmission)
- how to manage and provide service level agreements in the fronthaul service [32,37]

Recently, in both research community and industry it has been discussed if Ethernet networks can be used to transmit the physical layer signals, which would initially imply some more framing overhead and struggle to meet the requirements. However, on the other hand, using Ethernet links to encapsulate sample base band signals brings several advantages [23]:

(a)   Lower cost-industry standard equipment
(b)   Sharing and convergence with fixed networks
(c)   Enables statistical multiplexing gains when signal has a variable bit rate
(d)   Enables the use of virtualisation and orchestration
(e)   Allows network monitoring
(f)   Allows managing the fronthaul network (i.e., Fronthaul as a service). Path management enables the use of virtualisation and software defined networks (SDNs)

Originally, Ethernet is the best effort-based technology, and it is not designed to meet the low jitter and latency requirements for base band signals transmission, i.e., CPRI. In general, works considering CPRI over Ethernet [38] suggest dedicated links between RRH and BBU, and the Ethernet network is enhanced with additional features to satisfy stringent latency and jitter constraints.

In particular, the IEEE is actively working in new standardisation efforts. In particular, IEEE 1904.3 Radio over Ethernet (RoE) for encapsulation and mappers [39], the main objectives are: (a) to define a native encapsulation transport format for digitised radio signals and (b) a CPRI frame to RoE (i.e., Ethernet encapsulated frames) mapper. Also, several enhancements to allow the transport over time sensitive traffic have been presented in the IEEE 802.1CM Time-Sensitive Networking for Fronthaul Task Group [40]. Solutions such as frame pre-emption or scheduled traffic allows to better manage packets in Ethernet and reduce jitter [38].

Other solutions to support CPRI over Ethernet is the use of timing protocols, such as Precision Time Protocol (PTP) that provides synchronism through the exchange of time stamped packets; however, Gomes *et al.* [23] remark that bitrate as well as delay and delay variation requirements result in significant implementation challenges when using Ethernet in the fronthaul to transmit digitised radio signals, i.e., CPRI.

7.3.4   C-RAN integration in 5G: feasibility discussion

5G can be seen as a tool box of enabling technologies, where a specific set of these technologies can potentially provide the requirements of a given service. Therefore, when assessing the suitability of C-RAN with a fronthaul network in 5G, the main conclusion is that a flexible network architecture should be considered. Depending on the service and the underlying network infrastructure the RAN can be configured to employ any set of technologies and split functionalities. Cloud RAN and fronthaul/midhaul is all about heterogeneity. An overall conclusion of the transport network for fronthaul or midhaul and functional split trade-off discussion is that there is no configuration that can satisfy all the requirements for 5G simultaneously:

*   High levels of processing in the BBU may increase latency due to fronthaul links, and pose challenging capacity requirements for CPRI transmission, especially if higher number of antennas are employed;

- High levels of processing in the RRH may impair the cooperative capabilities and neglect the multi-connectivity features;
- Low latency or congested fronthaul networks may impair the correct function of reliability algorithms (i.e., HARQ in Layer 2), and increase latency and decrease user throughput;
- The lower layers are extremely latency critical whereas the higher layers are not;
- Network slicing capabilities and virtualisation outcomes are very much dependent on fronthaul transport networks;
- All these requirements in latency to maintain the high synchronicity between the lower layers are going to be more demanding when implementing centralisation with 5G radio access technologies.

Based on this, no rule of thumb can be proposed to efficiently implement partial fulfilment of fully centralised RAN. There is a strong trend from industry and academia to leverage packet switched networks, such as Ethernet, to provide a cost-effective transport network solution that allows to converge backhaul with fixed backbone networks. In the practical environment, Korean operator's KT big picture of 5G and re-design of the RAN involves the use of functionality split to support packet networks that deliver Ethernet frames. The lowest level functionality split option would involve the transmission of MAC PDUs over the fronthaul, meaning that the entire PHY layer is kept in the radio head side [41]. Table 7.2 summarises the main 5G enablers described in this document with its optimal Cloud RAN configuration.

Splits within the PHY layer, using either CPRI for sampled radio signals or packet switched networks after resource de/mapping, have the main advantage of maintaining the PHY, MAC and RLC layers operating together and keep full cooperation gains of the centralised architecture. Conversely, 5G latency is the key performance indicator, which may indicate that these layers need to stay as close as possible to the radio site.

In this sense, to be able to integrate all the 5G enablers and satisfy its individual requirements for implementation, base stations (RRH + BBU) functionalities should be completely flexible; allowing efficient configurations of slices that satisfy the services requirements and the UE needs at all times, considering each key performance indicator separately.

Table 7.2   5G enablers mapped to the optimal Cloud RAN configuration

5G Enabler	Cloud RAN Configuration
Low latency RAT	Lower layers close to radio unit
Multiple RAT integration	Common PDCP
LTE in unlicensed band	Common MAC
Massive MIMO	Lower layers close to radio unit
Multi-connectivity	High level of centralisation
Network slicing and service virtualisation	High dependency on transport network performance and availability

7.4　Market direction and real-world RAN split examples

Backhaul networks to connect mobile radio base stations to core networks are a key area of interest for both mobile operators (as they form a significant cost element) and fixed operators (who often provide the backhaul infrastructure). Over the lifetime of LTE, we have seen an evolution from the initial S1 backhaul interface to the CPRI-based "fronthaul" technology adopted by Cloud RAN (CRAN) and finally the proposal of alternative base station splits to reduce the overhead that CRAN fronthaul imposes.

7.4.1　Mobile backhaul

The architecture of a 4G mobile network as standardised by 3GPP is a flat structure simply involving a network of radio base stations (EnodeBs) which are linked to the Evolved Packet Core network via backhaul connections (Figure 7.6). The EnodeB sites incorporate both radio and baseband processing functionality linked to antennas located at the top of the associated masts by means of co-axial feeders. However, these can exhibit high losses and therefore some vendors moved to deployments which placed a Remote Radio Head or Radio Unit (RRH or RU) incorporating digital-to-analogue and analogue-to-digital conversion and power amplification next to the antenna and connected it via optical fibre to the Baseband Unit or Central Unit (BBU or CU) providing the packet processing and scheduling functions at the base of the mast.

The backhaul connections linking the remote ENodeBs to the EPC form the important S1 reference point which is standardised by 3GPP RAN3 working group. In addition, eNodeBs may be connected to each other by means of the X2 interface. Although the RAN backhaul network would appear to be simply a collection of point-to-point links, the reality is usually much more complex. A practical radio mobile network architecture is illustrated in Figure 7.6, with key features including:

1. Shared RAN – various aspects of the RAN are shared between different mobile operators, in this particular example through a joint venture MBNL. This includes the 3G radio, plus the towers and backhaul at the majority of sites.
2. The shared RAN arrangement is facilitated by a fibre ring providing resilient high-capacity connectivity between the core network and RAN Connectivity Sites (in the order of 10–20 serving the UK).
3. From these connectivity sites, point-to-point fixed connections provide backhaul connectivity to major base station sites.
4. Finally, second-tier, smaller base station sites are server from the major sites – mainly using point-to-point microwave links.

7.4.2　Centralised or Cloud RAN

As discussed earlier, the C-RAN architecture was first proposed in 2010 by IBM [2], and then described in detail by China Mobile Research, [3] extends the BBU-RRH concept described above by moving the BBU from the base station site to a centralised site – and co-locating or pooling with BBUs from a number of other base stations to form a Centralised-RAN.

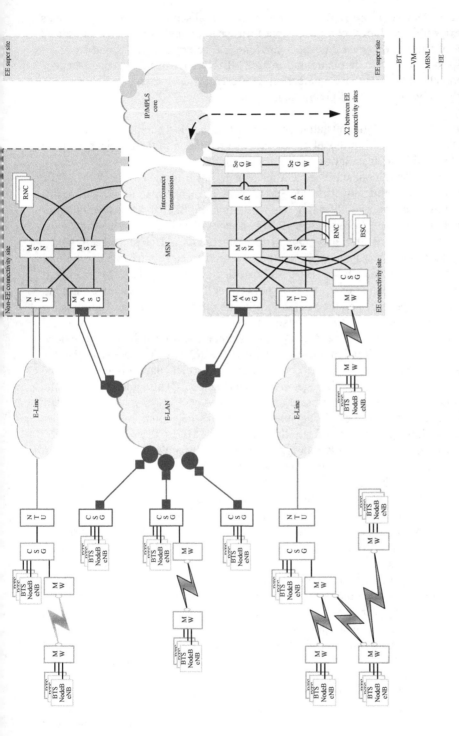

Figure 7.6 Practical end-to-end mobile network architecture. Less commonly used terms are: MSN – multi-service network, NTU – network termination unit, MW – micro/millimetre-wave backhaul, SeGW – security gateway. Diagram credit: Andy Sutton, Principal architect at BT

In a further step, the BBU functions may be realised in software components deployed within a shared computing platform or compute cloud. This virtualised BBU pool running on general purpose processors is referred to as a Cloud-RAN.

The key barrier to implementation of C-RAN is the bandwidth and latency requirements of the fronthaul connection. However, for operators, C-RAN approach offers operators a number of potential advantages:

1. Simplified base station installation: The BBU is the most complex element of the base station and it's removal can facilitate a reduction in the physical footprint and simplify the installation of the base station.

2. Reduced power consumption: China Mobile have estimated that 72% of the power used by the network is expended in the RAN and that, nearly half of this is consumed by air conditioning. Consolidating the BBU functions would allow most of this power to be saved as RRH elements can typically be air-cooled.

3. Increased spectral efficiency: Due to the characteristic of LTE that all cells generally operate on the same frequency (or set of frequencies), inter-cell interference is often the limiting factor on cell capacities and throughput. This manifests itself as a difference of up to $10\times$ between cell centre and cell edge throughput. Two approaches may be taken to mitigate interference effects – minimising interference and exploiting it constructively.

   One approach is to use inter-cell interference control which allows eNodeBs to co-ordinate with neighbouring cells over the X2 interface to ensure that the resources that are being used to communicate with a cell-edge mobile are not used by neighbours at their cell edge. eICIC (enhancedICIC) also addresses the time domain by introducing Almost Bland Subframes (ABS) which allows eNodeBs to negotiate (again via X2) an interval when it's neighbour cell will mute it's signal allowing it to send information to a cell-edge UE. CRAN does not directly improve either ICIC or eICIC performance – but it facilitates the establishment of the X2 interfaces on which they depend.

   An alternative is in utilisation of Co-Operative Multi-Point (CoMP) techniques which attempt to utilise interference. In CoMP several cells co-operate to serve cell-edge UEs – for example, by transmitting the data to a specific mobile from more than one cell so that the signals combine additively. However, this requires very tight synchronisation between the cells in the CoMP group, which can be achieved if all the cells are served from a centralised BBU pool.

4. Reduced upgrade and maintenance overheads: The CRAN architecture co-locates BBU functions at a much smaller number of locations and therefore repair and upgrade activities are also concentrated at these locations, significantly reducing overheads such as travel time. In addition, if equipment failure does occur, there are many more opportunities for absorbing the problem through re-configuration within the BBU pool with automation reducing further the need for human intervention.

5. Increased BBU resource flexibility and utilisation: The transition to a centralised, virtualised deployment makes it possible to pool BBU resources which can be shared across base stations. This can greatly increase the utilisation efficiency of these functions.

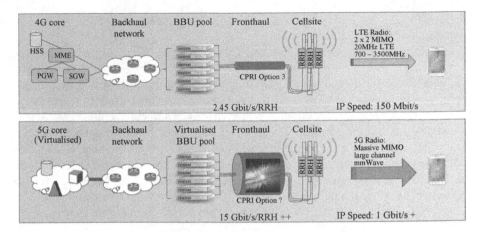

Figure 7.7 Correspondence between fronthaul and backhaul throughput requirements for 4G and 5G. High-speed radio ambitions to increase throughput by over 100× may result in similar increases needed on backhaul/fronthaul

7.4.3 Forward view to 5G

As shown in Figure 7.7, a typical deployment uses 20 MHz LTE bandwidth with 2T2R MIMO and therefore the CPRI rate to each radio head is 2.45 Gbit/s. Assuming that the base station site comprises three sectors, at a total of 7.37 Gbit/s is required (typically across 3 fibre pairs).

Compared to the usable data rate of 150 Mbit/s, the figures shown above indicate that CPRI introduces an overhead of over 1,600%. Taken together with the stringent latency requirements of CPRI, this make direct point-to-point optical fibre connections the only suitable technology for fronthaul links. Moreover, the introduction of 5G radio technologies such as massive MIMO and wider channel bandwidths will also make this situation even worse. Figure 7.7 also shows a possible 5G fronthaul scenario with a 5G radio systems supporting a real-world capacity between 1 and 3Gbit/s – translated into a CPRI fronthaul rate up to 45Gbit/s. Such speeds will drive up the costs of the termination devices for the optical fibre connections and delay or prevent the implementation of the required signal processing in virtualised software functions.

7.4.4 Industry 5G fronthaul initiatives

The bandwidth issues described above, taken together with the other challenges of current fronthaul protocols such as strict limits on latency and jitter and the inflexibility of the deployment architectures that they impose, have driven operators to look for alternatives which can deliver the benefits of CRAN without the costs.

An important vehicle delivery of this ambition is the work on the IEEE 1914 Next-Generation Fronthaul Interface (NGFI) specification. This will be based on Ethernet to deliver a packet-based, multi-point to multi-point interconnect using statistical multiplexing which will handle data security, quality of service and synchronisation.

The NGFI activity is supported by China Mobile, AT&T, SK Telecom, Nokia, Broadcom, Intel and Telecom Italia.

In addition, a number of operators have announced details of their plans covering C-RAN and fronthaul. These typically include the support needed for further study of RAN architectures to better understand both the trade-off between centralisation/virtualisation gains and fronthaul capacity and how efficient interworking between 4G, 5G and WLAN might be achieved in a CRAN architecture. In particular, investigations into the number of required splits between a software-based Control Unit (CU) and a remote hardware-based Access Units (AU) are ongoing. One important component from operators' perspective is that the interface between the CU and DU units is open and standardised to support multiple transport solutions required in a practical network deployment.

7.4.5 Split MAC trials

The BBU-RRH functionality separation in this case is within the MAC layer itself. Typically the MAC scheduler, i.e., the upper part of the MAC, is centralised within the BBU. This enables the possibility of coordinated scheduling amongst cells, whilst the lower MAC and HARQ functionality, given the 8 ms critical timing aspects of the HARQ cycle, are placed on the RRH in order to remove any additional latency from the fronthaul link. Having the lower MAC and HARQ on the RRH allows the remote scheduler to operate with the PHY layer in a semi-autonomous way and at a sub-frame level. Some tight coordination in the form of scheduling commands and HARQ reports is required for the communications between the MAC and HARQ scheduling. The required capacity and latency is 150/75 Mbps (similar to S1/X2 rate) for a 20 MHz carrier and <6 ms, respectively. Standard packetized technologies for fronthaul transport such as Ethernet can be sufficient to ensure good coordination between the two scheduling processes for some of the current basic LTE networks and not for LTE-A and 5G networks as the fronthaul requirements are typically higher (see Figure 7.8).

BT and Cavium have conducted trials on split cell options and on the suitability of various transport technologies for fronthaul since 2015. In an experiment that is believed to be a world first, BT has demonstrated Cavium's MAC split solution over LTE 2 × 2 MIMO 15 MHz carrier running successfully over 300 m of copper using G.FAST and 52 km of GPON under realistic load conditions with speeds of up to 90 Mbps (using standard UE devices) with only 10%–15% overhead associated with a CAL layer interface which is independent of the number of antennas, bandwidth and carriers. Further trials are ongoing on a further MAC/PHY split and denser deployments. The basic setup consists of Cavium's ThunderX blade, a virtual machine consisting of a complete EPC, eNB stack and application layer, capable of pooling up to 128 BBUs with varies splits running simultaneously. The connectivity between the BBU and RRH uses standard Ethernet protocol which can connect directly to G.FAST (see Figure 7.9) or GPON distribution point. The latency on both technologies was less than 3 and 1.5 ms, respectively, making them both suitable for MAC and MAC/PHY splits.

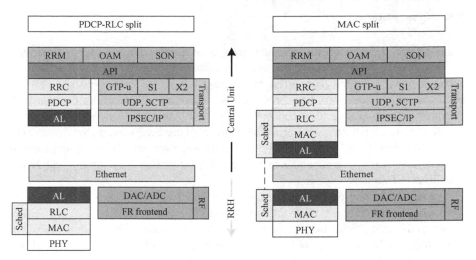

Figure 7.8 *Protocol architectures for experiments with base station PDCP-RLC and MAC splits. A split at higher layer typically has lower fronthaul requirements, while split at lower layers provides more coordination opportunities at the CU. AL here stands for an adaptation layer necessary to transport protocol data units over transport networks*

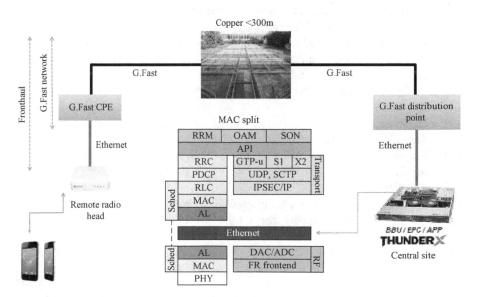

Figure 7.9 *Split MAC experiment setup*

7.5   Conclusion

This chapter focused on the practical aspects of fronthauling the next generation radio access networks. One of the important characteristics of 5G RAN is likely to be an increased degree of coordination between RAN nodes in networks that are becoming more dense. In this context CRAN becomes an interesting concept with overlaps with other emerging concepts such as NFV and MEC, however realisation of CRAN and associated requirements on fronthaul are dependent on the distribution of the RAN functions between central unit and the remote unit. This chapter provided an overview of the interdependence between RAN functions through the scheduling functions within a base station.

One of the main observations highlighted in this chapter is that many backhaul technologies can support 5G fronthaul requirements, however there is a significant opportunity for optimisation. In particular, while the use of CPRI can be viable for point-to-point links, associated transport overheads may be prohibitive for a dense network, especially where base stations utilise some of the proposed 5G air interface techniques such as higher order MIMO over large channel bandwidths. Provided examples of real-world split RAN operation illustrate the viability of alternative split RAN architectures that can be more efficient from deployment perspective, and invite further research in the area.

References

[1]   D. H. Ring, "Mobile telephony – wide area coverage – case 20564," *Bell Telephony Lab. Tech. Memoranda*, vol. 47-160-37, pp. 1–22, Dec. 1947.

[2]   Y. Lin, L. Shao, Z. Zhu, Q. Wang, and R. K. Sabhikhi, "Wireless network cloud: Architecture and system requirements," *IBM Journal of Research and Development*, vol. 54, no. 1, pp. 1–12, Jan. 2010.

[3]   C. M. R. Institute, "C-RAN the road towards green ran," White Paper, Oct. 2011.

[4]   E. Dahlman, S. Parkvall, and J. Skold, *4G: LTE/LTE-Advanced for Mobile Broadband*. Amsterdam: Elsevier Science, 2013.

[5]   S. Ahmadi, *LTE-Advanced: A Practical Systems Approach to Understanding 3GPP LTE Releases 10 and 11 Radio Access Technologies*. Amsterdam: Elsevier Science, 2013.

[6]   "Small cell virtualization functional splits and use cases," 2016.

[7]   "Fronthaul Requirements for Cloud RAN," White Paper, NGMN, Mar. 2015.

[8]   M. E. Forum, "MEF 22.1.1 Implementation Agreement Mobile Backhaul Phase 2 Amendment 1 – Small Cells," MEF, TR, Jul. 2014. [Online]. Available: https://www.mef.net/Assets/Technical_Specifications/PDF/MEF_22.1.1.pdf

[9]   3GPP, "Architecture enhancements for dedicated core networks; Stage 2 (Release 13)," 3rd Generation Partnership Project (3GPP), TR 23.707, Dec. 2014. [Online]. Available: http://www.3gpp.org/dynareport/23707.htm

[10] ETSI, "Network Functions Virtualisation (NFV); Architectural Framework," ETSI, TR GS NFV 002, Oct. 2013.

[11] "NGMN 5G White Paper," White Paper, NGMN, Feb. 2015.

[12] N. Nikaein, E. Schiller, R. Favraud, *et al.*, "Network store: exploring slicing in future 5g networks," in *Proceedings of the 10th International Workshop on Mobility in the Evolving Internet Architecture*. New York, NY: ACM, 2015, pp. 8–13.

[13] M. Iwamura, "Ngmn view on 5G architecture," in *2015 IEEE 81st Vehicular Technology Conference (VTC Spring)*, pp. 1–5, May 2015.

[14] A. Osseiran, F. Boccardi, V. Braun, *et al.*, "Scenarios for 5G mobile and wireless communications: the vision of the metis project," *IEEE Communications Magazine*, vol. 52, no. 5, pp. 26–35, May 2014.

[15] G. Berardinelli, K. Pedersen, F. Frederiksen, and P. Mogensen, "On the design of a radio numerology for 5G wide area," *ICWMC 2015*, p. 24, 2015.

[16] E. Lähetkangas, K. Pajukoski, J. Vihriälä, and E. Tiirola, "On the flexible 5G dense deployment air interface for mobile broadband," in *5G for Ubiquitous Connectivity (5GU), 2014 First International Conference on*, pp. 57–61, Nov. 2014.

[17] J. Vihriala, N. Ermolova, E. Lahetkangas, O. Tirkkonen, and K. Pajukoski, "On the waveforms for 5G mobile broadband communications," in *2015 IEEE 81st Vehicular Technology Conference (VTC Spring)*, pp. 1–5, May 2015.

[18] I. D. Silva, G. Mildh, J. Rune, *et al.*, "Tight integration of new 5G air interface and LTE to fulfil 5G requirements," in *2015 IEEE 81st Vehicular Technology Conference (VTC Spring)*, pp. 1–5, May 2015.

[19] 3GPP, "Study on small cell enhancements for E-UTRA and E-UTRAN-Higher Layer Aspects(Release 13)," 3rd Generation Partnership Project (3GPP), TR 36.842, Dec. 2014. [Online]. Available: http://www.3gpp.org/dynareport/36842.htm

[20] "LTE Aggregation & Unlicensed Spectrum," White Paper, 4G Americas, Nov. 2015.

[21] GPP, "Study on Licensed-Assisted Access to Unlicensed Spectrum," 3rd Generation Partnership Project (3GPP), TR 36.889, Mar. 2015. [Online]. Available: http://www.3gpp.org/dynareport/36889.htm

[22] E. G. Larsson, O. Edfors, F. Tufvesson, and T. L. Marzetta, "Massive MIMO for next generation wireless systems," *IEEE Communications Magazine*, vol. 52, no. 2, pp. 186–195, Feb. 2014.

[23] N. J. Gomes, P. Chanclou, P. Turnbull, A. Magee, and V. Jungnickel, "Fronthaul evolution: from CPRI to Ethernet," *Optical Fibre Technology*, vol. 26, pp. 50–58, 2015.

[24] S. Park, C. B. Chae, and S. Bahk, "Before/after precoding massive MIMO systems for cloud radio access networks," *Journal of Communications and Networks*, vol. 15, no. 4, pp. 398–406, Aug. 2013.

[25] J. Liu, S. Xu, S. Zhou, and Z. Niu, "Redesigning fronthaul for next-generation networks: beyond baseband samples and point-to-point links," *IEEE Wireless Communications*, vol. 22, no. 5, pp. 90–97, Oct. 2015.

[26]   "Further Studies on Critical Cloud RAN Technologies," White Paper, NGMN, Mar. 2015.

[27]   U. Dötsch, M. Doll, H. P. Mayer, F. Schaich, J. Segel, and P. Sehier, "Quantitative analysis of split base station processing and determination of advantageous architectures for lte," *Bell Labs Technical Journal*, vol. 18, no. 1, pp. 105–128, Jun. 2013.

[28]   D. Boviz, A. Gopalasingham, C. S. Chen, and L. Roullet, "Physical layer split for user selective uplink joint reception in SDN enabled cloud-ran," in *2016 Australian Communications Theory Workshop (AusCTW)*, pp. 83–88, Jan. 2016.

[29]   Huawei, "Handling of UL Traffic of a DL Split Bearer," 3GPP TSG-RAN, Tech. Rep. R2-140054, Feb. 2014.

[30]   ZTE, "Comparison of CP Solution C1 and C2," 3GPP TSG-RAN, Tech. Rep. R2-132383, 2013.

[31]   O. Simeone, A. Maeder, M. Peng, O. Sahin, and W. Yu, "Cloud radio access network: Virtualizing wireless access for dense heterogeneous systems," *Journal of Communications and Networks*, vol. 18, no. 2, pp. 135–149, Apr. 2016.

[32]   ETSI, "CPRI Specification v6.0," TR, Aug. 2013.

[33]   D. Samardzija, J. Pastalan, M. MacDonald, S. Walker, and R. Valenzuela, "Compressed transport of baseband signals in radio access networks," *IEEE Transactions on Wireless Communications*, vol. 11, no. 9, pp. 3216–3225, Sep. 2012.

[34]   B. Guo, W. Cao, A. Tao, and D. Samardzija, "LTE/LTE—a signal compression on the CPRI interface," *Bell Labs Technical Journal*, vol. 18, no. 2, pp. 117–133, Sep. 2013.

[35]   S.-H. Park, O. Simeone, O. Sahin, and S. Shamai, "Joint precoding and multivariate backhaul compression for the downlink of cloud radio access networks," *IEEE Transactions on Signal Processing*, vol. 61, no. 22, pp. 5646–5658, 2013.

[36]   S. H. Park, O. Simeone, O. Sahin, and S. S. Shitz, "Fronthaul compression for cloud radio access networks: signal processing advances inspired by network information theory," *IEEE Signal Processing Magazine*, vol. 31, no. 6, pp. 69–79, Nov. 2014.

[37]   "RAN Evolution Project Bachkaul and Fronthaul Evolution," White Paper, NGMN, Mar. 2015.

[38]   Huawei, "A Performance Study of CPRI over the Ethernet," Jan. 2015. [Online]. Available: http://www.ieee1904.org/3/meeting_archive/2015/02/tf3_1502_ashwood_1a.pdf

[39]   J. Korhen, "Radio over Ethernet Considerations," Feb. 2015. [Online]. Available: http://www.ieee1904.org/3/meeting_archive/2015/02/tf3_1502_korhonen_1.pdf

[40]   IEEE, "Time Sensitive Networking Task Group." [Online]. Available: http://www.ieee802.org/1/pages/tsn.html

[41]   Harrison., J. Son, and M. Michelle, "5G Network as envisioned by KT – Analysis of KT's 5G Network Architecture," Nov. 2015.

Chapter 8

Interference management and resource allocation in backhaul/access networks

L. Zhang[1], E. Pateromichelakis[2], and A. Quddus[3]

Abstract

The interface management and resource allocation in backhaul (BH)/access networks for the cloud/centralized radio access networks (C-RAN) are two of the largest challenges to enable the C-RAN architecture deployment successfully in cellular wireless systems. From the PHY perspectives, one major advantage of C-RAN is the ease of implementation of multicell coordination mechanisms to manage the interference and improve the system spectral efficiency (SE). Theoretically, a large number of cooperative cells leads to a higher SE; however, it may also cause significant delay due to extra channel state information feedback and joint processing computational needs at the cloud data center, which is likely to result in performance degradation. In order to investigate the delay impact on the throughput gains, we divide the network into multiple clusters of cooperative small cells and formulate a throughput optimization problem. We model various delay factors and the sum-rate of the network as a function of cluster size, treating it as the main optimization variable. For our analysis, we consider both base stations' as well as users' geometric locations as random variables for both linear and planar network deployments. The output signal-to-interference-plus-noise ratio and ergodic sum-rate are derived based on the homogenous Poisson-point-processing model. The sum-rate optimization problem in terms of the cluster size is formulated and solved.

From the radio resource management (RRM) perspective, we consider the problem of joint BH and access links optimization in dense small-cell networks with special focus on time division duplexing mode of operation in BH and access links transmission. Here, we propose a framework for joint RRM where we systematically decompose the problem in BH and access links. To simplify the analysis, the procedure is tackled in two stages. At the first stage, the joint optimization problem is formulated for a point-to-point scenario where each small cell is simply associated

[1] School of Engineering, University of Glasgow, UK
[2] Huawei Technologies, Germany
[3] Institute for Communication Systems, University of Surrey, UK

to a single user. In the second stage, the problem is generalized for multiaccess small cells. In addition, the chapter addressed the joint routing and BH scheduling in a dense small-cell networks using 60 GHz multihop BH, coordinated by a local C-RAN central unit. The problem is formulated as a generalized vehicle routing problem and decoupled into two subproblems, channel-aware path selection and queue-aware link scheduling.

8.1 Introduction

As a promising candidate technology for next-generation wireless communications, cloud (or centralized) radio access network (C-RAN) has drawn significant attention by both academia and industry in the last few years. Apart from C-RAN's advantage of reducing radio site operations and capital costs, another benefit is related to ease in the implementation of multicell coordination mechanisms such as coordinated multipoint transmission and reception [1–4], thus promising higher system performance through efficient interference management.

In addition, the cloud-based architecture provides the flexibility of splitting the radio access functionalities between the cloud and the remote sites depending on the backhaul (BH) link capacity and software/hardware processing capability of the access and cloud entities in the network [2,5,6]. One of the most popular functional split options in a C-RAN is to consider a high computational capability central processor taking high-complexity tasks in the cloud, and a set of densely deployed, low power, low-complexity radio remote heads (RRHs) [7,8]. This option can harness the benefit of deploying a low-cost dense small-cell network, whereas at the same time, efficient interference avoidance and cancelation algorithms across multiple small cells can be realized through centralized processing in order to improve network spectral efficiency (SE).

Theoretically, larger cooperation cluster size (i.e., number of cooperating cells) leads to better interference cancelation and higher system SE. However, this is in practice not true if real-world implementation factors, such as latency, are taken into account. Larger number of cooperating cells/antennas results in more complex channel estimation and precoding implementation; this is especially true for advanced channel estimators such as minimum mean square error [9,10] and zero-forcing (ZF) precoders [11] whose complexity is in cubic order of the number of involved (transmitting or receiving) antennas. In addition, more antennas/cells/users in a cluster imply more channel state information (CSI) required for precoding, bringing further CSI feedback delay into the system. Furthermore, due to general-purpose hardware processing in the cloud data center, and also due to the uncertainties in availability of computational resource, significant processing delay may get added.

All these delays can cause mismatch between actual channels and the channel used for calculating precoder matrix, consequently, performance degradation results. Therefore, we conjecture that there must be an optimal cluster size, large enough to mitigate interference into a reasonable level yet small enough to save the performance loss due to the delay-caused channel mismatch.

Another practical challenge of C-RAN architecture is the requirement for high capacity and ultra-low-latency BH/front-haul, which can be achieved mainly using ideal BH (fiber) for the small-cells connectivity. This can make the employment of C-RAN impractical and cost inefficient especially for ultra-dense urban scenarios. In this chapter, an alternative to wired BH technology is the millimeter wave radio (mmW) which operates in 58–80 GHz bands. In this context, several GHz-wide chunks of spectrum are available and can provide multiple Gbps even with low-order modulation schemes. In addition to these high-data rates, mmW radio band can offer excellent immunity to interference, high security, and the reuse of frequency. mmW radio requires clear line of sight (LoS) propagation and its range is restricted by the oxygen absorption which strongly attenuates ≥ 60 GHz signals over distances. To combat this, high gain directional antennas are used in order to compensate for the large free space propagation losses. Moreover, data can be transferred via multiple low-distanced hops to ensure good BH link channel qualities. To this end, one of the major challenges that need to be addressed is the routing and scheduling the traffic in BH links by taking into account the BH channel conditions and the traffic demand.

Furthermore, for efficient spectrum utilization in cloud-based RAN assuming heterogeneous BH, a key scenario involves the coupling between BH and access resources (usually termed as in-band backhauling), which can theoretically provide high gains by enhancing the spatial reuse. Nevertheless, the coupling of BH and access can limit the multiuser diversity gain of radio resource allocation in the access link. In case of time division duplexing (TDD) between the transmission links, each packet would be received after two consecutive transmissions, i.e., on BH link to small-cell station and from small-cell station to the end user (access link). Therefore, the efficiency of system depends upon the balance of resources between the two links for each small cell. This requires efficient resource partitioning on BH or access links and rate balancing strategies between the two.

8.2 Optimal cooperative cluster size

8.2.1 System model

To optimize the cluster size of cloud-based small-cell networks, building a mathematic link between the optimizing criterion in terms of sum-rate and the cluster size is the key. Two problems arise in that building process: (1) how to model the signal model of the network as a function of the cluster size and (2) how to map the cluster size as a function of latency.

For the first problem, we start our analysis from the deployment of cellular system, where typically, the base stations (BSs) are fixed in homogeneous grid. However, the most significant change that has to be taken into consideration for cooperation in small cells (in comparison with a point-to-point MIMO system) is the geometric location randomness which leads to the uncertainty of the large-scale fading. Toward this end, several papers have considered large-scale fading as well as small-scale fading in analyzing the ergodic capacity of cooperative systems for uplink [12] and downlink [13–16], considering BSs locations fixed while treating user equipments (UEs) locations as random variables.

However, this model is likely to be inaccurate for heterogeneous networks consisting of small-cell deployments both in urban and suburban areas, where cell radius varies significantly and should be modeled as a random variable in itself. In a befitting direction, in [17,18], an analysis was presented by introducing an extra source of randomness, i.e., modeling the position of the BS as a homogeneous Poisson point processing (PPP), which will be used as a framework in our analysis. In addition, a tractable model for noncoherent joint transmission BS cooperation is established and closed-form for signal-to-interference-plus-noise ratio (SINR) distribution by considering a single UE is proposed in [19].

Regarding the second problem of mapping cluster size with latency, few works in literature so far have considered the impact of latency in multicell cooperation systems. The authors in [20,21] considered CSI feedback delay in their analysis for distributed antenna systems. Moreover, BH latency models for various BH topologies and technologies were only introduced in [22,23]; however, the performance analysis and proposed algorithms are based on a single cooperative cluster instead of a network composed of multiple clusters that may interfere to each other.

In this first part of the chapter, we consider a set of clusters in a cloud-based network sharing the same cloud resources, where each cluster is composed by a number of RRHs performing joint processing and operating as multiantenna BS. In our work, we consider both RRHs' and users' geometric location as random variables based on homogeneous PPP [17,18] as well as the effects of processing and CSI feedback delay. This approach is not only more generic but also more realistic considering the dynamic deployment nature of small cells in the future. The output SINR is derived in terms of the RRH and UE density in the presence of delay-caused channel mismatch. All of the parameters are converted into a function of cluster size to formulate the optimization problem.

As shown in Figure 8.1, we consider the downlink of a network comprising a large number of RRHs and we suppose that each RRH is connected with the cloud data center through fiber or other capacity unlimited BH links, whereas there is no direct physical link between RRHs. Therefore, each RRH in a cooperative cluster can only acquire a local CSI and global CSI is accumulated at the cloud by RRH feedback via the BH links. The precoding matrix calculations will be done in the cloud. However, there are two main options for the precoding implementation: (a) the implementing at the cloud and then forwarding the precoded I/Q signals to individual RRH for transmission; (b) cloud-assisted implementation at each individual RRH, i.e., the modulated I/Q signal (before precoding) and relevant precoding coefficients will be sent from the cloud to the each RRH and the rest of physical layer processing will take place in RRH. There are pros and cons for each architecture [24]. Here, we focus our investigation on case (a) only, which is a more popular cloud architecture.

To mitigate the expected delay under joint transmission operation (mainly due to CSI feedback and precoding matrix calculation), we need to divide the network into a set of clusters with each one consisting of reasonable number of cooperative RRHs as shown in Figure 8.1. In this case, each cluster only requires CSI between

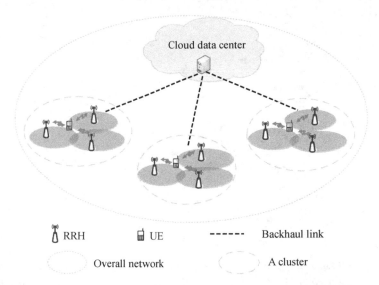

Figure 8.1   Considered C-RAN architecture and RRH clustering. (© [2017] IEEE. Reprinted, with permission, from [25])

the RRHs and the UEs within the cluster and the amount of feedback will drop as low dimension precoding requires less CSI [26]. Meanwhile, precoding can be performed in the cloud separately for each cluster; therefore, the computational complexity will be reduced. Consequently, the introduced delays could be mitigated and performance loss due to the channel mismatch will be reduced. On the other hand, extra inter-cluster interference due to small cluster size may diminish throughput in each cluster. Therefore, there must be an optimal cluster size to trade off delay and interference for maximizing system performance.

8.2.1.1   Clustering model

Consider a network served by a cloud in a d-dimensional space with a volume of V, where d could take the value of 1 or 2, corresponding to the linear or planar deployments, respectively. Here, we assume that the space is centrally symmetric, and therefore, the volume of the space could be generally expressed by $c_d R_t^d$, where c_d is the volume of the d-dimensional unit ball and R_t^d is the distance from the center to an arbitrary point on the bound of the d-dimensional space. Obviously, for linear and planar deployments, $c_1 = 1$ and $c_2 = \pi$, respectively.

Focusing on the planar deployment for demonstration purposes, we consider the network is divided into N_c same area and same shape clusters, i.e., each cluster has an area of $v = V/N_c$. One practical cluster shape in order to avoid adjacent cluster overlap and to maximize the density of packing in this 2D space is the hexagonal shape, as shown in Figure 8.2. However, as hexagonal boundary is relatively difficult to analyze, the hexagonal cluster can be replaced by an equivalent circular cluster

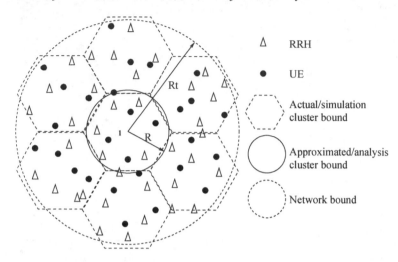

Figure 8.2   Cluster division for planar deployment with each cluster being shaped as a hexagon. (© [2017] IEEE. Reprinted, with permission, from [25])

having the same central point and the same area as the hexagon. The approximation is practical and accurate as the sum-rate contribution from the edge of a cluster is marginal (see Figure 8.2). In addition, we consider a network comprising of b tiers clusters, i.e., the number of clusters in a network can only be $1 + 3b(b + 1)$ [27], where b denotes the number of tiers. Figure 8.2 gives an example of one-tier network comprising seven clusters. Note that under the assumption of network bounding with a circular one, the network is not completely overlapped by the clusters at the edge of the network. However, larger number of tiers leads to a negligible model error. In addition, the approximation is made at cluster edges, which contributes marginal sum-rate to the network. Note that the above approximation for the planar deployment is not required for the linear case, where each cluster evenly occupies the same portion (length) of the network.

Next, we consider the active RRHs and UEs randomly scattered into the d-dimensional space with density ρ_R and ρ_U, respectively (see Figure 8.2). We assume that each RRH is equipped with M antenna and each UE is equipped with single antenna. Thus, the number of UEs $n_{U,i}$, RRH $n_{R,i}$ and transmission antennas $n_{A,i} = Mn_{R,i}$ in each cluster (for $i = 1, 2, \ldots, N_C$) are variables. As the UEs and RRHs are dropped based on the homogeneous PPP model, we can express the expected number of RRHs N_R, UEs N_U and total transmission antennas N_A in any cluster i with volume υ as [18]:

$$N_U = E(n_{U,i}) = \upsilon\rho_U, N_R = E(n_{R,i}) = \upsilon\rho_R, N_A = E(n_{A,i}) = E(Mn_{R,i}) = M\upsilon R \quad (8.1)$$

Apparently, the expected number of UEs and RRHs in the region outside of the considered cluster i is $(V - v)\rho_U$ and $(V - v)\rho_R$, respectively.

8.2.1.2 Channel model

By considering both path-loss and small-scale fast fading effects of the wireless access channel between RRH and UE, the baseband channel at time index t can be expressed as

$$\overline{\mathbf{h}}_{i,j,k}[t] = \mathbf{h}_{i,j,k}[t]\mathbf{G}_{i,j,k} \tag{8.2}$$

where $\overline{\mathbf{h}}_{i,j,k}[t] \in \mathbb{C}^{1 \times n_{A,j}}$ is the compound channel between the jth RRH and kth UE in the ith cluster. $\mathbf{G}_{i,j,k} = diag[g_{i,j,k,1}, g_{i,j,k,1}, \cdots g_{i,j,k,n_{A,j}}] \in \mathbb{R}^{n_{A,j} \times n_{A,j}}$ is a diagonal matrix with $g_{i,j,k,l}^2$ corresponding to the channel path-loss. $\mathbf{h}_{i,j,k}[t] \in \mathbb{C}^{1 \times n_{A,j}}$ is the corresponding fast fading factor of the channel with complex Gaussian distribution $\mathcal{CN}(0, 1)$.

The precoding matrix calculation is based on a delayed (outdated) version of the channel, at time $t - \Delta t$, which can be written as follows:

$$\overline{\mathbf{h}}_{i,j,k}[t - \Delta t] = \mathbf{h}_{i,j,k}[t - \Delta t]\mathbf{G}_{i,j,k} \tag{8.3}$$

where we have assumed that RRHs and UEs are essentially static and the path-loss $\mathbf{G}_{i,j,k}$ is kept as a constant during the Δt period. Here, we suppose that the channel is spatially uncorrelated but time correlated as follows [28]:

$$\mathbf{h}_{i,j,k}[t] = \mathbf{h}_{i,j,k}[t - \Delta t]\mathbf{\Lambda}_{i,j,k} + \mathbf{e}_{i,j,k}[t] \tag{8.4}$$

where $\mathbf{\Lambda}_{i,j,k}$ is a diagonal matrix defined as $\mathbf{\Lambda}_{i,j,k} = diag[\lambda_{i,j,k,1}, \lambda_{i,j,k,2}, \ldots, \lambda_{i,j,k,n_{A,j}}] \in \mathbb{R}^{n_{A,j} \times n_{A,j}}$ with $\lambda_{i,j,k,l} = \mathcal{J}_0(2\pi f_{D,i,j,k,l}\Delta t) < 1$ being the temporal correlation factor. $\mathcal{J}_0$ is the zero-order Bessel function of first kind, $f_{D,i,j,k,l}$ is the Doppler spread from the lth antenna that belongs to the jth cluster to kth UE of the ith cluster. As the cooperative RRHs in one cluster are assumed static, we will assume that $\lambda_{i,j,k,1} = \lambda_{i,j,k,2} = \cdots = \lambda_{i,j,k,n_{A,j}} = \lambda_{i,j,k}$ in the rest for brevity. Finally, $\mathbf{e}_{i,j,k} \in \mathbb{C}^{1 \times n_{A,j}}$ denotes the channel mismatch vector with each element being modeled as complex Gaussian distribution as $\mathcal{CN}(0, 1 - \lambda_{i,j,k}^2)$ [28].

Substituting (8.4) into (8.2) and considering (8.3), we obtain following relationship between the current and outdated channels:

$$\overline{\mathbf{h}}_{i,j,k}[t] = \lambda_{i,j,k}\mathbf{h}_{i,j,k}[t - \Delta t]\mathbf{G}_{i,j,k} + \mathbf{e}_{i,j,k}\mathbf{G}_{i,j,k} = \lambda_{i,j,k}\overline{\mathbf{h}}_{i,j,k}[t - \Delta t] + \mathbf{e}_{i,j,k}\mathbf{G}_{i,j,k} \tag{8.5}$$

It is apparent that as the calculation of the precoding matrix is based on the outdated channel $\overline{\mathbf{h}}_{i,j,k}[t - \Delta t]$ instead of the actual channel $\overline{\mathbf{h}}_{i,j,k}[t]$, the performance of the joint transmission process will be affected. Note that in the following, for simplification, we define:

$$\widehat{\mathbf{h}}_{i,j,k}[t] = \overline{\mathbf{h}}_{i,j,k}[t - \Delta t] \tag{8.6}$$

and we omit the time index for brevity.

Thus, the received signal of the kth UE in the ith cluster can be written as follows:

$$y_{i,k} = \sqrt{\gamma_i}\overline{\mathbf{h}}_{i,i,k}\mathbf{w}_{i,k}x_{i,k} + \sqrt{\gamma_i}\sum_{l=1,l \neq k}^{n_{U,i}}\overline{\mathbf{h}}_{i,i,k}\mathbf{w}_{i,l}x_{i,l} + \sum_{j=1,j \neq i}^{N_C}\sqrt{\gamma_j}\sum_{l=1}^{n_{U,i}}\overline{\mathbf{h}}_{i,j,k}\mathbf{w}_{j,l}x_{j,l} + n_{i,k} \tag{8.7}$$

where $n_{i,k}$ is the Gaussian noise with distribution $\mathcal{CN}(0, \sigma_{i,k}^2)$, and $\mathbf{w}_{i,k} \in \mathbb{C}^{n_{A,i} \times 1}$ and $x_{i,k}$ stand for the precoding vector and transmit signal for the kth UE in the ith cluster, respectively. $x_{i,k}$ and $x_{j,l}$ are assumed uncorrelated for $(i,j) \neq (k,l)$ (for $i, j = 1, 2, \ldots, N_c$ and $k, l = 1, 2, \ldots, n_{U,i}$) and being zero mean and unit power variables, i.e., $\mathcal{E}\{|x_{i,j}|^2\} = 1$. Finally, γ_i denotes the transmitting power of the signal $x_{i,k}$.

Note that here we have supposed power is evenly allocated to each UE in a cooperative cluster. The second and third items are intra-cluster and inter-cluster interference, respectively.

By substituting (8.5) into (8.7), the desired signal power, P_x, and the interference power, P_I, for the kth UE in the ith cluster can be expressed as:

$$P_x = \gamma_i |\lambda_{i,i,j} \mathbf{h}_{i,i,k} \mathbf{G}_{i,i,k} \mathbf{w}_{i,k} x_{i,k}|^2 + \gamma_i |\mathbf{e}_{i,i,k} \mathbf{G}_{i,i,k} \mathbf{w}_{i,k} x_{i,k}|^2$$

$$P_I = \left| \sqrt{\gamma_i} \sum_{l=1, l \neq k}^{n_{U,i}} (\mathbf{h}_{i,i,k} \lambda_{i,i,j} \mathbf{h}_{i,i,k} \mathbf{G}_{i,i,k} + \mathbf{e}_{i,i,k} \mathbf{G}_{i,i,k}) \mathbf{w}_{i,l} x_{i,l} \right|^2 + \left| \sqrt{\gamma_i} \sum_{j=1, j \neq i}^{N_c} \sum_{l=1}^{N_{U,j}} \overline{\mathbf{h}}_{i,j,k} \mathbf{w}_{j,l} x_{j,l} \right|^2$$

$$(8.8)$$

where we have used the fact that $\mathbf{e}_{i,i,k}$ is independent of the channel vector $\mathbf{h}_{i,i,k}$, and $\mathbf{h}_{i,j,k}$ is independent of $\mathbf{h}_{i,j,k}$ when $i \neq j$. Thus, the expectation of the output SINR in the presence of delay impact can be written as follows:

$$\overline{\text{SINR}}_{i,k} = \mathcal{E}\left\{ \frac{P_x}{P_I + \sigma_{i,k}^2} \right\} \tag{8.9}$$

8.2.2 Desired signal and interference power

8.2.2.1 Desired signal power

The SINR expression in (8.9) is very difficult to analyze theoretically as it is a compound function of multiple variables including large-scale and fast fading of the channels, delay and precoding coefficients as well as multiple random deployed RRHs and users. As a solution, we will first derive the desired signal and interference power for single user case without considering delay impact and any specific precoding algorithm. In other words, we focus on user k and set $\mathbf{e}_{i,i,k} = \mathbf{0}$ and $\mathbf{w}_{i,l} = \mathbf{1}, \forall i, l$, in (8.9). However, we will consider all of these factors in the next section and we will demonstrate that the fast fading, and specific precoding coefficients along with delay-caused error, can be treated independently.

When the UE of interest is assumed to be located at the center of the cluster, the received signal power can be given by [18]:

$$\overline{P}_x = M \rho_R c_d d \int_0^R g(r) r^{d-1} \, dr = 2\pi M \rho_R \int_0^R r^{-\eta} r \, dr \quad \text{(when } d = 2) \tag{8.10}$$

Note that in this work, we consider that path-loss coefficients, $g(r)$, are derived from the following model:

$$g(r) = \begin{cases} R_0^{-\eta} & \text{if } r \leq R_0 \\ r^{-\eta} & \text{if } r > R_0 \end{cases} \tag{8.11}$$

where η is the path-loss exponent and R_0 is a minimum distance between the UE and RRH to bound the path-loss [18], i.e., the path-loss during the distance $[0, R_0]$ is assumed constant.

Equation (8.10) gives a general calculation method of the desired signal power for a UE located at the center of the network. In essence, this stands for the best case scenario on average as the UE receives the largest power from the distributed RRHs and smallest interference from outside of the cluster. However, the UE could be located at arbitrary point in the cluster, i.e., the location of UE is another random variable. To solve the above problem, we will take a step-by-step approach. First, the desired signal and interference power will be derived considering the UE located at an arbitrary but fixed point in a cluster. Then, by treating the UE location as a random variable, the generic expressions will be derived as follows:

$$\overline{P}_x = \overline{P}_{x1} + \overline{P}_{x2} \tag{8.12}$$

where [25]

$$\overline{P}_{x1} = \int_0^R \frac{2a}{R^2} \overline{P}_{x1}(a)da = \int_0^{R-R_0} \frac{2a}{R^2} \overline{P}_{x1}(a)da + \int_{R-R_0}^R \frac{2a}{R^2} \overline{P}_{x1}(a)da$$

$$= \frac{\pi M \rho_R (R-R_0)^2 R_0^2}{R^2} \left(1 - \frac{2R^{-\eta}}{2-\eta}\right) + \frac{2\pi M \rho_R [R_0^{3-\eta} - R^{3-\eta}](R-R_0)^2 R_0}{R(2-\eta)(3-\eta)}$$

$$+ \frac{2\pi M \rho_R (-R_0)^{4-\eta} - (R)^{4-\eta}}{R^2(2-\eta)(4-\eta)} \pi M \rho_R R_0^2 \left[1 - \frac{R - R_0^2}{R^2}\right] \tag{8.13}$$

and

$$\overline{P}_{x2}(a) = 2M\rho_R \int_0^R \int_{R-a}^{R+a} g(r)r \arccos \frac{r^2 + a^2 - R^2}{2ar} drda \tag{8.14}$$

Unfortunately, there is no close-form solution for (8.14) when η takes most of the possible values. For that reason, numerical methods are adopted to verify the effectiveness of the derivation by simulations.

8.2.2.2   Interference power

Unlike the desired signal power, the expression of the interference power cannot be derived straightforwardly; the integral region is irregular and also depends on the location of the cluster within the network. However, it can be obtained indirectly by deriving the received power from the whole network and subtracting the desired signal power part. For a UE at a random location inside the cluster, the interference power contributed by the area outside of the cluster is

$$\overline{P}_I = \overline{P}_{tot} + \overline{P}_x \tag{8.15}$$

Equation (8.15) is very accurate approximation when the cluster is in the center of the network and the interfering area is much larger than the cluster size. In most cases, it is a practical assumption as the defined network with limited radius R_t will be surrounded by other networks and hence receive interference from them.

8.2.3 Cluster size optimization

In this section, we first derive the expectation of the SINR in the presence of small-scale fast fading and the delay for two representative precoders: maximum ratio combining (MRT) and ZF, respectively. The CSI latency model in terms of cluster size and cloud/RRH configuration will be built in the next section. Then, the optimization problem in the criterion of maximizing ergodic sum-rate in terms of cluster size will be formulated.

8.2.3.1 Output SINR

Let's define the precoding matrix and observed channel matrix for the ith cluster as $\mathbf{W}_i = [\mathbf{w}_{i,1}, \mathbf{w}_{i,2}, \ldots, \mathbf{w}_{i,n_{U,i}}]$ and $\widehat{\mathbf{H}}_i = [\widehat{\mathbf{h}}_{i,i,1}, \widehat{\mathbf{h}}_{i,i,2}, \ldots, \widehat{\mathbf{h}}_{i,i,n_{A,i}}]$, respectively. Then, the MRT and ZF precoders can be expressed as follows [11]:

$$\mathbf{w}_{i,k}^{\text{MRT}} = \frac{\widehat{\mathbf{h}}_{i,i,k}^{H}}{\|\widehat{\mathbf{h}}_{i,i,k}^{H}\|} \tag{8.16}$$

$$\mathbf{w}_{i,k}^{\text{ZF}} = \frac{\widehat{\mathbf{H}}_i^H (\widehat{\mathbf{H}}_i \widehat{\mathbf{H}}_i^H)^{-1} \mathbf{1}_k}{\|\widehat{\mathbf{H}}_i^H (\widehat{\mathbf{H}}_i \widehat{\mathbf{H}}_i^H)^{-1} \mathbf{1}_k\|} \tag{8.17}$$

where $\mathbf{1}_k = [0, \ldots, 0, 1, 0, \ldots]^T$ refers to a vector with its kth element being 1 and all other elements being 0. Thus, $\widehat{\mathbf{H}}_i^H (\widehat{\mathbf{H}}_i \widehat{\mathbf{H}}_i^H)^{-1} \mathbf{1}_k$ refers to the kth column of matrix $\widehat{\mathbf{H}}_i^H (\widehat{\mathbf{H}}_i \widehat{\mathbf{H}}_i^H)^{-1}$.

8.2.3.2 Output SINR with MRT precoder

By substituting (8.16) into (8.9) and considering the random vector $\mathbf{e}_{i,j,k}$ independent of $\mathbf{h}_{i,j,k}$, along with assumption that each cluster consumes the same power for transmission, i.e., $\gamma_1 = \gamma_2 = \ldots = \gamma_{N_c}$, we get [25,29]:

$$\overline{\text{SINR}}_{i,k}^{\text{MRT}} \approx \left[\lambda_{i,i,k}^2 + \frac{1 - \lambda_{i,i,k}^2}{MN_R} \right] \overline{P}_x \overline{P}_{I,\exp}^{\text{MRT}} \tag{8.18}$$

where

$$\overline{P}_{I,\exp}^{\text{MRT}} = \sum_{l=1}^{N_U-2} (-1)^{N_U-2-l} \sum_{n=0}^{l-1} \frac{P_{I,2}^{N_U-2-l+n} \xi_1^{N_U-1-l+n}}{l(N_U-2-l)n!}$$

$$+ \int_0^\infty e^{P_{I,2}\xi_1} (-P_{I,2})^{N_U-2} \frac{e^{(P_{I,2}+u)\xi_1}}{(u+P_{I,2})(N_U-2)!} du \tag{8.19}$$

with $\xi_1 = \frac{MN_R}{\gamma_1 \overline{P}_x}$ and $P_{I,2} = \gamma_1 \frac{N_U}{MN_R} \overline{P}_I + \sigma_{i,k}^2$

8.2.3.3 Output SINR with ZF precoder

Similarly, considering ZF precoder, the output SINR in the presence of latency can be expressed as [25]:

$$\overline{\text{SINR}}_{i,k}^{\text{ZF}} \approx \left[\lambda_{i,i,k}^2 + \frac{1 - N_U \lambda_{i,i,k}^2}{MN_R} \right] \overline{P}_x \overline{P}_{I,\exp}^{\text{ZF}} \tag{8.20}$$

where $\overline{P}_{I,\exp}^{\text{ZF}}$ has the same expression as $\overline{P}_{I,\exp}^{\text{MRT}}$ except replacing ξ_1 by ξ_2 in (8.19) with $\xi_2 = MN_R/\gamma_1(1 - \lambda_{i,i,1}^2)$.

Comparing the first part (before the multiplication sign) of (8.18) and (8.20), we observe that, when $N_U > 1$, MTR-based desired signal power will be always larger than the ZF-based one. This is due to the fact that ZF uses the spatial (i.e., antenna) degrees of freedom (DoF) to eliminate interference, whereas MRT explores all DoF to maximize the desired signal power. Comparing the second part (after the multiplication sign) of (8.18) and (8.20), the only difference is that an extra term $(1 - \lambda_{i,i,1}^2)$ is multiplied with ξ_1 in the case of ZF precoding. This term is essentially the residual intra-cluster interference due to the delay caused mismatch; larger delay leads to smaller $\lambda_{i,i,1}^2$ and higher the residual intra-cluster interference. Note that in case of MRT precoding, this second part is not affected by the introduced delay.

To express the output SINR as a function of the cluster size R, taking planar case as an example, we can substitute $v = 2\pi R^2$, $V = 2\pi R_t^2$, $\lambda = \mathcal{J}_0(2\pi f_D)$ and (8.1) into (8.18) and (8.20). Then, the output SINR for the random distributed RRHs and UEs in the presence of delay for MRT precoding could be expressed as follows:

$$\overline{\text{SINR}}_{i,k}^{\text{MRT}} \approx \left[\lambda(R^2) + \frac{1 - \lambda(R^2)}{2\pi R^2 M \rho_R} \right] \overline{P}_x \overline{P}_{I,\exp}^{\text{MRT}}(R) \tag{8.21}$$

and for ZF precoding

$$\overline{\text{SINR}}_{i,k}^{\text{ZF}} \approx \left[\frac{2\pi R^2(M\rho_R - \rho_U) + 1}{2\pi R^2 M \rho_R} \right] \overline{P}_x \overline{P}_{I,\exp}^{\text{ZF}}(R) \tag{8.22}$$

where the temporal correlation factor λ is expressed as a function of R, and its subscripts are omitted for brevity.

8.2.3.4 Delay model

In general, the total delay of the precoding process is caused by several factors such as the pilot estimation and processing delay at the UE, propagation delay from UE to RRH and from RRH to cloud, CSI feedback (and scheduling) delay, RRH processing delay, cloud data center processing delay and BH latency. Thus, total delay can be generally modeled as follows:

$$\Delta t = \varpi_1 \Delta t_{\text{chan-est}} + \varpi_2(\Delta t_{fb} + \Delta t_{\text{prop-tot}}) + \Delta t_{\text{process-cloud}}$$
$$+ \Delta t_{\text{process-RRH}} + \varpi_3 \Delta t_{\text{BH}} \tag{8.23}$$

where

$$\varpi_1 = MN_R = vM\rho_R; \quad \varpi_2 = \frac{MN_U N_R N_C}{q_{fb}} = \frac{VvM\rho_R}{q_{fb}} \tag{8.24}$$

The physical meaning of each item in (8.23) is explained one-by-one in the following subsections.

The first item in (8.23), $\Delta t_{\text{chan–est}}$, denotes the channel estimation delay at the UE and ϖ_1 stands for the number of channel coefficients to be estimated for one UE. Apparently, the more channels to be estimated, the larger the delay is likely to be.

Δt_{fb} and $\Delta t_{\text{prop–tot}}$ in (8.23) denote the average per channel coefficient feedback delay and total propagation delay, respectively. $MN_U N_R N_C$ stands for the total number of channel coefficients and q_{fb} is a factor denoting how many channels can be fed back each time. ϖ_2 stands for the total number of times CSI is to be fed back for the whole network.

Assuming that CSI feedback from UE to RRH has a capacity of C_{fb} and considering that each CSI is quantized to B_1 bits, the feedback delay can be written as $\Delta t_{fb} = B_1/q_{fb}$.

Moreover, the total propagation delay can be expressed as $\Delta t_{\text{prop–tot}} = 2(s_{u2r} + s_{r2c})/cq_{fb}$, where c is the speed of light, and s_{u2r} and s_{r2c} are the distances from UE to RRH and from RRH to cloud, respectively.

$\Delta t_{\text{process–cloud}}$ in (8.23) denotes the cloud processing delay, which is composed of two factors and can be written as $\Delta t_{\text{process–cloud}} = \Delta t_{Tx1} + \Delta t_{\text{precoder–cal}}$. Δt_{Tx1} is attributed to the (part of) baseband processing (such as coding, modulation, precoding, IFFT etc.) depending on the transmission chain functionality split between cloud and RRH [1,2]. In general, the total delay caused by baseband processing at the transmitter $\Delta t_{Tx} = \Delta t_{Tx1} + \Delta t_{Tx2}$ is assumed to be constant, where Δt_{Tx2} refers to the respective delay at RRH.

$\Delta t_{\text{process–cal}}$ stands for the precoder calculation delay, which is a dominating factor when the cluster size is relatively large and the available computational resource is limited. Note also that different precoding algorithms lead to dramatically different computational complexity. For example, ZF has significant larger complexity than MRT precoding. Taking ZF as an example, the delay caused by the precoding matrix calculation can be written as:

$$\Delta t_{\text{precoder–cal}} = \frac{(K_{\text{add}}^{\text{ZF}} + \zeta_2 K_{\text{multi}}^{\text{ZF}})V}{vC_{\text{com}}q_c} \tag{8.25}$$

where $K_{\text{multi}}^{\text{ZF}}$ and $K_{\text{add}}^{\text{ZF}}$ denote the required real-time operations of multiplication and addition, respectively; ζ_2 is the equivalent addition operation times for each multiplication; C_{com} denotes the cloud computational capability; and q_c is the resource division factor (as the computational resources are likely shared by multitasks, and assuming a uniform distribution of resources, $1/q_c$ of the total available resources will be allocated to the precoding matrix calculation). Thus, $C_{\text{com}}q_c$ available computational capability will be allocated in total to the precoding matrix calculation. Note also that q_c could be set smaller than 1 corresponding to the case where multiple processors could contribute to the computation in parallel. The values of $K_{\text{multi}}^{\text{ZF}}$ and $K_{\text{add}}^{\text{ZF}}$ depend on the number of users and RRHs in the cluster and can be calculated as:

$$K_{\text{multi}}^{\text{ZF}} = 8N_{UE}^2 MN_R + \mathcal{O}(4N_U^3) + 8N_U MN_R = 8v^3 \rho_{UE}^2 M\rho_R + \mathcal{O}(4v^3 \rho_U^3)$$
$$+ 2v^2 \rho_U M\rho_R \tag{8.26}$$

and

$$K_{\text{add}}^{\text{ZF}} = 8N_{UE}^2 MN_R - 2N_{UE}^2 + \mathcal{O}(4N_U^3) - 2N_U + 2\zeta_1 N_U MN_R$$
$$= 8v^3 \rho_{UE}^2 M\rho_R - 2v^2 \rho_U + \mathcal{O}(4v^3 \rho_U^3) + 2v^2 \rho_U M\rho_R \tag{8.27}$$

where the term $\mathcal{O}(4N_U^3)$ arises from the matrix inversion process and its complexity depends on the specific implemented algorithm (its typical value takes 8/3 [30]). Moreover, ζ_1 factor indicates how much time the multiplication process consumes compared to the addition process.

$\Delta t_{\text{process–RRH}}$ in (8.23) denotes the RRH processing delay which also comprises of two parts, i.e., $\Delta t_{\text{process–RRH}} = \Delta t_{Tx2} + \Delta t_{\text{CSI–}fw}$. The first item, Δt_{Tx2}, as already mentioned before is attributed to (part) of the baseband and RF implementation at the RRH. The second item, $\Delta t_{\text{CSI–}fw}$, stands for the delay due to CSI feedback from RRHs to the cloud in the uplink.

Apparently, the total processing delay (both at the cloud and RRH) can be given by $\Delta t_{Tx} + \Delta t_{\text{CSI–}fw} + \Delta t_{\text{precoder–cal}}$.

The last item contains the BH latency given by the BH latency Δt_{BH} per hop multiplied with the number of BH hops ϖ_3 from the cloud data center to the small cells. The values for Δt_{BH} of various BH technologies can be found in [31–33].

8.2.4 Discussion on backhaul load

Though has not been considered into the system model and cluster size optimizations, BH load is one of the key factors that may significantly affect the optimization results. The required BH load depends on the baseband signal processing chain functionality split between RRHs and the data center. By utilizing the data center's computational capability, more functions (i.e., lower PHY processing layers) can be moved to the cloud to simplify the RRH processing; in addition, it can also significantly reduce the required data rate between RRH and data center. However, it also leads to a smaller joint processing gain. For more information on the functionality split for CRAN system, please refer to [2,34] for more details.

8.2.5 Ergodic sum-rate and optimization formulation

The optimization problem for maximizing the ergodic sum-rate of the network in terms of cluster size can be $\text{SINR}_{i,k}$ expressed as:

$$\max_R C_s \quad \text{subject to} \quad 0 < R \le R_t \tag{8.28}$$

where ergodic sum-rate C_s in our system model can be given by:

$$C_s = \mathcal{E}\left\{\sum_{i=1}^{N_c}\sum_{k=1}^{N_{U,i}}\log_2(1 + \text{SINR}_{i,k})\right\} = N_c\mathcal{E}\left\{\sum_{k=1}^{N_{U,i}}\log_2(1 + \text{SINR}_{i,k})\right\}$$

$$= \rho_U V\mathcal{E}\{\log_2(1 + \text{SINR}_{i,k})\} \leq \rho_U V\log_2(1 + \mathcal{E}\{\text{SINR}_{i,k}\}) \qquad (8.29)$$

where $\text{SINR}_{i,k}$ is instantaneous SINR of the kth UE at the ith cluster. The second step of (8.29) is based on the assumption that the cellular network is surrounded by other nonoverlapping and same configured networks. In that case, each cluster can be effectively considered at the center of a network, therefore, equally contributing to the sum-rate. The third step in (8.29) holds when each UE is randomly and independently distributed within the network, where the total number of the UEs can be given in terms of the UE density and network volume as $\mathcal{E}\{N_c N_U\} = \rho_U V$. As it is very difficult to solve the ergodic sum-rate directly due to the expectation operation implemented outside of the logarithm, the well-known Jensen's inequality is used in the fourth step of (8.29) to obtain an upper bound. Although the ergodic sum-rate upper bound can be a loose bound in some cases, we will show that the optimal cluster size obtained from the proposed analytical framework match very well with the simulation results.

Thus, by considering MRT precoding and writing $\mathcal{E}\{\text{SINR}_{i,k}\} = \overline{\text{SINR}}_{i,k}^{\text{MRT}}$, the optimization problem in (8.28) is approximately equivalent to

$$\max_R \overline{\text{SINR}}_{i,k}^{\text{MRT}} \quad \text{subject to} \quad 0 < R \leq R_t \qquad (8.30)$$

and similarly for ZF precoding, we have

$$\max_R \overline{\text{SINR}}_{i,k}^{\text{ZF}} \quad \text{subject to} \quad 0 < R \leq R_t \qquad (8.31)$$

Note that, as the cost function is complex in terms of R, it is difficult to obtain the closed-form optimal solution, and in the next section, the numerical methods are adopted to verify its effectiveness.

8.2.6 Simulation results

In this section, we use Monte-Carlo simulations to investigate the proposed clustering optimization problem with ZF and MRT precoding algorithms for linear and planar dense small-cell deployments. 1000 small cells and 1000 active UEs are uniformly distributed in (a) a circular network area (planar deployment) with radius $R_t = 500$ m and (b) a linear network segment (linear deployment) of length $R_t = 1000$ m. We assume the number of antennas at each RRH $M = 2$.

For planar deployment, in order to approximate a circle bounded network we consider clusters formed by 1 to 7 tiers of cells, i.e., by considering that tier-1 consists of 7 cell, tier-2 of 19 cells and so on; cluster size will take values from the range set [7, 19, 37, 61, 91, 127, 169]. The input signal-to-noise power is set to 30 dB. The path-loss exponent is kept $\eta = 2$ and Rayleigh fast fading is considered to model channels between RRHs and UEs, as given by (8.3). The temporal correlation of the channel is modeled by (8.4) with Doppler spread $f_D = 10$ Hz for all links. In either linear or planar deployment, $R_0 = 10$ m. Without loss of generality, we only consider the performance of the UEs in the central cluster and assume interference from outside of the network to these UEs is negligible.

8.2.6.1 $\overline{P}_x, \overline{P}_I$ and output SINR in the absence of latency

To investigate the effectiveness of our theoretical analysis (i.e., proposed system, channel and clustering model), we first evaluate $\overline{P}_x$, $\overline{P}_I$ and output SINR in the absence of latency, i.e., $\Delta t = 0$ and $\lambda^2 = 1$.

$\overline{P}_x$ and $\overline{P}_I$ are evaluated using (8.12) and (8.15) for planar deployment and compared with simulation results in Figure 8.3, for different number of clusters within the network. It can be seen that the analytical results for both the desired signal and interference power match the simulation results perfectly for both planar and linear deployments. As expected, the desired signal power reduces with cluster size (i.e., larger number of clusters in the network) as less number of cooperating small cells contribute to the desired power. On the other hand, the interference power becomes larger as the number of clusters increases as the total number of interfering RRH outside the cluster is increased.

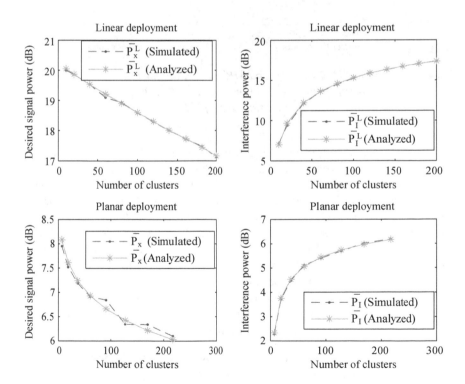

Figure 8.3 Desired signal and interference power for planar and linear small-cell deployments in the absence of delay. (© [2017] IEEE. Reprinted, with permission, from [25])

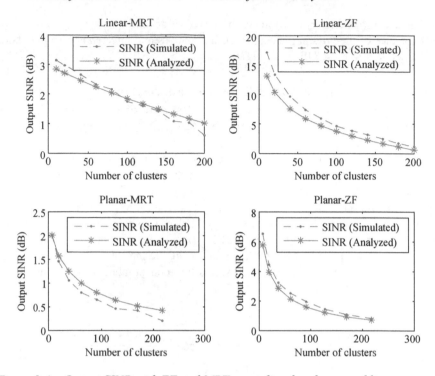

*Figure 8.4   Output SINR with ZF and MRT precoding for planar and linear
small-cell deployments in the absence of delay. (© [2017] IEEE.
Reprinted, with permission, from [25])*

Furthermore, the output SINR is evaluated using (8.18) and (8.20) for MRT and ZF precoding, respectively, and compared with simulation results in Figure 8.4. We observe that all four graphs (Linear-MRT, Linear-ZF, Planar-MRT and Planar-ZF) show good consistency between analytical and simulation results. For all cases, the output SINR decreases with increasing number of clusters due to the fact that a smaller cluster tends to obtain less desired signal power, whereas more interference is caused from outside the cluster. Note that small gaps can be observed between theoretical and simulated results due to the assumptions used in clustering modeling.

In the following, we only focus on the more complex planar deployment due to similar observed behavior of planar and linear case in terms of output SINR. Nevertheless, analogous results can be provided for the linear case as well, following the respective analytical expressions and simulation model.

8.2.6.2   Performance optimization in the presence of latency

To evaluate performance in the presence of latency, we first set some practical values for the parameters in the delay model. To this end, we consider the worst-case of Tx processing time defined in 3GPP, i.e., $\Delta t_{Tx} = 2.3$ ms; the channel estimation and RRH processing delay are set to zero as they are not as sizeable factors as

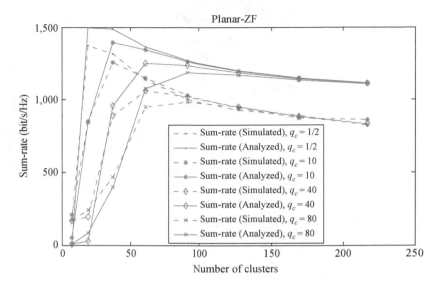

Figure 8.5 Sum-rate in the presence of delay for ZF-based planar deployment under various available computational resource division factors q_c. (© [2017] IEEE. Reprinted, with permission, from [25])

the processing or feedback delay. The factors for multiplication and division over addition are set to the typical values of $\zeta_1 = 1$ and $\zeta_2 = 10$, respectively. The average distance between UE and RRH plus the distance between RRH and cloud data center is considered to be $s_{u2r} + s_{r2c} = 1000$ m. The cloud is assumed to contain an Intel Xeon Processor E5-2680 with processing capacity of 1.73×10^{11} double-precision floating-point operations per second (DP-FLOPS). Assuming the addition operation is double-precision floating (64-bit), the available processing capacity for precoding calculation becomes $C_{com} = 1.73 \times 10^{11}/q_c$ addition operations per second. Furthermore, the computational resource division factor q_c is set to be 10, unless specified otherwise. The feedback capacity of the BH link from RRHs to cloud is set to $C_{fb} = 10^7$ bips and, unless specified otherwise, we set $q_{fb} = 1$, i.e., feedback of one channel coefficient each time. We assume that each small cell is connected to the cloud data center by one hop Dark fiber with latency $\Delta t_{FB} = 10$ μs/km per hop as defined by [31].

Figure 8.5 investigates the delay impact on sum-rate of the ZF-based algorithm in terms of cluster size. Various cases of cloud processing capability are considered with the computational load changing from 2, 1/10, 1/40 to 1/80 (i.e., q_c changing from 1/2, 10, 40 to 80). We observe that the analytical results roughly match the corresponding simulation results. More importantly, the peak points denoting the optimal cluster size are strictly overlapping with each other, for any specific configuration; thus, the optimal cluster size evaluation is not affected by the approximation in (8.29) where Jensen's inequality and the upper bound of the sum-rate have been considered

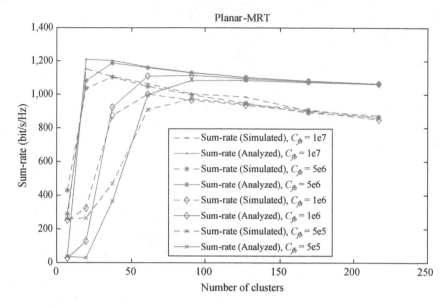

Figure 8.6   Sum-rate in the presence of delay for MRT-based planar deployment under different feedback capacity C_{fb}. (© [2017] IEEE. Reprinted, with permission, from [25])

(obviously, the theoretic sum-rate for each q_c is generally higher than the simulated results). We also note that the optimal cluster size decreases (i.e., optimal number of clusters increases) as the computational capability factor becomes larger. This is due to the fact that when q_c increases, less computational resources become available; therefore, smaller cluster size is needed to keep the delay-caused channel mismatch at low levels.

Figure 8.6 illustrates the impact of feedback delay on the sum-rate for various cluster sizes under MRT-based precoding, where the precoding matrix calculation delay is negligible. Compared to the results shown in regarding the processing delay, the curve slopes are more flat for increasing number of clusters. This is due to the fact that the feedback-caused latency is only in the second order of the number of cooperative antennas; on the other hand, latency caused by the processing capability is in the cubic order of the number of cooperative antennas; thus, decreasing cluster size will lead to faster reduction of latency.

8.3   Joint routing and backhaul scheduling

8.3.1   *System model*

In the multihop wireless BH consisting of small cells as can be seen in Figure 8.7, the routing information is required at the central unit (CU) (also called cloud data center)

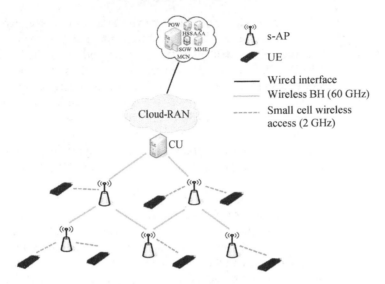

*Figure 8.7   Backhaul link scheduling and QoS-aware flow forwarding. (© [2017]
IEEE. Reprinted, with permission, from [35])*

from network layer so as to select paths (in longer time-scale) based on traffic load, spectral efficiency and other criteria. Nevertheless, this happens in longer time-span without taking into account the channel condition. Here, one of the major problems in mmW radio is the heavy reliance on LoS condition, which, in the case of dense urban small cells, could be an issue of major importance, significantly degrading the small-cell's network performance.

This chapter also addresses the problem of joint path selection and BH link scheduling in a dense small-cell network assuming 60 GHz multihop BH, coordinated by a local Cloud-RAN CU. Our objective is to optimally select the BH links and paths to be activated in time domain to enhance the cell throughput and to minimize the total delay, considering the BH channel conditions and the QoS requirements. This problem is formulated as a generalized vehicle routing problem and decoupled into two subproblems: the channel-aware path selection and the queue-aware link scheduling. The path selection subproblem is solved optimally using Branch-and-Cut exact algorithm for a given number of paths and per-link delay bounds. Subsequently, the scheduling subproblem is solved using back-pressure scheduling.

From the radio resource management (RRM) perspective, the ultra-dense heterogeneous networks with multiple small cells can be deployed under the macrocell umbrella, wireless backhauling is a cost efficient way to meet the high capacity and coverage demands, as expected in 5G radio access networks and to efficiently utilize spectrum in dynamic changing environments. Wireless backhauling for small cells and relay cells can introduce several challenges when it comes to RRM between the BH and the access links. Here, coupling between the two can limit the multiuser diversity gain of radio resource allocation in the access link. In case of TDD between the transmission links, each packet would be received after two consecutive transmissions, i.e., on BH link to small-cell station and from small-cell station to the end user (access link). Therefore, the efficiency of system depends upon the balance of resources between the two links for each small cell. This requires efficient resource partitioning on BH or access links and rate balancing strategies between the two.

The system consists of a CU, which serves as a controller and traffic aggregator for the dense small-cell network. This small-cell network encloses $l = 1, 2, \ldots, L$ s-APs, equipped with antennas for the access (s-AP-to-user), as well as directional antennas for the small-cell BH (which operate in higher frequencies). The latter is essential to exchange signaling and data with the CU (or other s-APs), using wireless BH (in mm-Wave band).

Let $G(V, E)$ be the graph consisting of a set of V nodes (s-APs) and a set of E edges. An edge $e \in E$ is a connection between any two nodes $v1, v2 \in V$. The presence of an edge $e \in E$ indicates that data can be exchanged between $v1$ and $v2$. We assume m links in the network, where the links are considered un-directional. Here, we also introduce a set of M demands where the mth demand originates from the source node s_m and terminates to the destination node d_m with a required rate r_m and maximum delay τ_m. Each link e has a desired data rate which corresponds to the summation of all the demands traversing it. We define the load of link e due to demand m as $l(e, m)$. Hence, the total load of this link is defined as follows:

$$l(e) = \sum_{m \in M} l(e, m) \tag{8.32}$$

Another important parameter is the link capacity, defined as $c(e) \; \forall e \in E$. Here, high directional antennas can be used to compensate for the high path attenuation. In this case, interference by other links is assumed to be negligible due to the high directivity of the antennas and the half duplex constraint (nodes only transmit or receive per time-instance). Hence, the channel is only affected by the path attenuation. It is high probable that we may have LoS between s-APs; however there is the chance that LoS is not available for an unplanned dense s-AP deployment.

As per the following equation, c_e can be computed using the LoS probability P_l, where c_e^{LoS} is the link capacity having LoS and c_e^{nLoS} is the case when we have non-LoS. Here, we introduce a Bernoulli random variable $I(x)$ based on the probability of LoS $(x = P_l)$ [36]:

$$c_e = I(P_l)c_e^{LoS} + (1 - I(P_l))c_e^{nLoS} \tag{8.33}$$

Both c_e^{LoS} and c_e^{nLoS} are computed using Shannon capacity formula. For the capacity computation, the 60-GHz path-loss models in LoS and nLoS are as follows [37]:

$$PL(d) = \begin{cases} A_{LoS} + 20 \log_{10}(f) + 10 n_{LoS} \log_{10}(d) \\ A_{nLoS} + 20 \log_{10}(f) + 10 n_{nLoS} \log_{10}(d) + X_\sigma \end{cases} \quad \text{or} \tag{8.34}$$

where f is the carrier frequency in GHz, d the distance in meters, $n_{LoS/nLoS}$ accounts for the path loss coefficient (typically 2 for LoS and 0.6 for nLoS). Moreover, A_{LoS}

and A_{nLoS} are the antenna-specific parameters associated with the beamwidth for LoS and nLoS, respectively (typically 32.5 dB for LoS and 51.5 dB for nLoS).

Using the definitions of the link capacity and the link load, we can introduce a new parameter which captures the number of timeslots required for a link to satisfy its demand:

$$f_e = \left\lceil \frac{\sum\limits_{m \in M} l(e, m)}{c_e} \right\rceil \tag{8.35}$$

where $\lceil x \rceil$ accounts for the ceiling function, i.e., the least integer that is greater than or equal to x. In other words, f_e captures how many timeslots are required for the traffic to be carried on each link, so as to meet the data rate requirements. The worse the channel conditions, the higher the number of timeslots needed to reserve one BH link.

Another key parameter, which is going to be used for the scheduling part of the generic formulation presented in III, is the set of all bipartite subgraphs of the graph $G(V, E)$, denoted as $\boldsymbol{S}$. Each of these S_i subgraphs represents a combination of link activations (one set of the bipartite graph is the transmitter nodes and the other set is the receiver nodes). Each of these subgraphs is associated with a weighting factor w_{Si} which represents the fraction of time that this combination of active links is live. Below, we illustrate a simple example of two different combinations with 6 random nodes. Note that the total number of these combinations is $2^{|V|} - 2$. We also define a binary indicator variable $1_{e,S}$ which is 1 if the edge e is included in subgraph S_i (otherwise will be 0).

8.3.2 Problem formulation

The joint routing and scheduling problem is formulated as a capacitated vehicle routing problem, where a central depot allocates passengers to vehicles and assigns paths that reach their destinations with the minimum time [38]. In similar manner, a CU centrally allocates path according to each small-cell's traffic requirements and BH channel constraints. The physical interpretation of a path resembles a tunnel, which encloses a group of nodes through which the traffic is forwarded by a variable number of hops. Each path can encapsulate multiple flows with different destinations and characteristics. The objective is how to select the best links and paths to be used from the CU to each small cell in order to satisfy its data rate demand and delay constraints.

The joint path selection and scheduling problem can be written as the following optimization problem. The maximization of total BH throughput is equivalent to the minimization of the total number of timeslots, defining the ratio of the demand over the BH link capacity toward an s-AP. In other words, the objective is to find paths the traffic should follow and links to be activated so as to maximize the system performance, i.e.:

$$\min \sum_{e \in E} f_e x_e \tag{8.36}$$

Subject to:

$$\sum_{e=\{0,j\}\in E} x_e = k \tag{8.37}$$

$$\sum_{e=\{i,j\}\in E} x_e \leq 1, \forall i \in V \tag{8.38}$$

$$\sum_{e=\{j,i\}\in E} x_e = 1, \forall i \in V \tag{8.39}$$

$$\sum_{e=\{i,j\}\in S_k} f_e x_e \leq D_{\max}, \forall k \tag{8.40}$$

$$f_e \leq T \sum_{i=1}^{|B|} w_{B_i} 1_{e,B_i}, \forall e \in E \tag{8.41}$$

$$\sum_{i=1}^{|B|} w_{B_i} = 1, w_{B_i} \geq 0, \forall B_i \in B \subseteq S \tag{8.42}$$

The first 4 constraints (8.37)–(8.40) are the routing constraints, whereas constraints (8.41) and (8.42) are the scheduling constraints. In (8.37), the number of links between CU (denoted as node 0) and all the s-APs depend on the number of paths and is equal to the variable k. The higher we set this value, the lower hops are expected in total. In (8.38) and (8.39), the number of incoming edges and outgoing edges to/from each s-AP is set exactly or less than one. By this, all the s-AP must be able to receive traffic; however, it is optional to have outgoing traffic to other links. Constraint (8.40) is the maximum delay constraint which has to be considered when creating a path. This constraint might be variable depending on the traffic (i.e., lower threshold for real time and higher one for nonreal time traffic). Moreover, (8.41) shows that the cost of the link shall not exceed the predefined time window (T); finally, (8.42) implies that the summation of the weights (i.e., the fraction of time each subgraph is active) is set to one.

8.3.2.1   Solutions framework

The problem of (8.36)–(8.42) is a NP-hard combinatorial optimization problem. Therefore, our proposed framework decouples the initial problem in two subproblems. At the first stage, we target to solve the path selection problem (constraints (8.37)–(8.40)) which has the form of an Integer Programming problem. By this, we identify links to activate and the number of slots to dedicate per set of links, such that the BH throughput is optimized. The solution of this subproblem is found using the Branch-and-Cut exact approach [39]. The next stage is the selection of the packet to be forwarded from the queues in a way to minimize the delay, taking into account the half duplex constraint, the multihop requirements and the queue buffers. This problem is solved using a variant of back-pressure scheduling algorithm [8].

8.3.2.2   Path selection algorithm

The objective of this problem is to deliver to a set of s-APs with known traffic demands on minimum cost paths originating from CU (given a predefined number of paths k). As discussed above, this is an Integer programming problem that can be solved by Branch-and-Cut algorithm.

The algorithm follows a branch-and-bound scheme, where lower bounds are computed by solving a linear program (LP) relaxation of the problem. This relaxation is iteratively tightened by adding valid inequalities to the formulation according to the cutting plane approach. The exact method is known as a Branch-and-Cut algorithm and is thoroughly described in [39] for the case of the IP problem.

8.3.2.3   Scheduling algorithm

After obtaining the paths and the number of timeslots that each link is going to be used for all destinations, the next phase is to find how to forward the packets from CU to all the s-APs, having a variable number of hops per path with the minimum delay. The packets are stored in separate queues per destination at the traffic aggregator. The target is to empty all the queues by the end of the given time window. Here, one constraint is related to half-duplexing, i.e., each node can either transmit or receive per time instance. Furthermore, the traffic that is forwarded via more than one hop should be stored in separate queues in the intermediate nodes. In each queue, First-In–First-Out policy is applied. For the solution of this problem, we propose a throughput optimal algorithm which follows the back-pressure concept [40]. Assuming slotted time, the basic idea of backpressure scheduling is to select a set of noninterfering links for transmission at each slot. Noninterfering links refer to links that do not have the same transmitting and/or receiving end, such that the half duplex constraint is maintained. Here, the objective is to serve the flow f with the maximum differential backlog. The differential backlog for each node i,j is defined as $\Delta Q_{i,j}^f = Q_i^f - Q_j^f$. The steps of this algorithm are as follows:

Step 1:   Compute the weight of each link (i,j) as $w_{i,j,t} = \max_f (Q_{i,t}^f - Q_{j,t}^f)$

Step 2:   Select links to maximize: $x^*(t) = argmax_x \sum_{(i,j)} w_{i,j,t} x_{i,j,t}$, where $\sum_t x_{i,j,t} = x_e : e = \{i,j\} \in E$

Step 3:   Transmit the chosen flows on the selected links.

8.3.3   Simulation results

To evaluate our work, Monte Carlo simulations for a 9-cell deployment are performed in a wide area scenario. In particular, Table 8.1 provides a summary of the simulation parameters.

The metrics used for the evaluations are the average BH link throughput, the average delay from the CU to reach each destination s-AP. Moreover, we provide another metric termed as satisfaction ratio, reflecting how well the solution meets the target delay threshold, i.e., the ratio of the successful snapshots over the total number of snapshots.

Table 8.1   *Simulation parameters*

s-APs	9
ISD	20 m
Users	Poisson arrivals per cell ($\lambda = 2.5$)
Traffic	Random traffic demand per user (10–50 Mbits)
Radio access channel	ITU [41]
Carrier	2 GHz (access), 60 GHz (BH)
Bandwidth	10 MHz (access), 100 MHz (BH)
Snapshots	5,000

The implementation of this scheme comprises two stages. The first stage is the extraction of results for the path selection problem. Here, we adjust the number of paths (k), so as to find the optimal path selection in different cases. The path selection algorithm was analyzed for all the possible number of paths. As shown in Figure 8.8 (top), the average BH link spectral efficiency drops when we increase k. This is due to the fact that the higher the number of paths, the lower the number of hops. In other words, long-distanced links with NLOS will impact the performance. On the other hand, in low-k regime with more hops, short-distanced LoS links increase the throughput performance.

In the second stage of our solution, the backpressure scheduling is used. As can be seen in Figure 8.8 (bottom), we evaluate the average delay (the average number of timeslots till each s-AP serves the required traffic) in this scenario. As shown, the higher the number of paths, the lower will be the delay. In other words, we may achieve higher throughput with more paths; however, this comes at the price of higher levels of delay. Another important observation comes from the impact of delay bound. As shown, by increasing the number of paths, the average delay with lower threshold bound (10 timeslots) decreases steadily. In the higher threshold (20 timeslots), also, the average delay drops; however, the pace of drop is relatively slower compared to the low threshold due to the fact that links with higher delays are admitted as feasible solutions. Here, both BH link throughput and delay are improved for lower delay threshold.

8.4   Evaluation of the joint backhaul and access link design

In the final topic of the chapter, we propose a framework for joint RRM where we systematically decompose the problem in BH and access links. For both single user and multiuser scenarios, it is shown that the optimization can be decomposed into separate power and subchannel allocation in both BH and access links where a set of rate-balancing auxiliary variables (pricing variable θ, as depicted in Figure 8.9) in conjunction with duration of transmission governs the coupling across both links (BH and access link phase durations (α) as can be depicted in Figure 8.9). In this context, novel algorithms were proposed based on grouping the cells to achieve rate-balancing

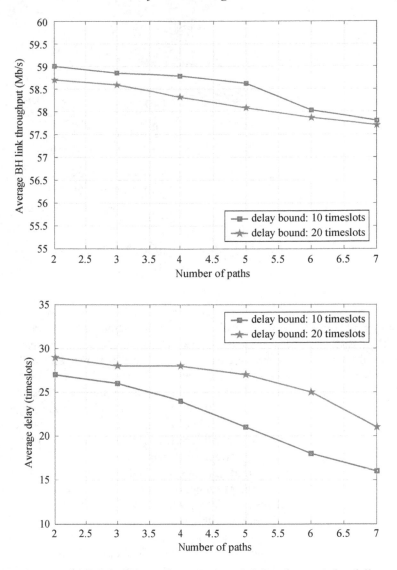

Figure 8.8 Average BH link throughput (top) and delay (bottom) for different delay bounds vs. number of paths. (© [2017] IEEE. Reprinted, with permission, from [35])

in different small cells. Based on the traffic demand and the channel conditions, the auxiliary variables are iteratively exchanged between small cells and a CU (Cloud RAN) and based on the aforementioned optimization problem the optimal phase duration between BH and access is extracted. On top of that, we perform dynamic clustering of users, based on [42], to orthogonally allocate resources to noninterfering users in order to avoid inter-cell interference.

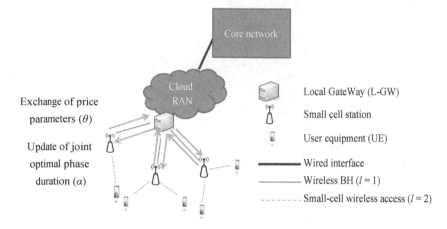

Figure 8.9   Proposed solution in dense small-cell network. (© [2017] IEEE. Reprinted, with permission, from [43])

Here, we extend our previous works in [42,44]. In [44], we proposed a low-complexity generic framework for resource allocation based on time-sharing applicable to different scenarios in downlink and uplink. On the other hand, in [42], we proposed an efficient graph-based dynamic clustering framework to control the level of cotier interference between small cells.

Combining the two solutions lead us to decouple the problem of resource allocation for small-cells networks across BH and access links by introducing a set of governing variables as a priority factor. The decoupled elements facilitate efficient RRM strategies that are flexible and applicable to a wide set of cases from full orthogonalization to full spectrum reuse between small cells. In particular, we present a novel framework to update the duration of transmissions in downlink between the BH and access links in TDD mode in conjunction with resource allocation in both links. Here, the time coupling element across the links is challenging yet presents another degree of freedom to jointly optimize the system performance.

The performance of the proposed algorithm is evaluated in a more detailed system-level topology consistent with 3GPP specifications [45]. In particular, the impact of different interference coordination strategies is examined from dynamic clustering to full reuse. Here, we have an outdoor random cell and user deployment in a cluster in line with scenario 2 of [46]. In particular, four small-cell stations are randomly dropped in a cluster area (ring) with minimum distances as outlined in [4]. Concerning the intra-cell scheduling, proportional fair (PF) scheduling is used for a multi-channel system per small cell to provide a fair allocation of resources between multiple users. Therefore, user weights are tuned based on this algorithm and are normalized to total weights per cell for a fair comparison. The samples are averaged over 5000 independent snapshots.

As it can be seen in Figure 8.10, the proposed rate-balancing scheme can provide significant improvement in both user spectral efficiency and weighted sum rate in

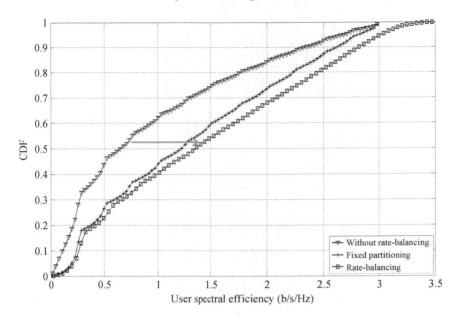

Figure 8.10   CDF of user spectral efficiency. (© [2017] IEEE. Reprinted, with permission, from [43])

this case study for both performance indicators. It is also interesting to note that fixed partitioning on the BH can be generally a better strategy compared to independent but dynamic RRM in BH and access links considering the results without rate balancing. However, fixed partitioning can be an expensive solution for small-cell operation as it yields low BH utilization when it comes to small cells in low-load scenarios.

Another interesting result is the evolution of α as the adjusting parameter for the duration of phases in the dynamic clustering case. Figure 8.11 shows the histogram of optimal value of across different snapshots. As it can be seen, the value is dynamically adjusted based on the instantaneous channel conditions in both links across different snapshots.

As shown, joint optimization with rate-balancing could provide significant improvement over independent resource allocation across BH and access links and also fixed BH partitioning strategies. In particular, the proposed algorithm could deliver significant performance improvement in conjunction with efficient interference coordination strategies like dynamic clustering.

8.5   Conclusions

Based on commonly used linear precoding algorithms (ZF and MRT) and linear and planar small-cell deployment configurations, the chapter proposed an approach for cluster size optimization in cloud-based distributed cooperative small-cell networks

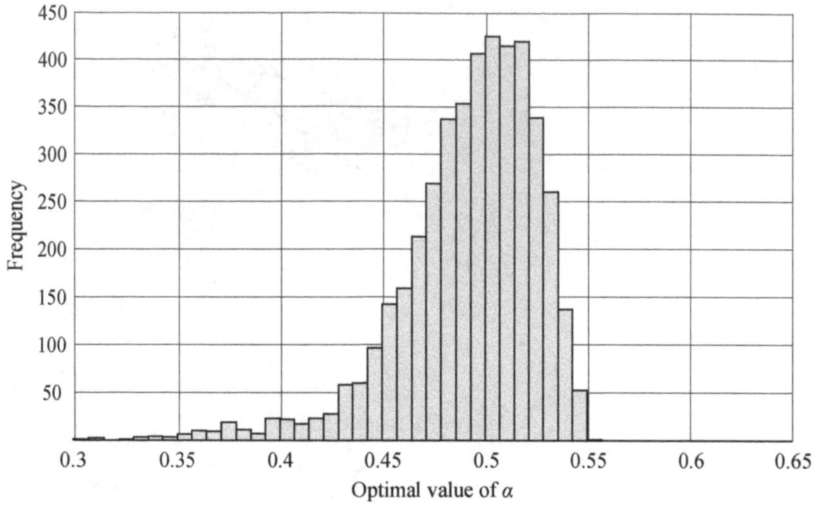

Figure 8.11   Histogram of optimal value of α. (© [2017] IEEE. Reprinted, with permission, from [43])

in the presence of CSI latency. An optimization problem is formulated in the afore-mentioned framework and desired signal and interference signal are calculated, which is followed by derivation of the output SINR by taking into consideration the channel mismatch caused by latency due to small-cells cooperation. Both delay and output SINR have been derived as a function of cooperation cluster size and an optimization problem to trade off the interference and channel mismatch has been formulated for maximizing network sum-rate. Simulations reveal a small gap with the analytical results in terms of SINR and sum-rate evaluations, and the proposed concise analyt-ical framework can be safely used in order identify the optimal cluster size for any specific deployment.

In addition, a novel framework for joint path selection and scheduling problem was provided, for a cluster of small cells assuming mm-Wave BH. Unlike the con-ventional routing and scheduling problems, this work aimed to find the best path selection and link scheduling taking into account the mmW channel at the BH and the traffic demand at the small-cell network. To this end, we provided a tune-able and adaptive solution framework by decoupling the generic problem into two stages: the channel-aware path selection and the queue-aware scheduling subproblem. Therein we provided an efficient solution to achieve higher spectral efficiency and utilization efficiency according to the QoS requirements.

Furthermore, the problem of joint resource allocation between BH and access links in TDD was addressed for dense small-cell networks. In this context, novel algorithms were derived for rate-balancing by employing the concepts of small-cell grouping and resource slicing. As shown, joint optimization with rate-balancing could provide significant improvement over independent resource allocation across BH and access links and also fixed BH partitioning strategies. In particular, the proposed algorithm could deliver significant performance improvement in conjunction with efficient interference coordination strategies like dynamic clustering.

Acknowledgements

The research leading to the results presented in this chapter received funding from the European Community's Seventh Framework Programme FP7/2007-2013 under grant agreement no. 317941 – project iJOIN, http://www.ict-ijoin.eu. The European Union and its agencies are not liable or otherwise responsible for the contents of this document; its content reflects the view of its authors only. The authors gratefully recognize the great contributions of many colleagues from iJOIN, who in fruitful cooperation, contributed with valuable insight, surveys and vision. The author would also like to acknowledge the support of the University of Surrey 5GIC (http://www.surrey.ac.uk/5gic) members for this work.

References

[1] U. Doetsch, M. Doll, H.-P. Mayer, F. Schaich, J. Segel, and P. Sehier, "Quantitative analysis of split base station processing and determination of advantageous architectures for LTE," *Bell Labs Technical Journal*, vol. 18, pp. 105–128, May 2013.

[2] D. Wubben, P. Rost, J. Bartelt, *et al.*, "Benefits and impact of cloud computing on 5G signal processing," *IEEE Signal Processing Magazine, Special Issue on 5G Signal Processing*, vol. 31, pp. 35–44, Nov. 2014.

[3] P. Rost, C. Bernardos, A. Domenico, *et al.*, "Cloud technologies for flexible 5G radio access networks," *IEEE Communications Magazine*, vol. 52, no. 5, pp. 68–76, May 2014.

[4] R. Irmer, H. Droste, P. Marsch, *et al.*, "Coordinated multipoint: concepts, performance, and field trial results," *IEEE Communications Magazine*, vol. 49, no. 2, pp. 102–111, Feb. 2011.

[5] L. Li, J. Liu, K. Xiong, and P. Butovitsch, "Field test of uplink CoMP joint processing with C-RAN testbed," in *IEEE Communications and Networking in China (CHINACOM)*, pp. 753–757, 8–10 Aug. 2012.

[6] J. Li, D. Chen, Y. Wang, and J. Wu, "Performance evaluation of cloud-ran system with carrier frequency offset," in *IEEE Globecom Workshops (GC Wkshps)*, pp. 222–226, 3–7 Dec. 2012.

[7] 3GPP TR 36.872, "Small cell enhancements for E-UTRA and E-UTRAN – physical layer aspects," 3GPP, Tech. Rep., Sep. 2013.

[8] 3GPP TR 36.842, "Small cell enhancements for E-UTRA and E-UTRAN – higher layer aspects," 3GPP, Tech. Rep., May 2013.

[9] J.-J. van de Beek, O. Edfors, M. Sandell, S. Wilson, and P. Ola Borjesson, "On channel estimation in OFDM systems," in *IEEE Vehicular Technology Conference*, vol. 2, pp. 815–819, Jul. 1995.

[10] S. Colieri, M. Ergen, A. Puri, and A. Bahai, "A study of channel estimation in OFDM systems," in *IEEE Vehicular Technology Conference Fall*, pp. 894–898, 2002.

[11] M. Joham, W. Utschick, and J. Nossek, "Linear transmit processing in MIMO communications systems," *IEEE Transactions on Signal Processing*, vol. 53, no. 8, pp. 2700–2712, Aug. 2005.

[12] M. Matthaiou, C. Zhong, M. McKay, and T. Ratnarajah, "Sum rate analysis of ZF receivers in distributed MIMO systems," *IEEE Journal on Selected Areas in Communications*, vol. 31, no. 2, pp. 180–191, Feb. 2013.

[13] D. Kaltakis, M. Imran, and C. Tzaras, "Information theoretic capacity of cellular multiple access channel with shadow fading," *IEEE Transactions on Communications*, vol. 58, no. 5, pp. 1468–1476, May 2010.

[14] S.-R. Lee, S.-H. Moon, J.-S. Kim, and I. Lee, "Capacity analysis of distributed antenna systems in a composite fading channel," *IEEE Transactions on Wireless Communications*, vol. 11, no. 3, pp. 1076–1086, Mar. 2012.

[15] W. Feng, X. Zhang, S. Zhou, J. Wang, and M. Xia, "Downlink power allocation for distributed antenna systems with random antenna layout," in *IEEE Vehicular Technology Conference*, pp. 20–23, Sep. 2009.

[16] F. Heliot, R. Hoshyar, and R. Tafazolli, "An accurate closed-form approximation of the distributed MIMO outage probability," *IEEE Transactions on Wireless Communications*, vol. 10, no. 1, pp. 5–11, Jan. 2011.

[17] J. Andrews, F. Baccelli, and R. Ganti, "A tractable approach to coverage and rate in cellular networks," *IEEE Transactions on Communications*, vol. 59, no. 11, pp. 3122–3134, Nov. 2011.

[18] M. Haenggi and R. K. Ganti, *Interference in Large Wireless Networks*. NOW, 2009.

[19] R. Tanbourgi, S. Singh, J. G. Andrews, and F. K. Jondral, "A tractable model for noncoherent joint-transmission basestation cooperation," *IEEE Transactions on Wireless Communications*, vol. 13, no. 9, pp. 4959–4973, Sep. 2014.

[20] P. Marsch and G. Fettweis, "Rate region of the multi-cell multiple access channel under backhaul and latency constraints," in *IEEE Wireless Communications and Networking Conference*, pp. 830–834, Mar. 2008.

[21] P. Marsch and G. Fettweis, "A framework for optimizing the uplink performance of distributed antenna systems under a constrained backhaul," in *IEEE International Conference on Communications*, pp. 975–979, Jun. 2007.

[22] J. Oueis, E. C. Strinati, and S. Barbarossa, "On the impact of backhaul network on distributed cloud computing," in *IEEE Wireless Communications and Networking Conference*, pp. 12–17, Apr. 2014.

[23] X. Ge, K. Huang, C. Wang, X. Hong and X. Yang, "Capacity analysis of a multi-cell multi-antenna cooperative cellular network with co-channel interference," *IEEE Transactions on Wireless Communications*, vol. 10, no. 10, pp. 3298–3309, Oct. 2011.

[24] iJOIN Project, "D2.1 – state-of-the-art of and promising candidates for PHY layer approaches on access and backhaul network," Tech. Rep., Nov. 2013. Available: www.ict-ijoin.eu/wp-content/uploads/2014/01/D2.1.pdf.

[25] L. Zhang, A. U. Quddus, E. Katranaras, D. Wubben, Y. Qi, and R. Tafazolli, "Performance analysis and optimal cooperative cluster size for randomly distributed small cells under cloud RAN," *IEEE Access*, vol. 4, pp. 1925–1939, 2016.

[26] L. Zhang, W. Liu, A. U. Quddus, M. Dianati and R. Tafazolli, "Adaptive distributed beamforming for relay networks based on local channel state information," *IEEE Transactions on Signal and Information Processing over Networks*, vol. 1, no. 2, pp. 117–128, Jun. 2015.

[27] D. S. Sesia, M. M. Baker, and M. I. Toufik, *LTE: The UMTS Long Term Evolution: From Theory to Practice*, 2nd ed. New York: Wiley-Blackwell, 2011.

[28] M. K. Simon and M.-S. Alouini, *Digital Communication over Fading Channels*, 2nd ed. New York: Wiley, 2005.

[29] L. Zhang, W. Liu, and L. Yu, "Performance analysis for finite sample MVDR beamformer with forward backward processing," *IEEE Transactions on Signal Processing*, vol. 59, pp. 2427–2431, May 2011.

[30] V. S. Ryaben'kii and S. V. Tsynkov, *A Theoretical Introduction to Numerical Analysis, Section 5.4*. Boca Raton, FL: CRC Press, 2006.

[31] iJOIN Project, "D5.2-final definition of requirements and scenarios," Tech. Rep., Nov. 2014. Available: http://www.ictijoin.eu/wp-content/uploads/2012/10/D5.2.pdf.

[32] U. Siddique, H. Tabassum, E. Hossain, and I. Dong, "Wireless backhauling of 5G small cells: challenges and solution approaches," *IEEE Wireless Communications*, vol. 22, no. 5, pp. 22–31, Oct. 2015.

[33] X. Ge, H. Cheng, and M. Guizani, "5G wireless backhaul networks: challenges and research advances," *IEEE Network*, vol. 28, no. 6, pp. 6–11, Nov. 2014.

[34] iJOIN Project, "D2.2 – definition of PHY layer approaches that are applicable to RANaaS and a holistic design of backhaul and access network," Tech. Rep., Oct. 2014.

[35] E. Pateromichelakis, M. Shariat, A. U. Quddus, and R. Tafazolli, "Joint routing and scheduling in dense small cell networks using 60 GHz backhaul," *2015 IEEE International Conference on Communication Workshop (ICCW)*, London, pp. 2732–2737, 2015.

[36] T. Bai and R. W. Heath, "Coverage analysis for millimeter wave cellular networks with blockage effects," *2013 IEEE Global Conference on Signal and Information Processing*, Austin, TX, pp. 727–730, 2013.

[37] A. Maltsev, E. Perahia, R. Maslennikov, A. Lomayev, A. Khoryaev, and A. Sevastyanov, "Path loss model development for TGad channel models," IEEE 802.11-09/0553r1, May 2009.

[38] I. Gribkovskaia, G. Laporte, and A. Shyshou. "The single vehicle routing problem with deliveries and selective pickups," *Computers & Operations Research*, vol. 35, no. 9, pp. 2908–2924, 2008.

[39] D. Naddef and G. Rinaldi, "Branch-and-cut algorithms for the capacitated VRP", in *The Vehicle Routing Problem*. Philadelphia, PA: Society of Industrial and Applied Mathematics, 2001, pp. 29–51.

[40]   L. Tassiulas and A. Ephremides, "Stability properties of constrained queuing systems and scheduling policies for maximum throughput in multihop radio networks", *IEEE Transactions on Automatic Control*, vol. 37, no. 12, pp. 1936–1948, Dec. 1992.

[41]   3GPP TR 36.814, "Evolved Universal Terrestrial Radio Access (E-UTRA): further advancements for E-UTRA physical layer aspects (Release9)", V9.0.0, Mar. 2010.

[42]   E. Pateromichelakis, M. Shariat, A. U. Quddus, and R. Tafazolli, "Graph-based multicell scheduling in OFDMA-based small cell networks," *IEEE Access*, vol. 2, pp. 897–908, 2014.

[43]   M. Shariat, E. Pateromichelakis, A. U. Quddus, and R. Tafazolli, "Joint TDD backhaul and access optimization in dense small-cell networks," *IEEE Transactions on Vehicular Technology*, vol. 64, no. 11, pp. 5288–5299, Nov. 2015.

[44]   R. Hoshyar, M. Shariat, and R. Tafazolli, "Subcarrier and power allocation with multiple power constraints in OFDMA systems," *IEEE Communications Letters,* vol. 14, no. 7, pp. 644–646, Jul. 2010.

[45]   Evolved Universal Terrestrial Radio Access (E-UTRA): further advancements for E-UTRA physical layer aspects, 3GPP TR 36.814, Mar. 2010.

[46]   Small cell enhancements for E-UTRA and E-UTRAN-physical layer aspects, 3GPP TR 36. 872, Dec. 2013.

Part II

Fronthaul Networks

Chapter 9

Self-organised fronthauling for 5G and beyond

Mona Jaber[1], Muhammad Ali Imran[2],
and Anvar Tukmanov[3]

The connecting link between the baseband unit (BBU) and the remote radio unit (RRU), in a centralised radio access network (C-RAN), is the fronthaul, which is the topic of this chapter. The fronthaul is presented as a key disruptive technology, vital to the realisation of 5G networks, but one with stringent requirements in terms of capacity, latency, jitter, and synchronisation. This chapter explains the fronthaul paradigm and presents an overview of legacy and new solutions in backhauling/fronthauling with critical analysis of respective advantages and limitations. In view of the debilitating expectations of 5G fronthaul performance, hybrid RAN architectures are also explored and analysed from their respective RAN gain and fronthaul requirements, leading to promising alternative RAN functional splits and innovations in *X-hauling*. The chapter delves into the (centralised) C-RAN/fronthaul versus (distributed) D-RAN/backhaul dilemma, offering a joint backhaul/RAN perspective and tangible trade-off analysis of the available options. The study advocates the need for a fronthaul architecture that is dynamic, adaptable, flexible, and expandable. To this end, a key catalyst to the realisation of the 5G fronthaul is equipping the network with self-optimisation and organisation (SON) capabilities that would timely adapt the network following the dynamically changing conditions. In this context, SON operation in the fronthaul becomes essential for optimisation objectives such as energy efficiency, latency reduction, or load balancing. Moreover, the SON-enabled joint RAN/fronthaul optimisation has a pivotal role in adjusting the level of centralisation according to the fronthaul capabilities and vice-versa.

9.1 Introduction

Ultra-dense-networks (UDNs) are considered an imperative 5G solution [1,2]. They are often heterogeneous networks (HetNets), i.e., multi-tier including legacy high

[1]5G Innovation Centre, University of Surrey, UK
[2]School of Engineering, University of Glasgow, UK
[3]BT Research and Innovation, UK

power macro-cells and very dense cells with lower power (small cells). Small cells support multi-radio access technologies (multi-RAT-capable) and represent an essential part of UDNs. Sharing the spectrum in a UDN requires intelligent inter-cell interference coordination (ICIC), cancellation or exploitation. Accordingly, key radio UDN facilitators have been developed: coordinated multi-point processing (CoMP) and enhanced ICIC (eICIC). Both these features require tight coordination between concerned cells and higher processing power.

The increasing need for processing power coupled with the emerging small cells diversity in traffic patterns, both spatial and temporal, renders the concept of centralised radio access network (C-RAN) very attractive. C-RAN consists of splitting the functions of the traditional evolved Node B (eNB, i.e., the cellular radio station) and migrating them towards a distant shared pool of baseband resources, referred to as baseband unit (BBU). Basic radio functions remain at the radio site, hence the terminology remote radio unit (RRU). The C-RAN architecture capitalises on the diversity of traffic peaks, hence, improves the utilisation efficiency of the infrastructure. At the same time, it promotes the green aspect of 5G, owing to close proximity of cells and users and corresponding lower transmission power requirements. The benefits of the C-RAN architecture are detailed in Section 9.2.

Another challenge resulting from UDNs is the user mobility management; traditionally, users moving from one cell to another require a handover procedure managed by the mobility management entity in the RAN. If this model were applied to UDNs, it would generate a crippling signalling overhead due to the limited footprints of small cells, hence, frequent cell border crossing. Accordingly, splitting data and control planes is another essential 5G technology: Small cells are used as data offloading points, whereas mobility handover is triggered when users move between clusters of cells. Tight collaboration and excessive signalling exchange between the control radio cell and the data radio cell (small cell) provide reliable, robust, and updated information. However, this increases the backhaul overhead and requires a backhaul with high throughput and low latency [3].

Although such technologies can potentially address the greedy 5G capacity requirements and reduced RAN-related capital and operational expenditures (CapEX and OpEX), a new challenge has nonetheless emerged: the 5G backhaul. With the rise of C-RAN architecture, the 5G backhaul has evolved to a more complex network composed of *fronthaul*, *midhaul*, and *backhaul*. The backhaul section connecting the RRH to the BBU directly, or to an intermediate aggregation point, is labelled fronthaul. The basic fronthaul is assumed to run over a common public radio interface (CPRI) separating the RRH from the BBU. Based on the 3GPP terminology, the inter-eNB X2-based interface is called the midhaul; the term has recently been used to refer to the group of links connecting the fronthaul aggregator to the backhaul aggregation point (see Figure 9.1), whereas the network connections between aggregation points and the core, based on the S1-interface, have retained the term backhaul.

Inhibitive bandwidths greater than 10 Gbps and constraining latency in the orders of hundreds of microseconds render fibre optics the only fronthaul viable solution

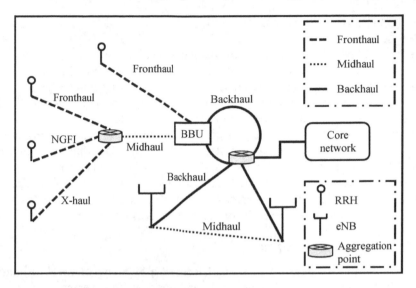

Figure 9.1 *Future cellular network's backbone composed of backhaul*
(S1-interface), midhaul (X2-Interface), fronthaul (CPRI),
NGFI, or X-haul

commercially available today, as detailed in Section 9.3. However, laying fibre to connect all envisaged RRH to the core network may be impossible in some cases and certainly very costly otherwise. In view of the immense challenge facing 5G deployment, the 5G backhaul research has been triggered, aiming at bridging the gap between the requirements stipulated by the 5G RAN and the realistic backhaul capabilities from two different perspectives. The first consists of evolving the current backhaul to meet 5G expectations and encompassing new wireless technologies. The other backhaul research perspective looks at adapting the 5G RAN to the available backhaul with realistic performance, such as investigating intermediate RAN architectures between the C-RAN and the D-RAN to fit the fronthaul (or X-haul) capabilities as detailed in Section 9.5.

Intermediate functional splits alleviate the exigent requirements of the X-haul to enable the usage of fronthauling solutions that are under-par with fibre optics. Wireless fronthaul links in the terahertz band are seen as a potential contender; however, they offer variable capacity, resilience, and latency according to changing propagation and interference conditions. A hybrid multi-hop backhaul network is, thus, inevitable and perhaps desirable for the launch of 5G. Such a network is expected to adapt to the dynamics of the radio network and the users' behaviour and to self-adjust to changes such as added cell, spectrum, radio feature, etc. Self-organised and optimised X-haul is thus a catalyst to enabling 5G and reaping the maximum benefits of C-RAN, as detailed in Section 9.6.

9.2 Merits of fronthauling

The *C* in C-RAN often stands for *centralised* or *cloud* but also *clean* and *cooperative* RAN. The focal concept is to redistribute functions, which are traditionally located in the eNB, towards a cloud-operated central processor. Such centralised intelligence would consequently enable cooperative operation among cells for greener and cleaner (i.e., fewer carbon emissions) communication. A fully centralised RAN consists of taking most of the base stations functionalities away from the eNB and leaving only the radio functions at the remote radio unit (RRU) or RRH. The BBU, which is traditionally located in the base station cabinet, is thus relocated to the cloud or central processor. Cloud-BBUs form jointly a shared BBU-pool that is connected to all RRHs. The fronthaul is the collection of interfaces linking the RRHs to the BBU–pool, and it may span any distance between tens of metres to tens of kilometres.

9.2.1 Capital expenditures

UDNs imply an invasive spread of small cells at a high cost of RAN equipment. The limited coverage of small cells results in a large peak-to-average ratio of users' traffic. In a D-RAN topology, this would necessitate large allocation of baseband resources, per small cell, in order to account for peak traffic. Admittedly, such a resource allocation would be underutilised most of the time. Alternatively, small cells may be dimensioned to cater for average traffic demand (instead of peak). This would improve the utilisation efficiency of baseband resources, but at the cost of quality degradation during peak traffic periods. The C-RAN reduces the complexity (hence cost) of small cells and improves the BBU utilisation efficiency without compromising users' quality, owing to the shared pooling. Standardised C-RAN solutions are not widely available commercially, nonetheless, based on information from proprietary solutions, the cost of an RRH is at most half the cost of a small cell. Consequently, given that the pooled baseband resources in a C-RAN deployment are less than half the cumulative baseband resources in D-RAN small cells, the equipment cost reduction is guaranteed.

The cost of radio planning and installation is also a factor in the capital expenditure calculation. Planning small cells typically involves two aspects: capacity and coverage. On the other hand, planning RRHs only entails coverage design as the capacity is accounted for by the shared BBU-pool, thus reducing the complexity and cost associated with the planning phase. Moreover, by virtue of their simplicity, robustness, and conciseness, RRHs are cheaper to install than small cells. Besides, in view of their discrete appearance, it is much easier and faster, hence, less costly, to acquire sites for installing RRHs.

9.2.2 Operational expenditures

Operational expenditures represent the monthly or yearly cost of operating a network including site rent, system upgrade, maintenance, electric bills, etc. RRHs are more robust than small cells owing to their simpler design; this has two operational advantages. First, RRHs may be installed without additional equipment to control the environment such as temperature and humidity. Consequently, the electric bills of an RRH will be considerably less than those of a small cell. Second, RRHs are less prone to failure than small cells because of their higher robustness. Accordingly, RRHs require less maintenance-related visits which are traditionally a major operational cost factor. As for the conciseness of RRHs, it entails smaller occupancy space, hence, cheaper site rental. In addition, most of the RRH optimisation and system upgrade procedures may be conducted from the BBU-pool location. Effectively, cost-inflating operational expenditures such as maintenance, optimisation, and environment-related electric bills are migrated and unified in the BBU-pool. In the D-RAN architecture, operational expenditures related to BBU functions scale linearly with the number of small cells.

9.2.3   How is C-RAN not DAS?

While the *fronthaul* technology seems to have been coined with the rise of fourth cellular generations and beyond, it brings to mind the question – Is it really any different to the incumbent distributed antennas systems (DAS) backhaul?

Indeed, DAS is a similar architecture to C-RAN and has been deployed, as early as the second generation (2G) of cellular networks (e.g., Global System for Mobile (GSM)). DAS is used to serve high user densities in areas with difficult indoor propagation characteristics (e.g., airports, shopping malls, and corporate building). It consists of breaking the traditional 2G radio site, called base transceiver station, into two parts: the BBU and a set of RRHs. These two parts are normally connected with optic fibre links inside the radio cabinet; thus, the solution requires removing and distributing the RRH by prolonging the fibre connection using radio over fibre transmission. Consequently, spreading diligently these RRHs around a building provides a continuous and close-to-uniform indoor coverage, irrespective of the traffic distribution. DAS may be considered as an implementation option of C-RAN in which quantised signals are exchanged among the RRHs to enable centralised or decentralised joint decoding. C-RAN is thus an evolution of DAS which introduces the novel concept of cloud BBU, whereby various BBUs may be located in different geographical areas while forming a cloud and connecting to more than on DAS. However, with the C-RAN architecture, the covered distances between RRH and BBU are larger than indoor solutions, and fibre is a luxury that is often unavailable.

9.3   Fronthaul solutions

5G targets are evidently ambitious and intensify the design challenges of the fronthaul. Current backhaul networks are mostly built with microwave links (often operator owned) and fibre/copper-based links (often leased) with different proportions per operator and country.

Furthermore, fronthaul design has unique requirements in addition to those that apply to the backhaul. For instance, a bandwidth of 100–1000 Mbps is sufficient for dimensioning small-cell backhaul links whereas RRH fronthaul links require 10 Gbps in order to be future proof. Similarly, fronthaul latency limitations, in the order of 100 μs, are significantly more stringent than traditional backhaul links which can safely operate with delays up to 30 ms. Besides, specific aspects of the CPRI link necessitate more careful planning such as a very low bit error rate in the order of 10^{-12} and high resilience.

Fronthaul designs are often tied to the existence of an end-to-end optical fibre network. Nevertheless, the vast majority of countries pushing towards 5G will not have nationwide optical fibre networks in the coming 10–20 years. Indeed, fibre to the home (FTTH) is scarce worldwide with only 16 countries exceeding 15 per cent FTTH penetration [4]. The need for innovation in backhaul provisioning is thus evident and becomes more vital in the dawn of 5G; to this end, a compilation of available and potential technologies is presented here.

9.3.1 Wired solutions

Wired backhaul/fronthaul solutions can be grouped into three categories, the first two being the most dominant. The first category is copper based and represents the family of digital subscriber lines (DSLs). The second and third categories are fibre based and encompass solutions employing partially or fully optical fibre technology. Fibre-based solutions that are complemented with copper-based, wireless, or optical fibre last-mile links are referred to passive optical networks, i.e., the second category referred to as any class of passive optical networks (PONs). The third category defines fibre-based solutions that use optical fibre as a trunk and coaxial cables as a last mile; the technology is termed data over cable service interface specification (DOCSIS).

9.3.1.1 Copper-based xDSL

Since its invention in 1988, the copper-based DSL connection has been the most popular solution for delivering broadband to the home owing to its relatively cheaper cost, compared to fibre. The latest statistics in 2014 show that global broadband connections still rely on copper more than 50 per cent of the time [5]. However, copper-based connections will not be viable for backhauling most 5G radio cells due to their limited capacity and high latency. According to Ericsson, the usage of copper-based backhaul for cellular sites will decrease from 40 per cent worldwide recorded in 2008 to less than 10 per cent forecast in 2020 [6]. It is envisaged that copper will be replaced by fibre-based connections because of their superior performance characteristics.

Advances in xDSL technology and recent field trials of XG-FAST[1] have achieved throughputs in par with optical fibre links in the order of 10 Gbps over a copper reach of 30 m [7]. The ideal channel capacity of an xDSL connection is a function of the copper line length and the bandwidth allocated. State-of-the-art xDSL technology is bringing copper-based throughput closer to the 5G targets by addressing two challenges.

The first consists of increasing the bandwidth by combining two (or more) regular very-high-bit-rate digital subscriber line 2 (VDSL2) lines into a single virtual pipe with, theoretically, double to capacity. This is referred to as bonding; however, such combinations generate crosstalk. The second challenge is, thus, crosstalk cancellation,

---

[1]A new technology prototype from Bell Labs (dubbed XG-FAST).

which is addressed using vectoring technology. It entails estimating the out-of-phase crosstalk signal and on each line in a cable, effectively making it disappear. This allows each individual line to operate at peak performance levels, unaffected by adjacent cables or bundles. Advances in DSL technology are looking at increasing further the bandwidth from 17.7 MHz with VDSL to 212 MHz with G.fast[2] and 500 MHz with XG-FAST.

Calculating the average packet delay over any link entails looking at four phenomena: the propagation delay, the transmission delay, the processing delay, and the queuing delay.

1. **The propagation delay**: It is the time difference between the moment a packet is placed on one end of the line and the time it is received at the other end. It is a function of the distance travelled and the refractive index of the propagation medium.

2. **The transmission delay**: It is the time taken to push all the packet bits on to the link, hence is the ratio of the average packet size over the nominal capacity of the channel.

3. **The processing delay**: The time the router takes to process a packet is referred to as the processing delay. The router needs to access the header of a packet and redirect it to the next path. The router processing may also include checking for bit errors in the packet. Processing delays are determined by the hardware (processing power) and are rarely dependent on the traffic conditions. Nevertheless, the processing delay varies according to the type of packet and the corresponding required task from the router.

4. **The queuing delay**: Once the router has identified the destination path of the packet, the path may not be immediately available due to preceding packets waiting in the queue. The time it takes a packet to stand in queue before getting access to the destination path is referred to as queuing delay; it may be null if no other packets are in queue and it may be very high if the queue is saturated.

The xDSL router acts also as a multiplexer to multiple xDSL lines and is called the digital subscriber line access multiplexer (DSLAM). For a DSLAM with a given processing power, the expected delay still varies drastically with respect to the number of connected lines and packet size. Consequently, xDSL latency is dominated by both processing and queuing delays as they jointly are one order of magnitude higher than propagation and transmission delays.

Nominal performance characteristics of the different members of the copper-based category of fronthaul solutions are summarised in Table 9.1 and relevant references are highlighted for further reading.

9.3.1.2   Fibre-based xPON

Optical fibre is hailed as the saviour and enabler of ultra-broadband networks, largely on the grounds of its exceptionally low loss and immunity to crosstalk when compared to copper-based links. These traits allow for long reaches, e.g., 80 km, without

---

[2]Fast Access to Subscriber Terminal; the letter G stands for ITU-T G series of recommendations.

Table 9.1   *Copper-based backhaul solutions with nominal performance characteristics*

xDSL technology	Upstream throughput	Downstream throughput	One-way latency (ms)	Reach	Note	References
VDSL2	15 Mbps	75 Mbps	5–15	1 km	Capacity decreases with distance	[8]
VDSL2 ph2	20 Mbps	100 Mbps	3[†]	1.5 km	Phantom mode; 2 pair bonding	[9]
VDSL2 ph4	40 Mbps	230 Mbps	3[†]	1.5 km	Phantom mode; 4 pair bonding	[9]
VDSL2 ph8	150 Mbps	750 Mbps	3[†]	1.5 km	Phantom mode; 8 pair bonding	[9]
G.fast 50 m	1 Gbps[‡]	1 Gbps[‡]	<1[‡]	50 m	TDD	[10,11]
G.fast 100 m	500 Mbps[‡]	500 Mbps[‡]	<1[‡]	100 m	TDD	[10,11]
G.fast 200 m	200 Mbps[‡]	200 Mbps[‡]	<1[‡]	200 m	TDD	[10,11]

[†]Realistic latency in the order of 10 ms.
[‡]These are theoretical values; however, realistic throughput may be sub-1 Gbps and latency closer to 3 ms. Moreover, G.fast is a TDD technology; thus, the throughput quoted assumes full occupation of either upstream or downstream.

the need of active repeaters, and ultra-high bandwidth which would have necessitated thousands of copper links to achieve the same capacity. Other practical benefits promote further the superiority of fibre such as its light weight and small cable size. In addition, communication is based on light, thus does not cause fire-igniting sparks and it is not prone to corrosion as the transmission medium is glass. On top of all that, it is much more difficult to tap into an optical fibre line compared to both copper-based lines and wireless connections and is therefore more secure.

However, optical fibre links are more expensive to install than copper links, and the majority of the cost is due to civil engineering work. This includes planning and routeing, obtaining permissions, trenching, creating ducts and channels for the cables, as well as installation and connection. Therefore, it has been the norm to lay significantly more fibre than required at the installation phase for potential future use and/or for redundancy purposes. Unused or unlit optical fibre links are often referred to as *dark fibre* and are often found in new buildings and structures that have been built in the last two decades. It is thus not surprising that the United Arab Emirates (with relatively new urbanisation) has the highest fibre-to-the-home penetration in the world with more than 70 per cent homes connected while the United Kingdom has less than 1 per cent (older urban centre) [4].

Whether dark fibre exists or new fibre is to be laid, the next step would be to connect that last mile to the optical fibre network/trunk. Many options are available

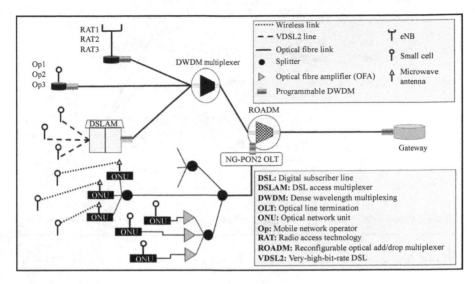

Figure 9.2 Fibre-based backhaul architecture with optional xDSL and wireless links used to cover the last mile

Table 9.2 Standardised xPON technologies are grouped into two classes: the ITU series and the IEEE series

Per line	ITU series (ATM)			IEEE series (Ethernet)		
	GPON	XG-PON	NG-PON2	EPON	10/1G-EPON	10/10G-EPON
Downstream (Gbps)	2.5	9.95	9.95	1.25	10.3	10.3
Upstream (Gbps)	1.2	2.5	9.95	1.25	1.25	10.3
Max reach (km)	60	40	40	20	20	20
Comment	XG-PON2 subsumed			Reach depends on splitter size and power budget class		

and often several of them are used within one network as shown in Figure 9.2. Each of these options would result in different link performance characteristics due to the processing characteristics of the components in the architecture.

Available throughput over an optical fibre link is commonly thought of as unlimited, indeed in a recent trial on a multi-core fibre link of 1 km, researchers were able to transmit 255 Tbps [12]. Although fibre lines are not free from power, the power budget remains amply relaxed to provision the required receiver sensitivity and bit error rate at standardised distances. Effectively, the achievable fibre link distance and throughput are governed by the specifications of the optical fibre components connecting it, which are standardised as listed in Table 9.2. Nonetheless, advances in the xPON technology are fast track ongoing; a recent article on the xPON roadmap

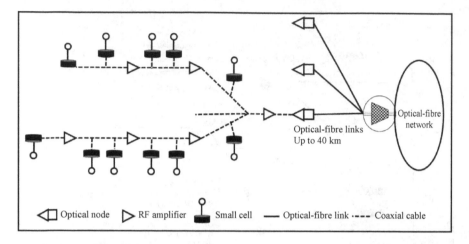

Figure 9.3   Typical DOCSIS solution using a hybrid fibre-coaxial infrastructure

described the newly standardised XGS-PON and introduces the concept of NG-PON3 which is 100 Gbps capable [13].

The propagation and transmission delays over xPONs are in the order of few microseconds. The queuing delay is in the order of tens to hundreds of microseconds when estimated separately. On the other hand, the processing plus queuing delay, incurred by different routers and multiplexers, is in the order of few milliseconds and is thus often considered the main source of delay in an xPON.

9.3.1.3   Fibre-based DOCSIS

DOCSIS is an international telecommunications standard that allows the transmission of high-bandwidth data over an existing cable television system. DOCSIS technology operates on hybrid fibre-coaxial infrastructure as shown in Figure 9.3. As frequency allocation bandwidth plans differ between the United States and European cable television systems, the European version of the DOCSIS series is published under the name EuroDOCSIS.

The coaxial portion of a DOCSIS network could typically connect 500 small cells in a tree-and-branch configuration. The interface between the fibre section and the coaxial section of the network is the fibre node. It has a broadband optical receiver, which converts the downstream optically modulated signal coming from the fibre section to an electrical signal going to the small cells. RF amplifiers are used at intervals to overcome cable attenuation and passive losses of the electrical signals caused by splitting or 'tapping' the coaxial cable. The performance characteristics of standardised DOCSIS technologies are summarised in Table 9.3.

9.3.2   Wireless solutions

A large toolbox of wireless links was originally promoted to enable fast and cost-effective connections for the pervasive small cells [8]. This included the following

Table 9.3   Standardised DOCSIS systems and corresponding performance characteristics [14]

	Upstream throughput	Downstream throughput	One-way latency	Reach
DOCSIS3.0	108 Mbps	304 Mbps	10–20 ms	1.5 km
DOCSIS3.1	1 Gbps	5–10 Gbps	N/A	N/A
EuroDOCSIS3.0	108 Mbps	400 Mbps	10–20 ms	1.5 km

technologies: microwave (point/multipoint to point/multipoint), millimetre wave (60 GHz or 70/80 GHz), sub-6 GHz (licenced or unlicensed), television white space, satellite, and free space optics [15]. Wireless links are effectively seen as the optimum solution to fill in the gap in the absence of optical fibre links, by virtue of the minimal required installation work and relative fast deployment. However, among all wireless technologies, the most viable ones are microwave and millimetre-wave, with various forms and topologies [8]. Although they are governed by very different regulations and propagation characteristics, these are often bundled under the terminology *microwave* with three different bands: the traditional band (6–13, 15–23, and 26–42 GHz), the V-band (often referred to as 60 GHz band), and the E-band (70/80 GHz). European Telecommunications Standards Institute (ETSI) has recently started looking at the D-band (141–174.8 GHz) with potential capacity of 50 Gbps [16]. Nonetheless, Extremely High Frequency (EHF) transmission industry is still in its early phase and faces many challenges [17]. The vast majority of countries pushing towards 5G will not be able (or willing) to timely deploy nationwide optical fibre networks in next two decades. Accordingly, the role of wireless backhaul links is pivotal in this transitory phase. Ericsson, for instance, predicts that 50 per cent of mobile network backhaul worldwide would still be using microwave solutions in the year 2020 [18].

Microwave technology is evolving and expanding to offer fibre-like throughput in the order of few Gbps. Evolution is based on improved spectrum efficiency, e.g., high adaptive modulation, improved system gains such as multiple-input–multiple-output (MIMO) technologies, and increase in throughput efficiency with advances such as multi-layer compression schemes. Currently, the average capacity of mobile backhaul microwave links is <200 Mbps but is expected to reach beyond 10 Gbps by the year 2020, partly as a result of the newly adopted millimetre-wave spectrum [18]. It is expected that the E-band would constitute 20 per cent of the microwave family of mobile backhaul in the year 2020 and the V-band 10 per cent.

The main impediment of wireless backhaul networks is their vulnerability to propagation conditions, not the least fading and loss of line of sight (LoS). There are two types of fading, small and large scale, that jointly result in a fluctuating received signal power level even in the case of static transmitter and receiver. Large scale fading, often referred to as shadowing or slow fading, may be compensated for by using power control and automatic gain control. On the other hand, small-scale fading is an outcome of multipath propagation caused by moving and static scattering objects in the propagation space between transmitter and receiver. These various obstacles result in reflections, refractions, and diffractions of the transmitted signal. As a result, several copies of the transmitted signal are received with different power levels, phases, and delays. Combining these multi-path signals may be constructive, i.e., the result is better than any individual path, but it can also be destructive, which causes the deep fades.

Besides, microwave radios require LoS deployment, specifically the traditional and E-band links. Traditional microwave radios are often above rooftops and use narrow-beam and high-gain antennae to secure high availability of LoS. However, E-band is often deployed below rooftops, often on street furniture, and suffers from frequent loss of LoS due to pole swaying and moving obstacles. More importantly, due to the E-band radio wavelength of few millimetres, the corresponding link is more sensitive to small deviations in the LoS.

A comprehensive list of available wireless fronthaul technical solutions is presented in Table 9.4. Based on fronthaul performance requirements highlighted earlier (>10 Gbps throughput <1 ms latency), the only viable solutions in Table 9.4 are the

Table 9.4   *Wireless backhaul technologies and corresponding performance*
           characteristics (US=upstream, DS=downstream, Tput=throughput,
           ch=channel(s))

Technology options		US Tput (Mbps)	DS Tput (Mbps)	Latency/ Jitter (ms)	Distance	Note
Microwave[†]	PtP	1,000	1,000	<1/hop	2–4 km	6–60 GHz
Microwave[†]	PtmP	1,000	1,000	<1/hop	2–4 km	6–60 GHz
Satellite[†]	LoS	15	50	300 one-way latency 5–30 jitter	Ubiquitous	Due to cost per Mbps realistic throughput 2–10 Mbps DL 1–2 Mbps UL
TVWS[†] television white space	NLoS	18/ch	18/ch	10	1–5 km	Up to 4 ch; up to 10 km at 10 Mbps using 2 ch with LoS
V-band[†]	LoS	1,000	1,000	0.2	1 km	Scalable
E-band[†]	LoS	10,000	10,000	0.065–0.350	3 km	Scalable
Sub-6 GHz[†] 800 MHz– 6 GHz	NLoS	170	170	5 single hop one way	1.5–2.5 km urban 10 km rural	Licensed (20 MHz TDD) expected to increase to 400 Mbps
Sub-6 GHz[†] 2.4, 3.5, 5 GHz	NLoS	150–450	150–450	2–20	250 m	Unlicensed data rate depends on MIMO
FSO[‡]	LoS	10,000	10,000	Low	1–3 km	

[†]Please refer to [8] for more information on these technologies.
[‡]FSO = Free space optics. Please refer to [22] for more information on this technology.

E-band and the free space optical communications (FSO). Commercial E-band solutions that are CPRI capable are readily available such as [19] with nominal throughput up to 2.5 Gbps over 5 km and latency less than 5 μs. Higher throughput E-band solutions are also offered for the backhaul in [20] with throughput up to 40 Gbps. On the other hand, commercial FSO solutions, such as [21], seem to be sub-par in comparison with the nominal performance shown in Table 9.4 (1.5 Gbps over a maximum range of 2 km). The solution provider claims that through their usage of a network management tool that gears the FSO multi-beam system according to distances and weather data, the offered solution provides carrier-class availability.

9.4 The dark side of fronthauling

The C-RAN is perceived as a disruptive technology which has created a new type of backhaul links, the fronthaul. The fronthaul connects essential parts of the VeNB; thus, the radio part and the fronthaul can only be designed jointly to ensure coordinated performance over the VeNB. As such, a disruption to the traditional network design is also incurred.

5G networks are expected to provide a step-change in performance; it is clear that more capacity, less latency, tight frequency and timing synchronisation, security, and resilience are all imperative requirements in the fronthaul design. The fronthaul bandwidth requirement scales with larger radio access bandwidth and becomes crippling with ≥100 MHz possible 5G spectrum allocations per small cell. The basic fronthaul is assumed to run over a CPRI separating the RRH from the BBU. CPRI was not designed for bandwidth efficiency and, effectively, full bandwidth occupancy occurs on the fronthaul even when the actual user traffic is null. Besides, the round trip time over the CRPI interface is 5 μs and the effective admissive delay over the fronthaul link, including propagation delay, is in the order of 100 μs. Admittedly, the C-RAN architecture inflates the effective fronthaul throughput such that links suitable for an eNB deployment act as a funnel for a VeNB, under the same user traffic load. Projected fronthaul throughput requirements of a radio cell with 500 MHz allocated bandwidth and 16 × 8 massive MIMO scale up to 2.25 Tbps which is impossible to reach with state-of-the-art fibre solutions [19]! Besides, it is rare to find dark-optical fibre links ready to be lit at the required location, which implies the need for new optical fibre links to be laid or an alternative RAN solution. Although E-band commercial products are available to cater for a fronthaul throughput of 2.5 Gbps (see Section 9.3.2), these are not future proof and may only be considered as transitory solutions awaiting the deployment of optical fibre links or advances in terahertz wireless technology. To this end, the fronthaul incurred capital and operational expenditures in this section are discussed on the grounds of fibre-based solutions.

9.4.1 Capital expenditures

Optical fibre networks offer the highest performance gain among all backhauling options but generate the highest capital expenditure. The crippling-associated deployment cost is due to the civil work required for trenching and laying fibre which adds up to >70 per cent of the cost whilst equipment's cost is about 8 per cent and the installation only 2 per cent [23]. The burden of civil work is because of digging trenches in busy roads which requires painstaking right-of-way negotiations, traffic management, and road reconstruction and hole backfilling, once the fibre ducts are laid. Micro-trenching is seen as a promising technique for reducing the current exorbitant traditional trenching cost associated with optical fibre networks [24]. It entails creating a narrow slit instead of the traditional open trenches. Micro-ducts are then pulled, pushed, or blown inside the slit and mini-cables are blown into those. Micro-trenching is said to be less disturbing than traditional trenching, thus facilitating the authorisation and realisation of the work and enabling a rate of 200 m of fibre per day. It is reported that this approach in civil engineering reduces the capital expenditure of fibre network deployment by 76 per cent. In a study conducted in 2011 by the UK government, micro-trenching is recommended as an acceptable method of installing communications cables, subject to certain caveats and conditions, mostly related to the depth of the trench and the structure and material of the road [25]. Alternatively, aerial optical fibre cables may be used for deploying fibre links without the need to dig up roads to bury cables or ducts [26]. They are hailed as the most cost-effective methods of deploying fibre links owing to the potential of reusing existing pole infrastructure. On the other hand, aerial fibre cables are fragile and prone to straining, sagging, and eventually breaking in extreme winds and temperature variations. They are also exposed to damage due to birds and other hazards. Moreover, often existing poles do not have the required strength to support the fibre cable span. These would need to be strengthened which increases the cost and time of deployment including getting the permits from the local authorities.

In summary, any of the techniques employed to spread fibre links, be it traditional trenching, micro-trenching, or aerial cables, contributes to the largest portion of the capital expenditure of optical fibre networks.

9.4.2 Operational expenditures

On the contrary, optical fibre networks require the least operational expenditure (excluding lease) because fibre degradation or breakdown is infrequent, thus rarely requires replacement. Moreover, bandwidth expansion (another general operational cost burden) can be realised via software upgrade rather than expensive hardware replacement [27].

Optical fibre cables were originally thought of as very fragile due to their intolerance to bending but proved as easy, if not easier, to handle than copper-based links, when regulations are respected. Intrinsic advantages of fibre that promote its resilience are many. First, fibre does not corrode and is not affected by moisture. Fibre can be installed over very long lengths without the need for joints nor amplifiers which would require power feed conductors. Moreover, fibre offers high resilience to radio radiation, power surges and lightning, thus improving the electromagnetic compatibility when compared to copper-based lines [28].

9.5   The emergence of X-haul

The fully centralised C-RAN, also referred to as baseline C-RAN configuration, employs a CPRI-based fronthaul, which dictates overwhelming capacity requirements that are independent of the actual traffic load in the RRH. Indeed, the CPRI-based C-RAN migrates both cell and user functions to the BBU, thus burdening the fronthaul with full load even when no users are served by the BBH. Moreover, the MIMO-related functions are also migrated to the BBU, resulting in a fronthaul traffic that scales with the number of MIMO antennae [29,30]. Therefore, researchers have been looking at ways for alleviating the requirements of the RRH–BBU interface, and as such, opening the door to non-fibre backhauling solutions, as those presented in Section 3.

To this end, the level at which the traditional base station functions should be split has become a prime research topic, termed *functional split*, which aims at finding an ideal split, by analysing the impact of different options on possible gains and fronthaul exigence [27,31]. Pioneering work in this domain has been conducted through iJOIN[3] and has resulted in essential quantification of latency and capacity requirements imposed on the fronthaul, for different eNB breaking points [32]. The challenge, however, is mediating between the limitations of the realistic fronthaul network, the exigent C-RAN needs, and the achievable centralisation gains by finding the optimum level of splitting the eNB functions.

Indeed, the optimum functional split can only be decided based on the available fronthaul solutions, and the required fronthaul performance can only be stipulated by determining the level of RAN centralisation. Diverse efforts in the industry are leading towards the convergence of using the common Ethernet packet backhaul for the fronthaul, motivated by the advantages of ease of deployment, interoperability, and cost. Delay and loss of synchronisation remain challenges for the adoption of Ethernet in the fronthaul and are currently being addressed by the iCIRRUS[4] project [33].

9.5.1   Functional split

Some cellular operators in Asia pacific have already started deploying outdoor C-RAN but are, however, facing challenging fronthaul issues, despite the broad availability of optic fibre in the country. As such, in [30], the usage of the next-generation fronthaul interface (NGFI), is proposed instead of the CPRI. The NGFI is Ethernet based, hence packet switched, and does not naturally support frequency and phase synchronisation; solutions are proposed but limitations still exist in the presence of inter-cell coordination techniques such as Coordinated Multi-Point (CoMP). Moreover, jitter and latency remain unsolved difficulties facing the realisation of NGFI. NGFI looks at a different functional split in order to alleviate the fronthaul

---

[3]Interworking and Joint Design of an Open Access and Backhaul Network Architecture for Small Cells based on Cloud Networks.
[4]Intelligent Converged network consolidating Radio and optical access aRound USer equipment is an EU Horizon 2020 project.

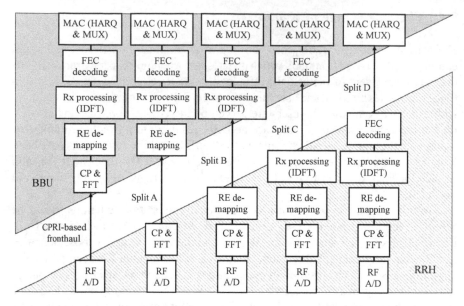

Figure 9.4 Different levels of splitting the functions of an eNB between the RRH and BBU-pool are proposed. The interface between the RRH and the BBU-pool is the CPRI-based fronthaul or the X-haul

requirements and decouple its dependency on the number of antennae, the number of users, etc. Authors in [29] present a study on backhaul requirements and impact on signalling processing for different centralisation levels. In [34], the authors compare the set of requirements to the various backhaul solutions, putting forward guidelines for fronthaul dimensioning. 'Fronthaul-lite' is proposed in [33], which transports base-band signals as opposed to sampled radio waveforms. With fronthaul-lite, exorbitant throughput requirements are alleviated and statistical multiplexing is enabled.

The block diagram shown in Figure 9.4 represents the functions that, traditionally, take place in an eNB. The different shown functional split levels indicate the breaking point of the function chain that determines which functions take place at the RRH and which occur at the BBU-pool. For each functional split level, the interface between these two entities is different. The term *X-haul* is used to designate these interfaces [30].

A key aspect is that the backhaul bandwidth requirement scales with larger radio access bandwidth and becomes crippling with the 5G target bandwidth exceeding 100 MHz per small cell. It should be highlighted that CPRI-based fronthaul depicts the baseline C-RAN architecture whereby the RRH and the BBU are connected over a CPRI interface. The round trip time over the CRPI interface is 5 μs and the effective admissive delay over the fronthaul link, including propagation delay is in the order of 100 μs.

The CPRI-based C-RAN and the Splits-A and B represent the physical layer functional split and enable full exploitation of spatial diversity. This may be done through advanced signal processing mechanisms that allow mitigating (even capitalising on) inter-cell interference. This level of cooperation requires tight synchronisation, hence, very low latency and ultra-high bandwidth. Other solutions may be employed to alleviate these requirements such multi-user-detection [35] and in-network processing [36]. In Split-C, equalisation and demodulation/precoding are performed at the RRH while forward error correction and decoding take place at the pooled BBU. This split hinders joint detection techniques but allows joint decoding. Only functions belonging to layers two and three are centralised in Split-D (medium access control MAC layer); i.e., the physical layer functions happen at the RRH. Subsequently, inter-cell coordination is limited to higher-level techniques such as joint scheduling, interference coordination, and path management.

All split options above the CPRI-based scale with the actual traffic, hence, allow exploiting the statistical multiplexing gain based on occupied physical resources. Moreover, the backhaul throughput requirement becomes flexible and more relaxed as it depends on the actual user throughput related to the user channel quality. On the other hand, latency requirements remain critical and are determined by the channel coherence time, hence, the user speed of movement. In these functional splits, latency requirements are governed by the hybrid automatic repeat request (HARQ), link adaptation, and scheduling processes. Opportunistic HARQ is proposed in [37], which divides the HARQ process into a time-critical part conducted at the RRH and computationally intense part that takes place at the BBU; such an approach relaxes the latency requirements over the backhaul.

9.5.2   Which X-haul level to choose?

From a RAN-gain perspective, the answer is clearly the physical layer split. However, from a fronthaul performance requirements point of view, MAC splits and above are more suitable. This trade-off is studied in [38] is a complexity analysis approach of various functional splits versus the fronthaul cost, using graph theory and genetic algorithm to find the optimised centralisation option. The C-RAN gain and incurred cost are also modelled and studied in [39]. These studies promote the development of a flexible and dynamic functional split that is orchestrated by the cloud RAN and that is designed jointly with the backhaul network [29].

A recent case study presents a quantitative take of the trade-off analysis of various functional splits and backhaul technologies by looking at incurred cost and achieved gains [40]. The starting point is a comparative study of the two extreme RAN architectures: fully centralised C-RAN and traditional D-RAN, as shown in Table 9.5.

The case study looks at six possible RAN architectures, as shown in Figure 9.4, and three backhaul technologies: G.fast, microwave, and fibre-based. The system model shown in Figure 9.5 assumes that the macro-cell acts as a backhaul aggregator connecting all small cells in the area back to the backbone through an ideal connection (high bandwidth and low latency). Nine realistic deployment scenarios

Table 9.5   Comparison between D-RAN and C-RAN architectures

Factor	D-RAN	C-RAN
Cost of RRH/small cell	High	Low
Planning, deployment, maintenance of RRH	High	Low
Energy efficiency of RRH	Low	High
Cost of BBU	N/A	High
Planning, deployment, maintenance of BBU	N/A	Low
Energy efficiency of BBU	N/A	High
Potential for resource pooling	Limited	High
Fronthaul requirements	Relaxed	Exigent
Cost of backhaul/fronthaul	High	Higher
Level of inter-cell coordination	Limited	Maximum

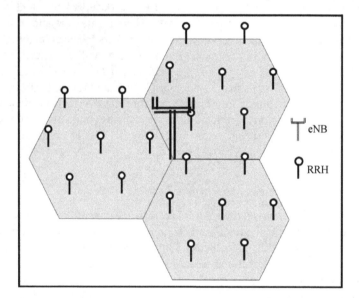

Figure 9.5   System model of a heterogeneous network consisting of 3-sectored macro-cell and seven RRHs (or small cells) per sector. The macro-cell is assumed to have an ideal backhaul link that aggregates the traffic of the 21 RRHs

are selected for the study as detailed in Table 9.6. Each scenario is simulated in MATLAB® and the total effective throughput of each is shown in Figure 9.6. There are two factors impacting the effective throughput: the functional split and the backhaul technology. The functional split results in different backhaul overhead, hence, may limit (or ease) the effective throughput, due to backhaul constraints. This is evident when we compare Scenario 3 with Scenarios 5–7, all employing microwave links

Table 9.6 *Simulated scenarios using different combinations of*
functional split options and X-Haul technologies

Scenario	Functional split	Backhaul technology
Sc1	D-RAN – Traditional eNB	G.fast
Sc2	Split D – MAC forwarding	G.fast
Sc3	Split D – MAC forwarding	Microwave
Sc4	Split C – Soft bit forwarding	G.fast
Sc5	Split C – Soft bit forwarding	Microwave
Sc6	Split B – Rx Data forwarding	Microwave
Sc7	Split A – Sub-frame forwarding	Microwave
Sc8	Split A – Sub-frame forwarding	Fibre-based
Sc9	C-RAN – CPRI – I/Q forwarding	Fibre-based

for backhaul with increasing overhead due to the functional split. Scenarios 5–7 are crippled with the backhaul overhead, resulting in 25 per cent less effective throughput in Scenario 7 compared to Scenario 3. The backhaul technology is another factor limiting the effective throughput, due to its capacity constraints, such as scenarios seven and eight which have the same RAN architecture with microwave backhaul and fibre, respectively. Scenario 7 is severely limited by the backhaul capacity whereas Scenario 8 allows 36 per cent more effective throughput.

The factors that drive the RAN cost down when functions are centralised are many-fold. First, the complexity and size of the RRH decrease with more centralisation, implying lower equipment cost, alleviated site requirements (e.g., cooling, security, rectifier backup, etc.), and less operational costs, such as maintenance, power bills, and site lease. Also, planning, installation and commissioning costs decrease with less complex RRHs, because they are easier to plug-and-play, and are often part of an area-wide deal (e.g. lamp posts, gas stations, billboards, etc.), hence, do not require individual planning and site acquisition efforts. Moreover, centralisation improves resource pooling, especially in a situation where the peak traffic of small cells occurs at different times. The difficulty in the cost analysis stems from tagging a realistic relative cost to each of these separate factors and eNB functions. The total cost of ownership (TCO) of each scenario is computed by adding the capital and operational expenditure over 10 years.

The case study, then, compares the gains/losses of each of the eight deployments scenarios featuring variable levels of centralisation, compared to the D-RAN, as shown in Figure 9.6. Scenario 1, D-RAN with G.fast, acts as a benchmark; the capacity gains/losses of all other scenarios are derived by comparing their respective cumulative effective throughput to the baseline. In parallel, the increase/decrease in TCO of each scenario is defined with respect to Scenario 1. The diagonal line $\Delta_{Tput} = \Delta_{TCO}$ separates the region of advantageous from the unprofitable scenarios. Those that fall on the line incur comparable cost increase and capacity gain, those below the line are dominated by cost, whereas those above have higher capacity gains. It can be observed from Figure 9.6 that the gain of centralisation in Scenarios 8 and 9 is more

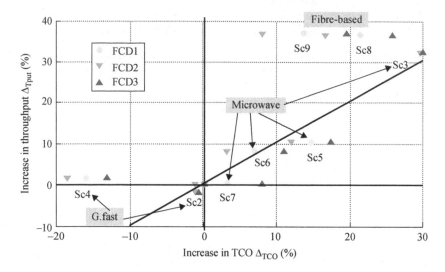

Figure 9.6   *The increase/decrease in effective throughput of each scenario, relative to Scenario 1, is compared to the corresponding increase/decrease in TCO for the three different FCD assumptions. The diagonal line separates the profitable and non-profitable regions; scenarios that fall below the line indicate higher increase in TCO than in throughput*

important than the extra cost of fibre. Thus, these results answer our open question partly: a joint perspective of C-RAN with fibre-based fronthaul reveals that the cost reduction of function centralisation overtakes the cost of fibre, hence is cost effective as opposed to the common belief that the C-RAN gain was met with debilitating fibre costs. Moreover, profitable scenarios are identified, where a scenario is considered profitable if the effective throughput gain is higher than the TCO increase, relative to the benchmark Scenario 1. Scenarios 2, 5, and 7 are not profitable as the incurred additional cost compares with the increase in throughput, hence are not justified. Scenario four (Split-C) is the most cost-efficient among the G.fast-based solutions because it offers a reduction in cost due to centralisation, while compensating for reduction is small cell capacity by shifting users to the available macro layer, thus resulting in unaltered cumulative throughput. Moreover, functional splits B, A and C-RAN may not be supported with G.fast; whereas with microwave, the only RAN options are D-RAN and Split D due to the high cost of the microwave that cancels out the centralisation gain.

The main message that can be drawn from this study is that RAN centralisation is advantageous for reducing the RAN cost, on one hand, and that fibre, while incurring the highest TCO among backhaul technologies, allows nonetheless a significant increase in throughput, rendering the C-RAN with fibre fronthaul (Scenario 9) the most cost-effective solution from a joint backhaul-RAN perspective. On the other hand, the results advocate the need for a heterogeneous backhaul/fronthaul with variable performance and cost to cater for different small cell needs.

9.6   Solving the X-haul challenge with SON

We have demonstrated in this chapter that the optimum X-haul option depends on many factors, some static and others dynamic. To this end, it is critical to design a dynamic X-haul solution that continuously adjusts to the changing demands and conditions of the radio and backhaul networks.

The rapidly rising number of radio cells and corresponding X-haul links increases the complexity bar of network optimisation. Meanwhile, optimisation objectives are also broadening with the emergence of new services and the explosive growth of connected devices. Henceforth, an optimised network is one that offers users' expected throughput, latency, availability, security, etc. with minimum energy consumption and minimum cost. Often, these optimisation objectives are conflicting and different users associate different priorities to each performance target. Thus, the most challenging aspect of future network optimisation is to dynamically adapt the optimisation technique to mediate between these conflicting goals and the diversity of users' requirements and network availability. This recent dimension renders the 5G network (both RAN and backhaul) optimisation highly complex with multi-objectives and multi-constraints.

To this end, the overwhelming growing number of cells and devices, a profusion of relevant parameters, and the mesh-like growth of the 5G backhaul dictate employing SON to automate the organisation and optimisation of the network in a distributed manner. The main advantage of SON is its fast adaptability to the dynamic network, relative to a centrally optimised solution. The challenge is to design efficient SON algorithms with low complexity in order to avoid an increase in cost and energy consumption of network elements that employ these algorithms.

9.6.1   Why SON for the fronthaul?

SON schemes have proved to be successful in addressing complexity issues and preventing rising operational costs, when applied to the RAN, why not extend their usage to the X-haul? By applying similar and complementary SON concepts to all parts of the network, synergies between the radio and backhaul networks can be created, providing benefits to operators in terms of reduced operational costs, and to users in terms of an optimised individual end-to-end experience.

SON is a broad terminology that encompasses three mechanisms: (i) self-organisation (or self-configuration), (ii) self-healing, and (iii) self-optimisation. These mechanisms jointly allow for accelerated deployment of radio cells and corresponding X-haul and for reduced total cost of ownership. Self-organised X-haul solutions speed up the deployment by virtue of their minimal-touch approach for installation and configuration. Similarly, the complex planning phase is avoided which results in faster rollout at lower capital expenditure. Moreover, the self-healing and self-optimising features of an X-haul solution reduce the optimisation and maintenance efforts of a network operator which effectively harnesses the operational cost.

9.6.2   State-of-the-art SON for the X-haul

The benefits of SON-capable backhaul networks have recently caught the attention of key industry players such as Ericsson [41], Nokia [42], and Fujitsu [43]. In view of the sheer number of small cells in future networks, a plug-and-play method is needed where the cells recognise each other and self-build the network topology. It is anticipated that these networks would be deployed in an unplanned incremental approach, following traffic demand. Besides, it is imperative that the fronthaul be adaptive and flexible in an organic way to allow for extra capacity injection where and when needed.

SON is a promising technique that allows for ease of fronthaul deployment with cost-effective scalability, dynamic expandability and optimisation, automated management, and energy efficiency while harnessing the incurred TCO.

9.6.2.1   Research body

The research community jointly with the telecommunication industry has started looking at solving the X-haul problem using SON in different projects. For instance, BuNGee[5] proposes a joint design of backhaul and access networks, using heterogeneous radio elements and a cognitive radio backhaul approach enabled by SON capabilities [44]. In one of their publications, they offer a green-oriented implementation of the cognitive backhaul [45] in which user association is geared towards prioritising RRHs with higher load, when possible, to allow a higher number of RRHs to be in sleep mode, thus economising energy. Self-optimisation is frequently employed in resource allocation schemes when the radio and backhaul share (fully or partially) the same spectrum. Such a SON solution is presented in a joint radio access and in-band backhaul scenario in [46]. The authors propose to use cognitive self-organising networking mechanisms in the form of reinforcement learning, in order to autonomously optimise the resources allocated to and within each carrier. SON is also used in the optimisation of user-cell association such as [47]. In this work, Q-learning is employed by cells to adjust dynamically their corresponding cell range extension offset values and the ICIC mechanism leading to improvement in users' throughput.

Authors in [48] propose a self-optimising wireless mesh network (SWMN) to cater for the fronthaul. SWMN assumes very narrow beam-steerable point-to-point millimetre wave links connecting the RRH/small cells. Network configuration is automated whereby complex and processor-intensive tasks (e.g., route and link schedule computation) are centralised while time-critical operations (e.g., failure recovery and congestion management) are locally performed in a distributed manner.

---

[5]BuNGee: Beyond Next Generation Mobile Broadband. A project part of the 7th Framework Programme funded European Research and Technological Development.

The user-centric-backhaul scheme is a novel SON feature that optimises the association of users with candidate cells based on users' quality of service (QoS) targets and network end-to-end capabilities and constraints [49]. This is implemented in a distributed manner and takes place at the small cells with minimum complexity. The small cells are reinforcement learning agents that receive dynamic information regarding network end-to-end conditions and optimise accordingly a set of bias factors that are broadcast. Each bias factor reflects the cell's end-to-end capabilities with respect to a given QoS. The users get the signal-to-noise ratio of candidate cells, read the broadcast bias factors, and develop accordingly a user-centric virtual view of the candidate cells' coverage that guides these users to select the cell that is more likely to deliver their required QoS.

9.6.2.2 Commercial solutions

A multi-point-to-multi-point microwave-based SON-enabled small-cell backhaul solution is commercially available since 2014 [50]. CCS Metnet nodes connect autonomously to form self-organising, self-healing links that dynamically reconfigure to optimise performance and spectral efficiency as traffic and LoS conditions change whilst minimising interference. The CCS Metnet system polls the network continually in order to determine the optimal topology that offers tailored capacity following the dynamic capacity needs. Another similar SON-enabled solution is presented in [51]. The V-60 version of this solution employs the millimetre-wave V-band (57–64 GHz) with 250 MHz channelization.

Another SON-enabled solution employs flexible optical fibre backhaul networks to provide a self-optimised mobile backhaul to small cells [42]. A proof of concept was demonstrated at the Mobile World Congress in 2015, showing fully automated capability for self-configuration and self-optimisation of multi-vendor mobile backhaul networks using SDN as the management layer.

A similar xPON-based SON-enabled backhaul solution is offered in [43]. It is called the Smart CPRI Transport solution and is developed to enable C-RAN architectures and corresponding fronthaul network. The solution is built around a smart WDM[6] system with integrated diagnostics, remote visibility, self-healing functionality, and ultra-low latency.

9.6.2.3 The FrontHaul-as-a-Service

Fronthaul as a service has been proposed in [52] as means of building a cost-efficient C-RAN solution (often referred to as *managed CPRI service*). The idea is that a FrontHaul-as-a-Service (FHaaS) would widen the C-RAN's addressable market by avoiding the mobile operators' need for self-owned fibre, and as such improving the business case of C-RAN. It is claimed that the first such simplified concept, referred to as FHaaS, has been deployed in a cellular network operator in Asia Pacific [53]. The RaaS Fronthaul-over-IP concept was demonstrated, recently, with the capability of virtualisation of the LTE RAN employing a patented transport and

---

[6]Wavelength Division Multiple access – a low-latency flexible-bandwidth xPON technology.

application protocol stack that replaces CPRI and can operate over a non-fibre-grade fronthaul [54].

9.7 Challenges of SON in the fronthaul

SON seems to be the ideal tool for setting-up, maintaining, upgrading, expanding, configuring, and optimising the evolving fronthaul network. However, there are important challenges ahead that need to be considered and addressed for such a successful usage of SON, some are listed here:

- Conflicting SON: Different SONs may be deployed at different stages of the network, simultaneously. For instance, a SON mechanism at the cloud RAN may be employed to control the inter-cell interference, while a second is installed at the fronthaul aggregation point for dynamic routing and scheduling, and a third runs at the network edge (small cell/RRH) responsible of power control and antenna azimuth/tilt tuning. The outcome of each of these mechanisms affects the other two and their interaction requires challenging orchestration to ensure network stability and optimum performance.
- SON inter-operability: Following the commercial development of SON, it is quite possible that SON solutions (or SON-enabled small cells/RRHs) would be sold as black boxes. Such an approach would lead to various types of black boxes co-existing in the same network or the same geographical area. It is then essential that inter-operability measures between SON mechanisms be standardised or regulated to ensure proper behaviour of the network(s). This is a major challenge that requires immediate attention assuming little (or none) visibility to the details of the SON mechanisms in each box, as these may be proprietary.
- SON TCO: SON is meant to alleviate the total cost of ownership of the network by automating network costly tasks, but SON has an inherent cost that needs to be contained as well. The design of SON algorithms, their tuning and adaptation to evolving network features, and their synchronisation to other running SONs in the network are all delicate and costly tasks. It is, hence, a challenge to design SON solutions with the least complexity possible but with enough intelligence and awareness to survive other entities' conflicting SON efforts and limit their own negative impact on the rest of the network. The complexity of a SON solution is reflected in its capital and operation expenditures (i.e., TCO).

Acknowledgements

The views expressed here are those of the authors and do not necessarily reflect those of the affiliated organisations. The authors would like to thank the UK Engineering and Physical Science Research Council (EPSRC) and BT Research and Innovation for funding this research through an Industrial Cooperative Awards in Science & Technology (iCASE) studentship. We would also like to acknowledge the support of the University of Surrey 5GIC (http://www.surrey.ac.uk/5gic) members for this work.

References

[1]   NGMN-Alliance, "5G white paper," [Online] Available: http://www.3gpp.org/technologies/presentations-white-papers, 2015.

[2]   D. Soldani and A. Manzalini, "Horizon 2020 and beyond: on the 5G operating system for a true digital society," *IEEE Vehicular Technology Magazine*, vol. 10, no. 1, pp. 32–42, 2015.

[3]   A. Mohamed, O. Onireti, M. A. Imran, A. Imran and R. Tafazolli, "Control-data separation architecture for cellular radio access networks: a survey and outlook," *IEEE Communication Surveys and Tutorials*, vol. 18, no. 1, pp. 446–465, 2016.

[4]   F. Barros, K. Ahl and V. Chaillou, "Creating a connected continent: press Conference Warsaw ([Online]. Available: http://www.ftthcouncil.eu/)," Warsaw, 2015.

[5]   T. French, "Global broadband subscriber numbers – Q1 2014," [Online] Available: http://point-topic.com/free-analysis/global-broadband-subscriber-numbers-q1-2014/, 2014.

[6]   Ericsson, "Microwave towards 2020: Delivering high-capacity and cost-effective backhaul for broadband networks today and in the future," [Online] Available: https://www.ericsson.com/res/docs/2015/microwave-2020-report.pdf, Sep. 2015.

[7]   J. Maes and C. J. Nuzman, "The past, present, and future of copper access," *Bell Labs Technical Journal*, vol. 20, pp. 1–10, 2015.

[8]   Small Cell Forum, "049.01.01 – Backhaul technologies for small cells: use cases, requirements and solution," *Small Cell Forum*, [Online] Available: http://scf.io/en/documents/all_documents.php, 2013.

[9]   J. Segel and M. Weldon, "Light radio portfolio: technical overview," *Alcatel-Lucent*, [Online] Available: https://www.scribd.com/document/49017733/Light-Radio-by-ALU, 2011.

[10]   S. Lins, P. Figueiredo and A. Klautau, "Requirements and evaluation of copper-based mobile backhaul for small cells LTE networks," *SBMO/IEEE MTT-S International Microwave & Optoelectronics Conference (IMOC)*, Rio de Janeiro, pp. 1–5, 2013.

[11]   D. Bladsjo, M. Hogan and S. Ruffini, "Synchronization aspects in LTE small cells," *IEEE Communications Magazine*, vol. 51, no. 9, pp. 70–77, 2013.

[12]   R. G. H. Van Uden, R. A. Correa, E. A. Lopez, *et al.*, "Ultra-high-density spatial division multiplexing with a few-mode multicore fibre," *Nature Photonics*, vol. 8, no. 11, pp. 865–870, 2014.

[13]   D. Nesset, "PON roadmap [invited]," *Journal of Optical Communications and Networking*, vol. 9, no. 1, pp. A71–A76, 2017.

[14]   R. Parker, A. Slinger, M. Taylor and M. Yardley, "Final report for Ofcom: future capability of cable networks for superfast broadband," www.analysismason.com, 2014.

[15]   M. Jaber, M. Imran, R. Tafazolli and A. Tukmanov, "5G backhaul challenges and emerging research – a survey," *IEEE Access*, vol. 4, no. 2016, pp. 1743–1766, 2016.

[16]   M. Coldrey, "Maturity and field proven experience of millimetre wave transmission," ETSI – European Telecommunications Standards Institute, 2015.

[17]   ETSI, "Introduction to millimetre wave transmission industry specification group," Millimetre Wave Transmission Industry Specification Group (mWT ISG), 2015.

[18]   Ericsson, "Microwave towards 2020," *White Paper* [Online]. Available: https://www.ericsson.com/res/docs/2015/, 2015, Accessed on 30 June 2016.

[19]   E-band Communications, "E-Link 1000Q-CPRI," [Online]. Available: http://www.e-band.com/E-Link-1000Q-CPRI. Accessed on 15 September 2016.

[20]   CableFree, "MMW: CableFree 10Gbps Links," [Online]. Available: http://www.cablefree.net/cablefree-millimeter-wave-mmw/10g/. Accessed on 15 September 2016.

[21]   CableFree, "FSO Gigabit," [Online]. Available: http://www.cablefree.net/pdf/CableFree%20FSO%20Gigabit%20Datasheet.pdf. Accessed on 15 September 2016.

[22]   M.-C. Jeong, J.-S. Lee, S.-Y. Kim, *et al.*, "8 × 10-Gb/s terrestrial optical free-space transmission over 3.4 km using an optical repeater," *IEEE Photonics Technology Letters*, vol. 15, no. 1, pp. 171–173, 2003.

[23]   T. Martin, "Fibre to the home," [Online]. Available: http://www.cisco.com/, 2007, Accessed on 27 June 2016.

[24]   V. Diaz, "Backhauling with fibre?," [Online]. Available: http://www.corning.com/, 2015, Accessed on 27 June 2016.

[25]   P. McDougall, "Microtrenching and street works: an advice note for local authorities and communications providers," [Online]. Available: https://www.gov.uk/government/publications/, 2011, Accessed on 27 June 2016.

[26]   K. Ahl, "FTTH handbook," [Online]. www.ftthcouncil.eu/documents/Publications/, 2014, Accessed on 27 June 2016.

[27]   D. Mavrakis, C. White and F. Benlamlih, "Last mile backhaul options for west {European} mobile operators," [Online]. Available: http://cbnl.com/sites/all/files/userfiles/files/, 2010, Accessed on 22 July 2016.

[28]   Alcoa Fujikura Ltd., "Reliability of fibre optic cable systems," [Online]. Available: www.southern-telecom.com/solutions/afl-reliability.pdf, 2011, Accessed on 17 July 2016.

[29]   D. Wubben, P. Rost, J. Bartelt, *et al.*, "Benefits and impact of cloud computing on 5G signal processing," *IEEE Signal Processing Magazine*, vol. 31, no. 6, pp. 35–44, 2014.

[30]   C. L. I, J. Huang, Y. Yuan, S. Ma and R. Duan, "NGFI, the xHaul," in *IEEE Globecom Workshops*, San Diego, CA, 2015.

[31]   U. Dotsch, M. Doll, H.-P. Mayer, F. Schaich, J. Segel and P. Sehier, "Quantitative analysis of split base station processing and determination of advantageous architectures for LTE," *Bell Labs Technical journal*, vol. 18, no. 1, pp. 105–128, 2014.

[32]   A. Maeder, M. Lalam, A. de Domenico, *et al.*, "Towards a flexible functional split for cloud-RAN networks," in *European Conference on Networks and Communications (EuCNC)*, Bologna, 2014.

[33] N. J. a. C. P. Gomes, P. Turnbull, A. Magee and V. Jungnickel, "Fronthaul evolution: from CPRI to Ethernet," *Optical Fiber Technology*, vol. 26, pp. 50–58, 2015.

[34] J. Bartelt, P. Rost, D. Wubben, J. Lessmann, B. Melis and G. Fettweis, "Fronthaul and backhaul requirements of flexibly centralized radio access networks," *IEEE Wireless Communications*, vol. 22, no. 5, pp. 105–111, 2015.

[35] INFSO-ICT-317941 iJOIN, "D2.1 – State-of-the-art of and promising candidates for PHY layer approaches on access and backhaul network," www.ict-ijoin.eu/wp-content/uploads/2014/01/D2.1.pdf, 2013.

[36] P. Henning, F. Jorg and D. Armin, "In-Network-Processing: distributed consensus-based linear estimation," *IEEE communication letters*, vol. 17, no. 1, pp. 59–62, 2013.

[37] P. a. P. A. Rost, "Opportunistic hybrid ARQ-enabler of centralized-RAN over nonideal backhaul," *IEEE Wireless Communications Letters*, vol. 3, no. 5, pp. 481–484, 2014.

[38] J. Liu, S. Zhou, J. Gong, Z. Niu and S. Xu, "Graph-based framework for flexible baseband function splitting and placement in C-RAN," arXiv preprint arXiv:1501.04703, 2015.

[39] V. Suryaprakash, P. Rost and G. Fettweis, "Are heterogeneous cloud-based radio access networks cost effective?," *IEEE Journal on Selected Areas in Communications*, vol. 33, no. 10, pp. 2239–2251, 2015.

[40] M. Jaber, D. Owens, M. A. Imran, R. Tafazolli and A. Tukmanov, "A joint backhaul and RAN perspective on the benefits of centralised RAN functions," in *IEEE International Conference on Communications (ICC)*, Kuala Lumpur, 2016.

[41] S. Khan, J. Edstam, B. Varga, J. Rosenberg, J. Volkering and M. Stumpert, "The benefits of self-organizing," 27 September 2013. [Online]. Available: https://www.ericsson.com/res/thecompany/docs/publications/ericsson_review/2013/er-son-transport.pdf. Accessed on 13 September 2016.

[42] Nokia, "Nokia Networks, Coriant first to extend self-organizing networks to mobile backhaul #MWC15," 12 February 2015. [Online]. Available: http://company.nokia.com/en/news/press-releases/2015/02/12/nokia-networks-coriant-first-to-extend-self-organizing-networks-to-mobile-backhaul-mwc15. Accessed on 13 September 2016.

[43] Fujitsu, "Fujitsu unveils intelligent mobile fronthaul solution for C-RAN Architectures," 17 March 2016. [Online]. Available: http://www.fujitsu.com/us/about/resources/news/press-releases/2016/fnc-20160317-02.html. Accessed on 13 September 2016.

[44] Z. Roth, M. Goldhamer, N. Chayat, *et al.*, "Vision and architecture supporting wireless Gbit/sec/km2 capacity density deployments," in *Future Network and Mobile Summit*, Florence, 2010.

[45] J. Lun and D. Grace, "Cognitive green backhaul deployments for future 5G networks," in *First International Workshop on Cognitive Cellular Systems (CCS)*, Rhine River, 2014.

[46]   P. Blasco, M. Bennis and M. Dohler, "Backhaul-aware self-organizing operator-shared small cell networks," in *IEEE International Conference on Communications (ICC)*, Budapest, 2013.

[47]   M. Simsek, M. Bennis and A. Czylwik, "Dynamic inter-cell interference coordination in HetNets: a reinforcement learning approach," in *IEEE Global Communications Conference (GLOBECOM)*, Anaheim, CA, 2012.

[48]   P. Wainio and K. Seppanen, "Self-optimizing last-mile backhaul network for 5G small cells," in *IEEE International Conference on Communications (ICC)*, Kuala Lumpur, 2016.

[49]   M. Jaber, M. A. Imran, R. Tafazolli and A. Tukmanov, "A distributed SON-based user-centric backhaul provisioning scheme," *IEEE Access*, vol. 4, pp. 2314–2330, 2016.

[50]   Cambridge Communication Systems (CCS), "Metnet system," 2014. [Online]. Available: http://www.ccsl.com/metnet-system/. Accessed on 13 September 2016.

[51]   Intracom Telecom, "SON automation in small-cell backhaul," 2014. [Online]. Available: http://www.intracom-telecom.com/downloads/pdf/products/wire less_access/StreetNode_Solution_Paper-SON.pdf. Accessed 13 on September 2016.

[52]   R. Avital, "Cloud-RAN – fronthaul perspective," 4 September 2013. [Online]. Available: http://blog.ceragon.com/blog/blogs/backhaulforum/cloud-ran-front haul-perspective/. Accessed on 16 September 2016.

[53]   C. Gabriel, "Malaysia gets first 'fronthaul as a service' option for C-RAN," 8 April 2016. [Online]. Available: http://rethinkresearch.biz/articles/malaysia-gets-first-fronthaul-service-option-c-ran/. Accessed on 16 September 2016.

[54]   Phluido, "Radio-as-a-Service," 6 September 2016. [Online]. Available: http://www.phluido.net/main/. Accessed on 16 September 2016.

Chapter 10

NFV and SDN for fronthaul-based systems

Maria A. Lema[1], Massimo Condoluci[1], Toktam Mahmoodi[1],
Fragkiskos Sardis[1], and Mischa Dohler[1]

10.1 Introduction

The trend to employ softwarisation in order to introduce virtualisation of network functions in the mobile core is driven by the increasing need for *flexible configuration and operation* of the network. Such needs are stipulated by the wide and dynamic requirements of new services that next-to-come fifth generation (5G) of mobile networks will be supporting. Traditional hardware-based network deployments would incur high capital and operation expenditures (CAPEX and OPEX, respectively) as a result of network expansion or modification requiring hardware addition or replacement. Flexibility is thus one of the key drivers in the design of 5G networks in order to support the heterogeneous requirements of all 5G applications, and to enable multiple network embodiments within one single network deployment.

In the design of Radio Access Network (RAN) architecture, there are similar trends of including softwarisation and virtualisation to introduce flexibility, where flexibility means the possibility of configuring and deploying network functions according to the needs of the traffic to be supported. This dictates for a modular architecture, where different building blocks can be configured and inter-connected to satisfy heterogeneous requirements [1]. One example in this direction is the Centralised/Cloud RAN (C-RAN) architecture [2], which splits the functionalities of a base station in two parts. In a C-RAN, the remote radio head (RRH, transmitting/ receiving radio signals) is decoupled from the digital function unit, i.e., the Base Band Processing Unit (BBU). This means that some of lower layer functionalities are close to the radio side and some of the higher layer functionalities can be now managed by a centralised entity which may also run in a cloud environment.

From a more general point of view, flexibility means having different degrees of freedom when managing the traffic, especially when considering that network functionalities are spread across the whole network (i.e., some functions may be managed at the radio side and other in cloud environments). When coming to a C-RAN environment, the split between low and high layer functionalities involves

[1]Department of Informatics, King's College London, UK

the fact that the traffic is split into two segments, fronthaul and backhaul [3]. The fronthaul deals with the RAN segment of the mobile network and manages the traffic from the RRHs to BBU, and vice versa, where such a traffic asks for: (i) high capacity as it carries the whole radio frame which may require higher data rate than data carried by the radio frame and (ii) very low latency and jitter requirements to guarantee synchronisation between the baseband processing at BBU and radio frame at RRHs. The backhaul manages the traffic from the BBU to the core network, which has relaxed capacity and latency requirements compared to fronthaul as in this case the traffic is composed of packets instead of whole radio frames (thus meaning that the requested data rate depends only on the amount of data/control information to be transmitted). Traffic in fronthaul and backhaul networks also depends on how network functionalities are split. From this point of view, there have been discussions in the community that argue the feasibility of centralisation. A fully centralised RAN will typically have all the functionalities decoupled from the radio unit and this facilitates the use of cooperative radio features to optimise network performance. Nevertheless, this poses capacity issues and it certainly limits the use of multiple antenna systems asking for very high data rates, whose support depends on the available technology. This underlines the fact that fronthaul availability could pose stringent constraints to C-RAN configuration.

This dependency with the underlying fronthaul infrastructure motivates the development of solutions allowing flexibility in the level of centralisation according to fronthaul availability [4]. In this direction, some of RAN functionalities can be implemented as software and virtualised to develop a virtualised RAN [5]. In this sense, the level of centralisation can dynamically go from decoupling the RAN radio and baseband (i.e., split at physical layer) to splitting functionalities at higher levels.

This chapter discusses two enabling technologies that allow the efficient implementation of fronthaul-based radio access networks. Network Functions Virtualisation (NFV) allows for network functions to be decoupled from hardware, and each functionality can be represented as a virtual machine (VM) or container with the possibility to be migrated across the available infrastructure. Software Defined Networking (SDN) is a network architecture that decouples control and data plane to better manage traffic within the network. A Software-Defined (SD)-RAN controller can bring dynamic programmability into the control and data traffic. In general, the use of NFV provides a flexible architecture, orchestrated by a controller, which considers fronthaul network capabilities to separate RAN functionalities with the aim of maximising the quality of service (QoS) as well as utilisation of network resources.

10.2   Background: NFV and SDN in research and standardisation

10.2.1   Network functions virtualisation (NFV)

The proprietary nature of existing hardware as well as the cost of offering the space and energy for a variety of middle-boxes limits the flexibility into today's networks.

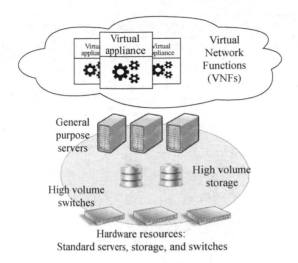

Figure 10.1   The Network Functions Virtualisation (NFV) paradigm

For instance, current 4G mobile networks suffer in terms of presence of proprietary black boxes, huge variety of expensive and proprietary equipment and inflexible hard-state signalling protocols. This drastically limits network programmability from a vendor point of view while involving high costs and limiting the time to market of introducing new services from an operator point of view [6,7].

NFV [8,9] brings a radical shift in the way network vendors and operators design and deploy their infrastructure by separating software instances from the underlying hardware. The main idea behind virtualisation is that Virtualised Network Functions (VNFs) are implemented through software virtualisation techniques and run on commodity hardware (i.e., industry standard servers, storage, and switches). This is exemplified in Figure 10.1.

The virtualisation concept is expected to introduce a large set of benefits to telco operators: (i) reduction of capital investment; (ii) energy savings by consolidating networking appliances; (iii) reduction in the time to market of new services thanks to the use of software-based service deployments; and (iv) introduction of services tailored to the customer needs. From a vendor point of view, virtualisation allows an easier network (re)configuration and high degree of freedom through the possibility of chaining functions according to operators' needs.

The NFV high level architecture from [10], shown in Figure 10.2, highlights the two major enablers of NFV, i.e., industry-standard servers and technologies developed for cloud computing. Being general-purpose servers, industry-standard servers have the key feature of a competitive price, compared to network appliances based on bespoke application-specific integrated circuits (ASICs). Using these servers may come in handy to extend the life cycle of hardware when technologies evolve (this is achieved by running different software versions on the same platform). Cloud computing solutions, such as various hypervisors, OpenStack, and Open vSwitch [11],

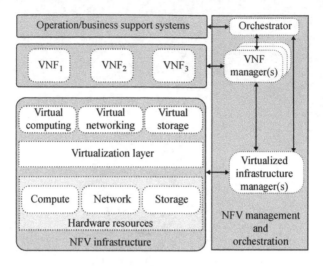

Figure 10.2   High level NFV architecture [10]

enable the automatic instantiation and migration of VMs running specific network services.

The NFV architecture is mainly composed of four different blocks:

- *Orchestrator*, responsible for the management and orchestration of software resources and the virtualised hardware infrastructure to realize networking services.
- *VNF manager*, in charge for: instantiation, scaling, termination, updating events during the life cycle of a VNF, supporting zero-touch automation.
- *Virtualisation layer*, abstracting the physical resources and anchoring the VNFs to the virtualised infrastructure. The virtualisation layer has the key role to ensure that the life cycle of VNFs is independent of the underlying hardware platforms through the use of VMs and their hypervisors.
- *Virtualised Infrastructure Manager (VIM)*, which has the role of virtualising and managing the configurable computing, networking, and storage resources, and control their interaction with VNFs.

Further reading about the NFV architecture can be found in [9,12] and more information on NFV in 5G networks can be found in [5,8].

10.2.2   Software Defined Networking

In traditional IP networks, control (e.g., commands for network devices configuration) and data packets are managed by the same networking devices. This was considered important for the design of the Internet in the early days as it was the best way to guarantee network resilience, which was a crucial design goal. The main drawback of this coupled paradigm is a very complex and relatively static architecture as, for instance, addressed in [13]. A further issue is related to the network management, which is typically handled through a large number of proprietary solutions with their own specialised hardware, operating systems, and control programs. This involves high OPEX/CAPEX as operators have to acquire and maintain different management solutions and the corresponding specialised teams and this further involves long return on investment cycles and limits the introduction of new services.

The *softwarisation* paradigm is useful to overcome above limitations as it introduces the following features:

- *Decoupling of control/data planes.* This means that network devices will act only as packet-forwarding units and will not handle control functionalities.
- *Per flow-based forwarding.* This means that all packets belonging to the same flow (identified by packet header information from layers 2 to 4) receive identical service policies at the forwarding devices, instead of having per-packet routing decisions based only on the packet destination address.
- *Network controller.* Control logic is moved to an external controller, which is a software platform that runs on commodity server technology and provides the essential resources and abstractions to facilitate the programming of forwarding devices based on a logically centralised, abstract network view. This allows to control the network by taking into consideration the whole state of the network.
- *Software-based network management.* The network is programmable through software applications running on top of the network controller that interacts with the underlying devices handling data packets. This allows a quick network reconfiguration and introduction of novel features.

This new vision finds its realisation in SDN [14], whose high-level architecture is depicted in Figure 10.3. SDN will enable an easier way of programming novel network functionalities, of achieving network balancing, and of availing same network information to all applications (i.e., global network view). The switches in Figure 10.3 are SDN network elements running OpenFlow, as such, they receive information by the SDN controller to configure link parameters (bandwidth, queues, metres, etc.) as well as intra-network paths. As a consequence, Figure 10.3 highlights that control and data traffic is now decoupled. Control traffic is now removed from the physical links between the switches and it is instead managed by the SDN controller.

The SDN architecture is depicted in Figure 10.4. Different components may be envisioned, which are presented in more detail in [15]. In the remainder of this section, we will provide a global overview of the SDN architecture.

- *Entities.* SDN is composed of forwarding devices and controllers. The former are hardware- or software-based elements handling packet forwarding, while a controller is a (potentially virtualised) software stack.
- *Planes.* SDN network is composed of three different planes. The data plane (DP) refers to the plane where interconnected devices manage data flows generated from end-users. The control plane (CP) can be considered as the "network brain", as it sends control messages to forwarding devices configure the DP. Finally, the management plane (MP) deals with the set of applications that leverage functions

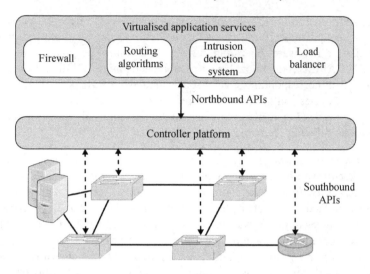

Figure 10.3   SDN with decoupled control and data planes

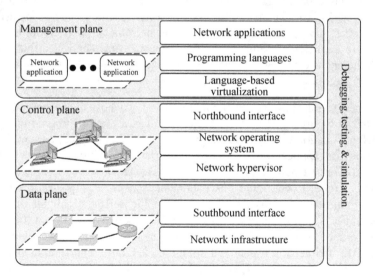

Figure 10.4   The SDN architecture with the related planes, layers, and main entities

such as routing, firewalls, load balancers, monitoring, and so on. Essentially, MP defines the policies, which are ultimately translated by the CP to specific instructions that program the behaviour of DP. From a mobile network point of view, the DP is usually referred to as user plane (U-plane) and it carries data flows. The CP is referred to as C-plane, and it carries control information handling, for instance, user mobility in addition to commands to configure the U-plane.

- *Interfaces.* SDN introduces the concept of southbound interface (SI), which defines the communication protocol and APIs between forwarding devices and CP. OpenFlow [16] is the most widely accepted and deployed open SI, and other protocols are being developed as complimentary to OpenFlow, e.g., ForCES, OVSDB, POF, OpFlex, OpenState, Revised Open-Flow Library (ROFL), Hardware Abstraction Layer (HAL), and Programmable Abstraction of Datapath (PAD) [15]. The northbound interface (NI) is a common interface exploited to develop applications, i.e., the NI abstracts the low-level instruction sets used by SIs to program forwarding devices. A common NI is still an open issue. Commonly, controller applications provide a Representational State Transfer (REST) API for the NI, where system administrators can forward the desired network configuration using a modelling language in XML or JSON format [17].

In a mobile network, SDN is a solution for vendors to manage in a dynamic way the links among the different network functions needed to support a given service and it is thus clear its close relationship with NFV. Indeed, NFV and SDN share the goal of accelerating innovation inside the network through the introduction of programmability. In this direction, SDN can support network virtualisation to enhance its performance and simplify the compatibility with legacy deployments. From an operator point of view, SDN is able to provide an overall abstracted view of the network in order to monitor (and react to) aspects such as congestion.

More information on SDN can be found in [14,15], and further details on SDN in mobile networks can be found in [18]. Finally, further readings on SDN/NFV cooperation can be found in [19].

10.2.3   Standardisation activities

The standardisation landscape on SDN and NFV is already wide. The *European Telecommunication Standards Institute (ETSI)* is managing the standardisation of VFNs and the NFV infrastructure. Separately, the orchestration and management of the architectural framework is carried out by *ETSI-NFV Management and Orchestration (MANO)*. Similarly, the *3rd Generation Partnership Project (3GPP)* is studying the management of virtualised networks. The main aspects of network virtualisation will be treated in the next section.

The *Open Networking Foundation (ONF)* is aiming at promoting the adoption of SDN. The main contribution has been the development of the OpenFlow protocol. The *Internet Research Task Force (IRTF)*, has created the *Software Defined Networking Research Group (SDNRG)* that is currently investigating SDN in both short- and long-term activities, i.e., aiming at identifying the approaches that can be defined, deployed and used in the near term, as well as identifying future research challenges. Similarly, the *International Telecommunications Union's Telecommunication sector (ITU-T)* has already started to develop recommendations for SDN. The *Institute of Electrical and Electronics Engineers (IEEE)* has started some activities to standardize SDN capabilities on access networks based on IEEE 802 infrastructure, for both wired and wireless technologies to embrace new control interfaces.

In the remainder of this chapter we will focus on how NFV and SDN can help in deploying the concept of a dynamic C-RAN to take into consideration fronthaul availability.

10.3 NFV: Virtualisation of C-RAN network functions

10.3.1 Virtualisation as a flexible C-RAN enabler

The heterogeneous nature of 5G use cases in terms of services, features, and traffic requirements requires high degrees of freedom when configuring the best combination of available options from either a network architecture and RAN configuration points of view. Especially from a RAN side, different solutions such as, e.g., Cooperative Multi-Point (CoMP, i.e., different coordinated base stations serve one user to improve his/her channel quality), Dual Connectivity (where a user is simultaneously attached to two or more base stations and it can aggregate traffic from both), Multiple Input Multiple Output (MIMO, i.e., a base station using multiple antennas to boost user throughput), result in demanding support from fronthaul connections, mainly in terms of capacity and latency. Supporting high capacity would require high costs due to the deployment of novel high-bandwidth fronthaul networks, while supporting low-latency would limit the distance from RRHs to BBU (and thus the benefits of having a centralised BBU). To reduce the fronthaul costs and mitigate its demands for higher capacity, solutions considered by the community include the so-called *functionality split* [4], which allows operators to choose the best "level of centralisation" for the C-RAN.

There has been recent work on analysing the inherent trade-off and identifying the ideal functional split. Conclusions mainly highlight that there is no such ideal split, since performance maximisation comes together as a combination of radio features, access techniques, and the available fronthaul/backhaul technology. Hence, a flexible centralised RAN architecture should be pursued, where different functional splits can be configured, and the architectural solution can be easily adapted to satisfy different requirements: service needs or network cost, for example.

This type of RAN configuration requires a complete virtualisation of the functionalities of the base station at all levels. In this context, NFV can provide the C-RAN with the flexibility needed to attend the diverse requirements depending on the specific service and available fronthaul technology. For example, a user located at cell-edge is more likely to be receiving/transmitting in a CoMP mode, whereas a cell-centre user might take more advantage of a Massive MIMO configuration. To this end, every RAN functionality needs to be modular, e.g., represented as a VM or container, which can be easily enabled for a base station according to its needs as well as moved from one machine to another.

Moreover, the increasing signalling and stringent latency required by inter-cell cooperation create additional challenges for network providers in C-RAN environments with heterogeneous deployments, e.g., co-existence of macro and small cells. A possible solution for signalling reduction is the integration at BBU level by grouping

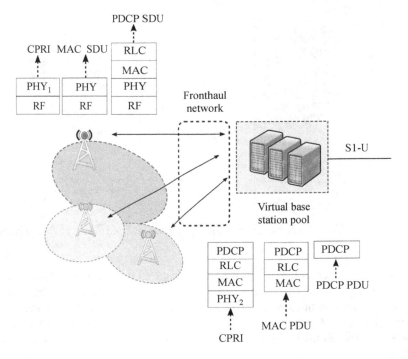

Figure 10.5   C-RAN architecture with multiple functional splits and BBU sharing

several small cells. This idea is popular in the literature, and several architectural solutions involving virtualisation and BBU sharing have been proposed [20–23]. In this way, e.g., mobility signalling due to transitions between the macro and small cells, or other signalling due to cell cooperation can handled locally at the same BBU. To this end, fronthaul networks need to be flexible and be easily reconfigurable, in order to allow path management/re-configuration and efficient BBU mapping.

Figure 10.5 shows an example of the virtualised C-RAN architecture. All three neighbouring cells may share the same BBU (and virtually be considered as the same base station from a core network point of view) and act as one cell for mobility or cooperative purposes. At the same time, different users may need different functionality split options, and the BBU may decide to migrate some functions (i.e., VMs or containers) from RRHs to the BBU, or vice-versa. This means that the fronthaul network should be able to support VM/container migration without affecting capacity and reliability of RRH/BBU traffic.

Signalling is not critical only in terms of inter-cell coordination, but more in general C-plane traffic represents an important ratio of the traffic within the core network, i.e., the Evolved Packet Core (EPC) by referring to 4G networks. This aspect becomes challenging when considering that all C-plane traffic is passed to the EPC in 4G networks and this presents a significant load. Traditional C-RAN can be enhanced by moving other EPC functionalities to the C-RAN in addition to RAN

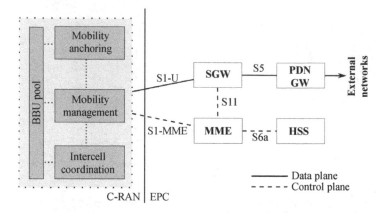

Figure 10.6   C-RAN architecture with core network functions and BBU sharing

functionalities. An example of a C-RAN architecture containing EPC functionalities is shown in Figure 10.6. In this architecture, the C-RAN is thus considered to provide a subset of functionalities of the Mobility Management Entity (MME, the C-plane anchor point in the EPC) and the Serving Gateway (SGW, the U-plane anchor point in the EPC). Examples of functionalities that could be moved from the EPC to the C-RAN are:

- *Control and user plane (C/U-plane) anchoring.* To provide a static endpoint on the C/U-planes for connections between cells belonging to the same C-RAN. This could offload C/U-plane traffic in the EPC and could potentially enable the support of ultra-low latency services by allowing U-plane traffic to be re-direct to the destination cell without going into the EPC.
- *Mobility management.* Several mobility functions can be migrated close to RAN. In particular, by placing those that allow handover between cells belonging to the same BBU pool, backhaul/EPC load can certainly reduce as well as delay and packet loss. Other functions such as those related to device tracking or paging can be also handled at the C-RAN instead of the EPC.

More information on flexible configuration for C-RAN can be found in [2–4].

10.3.2   Virtualisation of the network functions

The main challenge in the virtualisation of mobile networks is related to the physical layer functionalities of the base stations [24]. The main challenge identified is the virtualisation of compute-intensive baseband functions such as the physical (PHY) layer (layer 1), typically implemented on dedicated hardware or on general purpose hardware accelerators. Thus, physical layer virtualisation is discussed in terms of acceleration technologies. Some effort to centralize the functionalities of layer 1 of several base stations are currently in progress aiming at supporting multiple telecommunications technologies and adapting them for new releases.

Virtualisation is most commonly considered for implementation in the higher network stack layers. As an example, [25] is considering to introduce virtualisation in the layer 3 and then in layer 2 of the base stations: layer 3 hosts the functionalities of C/U-planes that connect to the mobile core network while layer 2 hosts the packet data convergence protocol (PDCP), radio link control (RLC), and media access control (MAC) network functions. This level of centralisation provides sufficient level of centralised computing infrastructure for multiple base stations. This may allow the effective deployment of C-RAN networks as service providers will benefit from sharing their remote base station infrastructure to achieve better area coverage with minimum CAPEX and OPEX investment. A more detailed overview on the state of the art in the virtualisation of mobile network can be found in [26].

It is important to note that such utilisation of visualisation and functionality migration will introduce additional delays due to the instantiation of VMs. VMware new family of cloud operating system, i.e., vSphere, is targeting ultra-low latency applications and can achieve near-native performance of 10 μs. However, to support dynamic association and migration of functionalities virtualised under NFV, the VM instances should be migrated while they are operating. There is a body of research in this domain exploring "live migration of VM", and latency of approximately 50 ms is reported in the literature [27]. The VM live migration studies often move the full VM with their associated memories to the new host, while in the case of NFV, a minimum/essential set of migrating elements should be identified so that latency can be kept within the acceptable threshold. Above analysis is interesting from operators point of view and can be used to understand the impact of VNF migration on user performance, while from a vendor side it focuses with a more general aspect on how migration can be optimised to avoid service interruption.

Nikaein et al. [28] presented a proof-of-concept prototype for a virtualised C-RAN built upon the OpenAirInterface Long-Term Evolution (LTE) software implementation [29], Ubuntu 14.04 with low latency kernel (3.17), Linux containers, OpenStack, Heat orchestrator, and Open vSwitch and National Instrument/Ettus USRP B210 RF front-end. As a reference scenario, authors considered a 20 MHz FDD system with single-input single-output and additive white Gaussian noise (AWGN) channel. For this scenario, the processing requirements for handling a base station, considering the LTE subframe of 1msec, are found to be two cores running at 3 GHz. By considering a fully loaded system, the exploitation of one processor core for the receiver processing assuming 16-QAM (quadrature amplitude modulation) on the uplink and 1 core for the transmitter processing assuming 64-QAM on the downlink is able to meet the hybrid automatic repeat request (HARQ) deadlines. By comparing different virtualisation environments, authors conclude that containers (LXC and Docker) offer near bare metal runtime performance, while preserving the benefits of virtual machines in terms of flexibility, fast deployment, and migration. This analysis can be used an inputs for vendors to build their NFV-based platforms.

Operators have shown keen interest in NFV technology in general, and virtualisation of the RAN in particular. This has pushed vendors to investigate different aspects of NFV realisation and create commercial solutions virtualised networks. Leading vendors like NEC, Ericsson, Nokia, Alcatel-Lucent, and Huawei have already started to adopt and upgrade their equipment to support NFV. Examples of this can be found in [20,30].

10.4   SDN: Towards enhanced SON

10.4.1   *C/U-plane decoupling in the mobile network*

It is also interesting to see the transition from fully coupled C/U-planes in the mobile network to a relatively decoupled architecture. The EPC, for the first time in the history of mobile networks, has a clear split into: (i) a packet-only U-plane, comprised of base station (namely the eNB), S-GW and packet data network gateway (PDNGW) and (ii) a C-plane comprised of MME, policy and changing rules function (PCRF) and home subscriber server (HSS) to manage mobility but also management plane functions such policies and charging rules. Although the LTE architecture yields to easier management compared to previous generations of mobile networks, it is still not as adaptable, flexible and programmable as needed.

Introducing SDN in the mobile core network has been so far discussed through integration of software agents (possibly Open vSwitch[1]), installed in all devices, that can be controlled by a SDN controller; examples can be found in [31] and [32]. The introduction of these agents is mainly to maintain the logically centralised nature of the SDN-controller, with the distributed solution, in-line with today's mobile architecture design. Considering 2G, 3G, and 4G networks are all simultaneously active in today's mobile networks, a clean slate approach is not justifiable. To this end, introducing SDN within the existing operational mobile network as in [33], where management plane is retained and significant flexibility and programmability are introduced through a newly introduced C-plane. From a logical point of view, management plane could be exploited by operators to enforce their rules on how to manage the networks and the supported applications. Since the management plane could also potentially be software-defined, the C-plane may, in the long term, subsume the functions offered by the management plane.

10.4.2   *Software defined RAN controller*

Softwarisation represents an important enhancement in effective C-RAN deployments by exploiting the novel features of SDN paradigm. In the previous section the need for virtualisation to enable a fully flexible and adaptive architecture was described. However, such a flexible architecture needs to be orchestrated by a controlling entity capable of managing resources and reconfiguring the network when required. Towards this end, a software defined (SD)-RAN controller can bring dynamic programmability into the C/U-planes, and provide:

- Accurate functional splits based on the QoS, service requirements and underlying fronthaul network.
- Rapid reconfiguration of fronthaul links to optimise traffic paths.

---

[1]OpenVSwitch: An Open Virtual Switch, http://openvswitch.org.

The programmability of the SDN architecture allows the U-plane to only deal with fast rule lookups and executes forwarding at fine timescales, whereas new rules can be pushed into longer timescales due to the latency involved in communicating with the controller. Self-Organised Network (SON) solutions already incorporated this idea of coordinating RAN functions at a higher time scale, with the main objective of achieving significant performance gains in the use of enhanced cooperative transmissions such as: CoMP, interference management, load balancing or power scaling. To do so, operators can exploit an SON controller to collect information about the network (user equipments (UEs) and eNBs), and to configure network parameters and inform the base stations about the configuration [34].

Software defined radio (SDR) and SDN both go from hardware based designed approaches to open programmable reconfigurable systems. Both SDR and SDN can converge into SON architectures enabling SON programmability [35]. Hence, an SD-RAN controller can support SON algorithms and expand its applications to handle other network operations; by using the SDN paradigm, operators can have a global network view through the centralised controller and improve RRH and BBU mapping functions, re-calculate paths along the fronthaul and hence improve load balancing.

10.4.3   Controlled network operations in C-RAN

A C-RAN architecture can be seen as a direct extension of SDN's C/U-plane separation principle to the RAN where the C-RAN and SDN complements each other. Some examples of the application of the SDN in the RAN SON functions can be found in the literature. Authors in [36] and [37] use this idea of reprogrammable SON to adjust dynamically the load balancing among small and macro cells, by setting the cell range expansion (CRE) value. The CRE was introduced by the 3GPP to address load balancing problems in heterogeneous networks: the small cell coverage or range is expanded by adding a cell selection offset to the reference signal received power (RSRP) or reference signal received quality (RSRQ) measured from the small cell. In this way, the users placed in the range expanded area are associated to the small cell. However, to manage the interference caused in the CRE area, an enhanced inter-cell interference coordination mechanism, known as almost blank sub-frames (ABS) must be considered. In this case, the SD-RAN controller can adjust the interference coordination mechanism together with the CRE offset to maximise the RAN performance and thus increase spectrum utilisation from an operator point of view.

Moreover, authors in [38] proposed to handle C-plane tasks such as radio resource management (RRM) or interference coordination logic in an SDN controller implemented in the C-RAN architecture in order to orchestrate the parameters of RRHs (U-plane). In this case, SDN brings some benefits in terms of distributing the control information triggered by the C-RAN to the involved network entities. Another example is for instance the activation/de-activation of RRHs which is decided by the C-RAN according to network load and interference level. In detail, SDN updates the path configuration in order to guarantee reliable communication for the new activated RRHs or to optimize data paths when some RRHs are switched off. This leads to another important benefit of SDN in managing the fronthaul from an operator point of view, i.e., optimising fronthaul utilisation and thus enabling higher capacity.

When considering in detail how to manage the fronthaul links, the C-RAN decouples the BBUs from the RRHs in terms of physical placement, but there is a one-to-one logical mapping between BBUs and RRHs. The path between a BBU and an RRH can be computed to satisfy different rules or policies, like for example: minimum distance or resource utilisation. Since fronthaul networks contribute to the accurate performance of the RAN (in terms of capacity, latency or outage) this path must be accurately calculated. SDNs expose to the orchestrator an abstracted view of the network making sure that there is enough information to make good decisions. In particular, this level of abstraction and path computation in a fronthaul network is studied in [39], and conclusions highlight what a delicate task this is, as there is a high level of information required from the network to accurately allocate RRH and BBUs.

Furthermore, it has been also analysed in the literature, particularly in [40], that this notion of fixed one-to-one mapping can potentially limit the performance of C-RANs. For instance, mobile users require handovers when moving from one RRH to another and in this scenario a one-to-many mapping on the fronthaul link could reduce the overhead and optimize network performance. Similarly, in multi-connectivity scenarios, such as the ones defined by the 3GPP in [41], the complexity of aggregating multiple flows of different base stations can be significantly lower if several RRHs share the same BBU, especially at a radio resource control (RRC) level [22]. Such an aggregation may be useful to reduce load in the RAN and could allow operators to improve RAN capacity.

Another aspect to be taken into consideration is that, with the one-to-one mapping, several BBUs are active and generate frames (and thus consume energy in the BBU pool) even if an enhanced capacity may not be needed in all parts of the network or at all times. As an example, when the traffic load is low in a region (e.g., coverage area of multiple small cell RRHs), a single BBU may be enough to serve the offered load. The SDN paradigm may come in handy to introduce this flexibility in the fronthaul management by treating fronthaul links as network links. This could bring operators to exploit heterogeneous technologies on the fronthaul managing the co-existence between old and new fronthaul links through SDN. Authors in [42] proposed a flexible C-RAN system for RRHs that is based on the introduction of an intelligent controller in the BBU pool that, similarly to an SDN controller, dynamically re-configures the fronthaul (at coarse time scales) based on network feedback to cater effectively to both heterogeneous user and traffic profiles. As a consequence, the amount of traffic demand satisfied on the RAN is maximised for both static and mobile users, while at the same time the computation resource usage in the BBU pool is optimised.

Another element controlled by the SD-RAN is the method for polling all available resources at the connected cells. So doing, the C-RAN can ensure that only the minimum required resources are active at any time in the whole C-RAN, instead of activating the minimum required resources in each cell. The SD-RAN controller would also provide slicing of the network resources to allow for RAN sharing between operators, in order to deploy and support multi-tenancy deployments which are attracting the interest of 5G research community as outlined for instance in [43].

10.5 Conclusions

Virtualisation and softwarisation are two important enablers of 5G networks as they provide the necessary flexibility and configurability to the network design. These novel enablers have been widely considered in the context of mobile core networks, where NFV and SDN have proven to improve the network efficiency by decoupling the network functions from the hardware, facilitating network management. On a separate matter, C-RAN has also shown to have potential to improve the RAN performance, by enabling better inter-cell cooperation and reducing complexity sharing BBUs. However, there is no rule of thumb on which of the possible C-RAN configuration is the best, given that several variables need to be taken into account, as for example the availability of fronthaul networks or the user profile. To ensure a correct C-RAN configuration, a flexible and dynamic topology should be considered. To this end, the community is considering the use of the aforementioned virtualisation and softwarisation technologies. This chapter has identified the main motivations to include these technologies, and it has highlighted the applicability of NFV and SDN in the context of C-RAN.

First, the virtualisation of network functions allows to satisfy the high level of heterogeneity in services and features considered in nowadays access networks. It helps to reduce fronthaul costs and accommodate several functional splits, ensuring that cell cooperation can be configured at any desired level. Moreover, since the use of virtualisation in core networks is already considered, RAN virtualisation allows to collocate RAN and core network functions, migrating some C/U-plane functionalities close to the edge. Finally, the use of SDN in virtualised C-RAN environment has been mainly driven to include a control entity that instructs the network how to configure functional splits, topology and any other RAN feature. Hence, SDN has enabled the programmability of the former SON architectures.

References

[1]   Z. Feng, C. Qiu, Z. Feng, Z. Wei, W. Li, and P. Zhang, "An effective approach to 5G: Wireless network virtualization," *IEEE Communications Magazine*, vol. 53, no. 12, pp. 53–59, Dec. 2015.

[2]   A. Checko, H. L. Christiansen, Y. Yan, *et al.*, "Cloud RAN for mobile networks – A technology overview," *IEEE Communications Surveys Tutorials*, vol. 17, no. 1, pp. 405–426, Firstquarter 2015.

[3]   M. Peng, Y. Sun, X. Li, Z. Mao, and C. Wang, "Recent advances in cloud radio access networks: System architectures, key techniques, and open issues," *IEEE Communications Surveys Tutorials*, vol. 18, no. 3, pp. 2282–2308, Thirdquarter 2016.

[4]   J. Bartelt, P. Rost, D. Wubben, J. Lessmann, B. Melis, and G. Fettweis, "Fronthaul and backhaul requirements of flexibly centralized radio access networks," *IEEE Wireless Communications*, vol. 22, no. 5, pp. 105–111, Oct. 2015.

[5] P. Rost, I. Berberana, A. Maeder, *et al.*, "Benefits and challenges of virtualization in 5G radio access networks," *IEEE Communications Magazine*, vol. 53, no. 12, pp. 75–82, Dec. 2015.

[6] K. Pentikousis, Y. Wang, and W. Hu, "Mobileflow: Toward software-defined mobile networks," *IEEE Communications Magazine*, vol. 51, no. 7, pp. 44–53, Jul. 2013.

[7] H. J. Einsiedler, A. Gavras, P. Sellstedt, R. Aguiar, R. Trivisonno, and D. Lavaux, "System design for 5G converged networks," in *2015 European Conference on Networks and Communications (EuCNC)*, pp. 391–396, Jun. 2015.

[8] S. Abdelwahab, B. Hamdaoui, M. Guizani, and T. Znati, "Network function virtualization in 5G," *IEEE Communications Magazine*, vol. 54, no. 4, pp. 84–91, Apr. 2016.

[9] R. Mijumbi, J. Serrat, J. L. Gorricho, N. Bouten, F. D. Turck, and R. Boutaba, "Network function virtualization: State-of-the-art and research challenges," *IEEE Communications Surveys Tutorials*, vol. 18, no. 1, pp. 236–262, Firstquarter 2016.

[10] European Telecommunications Standards Institute (ETSI), "Network Functions Virtualisation (NFV); Architectural Framework," ETSI GS NFV 002, Oct. 2013.

[11] A. Blenk, A. Basta, M. Reisslein, and W. Kellerer, "Survey on network virtualization hypervisors for software defined networking," *IEEE Communications Surveys Tutorials*, vol. 18, no. 1, pp. 655–685, Firstquarter 2016.

[12] B. Han, V. Gopalakrishnan, L. Ji, and S. Lee, "Network function virtualization: Challenges and opportunities for innovations," *IEEE Communications Magazine*, vol. 53, no. 2, pp. 90–97, Feb. 2015.

[13] H. Kim and N. Feamster, "Improving network management with software defined networking," *IEEE Communications Magazine*, vol. 51, no. 2, pp. 114–119, Feb. 2013.

[14] W. Xia, Y. Wen, C. H. Foh, D. Niyato, and H. Xie, "A survey on software-defined networking," *IEEE Communications Surveys Tutorials*, vol. 17, no. 1, pp. 27–51, Firstquarter 2015.

[15] D. Kreutz, F. M. V. Ramos, P. E. Veríssimo, C. E. Rothenberg, S. Azodolmolky, and S. Uhlig, "Software-defined networking: A comprehensive survey," *Proceedings of the IEEE*, vol. 103, no. 1, pp. 14–76, Jan. 2015.

[16] F. Hu, Q. Hao, and K. Bao, "A survey on software-defined network and OpenFlow: From concept to implementation," *IEEE Communications Surveys Tutorials*, vol. 16, no. 4, pp. 2181–2206, Fourthquarter 2014.

[17] R. T. Fielding and R. N. Taylor, "Principled design of the modern Web architecture," in *Proceedings of the 2000 International Conference on Software Engineering. ICSE 2000 the New Millennium*, pp. 407–416, Jun. 2000.

[18] Z. Cao, S. S. Panwar, M. Kodialam, and T. V. Lakshman, "Enhancing mobile networks with software defined networking and cloud computing," *IEEE/ACM Transactions on Networking*, vol. 25, no. 3, pp. 1431–1444, Jun. 2017.

[19] Y. Li and M. Chen, "Software-defined network function virtualization: A survey," *IEEE Access*, vol. 3, pp. 2542–2553, 2015.

[20] NEC Corporation, "NFV C-RAN for Efficient Resource Allocation," White Paper.

[21] A. W. Dawson, M. K. Marina, and F. J. Garcia, "On the benefits of RAN virtualisation in C-RAN based mobile networks," in *2014 Third European Workshop on Software Defined Networks*, pp. 103–108, Sep. 2014.

[22] M. A. Lema, T. Mahmoodi, and M. Dohler, "On the performance evaluation of enabling architectures for uplink and downlink decoupled networks," in *2016 IEEE Globecom Workshops (GC Wkshps)*, pp. 1–6, Dec. 2016.

[23] D. Pompili, A. Hajisami, and T. X. Tran, "Elastic resource utilization framework for high capacity and energy efficiency in cloud RAN," *IEEE Communications Magazine*, vol. 54, no. 1, pp. 26–32, Jan. 2016.

[24] European Telecommunications Standards Institute (ETSI), "Acceleration Technologies; Report on Acceleration Technologies & Use Cases," ETSI GS NFV-IFA 001, Dec. 2015.

[25] European Telecommunications Standards Institute (ETSI), "Network Function Virtualization: Use Cases," White Paper, 2013.

[26] H. Hawilo, A. Shami, M. Mirahmadi, and R. Asal, "NFV: state of the art, challenges, and implementation in next generation mobile networks (vEPC)," *IEEE Network*, vol. 28, no. 6, pp. 18–26, Nov. 2014.

[27] C. Clark, K. Fraser, S. Hand, *et al.*, "Live migration of virtual machines," in *Proceedings of the Second Conference on Symposium on Networked Systems Design & Implementation – Volume 2*, ser. NSDI'05. Berkeley, CA, USA: USENIX Association, 2005, pp. 273–286. [Online]. Available: http://dl.acm.org/citation.cfm?id=1251203.1251223

[28] N. Nikaein, R. Knopp, L. Gauthier, *et al.*, "Demo: Closer to cloud-RAN: RAN as a service," in *Proceedings of the 21st Annual International Conference on Mobile Computing and Networking*, ser. MobiCom'15. New York, NY, USA: ACM, 2015, pp. 193–195. [Online]. Available: http://doi.acm.org/10.1145/2789168.2789178

[29] N. Nikaein, R. Knopp, F. Kaltenberger, *et al.*, "Demo: OpenAirInterface: An open LTE network in a PC," in *Proceedings of the 20th Annual International Conference on Mobile Computing and Networking*, ser. MobiCom'14. New York, NY, USA: ACM, 2014, pp. 305–308. [Online]. Available: http://doi.acm.org/10.1145/2639108.2641745

[30] Ericsson, "Cloud RAN: The benefits of virtualization, centralization and coordination," White Paper, Sep. 2015.

[31] M. Amani, T. Mahmoodi, M. Tatipamula, and H. Aghvami, "Programmable policies for data offloading in LTE network," in *2014 IEEE International Conference on Communications (ICC)*, June 2014, pp. 3154–3159.

[32] L. E. Li, Z. M. Mao, and J. Rexford, "Toward software-defined cellular networks," in *2012 European Workshop on Software Defined Networking*, pp. 7–12, Oct. 2012.

[33] T. Mahmoodi and S. Seetharaman, "Traffic jam: Handling the increasing volume of mobile data traffic," *IEEE Vehicular Technology Magazine*, vol. 9, no. 3, pp. 56–62, Sep. 2014.

[34] M. Y. Arslan, K. Sundaresan, and S. Rangarajan, "Software-defined networking in cellular radio access networks: Potential and challenges," *IEEE Communications Magazine*, vol. 53, no. 1, pp. 150–156, Jan. 2015.

[35] C. Ramirez-Perez and V. Ramos, "SDN meets SDR in self-organizing networks: Fitting the pieces of network management," *IEEE Communications Magazine*, vol. 54, no. 1, pp. 48–57, Jan. 2016.

[36] K. Tsagkaris, G. Poulios, P. Demestichas, A. Tall, Z. Altman, and C. Destré, "An open framework for programmable, self-managed radio access networks," *IEEE Communications Magazine*, vol. 53, no. 7, pp. 154–161, Jul. 2015.

[37] A. Gopalasingham, L. Roullet, N. Trabelsi, C. S. Chen, A. Hebbar, and E. Bizouarn, "Generalized software defined network platform for Radio Access Networks," in *2016 13th IEEE Annual Consumer Communications Networking Conference (CCNC)*, Jan. 2016, pp. 626–629.

[38] Z. Zaidi, V. Friderikos, and M. A. Imran, "Future RAN architecture: SD-RAN through a general-purpose processing platform," *IEEE Vehicular Technology Magazine*, vol. 10, no. 1, pp. 52–60, Mar. 2015.

[39] M. Fiorani, A. Rostami, L. Wosinska, and P. Monti, "Transport abstraction models for an SDN-controlled centralized RAN," *IEEE Communications Letters*, vol. 19, no. 8, pp. 1406–1409, Aug. 2015.

[40] K. Sundaresan, M. Y. Arslan, S. Singh, S. Rangarajan, and S. V. Krishnamurthy, "FluidNet: A flexible cloud-based radio access network for small cells," *IEEE/ACM Transactions on Networking*, vol. 24, no. 2, pp. 915–928, Apr. 2016.

[41] 3rd Generation Partnership Project (3GPP), "Study on Small Cell Enhancements for E-UTRA and E-UTRAN; Higher Layer Aspects," TR 36.842.

[42] K. Sundaresan, M. Y. Arslan, S. Singh, S. Rangarajan, and S. V. Krishnamurthy, "FluidNet: A flexible cloud-based radio access network for small cells," in *Proceedings of the 19th Annual International Conference on Mobile Computing & Networking*, ser. MobiCom'13. New York, NY, USA: ACM, 2013, pp. 99–110. [Online]. Available: http://doi.acm.org/10.1145/2500423.2500435

[43] M. Condoluci, F. Sardis, and T. Mahmoodi, "Softwarization and virtualization in 5G networks for smart cities," in *Internet of Things. IoT Infrastructures. Proceedings of the Second International Summit, IoT 360*, 2015.

Part III

Backhaul Network

Chapter 11

Mobile backhaul evolution: from GSM to LTE-Advanced

Andy Sutton[1]

Abstract

Mobile backhaul describes the connectivity between cellular radio base stations and the associated mobile network operator's (MNO's) core network. Previously, this was known as 'transmission' but during the 1990s, the term backhaul was adopted. This chapter will review the development of mobile backhaul for the global system for mobile communications (GSM), universal mobile telecommunications system (UMTS) and long-term evolution (LTE), including LTE-Advanced radio access technologies. Original GSM terrestrial transmission interfaces were defined as time division multiplexing (TDM), within Europe and elsewhere these were based on 2.048 Mbps E1 circuits. These circuits could be multiplexed via the plesiochronous digital hierarchy (PDH) and synchronous digital hierarchy (SDH) standards to realise higher order transmission systems. The introduction of UMTS brought a new requirement to support asynchronous transfer mode (ATM) technology within the mobile backhaul domain. ATM is a fixed length cell switching technology which is carried over TDM transmission systems such as PDH and SDH. The evolution of mobile networks from predominately voice-centric to increasingly data-centric operation resulted in the development of High Speed Downlink Packet Access technologies which, together with advanced uplink technologies, resulted in the need to significantly scale the capacity of mobile backhaul networks. To address the need for scalability mobile backhaul evolved from $n \times$ E1 and TDM systems to Carrier Ethernet, there was however a few challenges to address to support this migration, firstly, the end points on base stations and network controllers were all TDM based and secondly, the E1 circuits provided a deterministic synchronisation signal via the native line code which ensured the base station operated within its allocated radio frequency channels. To address the first requirements, there was widespread adoption of pseudo-wire technology, whereas the second challenge was addressed by the introduction of Synchronous Ethernet or alternatively by the deployment of a local or packet-based synchronisation reference. Overtime, the base station and network

---

[1]School of Computing, Science & Engineering, University of Salford, UK

controllers migrated to native Ethernet interfaces and therefore supported end-to-end Carrier Ethernet transmission with an Internet Protocol (IP) transport network layer. LTE was introduced with native Ethernet and IP support, the LTE radio interface offers significantly higher peak and average data rates than previous generations of cellular radio access networks. To support the deployment of LTE technology, the need for combined GSM, UMTS and LTE backhaul and the growing trend towards network sharing between MNOs, resulted in 1 Gbps Carrier Ethernet backhaul solution being deployed. This increase in backhaul capacity requirements effectively ruled out copper twisted-pair-based technologies in favour of ever more optical fibre and high-capacity microwave and millimetre wave radio backhaul technologies. The on-going development of LTE-Advanced features, such as carrier aggregation, continues to drive the capacity of backhaul networks, whereas new products and services ensure constant evolution of performance with lower latency and reduced packet error loss rates becoming the norm.

11.1   Global system for mobile communications

Global system for mobile communications (GSM) was developed by the Confeder-ation of European Posts and Telecommunications (CEPT) Administration as a result of a project which was initiated in 1982, to design a pan-European mobile communi-cations technology. Initial GSM specifications focused on providing a digital mobile voice service which would work across Europe through the establishment of roaming agreements between MNOs. In addition to voice, GSM supported the Short Message Service and circuit switched data services.

The GSM interfaces which would be classed as backhaul differed somewhat between network operators and was, in the main, determined by the operator's choice of network architecture and network element placement. The key decision was whether to deploy centralised or distributed base station controllers (BSCs). In practice, it was quite common for operators to have a mix of centralised and distributed BSCs; however, the level of centralisation or distribution would influence the backhaul transmission strategy, target architecture, design and network planning.

11.2   GSM network architecture

GSM standards define the network architecture and interfaces between network ele-ments. The radio base station is known as a base transceiver station (BTS); this connects to the BSC via the Abis interface. The Abis interface, as with all original European GSM terrestrial interfaces, is based on the ITU-T standardised 2.048 Mbps frame, often referred to as an E1.

The interface between the BSC and core network is an interesting topic to review because whilst standardised was subject to equipment manufacturer's implementation of the GSM network architecture. The Transcoder and Rate Adaptation Unit (TRAU) is a network component within the GSM base station subsystem which connects to the core network of mobile telephone exchanges, generally known as the mobile switching centre (MSC). The MSC is a telephone switch with additional functionality to support subscriber mobility and authentication (visitor location register which connects to the main subscriber register, the home location register). The telephony switching function is based on E1 interfaces carrying traditional 64 kbps A-law encoded pulse code modulation (PCM)-based digital voice signals, the Abis interface transports GSM coded voice which is 13 kbps and mapped with overheads to a 16 kbps sub-rate timeslot (TS). The transcoder function of the TRAU converts between GSM-coded voice and A-law PCM-coded voice. The location of the TRAU determines where 16 kbps voice transmission or 64 kbps voice transmission is required; therefore, it's easy to understand the need for flexible placement of the TRAU to optimise a centralised or distributed BSC architecture.

A centralised BSC architecture is illustrated in Figure 11.1. The Abis interface (16 kbps sub-TSs), identified as reference point number 1, connects between the BTS and the core network site in which it terminates to the BSC. Interface number 2, the Ater interface between the BSC and TRAU (16 kbps sub-TSs) could be internal to the BSC equipment or could be via local E1 coaxial cables, cross connected manually via a digital distribution frame (DDF), likewise transmission for interface number 3, the A interface between the TRAU and MSC (64 kbps TSs) would be implemented with local E1 coaxial cables via a DDF.

The alternative to the centralised BSC architecture is the distributed BSC architecture. This relies on BSCs being distributed to suitable geographical locations and terminating a number of locally connected BTS sites. It also requires connectivity to the core network however rather than transmitting the Abis interface towards the core network site, the Ater interface is transported. There are pros and cons of both approaches, an operator will likely have deployed some centralised and some

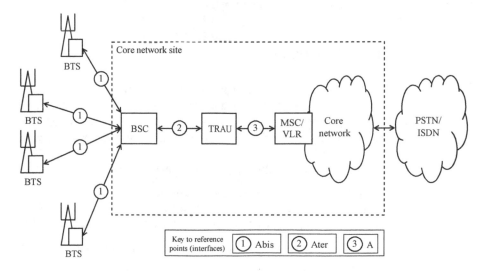

Figure 11.1   GSM network with centralised BSC architecture

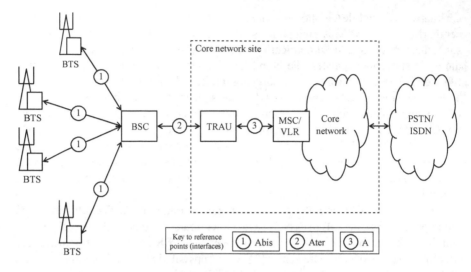

Figure 11.2   GSM network with distributed BSC architecture

distributed BSCs. Those who selected a mainly centralised BSC architecture would likely have a small total number of high capacity BSC platforms and have a need for careful management of Abis sub-TSs on the transmission links between the BSC and distributed BTS sites. Operators who selected a mainly distributed BSC architecture would have more BSCs, typically lower capacity units however would realise statistical multiplexing gain by having BSCs acting as efficient switched concentrators to minimise the backhaul requirements towards the core network site. The distributed BSC architecture is illustrated in Figure 11.2.

11.3   GSM mobile backhaul

To appreciate the backhaul transmission implications of a centralised or distributed BSC model requires a study of the Abis and Ater interfaces; there were some differences between equipment manufactures; however, the examples which follow highlight the principles on which GSM backhaul transmission networks were designed.

The Abis interface carries the following traffic types:

- Full rate voice traffic channels – mapped to 16 kbps sub-TSs
- Transceiver signalling links – mapped to 64 kbps TSs
- Operations and Maintenance (O&M) channel – mapped to a 64 kbps TS

The standard E1 frame is 2.048 Mbps and therefore supports 32 × 64 kbps TSs. TS0 is allocated for transmission link alignment and management [known as Frame Alignment Word/Not Frame Alignment Word (FAW/NFAW)] and is therefore not available to the Abis link. Consequently, this leaves 31×64 kbps (1.984 Mbps) for the Abis payload. The BTS requires a 64 kbps O&M channel to download configuration information and manage alarms, whereas the remainder of the payload is available for traffic channels and associated transceiver signalling. Each GSM transceiver has $8 \times$ TDMA TSs on the radio interface which map to 8×16 kbps sub-TSs on the Abis interface. To support call setup and other Transceiver (TRX) signalling activities, a 64 kbps TRX signalling link is assigned for each TRX. Therefore, a single TRX requires 192 kbps $[(8 \times 16) + 64]$ of Abis transmission capacity. A complete E1 frame can support a maximum of $10 \times$ TRX on a single BTS site (10×192 kbps $+ 64$ kbps O&M channel $+ 64$ kbps FAW/NFAW). As GSM has evolved equipment manufacturers have reduced the data rate required for TRX signalling to 32 kbps, even 16 kbps in certain cases, such that an increased number of TRX's can be supported on an E1 circuit; $12 \times$ TRX became a common number for a single E1 to a single BTS site. Sites with a requirement for > 12TRX will have $n \times$ E1 circuits delivered.

There are a number of physical technologies available to extend an Abis interface from the BSC site to the BTS location. These include copper-line-based technologies, optical fibre and microwave radio systems. The capacity required of these transmission backhaul technologies will depend on the provisioned GSM radio interface capacity and choice of centralised or distributed BSC architecture.

Starting with a review of a centralised BSC architecture, let us consider the technologies in turn and apply them to a typical GSM backhaul network of the 1990s. A new GSM cell site would be provisioned either with a single GSM TRX or with a three cell sector configuration (single GSM sector per 120 degrees of coverage via directional antennas); the backhaul requirements would therefore be 256 kbps $[(8 \times 16) + 64$ signalling $+ 64$ O&M] or 640 kbps $[((8 \times 16) + 64) \times 3 + 64]$. With the exception of mobile operators who were also the national incumbent fixed line operators, it's unlikely that a new GSM operator would have access to raw copper or fibre cables. Therefore, to consume fixed backhaul services, they would typically purchase leased lines from a fixed network operator, either part of the same company group, or in most cases from the national incumbent operator or a competitive challenger. Leased lines attract an upfront Capital Expenditure (CAPEX) and on-going Operational Expenditure (OPEX); however, it does effectively offload some of the technical challenges of implementing backhaul solutions to a third party.

Early copper line technologies could support $n \times 64$ kbps per copper pair, and therefore with copper bonding, a circuit of 2.048 Mbps could be delivered via High bit-rate Digital Subscriber Line (HDSL) technology over three copper pairs. As these copper line technologies developed, the number of copper pairs required to deliver a given data rate decreased and/or the distance over which the service could reach was extended. Copper delivery was very common for $n \times 64$ kbps and E1 based backhaul solutions when leased lines were ordered from national incumbents.

Optical fibre-based solutions can support significantly higher data rates than copper cables; however, the rollout of access and metro fibre transmission systems wasn't very extensive in the 1990s. Given the low data rate requirements of the new GSM networks, fibre wasn't essential in the backhaul transmission domain; however, there were some deployments in support of leased lines along with limited self-build from new GSM network operators. As traffic started to grow over the coming decade, the use of optical fibre-based communications would increase considerably.

Point-to-point microwave radio systems offer an alternative to wireline transmission and could be deployed by a fixed line provider as part of an end-to-end leased line or alternatively, directly by the mobile network operator (MNO) to enable self-management of their mobile backhaul network. The decision to self-deploy microwave radio backhaul transmission was generally driven by an economic analysis; the total cost of ownership was generally lower than the combined CAPEX and on-going OPEX of leased lines.

11.4 Leased lines

Figure 11.3 illustrates a centralised BSC deployment with all Abis backhaul transmission supplied via third party leased lines. The actual technology underpinning the leased lines could be copper, fibre or microwave radio, as long as the leased line meets the performance criteria set out in the service level agreement, the fixed operator can implement the circuit with whichever technology is most appropriate to the particular deployment scenario. Given the geographical reach of the fixed network providing the leased lines, there could be a significant distance between the BTS and centralised BSC location. The leased line service may provide for aggregation of low speed access circuits, typically 256 or 640 kbps per BTS site, to aggregate bearers based on $n \times E1$ with a higher fill ratio towards the BSC. This aggregation is provided by time division multiplexing (TDM) cross-connect equipment which involves mapping at 8 or 64 kbps.

A TDM leased line is effectively a transparent end-to-end connection over which the mobile service (i.e. Abis interface) is carried and the transmission network is

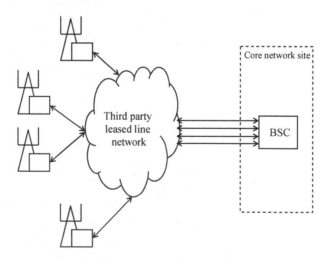

Figure 11.3   Centralised BSC architecture with leased line backhaul

under the operational control of the third-party-fixed network operator. Leased lines for GSM transmission may be $n \times 64$ kbps or full 2.048 Mbps E1 circuits; this was determined during the transmission planning phase and depended upon network operator's strategies and growth forecasts. Providing a full E1 would allow an operator to easily increase the number of TRXs on site by simply mapping the new sub-TSs to the Abis interface, if the site had an $n \times 64$ kbps circuit, it would require an upgrade which would incur additional costs and, therefore, the local market conditions would determine the best approach. Generally, a macro-cell base station, which provides wide area radio coverage, would be provided with a full E1, whereas micro-cells would often have $n \times 64$ kbps circuits. It wasn't unusual for early micro-cells to be limited to one or two TRXs with no upgrade path beyond this; therefore, the network planner could be fairly certain that an upgrade was unlikely in the short-to-medium term as this would likely involve significant costs to swap out cabinets, etc.

TDM leased lines could be implemented in support of a centralised BSC architecture, as shown in Figure 11.3, and also in support of a distributed architecture. In the case of a distributed BSC architecture, there is a need for wide area connectivity between the cell sites and remote BSC location (Abis interface) and also between the BSC site and core network site (Ater interface). The BSC is effectively a switched concentrator and therefore requires significantly less capacity on the Ater interface than the sum of the Abis connection. The Abis interface is a direct and static mapping of TRX capacity to the backhaul; hence, 128 kbps is required to support the $8 \times$ TDMA TSs from a single TRX which are mapped directly from the GSM radio interface to the backhaul. This direct mapping exists even though GSM TDMA TS0 on the air interface is typically used for radio interface broadcast-control information and therefore doesn't actually send anything over the Abis interface, and likewise regardless of whether the other TDMA TSs are carrying any actual traffic. The Ater interface in contrast is more dynamic in its design and implementation, only carrying actual live traffic channels along with an amount of signalling to support call setup and BSC operation. The actual capacity requirements on the Ater interface are calculated in accordance with Erlang B theory and, historically, it was not uncommon for GSM MNOs to see a statistical multiplexing gain of between 5 and 10 times when comparing the overall Ater load with that of the Abis. The distributed BSC-leased line architecture is illustrated in Figure 11.4.

The leased lines for Abis interface and Ater interface could be supplied by one or more fixed network providers with the MNO configuring the specific GSM interface mapping as appropriate. The connections between the BSC and core network will likely be implemented with higher availability through route diversity and redundant equipment to avoid a large-scale geographical outage which could occur if an entire BSC or BSC site was lost due to transmission equipment failure or cable break.

Whilst a leased line agreement allows an operator to offload the responsibility for many aspects of the mobile backhaul network, it is typically more expensive than a self-provided solution based on microwave radio. The relatively limited capacity requirements of GSM (in comparison with current LTE networks) meant that point-to-point microwave radio systems were a viable alternative to leased lines in many networks. Self-provided microwave radio may not address 100% of mobile backhaul

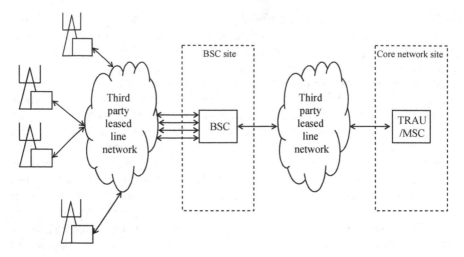

Figure 11.4   Distributed BSC architecture with leased line backhaul

requirements; however, it would certainly address a high percentage and therefore reduce the MNOs overall total cost of ownership. Consequently, GSM networks with up to 90% microwave connectivity were not uncommon.

11.5 Self-provide microwave backhaul

Point-to-point microwave radio links typically provided $n \times$ E1 of capacity where 'n' could be 1, 2, 4, 8 or 16. The multiplexing within the radio systems was based on the plesiochronous digital hierarchy (PDH) of 2 Mbps ($1 \times$ E1), 8 Mbps ($4 \times$ E1) and 34 Mbps ($16 \times$ E1) with additional rates of 4 Mbps ($2 \times$ E1) and 16 Mbps ($8 \times$ E1). Whilst PDH-type multiplexing took place within the radio baseband unit, the inter-faces were always $n \times$ E1, effectively implementing skip-multiplexing between the traditional PDH process of $4 \times$ E1 to an E2 (8.448 Mbps) interface and $4 \times$ E2 inter-faces to an E3 interface (34.368 Mbps). This enabled a simple and consistent network interface based around the primary building block of the 2.048 Mbps E1 circuit.

Microwave radio systems of the early GSM era had evolved from all indoor solutions (the active radio electronics was installed in the cabin next to the tower with an external waveguide connection to the antenna) for all frequency bands to split-mount systems with a component installed within the base station cabin (In-Door Unit or IDU) and an outdoor module (Out-Door Unit or ODU). The ODU could be installed at the base of the tower, on the tower close to the microwave antenna, or directly attached to the antenna. All indoor radios were still supplied for lower frequency bands and certain early implementations of higher capacity systems. The indoor component was either a baseband module or a combined baseband and modem module, the latter becoming the norm overtime. The interconnection between

Table 11.1   Microwave radio data rates, modulation
            schemes and channel bandwidths

Data rate	Modulation scheme	Channel bandwidth (MHz)
1 × E1	2 FSK	3.5
2 × E1	4 FSK/QPSK	3.5
4 × E1	4 FSK/QPSK	7
8 × E1	4 FSK/QPSK	14
16 × E1	4 FSK/QPSK	28

the two modules would be a pair of coaxial cables and quite often a separate DC power cable. Some manufacturers provided DC power over the coaxial cables and overtime, a single coaxial cable system would be adopted for all requirements; transmit and receive intermediate frequencies, DC power and telemetry.

The number of E1 circuits and modulation scheme would dictate the RF channel bandwidth of the microwave radio system. Typical access microwave radio systems of the 1990s used two- or four-level Frequency Shift Keying (FSK) or Quadrature Phase Shift keying (QPSK). International standards for RF channel bandwidths were set for the original PDH data rates with intermediate steps added as shown in Table 11.1.

In addition to data rate, modulation scheme and channel bandwidth, another important consideration when planning a microwave radio link, is the frequency band in which to operate. Prior to GSM, it is reasonable to say that most microwave radio systems operated in bands between 4 and 23 GHz. The introduction of a mass market for microwave radio-based mobile backhaul resulted in a significant investment in research and development in the field of microwave radio engineering. This investment returned higher frequency radios, improved Mean Time Between Failure (MTBF) as well as ever greater system capacity and spectral efficiency; all trends which continue today. The distance between cell sites can be quite short so higher frequency radios offering link lengths of several kilometres were well suited and offered a high-frequency reuse factor. Thirty-eight gigahertz became a popular band, and in some countries, this was allocated to MNOs for self-managed mobile backhaul implementation. Assignment of managed spectrum enabled a faster and often cheaper rollout of microwave transmission, the alternative being a per link licence application to the national authority responsible for assigning spectrum. Once a frequency channel assignment is granted, there is an associated annual licence fee which becomes an OPEX for the MNO. This OPEX along with any annual site rental plus operations and maintenance costs must be added to the cost of the microwave equipment, installation and commissioning costs, to derive the overall TCO of the microwave radio system. This overall TCO can be compared with the TCO of third-party-leased lines to enable an operator to set their backhaul transmission network strategy.

11.6   Planning the microwave backhaul transmission network

The fundamental decision of centralised or distributed BSC architecture will have an impact on the capacity requirements of the microwave radio backhaul network; however, it won't have a major impact on the microwave network topology. The frequency band to be used will be determined by a link planning process which will consider link length, data rate, modulation scheme, channel bandwidth, radio equipment specifications for transmitter output power and receiver sensitivity, structural loading of antenna supporting structure, proposed location of RF transceiver/ODU, target atmospheric availability and equipment configuration.

Link length is the direct line of sight distance between the two ends of a microwave radio link. Point-to-point microwave radio systems are said to require a clear line of sight between the two ends of the link; the actual technical requirement is based on achieving a minimum of 60% clearance of the first Fresnel zone. Links between two microwave antennas are often drawn as straight lines (as in the diagrams in this chapter) or as lightning bolts; both sufficiently represent connectivity; however, in reality, it is essential to appreciate that the transmission is in fact an expanding wavefront on which every point source conforms to Huygens' principles. As a result of this, it is necessary to maintain at least 60% Fresnel zone clearance to avoid significant diffraction which would result in a significantly attenuated received signal.

The data rate of the radio must meet or exceed the minimum demands of the Abis interface. In the case of a distributed BSC architecture; microwave radio may also be used for the Ater interface. During the 1990s, the modulation scheme was set on the basis of data rate, and this dictated the channel bandwidth; however, modern systems offer a greater flexibility to the network planner. The maximum transmit output power would typically be lower for higher frequency systems; a 38 GHz radio would typically have a maximum output power of $+16$ dBm or $+17$ dBm, whereas a 7.5 GHz radio system would have an output power of $+30$ dBm. The actual transmit power would be set to a level $\leq$ maximum transmit power. Receiver sensitivity gets worse as the channel bandwidth increases; typical receiver sensitivity for a 1×10^{-6} bit error rate on a 38 GHz radio was around -87 dBm in a 3.5 MHz channel with 2 FSK modulation, rising to about -72 dBm in a 28 MHz channel with 4 FSK modulation. This means that upgrading the capacity of a link from a lower capacity to a higher capacity may not be possible in the same frequency band, with the same sizes of antennas, depending on how much headroom was available to turn up the transmitter output power.

Microwave antennas are parabolic by design and often referred to as dish antennas; sizes vary from 0.2 up to 4.6 m although the smallest sizes are not possible in the lower microwave frequency bands. The objective was and is still today to balance the wind loading of the antenna on the supporting structure with the need to use as high a frequency band as possible, to ensure a channel is available and maximise frequency reuse.

The ODU is generally either installed outside, at the base of the tower, on a working platform close to the antenna or directly mounted on the back of the microwave antenna. There are pros and cons of these approaches. Mounting the ODU at the base of the tower allows easier and quicker access for maintenance in the event of a fault;

however, it comes at a technical and commercial cost. The ground mounted ODU will require a waveguide to provide connectivity to the antenna which is likely some tens of metres up the tower. This waveguide is expensive and introduces attenuation; the amount of attenuation increases with frequency and, therefore, it's very unlikely that higher frequency systems will be installed with waveguide. As an example, 20 m of 7.5 GHz will introduce 1.2 dB of attenuation, whereas 20 m of 23 GHz waveguide will introduce 5.6 dB of attenuation. Waveguide above 26 GHz isn't commonly available for long waveguide runs; however, short lengths of flexible waveguide are available to aid installation when an ODU is mounted very close to the antenna. From a microwave radio link engineering perspective, the integrated ODU/antenna mount arrangement is ideal; there's no need for any waveguide, as the ODU is directly coupled to the antenna, so system losses are minimised and, therefore, the maximum link length for a given frequency band with a given size of antenna is achievable.

Atmospheric availability is typically referred to as an uptime percentage such as: 99.99%, 99.995% or 99.999%. This refers to the duration in which the link will operate within a given atmospheric environment (usually per annum); the actual atmospheric effects fall into two main categories, atmospheric ducting and multipath for lower microwave frequency bands and attenuation due to precipitation in bands typically above 15 GHz. The various losses are calculated by the network-planning engineer for a given link location, and a fade margin is produced on the basis of the required atmospheric availability. This fade margin is the difference between the receive signal level under normal operating conditions and the point at which the background bit error rate reaches 1×10^{-6}. Beyond this point, a digital microwave radio system quickly drops off in performance and effectively fades out such that the link is no longer available. The fade margin effectively defines the level of headroom required in the receive signal in order for the link to adequately meet its designed atmospheric availability.

Microwave radio systems for GSM backhaul were typically deployed as $1 + 0$ or $1 + 1$ configurations. This refers to the equipment redundancy mode: $1 + 0$ is a single radio system with one IDU and one ODU, whereas $1 + 1$ is a protected or duplicated radio system with two IDUs and two ODUs. If a modem or RF transceiver failed on a $1 + 1$ system, the second unit would take over, often in a hitless or near hitless manner such that the GSM service carried over the microwave link was unaffected. $1 + 0$ links were the norm; however, $1 + 1$ links were deployed in certain circumstances.

Figure 11.5 can be used to discuss both a centralised and distributed BSC architecture with microwave radio backhaul. The diagram is of course a very simplified representation for the purpose of clarity; in reality, the core network site will sit at the centre of a large area of geographical coverage and have microwave radio links, or leased lines, connecting in all directions. There would be a large number of cell sites connected to a core network site, some hundreds or even thousands depending on the size of the network and ratio of BTS sites to core network sites. The diagram has three microwave links (represented by long/short dashed lines) connecting between the core network site and BTS sites; the first BTS site they connect to is numbered 1, 2 and 3.

BTS site 1 is on a direct point-to-point link to the core network site, a 1990s GSM BTS site would require $<1 \times E1$ and, therefore, this link could be a $1 \times E1$ radio or more likely, a $2 \times E1$ radio. The cost difference between the two capacity variants is

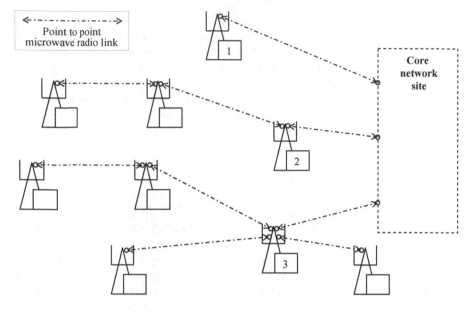

Figure 11.5   Microwave radio backhaul transmission topology

minimal; however, the $2 \times E1$ link will allow for a further site to be connected (sub-tended via a new microwave radio link) in the future, as network rollout continues or, to support capacity growth at site number 1. BTS site number 2 is the start of a chain of three links and is therefore acting as a relay site for the two sites behind it. This relay function could be required because the other sites have no direct line of sight to the core network site, or, because higher frequency radio equipment with smaller antennas can be used for the shorter hops, therefore saving cost in the overall TCO calculation. The link between the core network site and site number 2 could be a $2 \times E1$ radio if sub-multiplexing was used (where the E1 frame is shared between several sites). Sub-multiplexing was implemented by some MNOs; however, the cost of a $4 \times E1$ radio was not significantly higher than a $2 \times E1$ radio and would simplify site configuration and save the costs for the extra equipment and planning and configuration work associated with sub-multiplexing.

The topology associated with BTS site number 3 is rather more complex. This site supports four sub-tended sites and, therefore, the link from the core network site must support five Abis connections. A $4 \times E1$ radio could be used with sub-multiplexing at site number 3; this was quite a common approach during the early days of GSM with multiplexing equipment capable of mapping traffic at bit level between E1 frames. Another option would be deploy a radio with at least $5 \times E1$, in this case, the closest match would be an $8 \times E1$ radio although these didn't become available until the mid-1990s as the aggregate rate wasn't a standardised PDH interface. The only option before this would have been to deploy a $16 \times E1$ radio, and as these were significantly more expensive than a $4 \times E1$ radio, sub-multiplexing may well have been cheaper, particularly if the number of sites connected to site number 3 would increase as network rollout continued. Given that, five sites are dependent on the link between site number 3 and the core network site; the network-planning engineer would review the design of this link and likely increase the atmospheric availability and/or implement 1+1 equipment protection on the radio hardware.

From the previous discussion, it is clear that a centralised BSC architecture will require higher capacity microwave links (or leased lines) from the core network site than the distributed BSC architecture; it was also not uncommon for MNOs to use a mix of leased lines and microwave, picking the most suitable on the basis of a techno-economic analysis on a per site basis. The centralised architecture drove an early adoption of links operating in the synchronous digital hierarchy (SDH); these radio systems offered a significant capacity uplift as they operated at synchronous transfer module level 1 (STM-1) of 155.52 Mbps. Once overheads were removed, the STM-1 radio could carry 63 × E1 circuits; however, unlike the PDH radios, they required an external multiplexer to break the E1s out from the aggregate line rate.

A less obvious role of the backhaul network was the delivery of network synchronisation to support the cellular mobile radio network. A TRX must operate on a specific RF channel as allocated during the network-planning process, and it's essential that this radio transmission occurs at the correct frequency and, therefore, the oscillator in the TRX will require a source of reference. The simplest way to discipline the oscillator with a suitable reference is to use the deterministic 8 kHz clock source which can be derived from the incoming 2.048 Mbps signal. A 2.048 Mbps signal is by definition synchronous given the exact placement of TSs in time, within the frame structure. This reference signal is available due to the use of High Density Bi-Polar 3 (HDB3) line code on the standard E1 frame, and this is sufficient to deliver a frequency accuracy of ±16 parts per billion (ppb) which will ensure that the GSM TRX meets the radio interface requirements of ±50 ppb. The E1 sync signal will ultimately be traceable to an IUT-T G.811 frequency source in either the mobile operator's network or the leased line providers network. Whilst this results in multiple sources of frequency synchronisation, it works fine as all are conformant to the international specification, and this results in a network that is pseudo-synchronous.

11.7 Adding IP packet data to GSM

The introduction of General Packet Radio Service (GPRS) to GSM resulted in a significant change to the mobile network architecture. GPRS was developed during the late 1990s, and the world's first GPRS network was launched in the United Kingdom by BT Cellnet (now O2) in June 2000, quickly followed by many other operators who could see the potential to develop new data services over GPRS packet data bearers. From a mobile backhaul perspective, the Abis link would remain; however, it would now be modified to support packet data. It would also require mechanisms to enable the co-existence of circuit switched traffic and packet-based data traffic. The most significant architectural change occurred within the core network; however, the BSC would have a new architectural peer called the packet control unit (PCU). The PCU was implemented as a number of plug-in units in the BSC by most vendors although some did implement the functionality as a standalone module. Whilst GPRS traffic shared the Abis interface, it did not share the Ater interface, a new interface was specified between the PCU and new packet core network, the Gb interface was a narrow-band data networking interface based on frame relay (FR) technology. Very few operators actually implemented a FR network in support of Gb interface connectivity to the packet core network. In most cases, MNOs chose to simply map FR over $n \times 64$ kbps TSs or allocate $n \times$ E1s, depending on capacity requirements. As the demand for data communications on the mobile network increased, it was clear that GSM wouldn't scale to meet future demand. To address this concern, two things happened in parallel; a higher capacity radio interface was developed for GPRS, known as Enhanced Data rates for GSM Evolution (EDGE) which would further modify the Abis interface and result in greater capacity demands on the Gb interface, and, the specifications for the universal mobile telecommunications system (UMTS) were developed. UMTS would be commonly known as 3G and would result in significant changes to the mobile backhaul network.

11.8   Universal mobile telecommunications system

The original UMTS specification was issued by 3GPP in the year 2000 and known as Release 99. The standard was developed for broadband mobile data communications and introduced a new radio interface based on Wideband Code Division Multiple Access (WCDMA) along with new network nodes and backhaul requirements. The UMTS network architecture appears similar to that of GSM/GPRS; however, there are a number of significant differences. Mobility, which in GSM/GPRS is handled in the core of the network, is pushed to the radio network controller (RNC) level in UMTS; this requires RNCs to be interconnected. Referring to Figure 11.6, the radio base station is known as a NodeB or NB however often referred to as a WCDMA BTS, the Iub interface (1) provides the connectivity between NodeB and the RNC. The Iur interface (2) provides connectivity between RNCs in support of mobility, whereas the circuit switched core connects to the RNC via the Iu-cs interface (3) and the packet switched core via the Iu-ps interface (4). Early RNCs had significant capacity compared with BSCs and therefore, in most cases, the MNO would deploy RNCs on core network sites. The Iub interface would be supported on the backhaul network, whereas connectivity for Iur, Iu-cs and Iu-ps would generally be within the core transport domain.

Release 99 specified the use of asynchronous transfer mode (ATM) technology as the 3GPP transport network layer (TNL) for UMTS, and this applies to the Iub, Iur, Iu-cs and Iu-ps interfaces. ATM is a layer-2 technology and was selected because of its flexibility to handle multiple types of traffic as UMTS was designed from first principles to support voice, data and video traffic. ATM can utilise any layer-1 transmission technology and has the ability to logically combine lower speed circuits to form a higher speed link, a technique known as Inverse Multiplexing for ATM or IMA. ATM utilises a fixed 53 octet cell structure and can therefore be switched at high speed in hardware, a key advantage of ATM over the variable length packet switching technologies at the time. ATM enabled an appropriate quality of service (QoS) to be

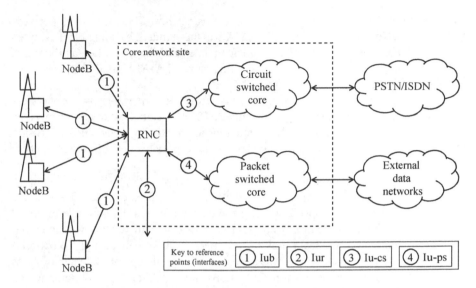

Figure 11.6 UMTS network architecture with centralised RNC

applied to the different traffic types and therefore ensured that voice, data and video traffic could coexist on the same network without any detrimental behaviour.

ATM standards define three layers: the ATM adaptation layer (AAL), the ATM layer and Physical layer. AAL is service specific; this layer includes a segmentation and reassembly (SAR) function which splits an incoming information stream and allocates blocks of 48 octets (or bytes) to the payload area of an ATM cell; the process is reversed at the receiving end of the link. There are five defined AALs known as: AAL-0, AAL-1, AAL-2, AAL-3/4 and AAL-5. AAL-2 is used to carry voice traffic, whereas AAL-5 is used for data traffic in R99 UMTS transmission. The ATM layer adds a five-octet header to the ATM cell; this contains addressing information in the form of virtual path identifier (VPI) and virtual channel identifier (VCI) along with QoS information and a header error checksum. The physical layer manages the connection to a transmission system which, in the case of UMTS backhaul, is typically E1 or SDH based.

11.9 UMTS mobile backhaul

A NodeB transport interface card would typically support $n \times$ E1 interfaces; the exact number represented by n is vendor specific; however, 4 and $8 \times$ E1 configurations were common. NodeB ATM E1 interfaces could be configured as a user to network interface (UNI) in which a single E1 circuit provides an ATM connection, or as an Inverse Multiplexing for ATM (IMA) group in which $n \times$ E1 circuits could be configured as an ATM connection. In the case of IMA, n could be anything from 1 to the maximum number supported by the NodeB, typically a maximum of 8. In this example, an Iub interface could scale from 2 to 16 Mbps, higher capacities were available with E3 and even STM-1 interface cards; however, these were not very common on eNodes and, therefore, E1 IMA was the standard building block adopted by most MNOs. An early RNC had significant interface capacity, often a mix of E1s and STM-1s. The STM-1 interfaces were unstructured ATM Virtual Container level 4 (VC-4) with a payload capacity of 149.76 Mbps.

In some countries, the 3G licencing process resulted in new 3G network operators and, therefore, there were some greenfield UMTS MNOs. In most cases, however, an operator would add UMTS backhaul demands to an existing GSM/GPRS network, and in doing so would look for as many synergies as possible between the two sets of requirements.

Reviewing UMTS backhaul evolution, as with GSM/GPRS before this, is important as the decisions taken here will likely influence, to a greater or lesser extent, the starting point for deploying LTE networks and developing an LTE mobile backhaul strategy and target architecture. Adding ATM to the backhaul network was an implementation challenge on a scale previously unseen in ATM networking. To optimise the design of UMTS backhaul, it was necessary for mobile operators to deploy ATM cross-connects/switches between their NodeBs and RNC; the level of manipulation at the ATM layer would differ between MNOs, some treating ATM E1 circuits in a TDM-like manner, whereas others would truly optimise the ATM layer to gain trunking efficiencies and realise economic benefits. In addition to the introduction of ATM, MNOs had to scale their backhaul networks as multiple RATs (Radio Access Technologies) and data growth started to generate significant traffic. In the base case, a second E1 would be added to the existing GSM site to support the newly deployed UMTS NodeB. Referring back to Figure 11.5, there would be a need to upgrade many of the microwave radios to support a greater E1 count, geographically distributed BSC sites and microwave radio hub sites would be ideal places to deploy ATM aggregation capability; however, the backhaul from these sites towards the core network would have to scale significantly, often from a few E1s to an STM-1. ATM supports circuit emulation via AAL-1, this enables a TDM E1 circuit, such as a GSM Abis or Ater interface, to be carried over an ATM network and presented at the far end as a standard TDM interface; this is known as circuit emulation service (CES). This approach can also be used for FR-based Gb interface; however, alternatively FR can be carried natively over ATM, but in most practical implementations, it was easier to transport the FR E1s as CES. Common ATM interfaces, data rates and corresponding cells per second are detailed in Table 11.2.

Table 11.2   ATM interfaces, data rates and cells per second

Interface type	Data rate in Mbps	ATM cells per second
STM-1 VC4	149.76	353,208
E1 CES	2.309	5,447
E1 UNI	1.920	4,528
E1 IMA	1.904	4,490

11.10   Planning the UMTS mobile backhaul network

Assuming an existing GSM/GPRS, network operator had an installed base of BTS equipment; it was most economical to deploy UMTS NodeBs to those sites in the first instance. If the GSM network was operating in the 900 MHz band, then a large number of in-fill sites would have been required, given that UMTS operates in the 2100 MHz band. Conversely, GSM 1800 MHz operators had a more densely packed cell site grid anyway and would therefore require less in-fill sites from a coverage perspective. Both types of operator added in-fill sites for capacity management as UMTS traffic grew. Overtime, the 900 MHz operators have reformed some of their 900 MHz spectrums for UMTS use.

Using Figure 11.7 as a use case, it is possible to review the steps required to upgrade a GSM/GPRS backhaul network to support UMTS. The site labelled as 'Hub' is either a microwave radio hub site in the case of a centralised BSC architecture, or a combined BSC site/microwave radio hub in the case of a distributed BSC architecture. As discussed previously, the link numbered 1 will either be a high-availability microwave radio link with equipment redundancy and high atmospheric availability, say 99.999%, or be implemented via diverse routes. Alternatively, this could be $n \times$ E1 leased lines from a fixed network provider.

The first requirement is to upgrade link number 1; there are only 13 cell sites illustrated in Figure 11.7 (including the hub site), this is simply for clarity of the diagram; however, in reality, the hub will likely support significantly more sites. In the case of a distributed BSC site, it may support somewhere in the region of 50 to 100+ sites. The most likely step is to upgrade link number 1 to STM-1; this could be via a self-provided microwave radio or leased line. Considering these two options in turn, self-provided microwave radio would require the implementation of a much

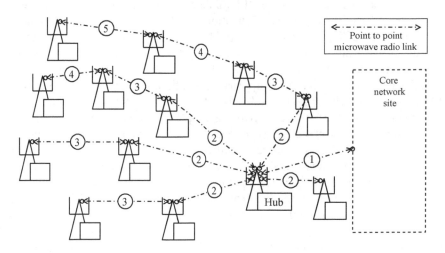

Figure 11.7   Upgrading GSM/GPRS backhaul to support UMTS

higher modulation scheme than the existing lower capacity PDH link, even if a wider RF channel was used. In many cases, an STM-1 radio of the day would be configured to use a 28 MHz channel with 128 QAM; this is the same channel bandwidth as a 16 × E1 radio with QPSK modulation. Implementing such a high-order modulation scheme means that the link is more likely to suffer from error vector magnitude issues than the previous QPSK link, this results in a higher (worse) receive signal threshold, whereas the linearization requirements of the transmitter amplifier will result in a lower output power. As a result of this, the overall system gain will be significantly reduced in comparison with the previous link. It is therefore unlikely that a like-for-like link swap-out could be implemented unless the link length was short; this itself is unlikely as the hub sites are generally a reasonable distance away from the core network site to extend the overall geographical footprint of the core site. Options for recovering the loss of system gain include dropping to a lower frequency band, which has lower free space pass loss, and/or deploying larger antennas with a higher gain. Finally, the use of ring protection is highly unlikely given the capacity requirements per hub; this would require $n \times$ STM-1 radio systems which were not commonly or cheaply available in the early 2000s. The probability of being able to complete the ring is also unlikely, given the constrains on system gain and link length discussed above.

Leased STM-1 circuits were readily available from fixed network operators and offered significant geographical reach in many markets. Given the geographical significance of the hub/BSC site, the upgrade to an STM-1 fibre-based leased line often involved the implementation of a diverse route to maximise availability in the event of a fibre break, which typically has a long mean time to repair. A leased STM-1 would offer a VC-4 payload which could be utilised as the MNO deemed appropriate, from mapping 63 × E1 circuits via an add/drop multiplexer to running a high-speed ATM connection which utilised the unstructured VC-4. The NodeB mapped all voice and data traffic to the Iub interface as AAL-2 while using AAL-5 for control and user signalling. Typically, a single VPI would contain a number of VCIs to identify end points such as O&M, NodeB Application Part (NBAP) signalling, user signalling and user plane traffic. 3GPP specified the Iub interface with ATM service category of Constant Bit Rate (CBR); this is implemented to ensure QoS however comes at a cost in terms of inefficient use of the STM-1 link between the hub site and core network site. An alternative to this approach is to utilise CBR between the NodeB and ATM hub then switch to real time-Variable Bit Rate (rt-VBR) between the two ATM cross-connects/switches; this effectively decouples the relationship between provisioned Iub circuits and the capacity utilisation on the STM-1 link, saving significant opex by negating or delaying the requirement of a second STM-1 circuit. The use of rt-VBR will require an additional level of traffic management; however, this is simple enough to manage and is justified by the associated OPEX savings. As previously mentioned ATM supports CES, therefore, if the hub site is also a BSC site, the Ater and Gb interface circuits could be emulated over the ATM STM-1. As a result of this, any parallel backhaul requirements can be cancelled/decommissioned in favour of a single-converged backhaul solution for GSM/GPRS and UMTS. As the CES uses AAL-1 and CBR and there is no option to switch these particular circuits to rt-VBR,

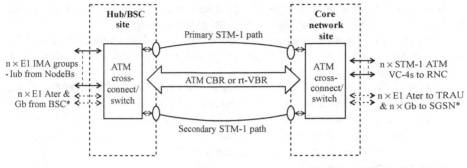

Figure 11.8   Converged ATM backhaul for GSM/GPRS and UMTS traffic

the sum of CES circuits is therefore subtracted from the available Iub capacity which can operate as rt-VBR.

Figure 11.8 illustrates the connectivity between the hub site and core network site (link number 1 in Figure 11.7). The E1 IMA groups from NodeBs arrive at the hub site via microwave radio systems or E1 leased lines; if microwave radio then it's likely that the radio will carry Abis and Iub circuits on different E1s of the same $n \times$ E1 link. GSM traffic will either pass straight through the hub site (possibly via the ATM cross-connect/switch), be TDM aggregated or, if a distributed BSC is on site, terminate to the BSC. The Iub traffic will terminate on the ATM cross-connect/switch. The connection between the hub site and core network site is fully protected which means there are two fibre-terminating equipment at each end of the link and two, diversely routed, optical fibre connections. It should also be noted that it is possible for the fixed network provider to provide one or more legs of the STM-1 leased line via microwave radio. The two diverse paths provide a primary working path and secondary backup path to ensure maximum availability. Control plane and switching redundancy in the ATM cross-connect/switch is usual to complement the link diversity, along with line card separation. The ATM cross-connect/switch at the core network site will have STM-1 ATM VC-4 connectivity to co-located RNCs and will provide a CBR interface as per 3GPP specifications. Any TDM circuits are handed off as E1 to either, a centralised BSC, or, as shown in Figure 11.8, from a distributed BSC to the circuit-switched TRAU and packet-switched Serving GPRS Support Node (SGSN).

The cell site facing links in Figure 11.7 will have to scale to meet the demand of an initial $1 \times$ E1 Iub to each new NodeB which is co-located with the GSM BTS. Overtime, the number of E1s will increase as data traffic grows, 2 then $3 \times$ E1 IMA groups were common with R99 UMTS. The same considerations apply to the scaling of these microwave links as discussed with regard to link number 1. As more links in a given chain results in a higher capacity demand on the first hop out of the hub site, it wasn't long before $16 \times$ E1 simply wasn't enough and STM-1 access radio was introduced, with ADMs enabling access to the E1 circuits. At a similar time, vendors started to develop a wider range of $n \times$ E1 microwave radio systems where n could be 21, 32, 40 or 48, depending on vendor and configuration.

Having removed the E1s between the hub/BSC site and core network site, there is no longer an E1-based synchronisation source and, therefore, MNOs required a new synchronisation strategy. UMTS frequency division duplex (FDD) radio interface operates with the same 50 ppb radio interface stability requirement of GSM; therefore, the same 16 ppb requirement must be delivered to the NodeB. Many mobile operators experimented with sync solutions based on ATM; however, they tended to have a number of issues. It was quite common for mobile operators to adopt global navigation satellite systems (GNSS)-based sync solutions at hub sites. At the time, the only solution available was based on the Global Positioning System (GPS), and whilst operators in North America and elsewhere had relied on GPS for many years, due to their cellular systems requiring phase alignment as well as frequency synchronisation, European operators in particular were nervous about over-reliance on a system which has vulnerabilities; including jamming, spoofing and space weather disturbances. Whilst a fairly simple and reasonably cost-effective system to install and operate, GPS requires a backup solution, and this was generally the STM-1 line rate from the transmission network which would ultimately be synchronous and traceable to a G.811 clock source.

11.11   High-speed packet access

3GPP Release 5 introduced high-speed downlink packet access (HSDPA) to UMTS; this was the first in a series of downlink and uplink enhancements which would see the peak data rates and overall system capacity of UMTS increase significantly, with the related knock on effect to backhaul capacity requirements. HSDPA first introduced 16 QAM on the UMTS radio downlink and then an enhanced uplink was introduced in Release 6. HSDPA enables data rates of 14.4 Mbps, whereas the enhanced uplink supports data rates of up to 5.76 Mbps. The realisation of HSDPA on devices was through a series of improvements, early devices supported 1.8 Mbps, then 3.6 Mbps, 7.2 Mbps and eventually 14.4 Mbps. Enhancements are still on-going with current commercial networks supporting 64 QAM which offers data rates up to 21 Mbps per 5 MHz WCDMA carrier and 42 Mbps with dual carrier operation (all of which adds more traffic to the backhaul networks). With peak rates of some tens of megabits per second, it's clear to see that continuing to scale, the Iub interface with E1s simply is not feasible, so a new approach is necessary.

11.12   Carrier Ethernet and pseudo-wires

Ethernet is a long-established local area network (LAN) technology which was invented in the early 1970s. The idea to evolve and extend the reach of Ethernet to the metropolitan area network (MAN) was championed by the Metro Ethernet Forum (MEF) which was formed in 2001. Metro Ethernet as it was originally known quickly established itself as a viable alternative to traditional PDH and SDH transmission technologies in the MAN and before long, researchers started to explore the possibility of building wide area and even global networks with the evolving Ethernet technologies. A name change to Carrier Ethernet followed, and the foundation for evolving mobile backhaul to much higher capacities with the use of lower cost Carrier Ethernet transport became a reality. Fixed line operators started to develop a portfolio of Carrier Ethernet-based products which included 100 Mbps symmetrical circuits, ideally suited to mobile backhaul given the HSPA data rates discussed above. There were however a couple of challenges, as all of the base stations and network controllers had E1 or STM-1 interfaces; in addition, how could operators deliver a frequency synchronisation signal over Carrier Ethernet? In parallel with developments on fibre-based solutions, microwave radio vendors introduced hybrid radio systems which could support E1s and/or Carrier Ethernet circuits, however as already discussed, the network controllers were increasingly centralised and few circuits landed directly at core sites via microwave radio. Even MNOs with a distributed BSC architecture were starting to centralise their BSCs as ever higher capacity platforms became available, and in any case, the amount of GSM/GPRS backhaul from a hub/BSC site, even for Abis, was now very small in comparison with Iub traffic.

Given the installed base of TDM interfaces on BTS, NodeB and network controller platforms, it became obvious that to start an evolution towards Carrier Ethernet backhaul would require a mechanism for transporting legacy interfaces over the new transmission technology. With ATM, it is possible to emulate circuits over a different technology, and this is exactly what happened with Carrier Ethernet. The Internet Engineering Task Force issued RFC3985 in 2005 which describes an architecture for Pseudo-Wire Emulation Edge-to-Edge (PWE3) which enables the transportation of protocols such as TDM, FR and ATM along with Ethernet over a generic Packet Switched Network (PSN). PWE3 is a mechanism that emulates the essential attributes of a telecommunications service and can therefore take an E1 circuit in, modify it to a form suitable for transmission over a PSN and then recreate the E1 frame at the far end; the most common form of PSN is based on Multi-Protocol Label Switching (MPLS) technology.

A PWE3 carries traffic between two Customer Edge (CE) devices. This is standard MPLS terminology for the customer's equipment which in the case of mobile backhaul is typically the base station (BTS or NodeB) and the network controller (BSC or RNC). The attachment circuit carries the native service such as a TDM Abis interface or IMA Iub interface which terminates to a Provider Edge (PE) router. The PE router is the edge of the MPLS network and the point at which the native circuit is terminated and the pseudo-wire is generated. The PE router establishes a label-switched path across the MPLS network, possibly passing through a number of Provider (P) routers which simply act on the label-switching header rather than interacting with the payload. At the far end, the pseudo-wire terminates to a PE router from which the attachment circuit containing the native service is extracted and connected to the CE. This process is illustrated in Figure 11.9.

The introduction of pseudo-wires enabled the migration from TDM transmission to Carrier Ethernet which ensured mobile backhaul could scale to meet the growing

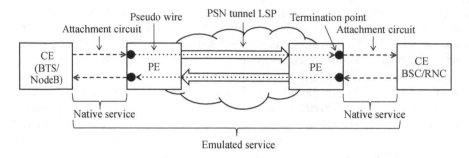

Figure 11.9   Pseudo-wire emulation edge-to-edge (PWE3)

demand for mobile broadband. Synchronisation was the second key challenge with Carrier Ethernet backhaul; however, this was successfully addressed through two parallel developments; Synchronous Ethernet (SyncE) from the ITU-T and a Precision Timing Protocol (PTP) known as 1588v2 from the IEEE, and standardised as IEEE 1588-2008. SyncE is a physical layer clock, similar to HDB3 on an E1, which can deliver frequency synchronisation. IEEE 1588-2008 is a layer-2 message-based timing protocol which can deliver frequency and, with suitable engineering, phase and time synchronisation, a capability which will become essential for many modern cellular network features.

The drive towards Ethernet interfaces on the cellular equipment came with 3GPP Release 5 (2002), this specified an IP TNL as an alternative to ATM for UMTS, and incidentally the same release specified IP for the GPRS Gb interface. Carrier Ethernet is more scalable and cost-effective than legacy TDM for carrying broadband IP-based traffic.

11.13   Long-term evolution

Long-term evolution (LTE) was a 3GPP project which explored the long-term evolution of UMTS and resulted in the development of new radio access and core network technologies. LTE was developed to be an all-IP technology and differed significantly from GSM and UMTS because of the lack of a circuit-switching capability. As an all-packet technology, it was optimised for the growing requirements of mobile broadband; however, it has integrated QoS mechanisms to support Voice over IP (VoIP) or what became known as Voice over LTE (VoLTE). The new radio interface is based on Orthogonal Frequency Division Multiple Access (OFDMA) in the downlink and can offer data rates of up to 150 Mbps in a 20 MHz channel with a 2×2 multiple input multiple output (MIMO) antenna system.

LTE is a hugely flexible system which can support a range of operating frequencies, channel bandwidths and two modes of operation [FDD and time division duplex (TDD)]. LTE was standardised in 2008 in 3GPP Release 8; in parallel with LTE

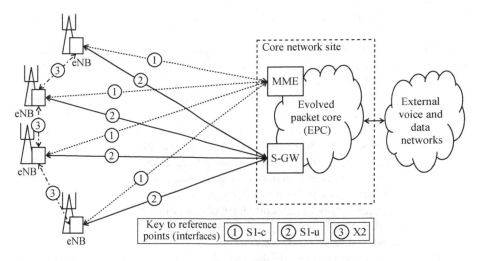

Figure 11.10   LTE network architecture

standardisation efforts, an influential group of MNOs formed the Next Generation Mobile Networks (NGMN) Alliance to contribute to standardisation activities and produce guidance on industry best practice. Amongst many documents produced by the NGMN are a number on mobile backhaul dimensioning, deployment and security. The 3GPP LTE network architecture is illustrated in Figure 11.10; the connectivity within the access network (between base stations) is new, whereas connectivity to the core is split across two network elements, one for control plane and one for user plane.

What is commonly known as LTE is strictly the Evolved Packet System (EPS); this consists of an evolved UMTS Terrestrial Radio Access Network (eUTRAN) and an Evolved Packet Core (EPC) based on the System Architecture Evolution (SAE) activity within 3GPP. The term LTE is now commonly used and accepted as the overall system name. The LTE base station is known as an evolved NodeB, often written as eNB, and connects to the core network via the S1 interface. Mobility management has moved closer to the edge of the network as there is no network controller in the LTE architecture; this necessitates the logical meshing of eNBs to enable signalling and user plane traffic flows in support of mobility management and network optimisation. All interfaces are Ethernet based with an IP TNL.

Referring to Figure 11.10, interface number 1 is the S1-C; this is a signalling plane interface between the eNB and the Mobility Management Entity (MME). This interface carries the S1-Application Part (S1-AP) signalling connection. Interface number 2 is the S1-U; this is the user data plane connection between the eNB and the Service GateWay (S-GW). Interface number 3 is the X2 interface which connects between adjacent eNBs. An eNB will have a large number of X2 interfaces to manage co-operative interworking and radio interface scheduling along with mobility with adjacent eNBs.

11.14   LTE mobile backhaul

LTE drives a significant uplift in mobile backhaul capacity requirements, driven by wideband radio channels and higher order modulation, up to 64 QAM in release 8. There is an additional backhaul requirement which isn't immediately obvious; this is network security and is due to the fact that an eNB terminates all radio protocols, as in GSM, however, given that an eNB has an Ethernet interface and runs IP; it is more susceptible to attacks such as man in the middle and interception of microwave radio transmissions, than the proprietary GSM Abis over TDM. This isn't a concern in UMTS as the higher layer radio protocols (including encryption) terminate in the RNC and as such are within a trusted network on the core site. To address the vulnerability in LTE, it is recommended by 3GPP that MNOs utilise IP Security (IPSec) as defined by the IETF. IPSec supports both authentication of end points and encryption of content with keys provided by a Public Key Infrastructure (PKI) Certificate Authority (CA) deployed within the operator core network.

The most common practical realisation of the S1 and X2 interfaces is via a single IPSec tunnel between the eNB and an IPSec gateway on the MNOs core network site, the latter being considered a trusted network as opposed to the base station site which could be vulnerable to physical access. The use of an IPSec Authentication Header (AH) and Encapsulating Security Payload (ESP) ensures integrity of the end station and associated IP traffic flows. In addition to S1 and X2, the eNB will have an operations and maintenance channel which will connect back to the Operational Support System (OSS) in the Network Operations Centre (NOC). It is not unusual for hundreds or even thousands of eNBs to connect to a single core network site and therefore the IPSec gateways must implement high-availability features and are usually deployed in clusters of at least two units, one working and one standby or both working with a load-sharing mechanism and each having enough capacity for one gateway to manage all traffic in the event of a failure of its peer node. The use of IPSec adds more overheads to the backhaul traffic, and when added to the GPRS Tunnelling Protocols (GTP), IP and Ethernet results in an overhead in the region of 20%–25%, depending on the average packet size and distribution. Whilst the X2 interface is a logical connection between adjacent eNBs, it is generally routed back to the IPSec gateway in a single security tunnel which it shares with S1. It is then decrypted, routed within the gateway, re-encrypted and transmitted back to the adjacent eNB.

Figure 11.11 is an evolution of the backhaul network architecture illustrated in Figure 11.7, and the same caveats apply in as much as this is a conceptual diagram. There would be a much higher number of cell sites connected to a core network site in a typical deployment. The diagram illustrates the concept of introducing fibre-based backhaul to what was previously a predominately microwave radio-based backhaul network. Some operators started to implement this architecture in support of UMTS/HSPA, whilst others waited until LTE, the key point being that microwave radio systems still have a role to play in LTE mobile backhaul; however, practical limitations on scalability means a hybrid of fibre and microwave will be required to avoid the backhaul network becoming a throughput bottleneck. It is common for a

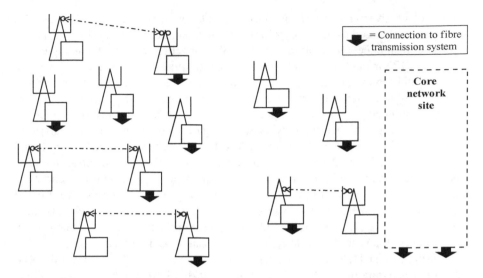

Figure 11.11   High-capacity backhaul network with fibre and microwave radio connectivity

single-cell site or hub supporting a small number of microwave-based sub-tended sites to be on a single-fibre cable connection; rings could be implemented, however, they are less common due to cost and added technical complexity. The fibre connectivity to the mobile core network site would typically be fully protected via two diverse cable routes and diverse equipment; this is important given the aggregate number of cell sites carried over these connections, and the consequences of a significant geographical service outage if a fibre cable is damaged, not an uncommon occurrence given the amount of general construction activity which takes place.

If the mobile operator is also a fixed network operator, as is the case with many national incumbent operators and some established competitive network operators, they could self-provide fibre cables and the associated electronics to enable this solution. In many cases, the MNO will have to purchase fibre-based services from third-party providers, and this will increase overall TCO; however, it's the only practical solution to enable LTE mobile broadband networks to realise their full potential. Fibre cables would be installed to a number of cell sites, some would operate in a standalone mode; in this scenario, the fibre connection supports a single site, whilst others would act as hub sites where the fibre connection supports the sum of traffic from the local site and any sub-tended sites connected by microwave radio. The actual design of the backhaul over the fibre network could differ considerably between network operators/third-party providers; from simple point-to-point virtual connections such as the MEF defined E-Line service, to more complex aggregation networks which offer a statistical multiplexing gain between the sum of access tails and the total connection capacity presented to the core network site. In addition, different solutions could result in two main scenarios: backhaul circuits which are effectively hard-wired connections between two points or a more flexible connectivity model which enables circuits to be changed from one core network site to another to manage load balancing or provide resilience. Quite often such flexibility is achieved through the implementation of an IP Virtual Private Network (VPN), also known as Layer 3 VPN. There are technical and economic arguments for and against both approaches. Microwave radio links are still used however are often deployed in shorter chains of links to prevent capacity bottlenecks. Systems supporting between 100 and 800 Mbps are increasingly common, the higher capacity systems requiring co-channel cross-polarisation operation, typically in 56 MHz channels. The 112 MHz channels have been proposed and will likely to be available shortly; however, it's a challenge to secure such a wide channel in the traditional microwave bands. Point-to-point radios in millimetre wave bands are increasingly common and offer capacity >1 Gbps in the E-band. E-band radios can scale to 10 Gbps and operate in 71–76 GHz paired with 81–86-GHz spectrum. At such a high frequency, the links lengths are limited; however, for many urban deployments, this is sufficient to replace traditional 32 and 40-GHz links.

Assuming an LTE FDD channel bandwidth of 20 MHz with 2 × 2 MIMO, the peak radio interface data rate is 150 Mbps, adding backhaul overheads, including IPSec, increases this by approximately 20%–25%; therefore, a minimum of 180 Mbps of backhaul is required. The 180 Mbps would support a single peak on a single 20 MHz radio channel, most cell sites are typically built with three cell sectors; therefore, in theory, the backhaul requirement could be 3 × 180 Mbps or 540 Mbps. In reality, the actual figure is somewhat lower than this as the probability of a single radio interface peak of 150 Mbps is very low given the signal to noise ratio and implementation of LTE. MNOs often take a pragmatic view based on theoretical analysis backed by field measurements when determining the actual backhaul capacity requirements. The average traffic on a 20 MHz LTE cell sector will be considerably lower than the theoretical peak on the basis of the actual achievable spectral efficiency. Practical measurements of release 8 LTE radio networks highlight that a spectral efficiency in the region of 1.1–1.3 bits per second per Hertz (bits/Sec/Hz) is common with a standard distribution of users within the cell coverage area. Users close to the cell centre will benefit from 64 QAM and low levels of coding overheads, whilst users at the cell edge will be operating with QPSK with a deep coding depth. The delta in spectral efficiency between cell centre and cell edge is considerable and as such the corresponding data rates vary considerably, impacting the overall average spectral efficiency. Assuming an operator wants to offer high data rates wherever possible, then a balance needs to be found between the peak rates which drive huge and otherwise unnecessary backhaul capacity requirements, and the average rates, based on average spectral efficiency, which would constrain the performance of users closer to the cell centre who experience better radio conditions. An approach referred to as 'all average-single peak' is often adopted in which the sum of a single-cell sector peak and two-cell sector averages is implemented.

An alternative approach to backhaul dimensioning is to use traffic forecasts, a MNO will maintain a detailed traffic plan for their network; this includes considerations such as subscriber numbers and growth projections, average usage per subscriber and anticipated growth along with any step changes which may occur due to new products, services and tariffs. Traffic distribution is also a key consideration as dense urban areas will have higher traffic demands due to greater subscriber density than sparsely populated rural areas. Given that, cellular radio access spectrum is an expensive and finite asset, MNOs don't want to be blocking services on their back-haul networks as this constrains spectrum utilisation; however, building non-blocking backhaul everywhere may not be commercially viable. An alternative is to use a mix of approaches, in urban areas adopt something along the lines of all average-single peak, whilst in rural areas focus more on actual traffic forecasts whilst allowing an additional margin to increase the probability of a subscriber getting a particular data rate on the basis of the marketing associated with the operator's mobile broadband products. Targeting non-blocking backhaul is a sensible approach and one which doesn't prevent operators from realising the benefits of statistical multiplexing gains in backhaul network; recall the example of rt-VBR in UMTS backhaul. Managed contention is sensible as it offers significant economic benefits, and if managed properly avoids congestion. However, contention resulting in congestion should be avoided.

11.15 Multi-RAT and multi-operator backhaul

Most LTE MNO will already operate GSM/GPRS and/or UMTS/HSPA networks; therefore, LTE has been deployed as an upgrade to existing cell sites, adding an eNB to the BTS and NodeB on site. Backhaul will have to support all traffic types including the Abis interface, Iub interface, S1 and X2 interfaces. The three RATs could be supported on physically different base station equipment and possibly from different vendors, the TNL could be TDM for GSM, ATM for UMTS and IP for LTE, therefore requiring the use of pseudo-wires to enable convergence to a single-Carrier Ethernet backhaul service, be this over fibre or microwave radio. Many MNOs have migrated UMTS to release 5 IP TNL; however, there was never a standard-based IP Abis specification. Consequently, vendors have developed proprietary solutions, and given that, the BTS and BSC are always from the same vendor; this works perfectly well and enables a move towards a truly all-IP TNL for the backhaul network, unified over a common Carrier Ethernet layer. IP backhaul interfaces from the different RATs will typically occupy different Virtual Local Area Networks (VLANs) within the Carrier Ethernet backhaul, enabling simple layer-2 switching where possible and layer-3 routing where necessary.

An alternative to multiple base stations per site is the multi-RAT or Single-RAN base station which has been developed and deployed over the last few years. This platform can support GSM/GPRS/EDGE, UMTS/HSPA and LTE in a single base station, sharing common functions such as power supplies, baseband processing, RF combiners and transmission interfaces whilst offering the flexibility to configure the radio interface to support any mix of radio access technologies across a range of frequency bands. A common backhaul interface can be configured which supports what's frequently known as 'shared', or, 'co-transmission'. This enables a single flow of IP packets which are addressed according to their end point, BSC for Abis,

RNC for Iub and EPC for S1, all via a common IP network which may include the IPSec gateways (which are required for GSM and LTE however not necessarily for UMTS).

RAN sharing is an increasingly common trend as MNOs look to reduce their overall TCO. This was discussed earlier with regard to UMTS, and the trend continues with LTE. Whilst operators have very different views on whether the actual base station equipment should be shared, it is very common for cell sites, enclosures, power supplies, antennas and backhaul to be shared. Sharing backhaul will increase overall capacity requirements whilst adding some technical complexity to enable the two operator's traffic to coexist on the same physical backhaul network. The simplest form of logical operation is to utilise VLANs, each partner in the sharing agreement having their own VLAN dimensioned against shared network principles. To ensure each partner operates within the guidelines, there will need to be some policing and traffic shaping. Per VLAN QoS is an important feature which offers the necessary management of allocated capacity and can deal with variable backhaul capacities which could result from the use of advanced features such as Adaptive Coding and Modulation (ACM) on microwave radios.

11.16   LTE-Advanced

LTE-A arrived with 3GPP Release 10 and has evolved with releases 11 and 12. LTE-A introduces a number of new concepts, several of which have implications for the backhaul network. The most common LTE-A feature is downlink Carrier Aggregation (CA) which refers to the aggregation of two, three, four or five carriers, each with a channel bandwidth of up to 20 MHz, in the LTE downlink (from network to mobile device). This feature enables an operator to offer higher peak and average data rates whilst increasing the overall area capacity density, it therefore drives higher capacity requirements on the backhaul network.

Evolved Multimedia Broadcast Multicast Service (eMBMS) supports a range of services which includes TV-like broadcasting of real-time linear audio and video content. This feature offers a significant optimisation of capacity on the LTE radio interface when supporting a large number of users in the same cell sector with the same real-time content. To enable this, the backhaul network must support IP-multicast and may well play a role in delivering the synchronisation necessary for this feature. With each generation of technology, we've discussed how frequency synchronisation is delivered, and this is no different for LTE, the requirements for the standard FDD radio interface is identical to that of GSM and UMTS. To this end, the eNB is connected with a backhaul connection which supports SyncE and provides the 16 ppb line feed reference to ensure the 50 ppb frequency stability of the radio interface. eMBMS is one of several LTE-A features which requires the addition of phase synchronisation with an accuracy of ± 1.5 μs on the radio interface. This phase alignment is achieved by adding a time reference, typically Universal Time Coordinated (UTC) which is most readily available from a GNSS. The vulnerabilities of GNSS can be mitigated by a holdover capability based on either network frequency assistance or a phase synchronisation signal delivered directly via the transmission link through the use of IEEE-1588-2008.

The last LTE-A feature to highlight here relates to the introduction of small cells and the concept of heterogeneous networks (het-nets). Small cells can mean many different things; however, in this context the term refers to public access pico-cells deployed by an operator to increase overall area capacity density. There are other use cases; however, this one highlights the need for a feature known as enhanced Inter-Cell Interference Coordination (eICIC). eICIC is required for co-channel operation of wide area macro-cells and small cells, and this presents another need for phase synchronisation. In addition to the LTE-A features mentioned, TDD systems will become increasingly common and by the very nature of the radio interface have a requirement for phase alignment. Overall, it is likely that phase synchronisation will be as common a requirement as frequency synchronisation in future mobile networks and, therefore, mobile backhaul must evolve to support this.

11.17   LTE-A backhaul

In most cases, the biggest impact on mobile backhaul from LTE-A to date is the increased capacity required to support CA. As already discussed, synchronisation is an evolving requirement; however, the most challenging requirement for LTE-A backhaul is small cell connectivity. Considering the small cell scenario for increasing area capacity density in urban areas, there is a need to connect a large number of small cells to the MNO's core network in as cost effective a manner as possible, whilst still maintaining performance and service availability levels. As LTE evolves and 5G is introduced the use of small cells in a het-net context will increase significantly; add to this, the need for ultra-reliable networks, and there's a whole new area of mobile backhaul to investigate and detailed strategies and target architectures to be developed.

11.18   Future RAN and backhaul evolution

The concept of Cloud-RAN (C-RAN) has resulted in the development of solution for fronthaul, carrying CPRI (Common Public Radio Interface) or OBSAI (Open Base Station Architecture Initiative) interfaces between remote radio units and centralised baseband modules. C-RAN enables tight coordination between radio units through the use of a shared baseband function; this enhances radio performance, particularly at cell edge. The major drawback of CPRI/OBSAI is the high capacity and stringent latency requirements placed on the fronthaul segment. Evolved LTE and 5G systems are considering alternative functional decompositions of the RAN such that wide area connectivity requirements are not as stringent, yet some benefits from a centralised component are realised. Such solutions will require an X-haul interface, likely similar in many cases to today's backhaul, although this depends on the actual functional split to be implemented.

11.19　Conclusion

Mobile backhaul is vital to the operation of cellular mobile communications networks. The correct mobile backhaul strategy and architecture will strike the right balance between technical and commercial considerations, whereas ensuring the long-term strategic goals of the business can be met in a cost-optimised manner. Optical communication systems operating over modern fibre optic cables represent the perfect solution; however, practical considerations mean that wireless backhaul, microwave and millimetre-wave, have a continued and evolving role to play. New copper line technologies such as G.Fast may offer some opportunities in certain use cases, whereas shared fibre-based systems such as wave division multiplexing passive optical networks (WDM-PON) offer new opportunities. The delivery of high capacity, resilient backhaul solutions with appropriate QoS will enable mobile networks to continue to evolve, support for frequency, and phase synchronisation will rely on backhaul networks to a certain extent to ensure service continuity. The evolution of the radio access network will include an alternative distribution of base station functionality over the wide area; this will lead to new challenges for not only backhaul but also fronthaul and other X-haul variants as described elsewhere in this book.

Acknowledgement

Thanks to Professor Nigel Linge, Julian Divett and Michael Ferris for their review of this chapter and constructive feedback which improved the final text.

Further Reading

[1]　Linge N. and Sutton A. *30 Years of Mobile Phones in the UK*, Amberley Publishing, Stroud, 2015.

[2]　Mouly M. and Pautet M.-B. *The GSM System for Mobile Communications* (self-published), 1992.

[3]　Sutton, A. "Building better backhaul", *IET Engineering & Technology*, vol. 6, no. 5, pp. 72–75, June 2011.

[4]　Agilent. Understanding General Packet Radio Service (GPRS) – Application Note 1377, June 2001.

[5]　Manning T. *Microwave Radio Transmission Design Guide* (second edition), Artech House, London, 2009.

[6]　Valdar A. *Understanding Telecommunications Networks*. Telecommunications Series, no. 52. London: IET; 2006.

[7]　Sexton M. and Reid A. *Broadband Networking – ATM, SDH and SONET*, Artech House, London, 1997.

[8]　Wilder F. *A Guide to the TCP/IP Protocol Suite* (second edition), Artech House, London, 1998.

[9]　Holma H. and Toskala A. *WCDMA for UMTS*, John Wiley & Sons, Hoboken, NJ, 2000.

[10] Jay Gou Y. *Advances in Mobile Radio Access Networks*, Artech House, London, 2004.

[11] Holma H. and Toskala A. *HSDPA/HSUPA for UMTS*, John Wiley & Sons, Hoboken, NJ, 2006.

[12] Kasim A. *Delivering Carrier Ethernet: Extending Ethernet Beyond the LAN*, McGraw-Hill, New York, NY, 2008.

[13] Grayson M., Shatzkamer K., and Wainner S. *IP Design for Mobile Networks*, Cisco Press, Indianapolis, IN, 2009.

[14] Holma H. and Toskala A. *LTE for UMTS*, John Wiley & Sons, Hoboken, NJ, 2011.

[15] Tipmongkolsilp O., Zaghloul S., and Jukan A. "The evolution of cellular backhaul technologies: current issues and future trends", *IEEE Communications Surveys and Tutorials*, vol. 13, no. 1, pp. 97–113, first quarter 2011.

[16] Robson J. *Guidelines for LTE Backhaul Traffic Estimation*, NGMN White Paper, 2011.

[17] Masala E. and Salmelin J. *Mobile Backhaul*, John Wiley & Sons, Hoboken, NJ, 2012.

[18] Alvarez M. A., Jounay F., and Volpato P. *Security in LTE Backhauling*, NGMN White Paper, 2012.

[19] Sutton A. "Mobile backhaul – from analogue to LTE", *The Institute of Telecommunications Professionals Journal*, vol. 7, no. 3, pp. 34–38, October 2013.

[20] Aweya J. "Implementing synchronous Ethernet in telecommunications systems", *IEEE Communications Surveys and Tutorials*, vol. 16, no. 2, pp. 1080–1113, second quarter 2014.

[21] Sutton A. and Linge N. "Mobile network architecture evolution", *The Institute of Telecommunications Professionals Journal*, vol. 9, no. 2, pp. 10–16, October 2015.

[22] Sutton A. "Microwave and millimetre wave radio systems", *The Institute of Telecommunications Professionals Journal*, vol. 7, no. 3, pp. 38–43, October 2013.

[23] Holma H. and Toskala A. *LTE-Advanced – 3GPP Solution for IMT-Advanced*, John Wiley & Sons, Hoboken, NJ, 2012.

[24] Ferrant J.-L., Jobert S., Montini L., *et al. Synchronous Ethernet and IEEE 1588 in Telecoms – Next generation Synchronisation Networks*, Hoboken, NJ, ISTE Wiley, 2013.

[25] Volpato P. *An Introduction to Packet Microwave*, CreateSpace (Amazon), Seattle, WA, 2015.

[26] Masala E. and Salmelin J. *LTE Backhaul Planning and Optimisation*, Hoboken, NJ, John Wiley & Sons, 2016.

Chapter 12

Wired vs wireless backhaul

Hirley Alves[1] and Richard Demo Souza[2]

Abstract

Wireless networks are becoming even more pervasive and an indispensable part of our daily life. In order to cope with a large increase in traffic volume, as well as in the number of connected devices, new technologies, practices and spectrum rearrangements are required. In this context, a key question arises: *how to provide extensive **backhaul** (and including fronthaul and midhaul) connectivity and capacity in a cost-effective and sustainable way for such pervasive networks?* The answer, if exists, is rather complex and not the goal of this chapter. However, this chapter aims at providing the reader with an overview of some of the most prominent solutions and shed some light into the technical challenges ahead. This chapter is organized as follows. The vision of a backhaul solution for future wireless networks is first introduced, then wired, and radio-frequency (RF) wireless candidate solutions are covered. Special attention is given to the advantages and disadvantages of each technology. Overall, this chapter provides extensive qualitative comparison and discussion of the key components of each available solution. The tone of this chapter is to emphasize that, due to the pervasiveness of wireless networks, an one-size-fits-all approach is not attainable and hybrid wired-wireless solutions will take place.

12.1 Introduction

In the recent years wireless networks have become ubiquitous and an indispensable part of our daily life due to a broad range of applications. We experience it through a myriad of connected devices that are changing the way we live, make business, gather into social groups and interact in an ever more connected society. As a result of this high demand for wireless data services, a 1000-fold increase in traffic volume is expected by 2020 [1,2]. Besides the increase on data traffic, it is also envisioned that by 2020 billions of devices composed of different sensors and actuators will be

[1]Centre for Wireless Communications, University of Oulu, Finland
[2]Department of Electrical and Electronics Engineering, Federal University of Santa Catarina (UFSC), Brazil

connected to the Internet, gathering all kinds of data and generating new business models with a potential economic impact in the order of trillions of dollars [3,4]. In this context, operators are compelled to cope with such demand and explore new ways to improve coverage and network capacity, while lowering their capital and operating expenditures (CAPEX and OPEX, respectively) [1,2,5,6].

12.1.1 Promising technologies for future wireless networks

Despite the recent advances in the field of wireless communications, current technologies are not able to cope with such large increase in traffic volume and in the number of connected devices, consequently new technologies and practices as well as a spectrum rearrangement are required [1,2,5–8]. Current generations of wireless systems have been designed as a single network architecture serving multiple purposes while addressing a narrow range of requirements and supporting limited interoperability. However, such one-size-fits-all approach does not apply to future wireless communications networks, whose architecture is evolving to become more scalable, flexible, heterogeneous and dynamic in order to offer tailored and optimized solutions [2,7,9], since the requirements imposed when designing the communication system shall be closely related to the target application.

Current wireless networks are already heterogeneous to some extent and such characteristic will be inherent of future generations [5,6]. By exploiting heterogeneity it is possible to cope at least in part with such galloping demand for mobile traffic so that the users can experience ubiquitous connectivity. In general, a heterogeneous network is composed of a set of small cells (ranging from femto to micro cells) access points, which are deployed either by an operator or by the user, in conjunction with the macro cell base stations. The small cell access points are low power base stations, and hence have limited coverage compared to a macro base station, but can be easily deployed. Nevertheless, such heterogeneous network presents the advantages of easy installation as well as reduced cost, which motivates a dense deployment of small cells, also known as network densification [9,10]. As pointed out in [10], there are still many open problems related to network densification, such as techniques to handle interference and aggressive frequency reuse.

Another feature of the current and future wireless networks is the coexistence of several radio access technologies (RATs) within small and macro base stations [2]. For instance, [11] evaluates the performance of dense small cell networks with co-located LTE and WiFi RATs, while a RAT selection algorithm is proposed in [12], showing that significant gains in throughput can be achieved. For instance, a base station can split the information flow into several RATs, and moreover each flow can be prioritized according to certain quality of service (QoS) requirements. Another advantage of multi-RAT is that one or even more of the air interfaces can be used for backhauling, which increases the flexibility of deployment along with a considerable reduction of costs to the operator. Ultra dense and multi-RAT deployments are paradigm shifts for wireless networks, and therefore there are many open problems in these areas [2,13]. In addition to interference issues mentioned above, decentralized RAT selection relying only on local information is a major issue, since centralized solutions would require additional overhead and may cause high handover from one air interface to another. Additionally, another issue that arises is how to prioritize traffic flows given the QoS and inherent constraints of each air interface, such as delay profile and achievable rates, for instance.

Besides the above-mentioned technologies, other promising enablers have been foreseen by academia and industry: massive multiple-input multiple-output (MIMO), coordinated multi-point processing, advanced inter-cell interference cancellation and decoupling user and control planes [1,2,5–9,14]. All the above technologies, have their own advantages and disadvantages, but one issue is clear from this point onward: there is no single solution for future networks that meet all the requirements in terms of capacity, latency, spectral efficiency, power consumption, reliability, security, and interoperability.

12.1.2 Challenges ahead

It is clear that ultra dense, multi-RAT, heterogeneous future networks will be composed of a wide range of technologies in order to meet the diverse requirements from each application. However, how to make future deployments feasible so that such large amount of diverse devices are connected? In other words, how to provide extensive *backhaul* connectivity and capacity in a cost-effective and sustainable way to small and macro base stations? This is already a relatively complex and open problem to current deployments, that are considerably less dense compared to what is expected of a future heterogeneous network. Therefore, this problem will become even more intricate as the number of cells and base stations grows.

In cellular networks, backhaul (also known as backnet, backbone, or transport network) is the network that connects, in general terms, the small and macro base stations to the core network [15]. In [15] authors provide an extensive survey about the evolution of the backhaul technologies for cellular networks up to its third generation (3G), while [8] visits the key aspects of the fourth generation (4G) and the trends envisaged for the fifth generation of cellular networks (5G). Often macro cell backhaul is provided via fiber, which has high capacity and large bandwidth, while providing low latency and delay. However, such gains come at a cost of elevated operating expenditures (about 15% on leased lines [10]), as well as scarcity of such resources due to the intensive use of the installed infrastructure. Moreover, the ownership of dedicated fiber becomes onerous, especially for dense deployment of small cells, which poses another challenge in terms of positioning. One of the major advantages of small cells deployment is the easy installation and reduced costs for the operator, or even the user. The installation may be indoors or outdoors, at the street level, in locations hard to be reached by dedicated fiber.

Alternatively, copper lines can be used as backhaul for the small cells; however, they offer limited bandwidth and large delay profile. Wired lines (for instance, xDSL family) offers data rates of hundreds of Mbps, while fiber achieves tens of Gbps. Nonetheless, similarly to fiber, but with lower capital expenditure, availability and locations hard to be reached by copper pose another issue due to the positioning of the small cells [7].

In this context, wireless backhaul appears as a way to complement the wired solutions. Wireless solutions present the advantage of ease installation and availability in locations that may be hard or expensive to reach by fiber or copper, and thus reducing capital and operational expenditures to the operator. Another attractive point of wireless solutions is that they may serve not only for backhaul purposes but as access links as well. Due to spectrum scarcity new frequency bands, commonly named millimeter wave (mmWave) [2,5], usually above 6 GHz, are being considered as a viable solution not only for access links but also for dedicated backhaul. In the mmWave region larger bandwidths are available, which allows high data rates, in the order of several Gbps [7,16].

The **backhaul** network is a major *enabler for future wireless networks*, such as 5G and beyond [7,10,16]. However, it may also be a major drawback if not addressed properly, as an inappropriate backhaul design and planning can hamper the development of future networks as discussed in [17].

In what follows we overview some of the most promising backhaul technologies to be used in the upcoming and future wireless networks. In Section 12.2.1 we discuss the pros and cons of the wired solutions, while in Section 12.2.2 we focus on the wireless and hybrid backhaul technologies.

12.2 Backhaul networks

Let us first introduce a general view of a future wireless network, as depicted in Figure 12.1, in order to discuss some of its key components. Therein, we contemplate the deployment of a large number of macro base stations, though in a smaller order of magnitude compared to small cell base stations. Additionally, both macro and small cells are likely to have several RATs available, even both air interface and wired (copper or fiber) connections. For example, some small cells may be connected to fiber or to a copper line (as xDSL) and also have two air interfaces such as WiFi and LTE [11]. On the other hand, due to positioning some base stations may have only a single air interface available (for instance, WiFi only). An example of such network is introduced in [11], where devices may both have WiFi and LTE air interfaces and performance gains are achieved when traffic is split into many RATs.

Figure 12.1 depicts part of a large urban network, where a macrocell is underlaid by some small cells. In this picture we depict only a snapshot of a small area, which can be then projected to a neighborhood or to a city-wide deployment. Notice that Figure 12.1 also illustrates possible connections of fronthaul, midhaul, and backhaul between small cells (which can be cells with reduced capability deployed at street level, as in lampposts [18]) and macro cells; these connections can be either wired or wireless, as we shall discuss in more detail in Sections 12.2.1 and 12.2.2, respectively.

In the recent years and with the evolution of the concept of centralized radio access network (C-RAN), the typical backhaul evolved to a complex network composed of: fronthaul, midhaul, and backhaul. Figure 12.2 provides a schematic example of deployment of this complex network. As the wireless technologies evolved, such as enhanced inter-cell interference coordination and distributed antenna systems, there

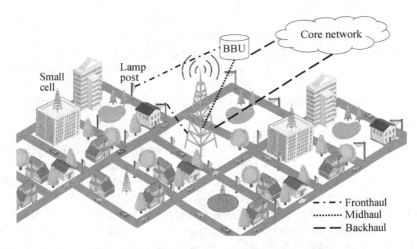

Figure 12.1   Envisaged deployment of a future wireless network composed of multiple tiers: macro cells and different types of small access points (e.g., femto, pico, and micro) and the core network. Each of these nodes may have one or more air-interfaces (or RATs) as well as wired connections, such as fiber or xDSL family. We also illustrate possible connections of fronthaul, midhaul, and backhaul, which may be wireless or wired

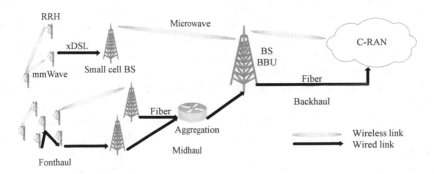

Figure 12.2   Schematic of a complete backhaul network, including fronthaul, midhaul, and backhaul. For each case we show some potential candidate solutions for wireless and wired backhauling

has been a crave for more processing power in order to process and manage the diverse traffic coming from the network, which renders the concept of C-RAN. The C-RAN aims to provide very low latency between multiple base stations, usually called RRH units in this context, by splitting its functions and moving them to a centralized shared pool of baseband resources, also known as baseband unit (BBU).

The fronthaul connects the RRH to the BBU, or to an aggregation point (see Figure 12.2), and has quite stringent requirements in order to meet latency and capacity requirements, and therefore fiber is a strong candidate solution capable of conveying tens of Gbps with low latency [19]; other aspects of the design are addressed in [20]. The midhaul, whenever needed, connects the fronthaul aggregation point to the BBU, and may present as stringent requirements as the fronthaul [21]. By its turn, the backhaul connects BBUs to the core network.

In this chapter, we focus on the backhaul network as whole (which encompass front, mid, and backhaul), and we provide an overview of the current and most promising technologies for backhauling with respect to their availability, cost, capacity, latency, and spectrum usage. It is worth mentioning that there are different requirements for each layer (front, mid, and backhaul), and therefore there is a wide range of candidate solutions that comply with some of those requirements [21,22] as we shall see herein. Current technologies used in backhaul networks include microwave links, fiber, and copper-wire links. As pointed out in [18,23,24] the resources available today are not able to cope with the exponential increase in volume of data traffic, high number of connected users, high demand for ultra-high quality video transmissions, low latency for real-time applications such as gaming and augmented reality. Future networks have to meet a wide range of requirements posed by an even wider range of applications. For instance, a smart meter application may allow for quite large latency in the order of minutes; on the other hand, a short-circuit protection network in a smart grid environment cannot tolerate latencies of more than a few milliseconds. Another example lies in the so-called Internet of Things (IoT), where a myriad of things (as sensors or actuators) will be connected to the Internet. Besides the large number of devices, the requirements vary from low throughput and moderate reliability (as in temperature monitoring), to real-time data transmission with high throughput and low latency, such as in video streaming. This wide range of requirements impacts directly in the design of the backhaul, which should be flexible and dynamic in order to efficiently comply with these large range of demands.

Current technologies cannot cope alone with the requirements imposed over the backhaul of future networks, thus the need for new technologies and even in-band backhauling solutions, which reuse the radio access spectrum for backhauling purposes. Current cellular networks operate at a carrier frequency under 6 GHz, though much attention has been given in the recent years to higher frequency bands, known also as mmWave, where high throughput can be achieved and more bandwidth is available, becoming one of the candidate solutions to deal with high throughput demands [5,9].

12.2.1　Wired solutions

Current backhaul networks rely heavily on fiber and copper-based links, which are often leased, increasing the operating expenditures [10]. Also, reports from 2014 show that there is a scarcity of wired links due to the intensive use of the installed infrastructure, besides the fact that fiber infrastructure is not available nationwide in

Europe [23,24]. These reports point out that in order to guarantee fair competition in the European mobile market, a fair access to fiber backhaul would be needed.

As discussed in the previous section, in future networks the radio unit and the BBU may be physically separated, by distances up to several kilometers [22], while low latency must be still met. As reported in [22], the separation distance between the radio unit and the BBU is then limited by the retransmission protocol currently employed in LTE, as illustrated in the following example.

Example: Latency of the Backhaul

Optical fiber transfers data at the speed of light in glass, which is typically around 200,000 km/s given the refractive index of the glass. Following the LTE constraint for retransmission protocols, a one way communication link is expected to have a delay of about 100 μs, which renders a maximal distance of 20 km [22]. However, as pointed out in [22], in order to support large C-RANs, the backhaul link is expected to be up to 50 km [25], which renders a one way latency of 250 μs just accounting for propagation delay. Such delays may hamper some applications that foresee 1 ms latency, such as in [26,27].

In what follows we provide an overview of some wired solutions that can be used in the future backhaul network.

12.2.1.1 Fiber

Direct fiber, also known as point-to-point fiber, provides high speed data rates, in the order to Gbps and low latency. Nonetheless, such benefits come at the cost of higher leasing fees. The distances covered vary with respect to the fiber mode, thus those that support multiple propagation paths are suitable for shorter distances, while single mode fiber are preferred for larger coverages [18].

Fiber to the x

Fiber to the x (FTTX) encompasses several variants of fiber optic access architectures, such as: fiber to the home (FTTH), fiber to the premises (FTTP), fiber to the cabinet (FTTC), and fiber to the node or neighborhood (FTTN). Each variant has its own nuances on technical aspects and nomenclature, but the common point is that FTTX is a system that allows the connection, via fiber cables and an optical distribution network, of an user (for instance, cabinet, home, node, which can be interpreted, in this context, as BBU, RRH, eNB) and a provider (for example, C-RAN and BBU) [18].

Example: FTTX Deployment

FTTX is often available in urban areas, specially cabinet from which fiber can be distributed. Let us assume that we have a network composed of a macro base station, which also acts as BBU, underlaid to this cell there are many small cells

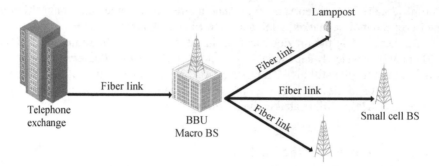

Figure 12.3   Example of FTTX connections. We depict a FTTC (Fiber to the Cabinet) connection from the core network (Telephone Exchange) up to the Cabinet, representing the BBU or a macro base station, which then lays Fiber to the Home (FTTH), perhaps a small cell or street lamppost or any other type of RRH

(considered to be RRHs here). Some of these small cells are close to a optical cabinet, therefore we can resort to a FTTC connection from the BBU to the cabinet, and from there are many alternatives to connect the cabinet to the RRHs depending on their positioning and available technologies, such as copper, fiber, or even wireless solutions.

In FTTX the distance and data rates attained vary from hundreds of Mbps up to some Gpbs with the technologies that are being used; however, leasing and renting charges are much lower than direct fiber [18]. An illustrative example of FTTX connections is shown in Figure 12.3.

Passive Optical Network
Passive Optical Network (PON) technology is a point-to-point architecture, where a single fiber serves multiple end-users via passive, or non-powered, switches. The most common variants are Ethernet passive optical networking (EPON), and ATM-based gigabit passive optical networking (GPON) [19,21]. As discussed in [28], it is possible to share resources of the existing PON networks to effectively serve as backhaul for future generations. An overview of the potential optical front/backhaul technologies to be used in large network deployments is given in [19,21]. PON architecture can offer high reliability and capacity up to 5 Gpbs in the downlink and 2.5 Gpbs in the uplink, large coverage up to tens of kilometers, supporting voice and data traffic as well. Further aspects and a comparison with the FTTX variants can be found in [8,18].

12.2.1.2   Digital subscriber line (xDSL) family
Digital subscriber line (xDSL) solutions dominate personal broadband applications. However, due to increased deployment of small cells, xDSL solutions have become an

attractive alternative for backhaul also in this context, given its ubiquitous presence and relative low cost of installation, operation and maintenance [18]. For instance, according to the Organization for Economic Co-operation and Development (OECD), in 2015, the DSL technology had more than 45% penetration worldwide [29]. The coverage range and data rates vary depending on the technology used, but in general the coverage is up to a few kilometers while the data rates range from few to hundreds of Mbps. For instance, the asymmetric digital subscriber line (ADSL) technology and its variants (2+ and 2+M) can provide up to 24 Mbps in the downlink and 3.3 Mpbs in the uplink, covering up to 5 km [18]. More recent standards, such as very-high bit-rate digital subscriber line (VDSL) and its variants (such as VDSL2) can provide downlink rates of up to 75 Mbps, while the peak uplink rate is around 15 Mbps [8]. However, latency may be an issue, being of more than 10 ms due to propagation speed and losses in the network equipment [8].

G.FAST is another variant of the xDSL family, whose recommendation was recently released by the International Telecommunication Union (ITU) as indicated in [30]. G.FAST provides data rates from 150 up to 1,000 Mbps, but highly dependent on the distances that range from tens to a few hundred meters. For instance, according to [30], 1 Gbps data rate is expected to be reached at 340 m range with low latency. Moreover, the work in [31] performs a cost assessment with respect to the deployment of this technology against fiber-based solutions (such as FTTX), and shows that G.FAST is suitable for indoor scenarios where the roll-out of fiber is cumbersome. All in all, depending on the application, a hybrid copper-fiber solution may be a cost-effective option.

Example: FTTC/VDSL

The Fiber to the Cabinet (FTTC) is often associated with VDSL (as well as with its variants) [8], and may reach 100 Mbps in downlink and 80 Mbps in uplink, with latency of a few milliseconds. FTTC may be used to cover hotspots, indoor areas and in some cases rural areas, though the distance is limited by the copper lines, reaching up to 1 km [18]. Figure 12.4 illustrates the use of FTTC.

12.2.1.3   Remarks on wired solutions

Other alternatives for wired backhaul are summarized in [8,18,32]. Herein, though, we focus on the most promising technologies that could potentially be employed in the next generation of wireless networks. Figure 12.5 gives a brief comparison in terms of some key characteristics of the different technologies discussed above. Notice that direct fiber offers the best compromise in capacity, distance, and latency, though in terms of deployment cost it is much behind other alternatives. On the other hand, VSDL2 has a low deployment cost, but comes short in terms of capacity, latency, and coverage.

As shown in Figure 12.5 there are several alternatives for wired backhauling, and even a few more are listed in [8,18,32]. From the presented wired alternatives, the fiber solution is currently often more suitable for fronthaul connections given

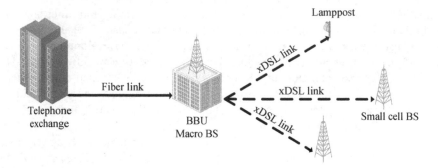

Figure 12.4   FTTC and VDSL connection. FTTC runs from the core network up to the cabinet, represented by a BBU or macro base station, and then from there is distributed via a xDSL connection to the small cell or any other type of access point or RRH. Notice that xDSL in the figure may be any of the solutions discussed above

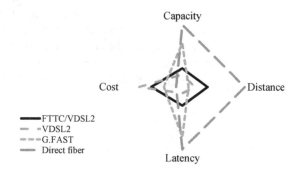

Figure 12.5   Comparative analysis of some key characteristics of wired technologies

their inherent low latency and high capacity characteristics which match well with the requirements imposed by the C-RAN architecture. Nevertheless, as pointed out in [8], fiber solutions for backhaul are the most attractive for future wireless networks, due to their superiority in capacity and data rates compared with other wired solutions. The use of fiber links poses challenges in terms of deployment costs and heterogeneity. The locations of the small cells may be hard to reach and to lay fiber [33]. In this case, hybrid solutions become a viable approach, though some restriction in coverage, latency, and data rates may apply. Technologies such as FTTX offer a good compromise. Besides, as pointed out in [23,24], there is a scarcity of fiber resources, whose infrastructure is not available nationwide in Europe, and also only few countries worldwide provide FTTH (Fiber to the Home) with more than 10% coverage [8]. As highlighted in [8], trenching and laying fiber comes at an exorbitant cost, which discourages operators to engage in such endeavors. To this end these hybrid solutions become even more attractive when associated with wireless solutions, since these can offer a good compromise in terms of coverage, latency, and data rates, as we shall see next.

12.2.2 Wireless solutions

Wireless backhaul solutions appear as an attractive alternative to wired ones due to the densification of network deployments, as discussed in Section 12.2.1.3, evincing the need for new solutions in order to cope with the stringent requirements of future wireless networks [34]. As pointed out in [22], under provisioning in backhaul can reduce considerably the rates perceived by the users; other requirements are also discussed in [22]. Herein, we focus on the wireless solutions that comply with some of these requirements. A wide range of solutions have been proposed for wireless backhaul, which provides the operator some flexibility in design, besides costs reduction and improved deployment speed [22]. These solutions can be sorted into large groups, for instance:

- Carrier frequency: from sub-1 up to 100 GHz;
- Propagation: line of sight (LOS) and non-line of sight (nLOS);
- Spectrum: licensing (link, area, light, exempt) and allocation (static, dynamic, shared);
- Topology and connectivity: point-to-point, point-to-multipoint, mesh, ring.

In wireless systems the carrier frequency has a great influence. For instance, propagation and spectrum occupancy play a great deal in defining the feasibility and capacity.

Spectrum and propagation

The available spectrum in the region below the frequency of 6 GHz is quite scarce due to the excessive use of those bands for broadcasting, military applications, cellular networks, among others [35], which limit wireless solutions for backhauling, since most resources are used to serve the users. On the other hand, in terms of propagation the sub-6 GHz portion renders flexibility since LOS and nLOS propagation is feasible. At lower frequencies, obstructions are often smaller compared to the wavelength, which gives better penetration through materials (for instance, walls and glass) and diffraction around obstacles. Generally speaking, as frequency increases, the harder it becomes to communicate in nLOS conditions—penetration and diffraction reduces and therefore the received signal becomes weak in very short distances [22]. Besides, in higher frequencies rain, fog, and oxygen absorption become an issue and lead to high propagation losses, which makes nLOS communication very difficult.

Coverage: Under nLOS conditions (e.g., sub-6 GHz band) coverage reaches up to tens of kilometers, which renders large cell sizes, and therefore the antennas deploy lower gains and wider beam widths [22]. Though, due to intense use of this frequency band, interference becomes an issue and sophisticated mechanisms need to be used to coordinate and mitigate interference from neighboring cells (either macro or small cells). In general, nLOS solutions have higher coverage in dense urban deployments and do not necessarily require antenna alignment, which simplifies deployment. On the other hand, in LOS conditions higher antenna gains are needed, as well as narrow beams that require alignment between transmit and receive antennas; however, higher throughput is attainable [22].

Spectrum licensing

The spectrum is regulated by national and international committees, defining the rules by which uses should abide. The most influential organ worldwide is the ITU, which is the United Nations specialized agency for information and communication technologies. One of the most significant rules covers the licensing to use a certain frequency band, and each license group has its own characteristics as summarized next.

- **Area:** allows the user to transmit in a given channel within a predefined region. Interference management is user responsibility and several mechanisms have been developed to optimize utilization and efficiency [18]. Examples are mobile radio access networks and multipoint microwave systems [22] (see Section 12.2.2.2).
- **Link:** regulates the use of a specific link, respecting predefined radiation patterns in order to manage the interference to other links. Point-to-point microwave links are an example of application [22]. A simplified version of the link license is the *light license*, which requires a registration to use the link, and the use is subject to a fee which may vary depending on the country own regulation as pointed out in [36,37]. Light license applications usually operate in the E-Band region from 71–76 to 81–86 GHz.
- **License exempt:** are the spectrum bands free for use, thus anyone can transmit within the band, given that the transmitter complies with in-band power limits and out-of-band interference. Bands under this license are often heavily used with high traffic demand and uncontrollable interference. These bands are at 2.4 and 5 GHz (commonly known as WiFi bands) and 60 GHz. Many of WiFi standards resort to these bands [22].

Dynamic spectrum management, sharing and licensing are blossoming research areas that will enable increased spectrum reuse and efficiency as pointed out in [35].

Topology and connectivity

In a wireless network the connectivity between nodes is established in a point-to-point or point-to-multipoint fashion. The first typically is established under LOS conditions, perhaps requiring directive antennas at the end of each link, in order to communicate over long distances [18]. On the other hand, the latter often operates under nLOS conditions and in some cases also in LOS conditions (as in microwave multipoint, see Section 12.2.2.2), omni or multisector antennas may be used, though the communication range is usually shorter than in point-to-point scenarios [18]. Other topologies are tree, ring, and mesh, whose main characteristic is the presence of some or several redundant links that are used to improve resiliency to link outages, though sophisticated routing algorithms are required [22].

Summary
It is worth mentioning that the throughput is taken as the peak rate, some restrictions may apply to each case since throughput may vary depending on the number of channels and/or antennas available, as well as bandwidth [8].

In what follows we discuss currently used and potential wireless technologies for future deployments.

12.2.2.1 Satellite

Satellite communication covers wide regions and is available worldwide, with relatively fast deployment times and high availability. On the other hand, data rates are not high enough and latency is extremely high for most of backhaul applications foreseen in future wireless networks. As indicated in [8], the data rates range from 15 to 50 Mbps, while latency is on the order of hundreds of milliseconds (mostly due to propagation delay [18]). Another drawback is the high cost associated with data transmissions since it is calculated based on the consumed Mbps. Usual satellite frequency bands are 4–6, 10–12 and 20–30 GHz, which may be subject to weather conditions (e.g., rain, fog), especially in high frequencies where up to 24 dB rain fade attenuation is expected [34].

Satellite communication becomes an option for locations where the cost of laying a wired solution, or having multiple wireless hops, overcomes the cost of the consumed Mbps. Another application is in high mobility scenarios, such as shipping vessels, airplanes, or even moving base stations (used for extended coverage) [34].

12.2.2.2 Microwave

Microwave is a mature technology and has already been used for backhaul applications for many years. Nonetheless, due to the rapid growth of wireless networks, microwave solutions are being exploited even more. As highlighted in [38] the microwave market is expected to grow over $500 million by 2017, due to the need for high capacity backhaul to cope with the thousand fold traffic demand expected for the upcoming years [3,9].

Microwave frequencies range from 6 up to 60 GHz, but are typically deployed at 10.5, 26, and 28 GHz [38,39], and present a good trade-off between capacity and available spectrum [38]. These solutions operate in LOS conditions, employing directional antennas with fixed alignments designed to be robust to weather-induced fading [38], and under licensed spectrum bands (see Table 12.1).

Example: Capacity of a Microwave Link

The link capacity (C_L) of a point-to-point microwave link can easily go up to 1 Gbps and can be calculated in Mbps as $C_L = \beta \times \eta$, where β is the channel size, while η is the spectral efficiency [18]. Channel sizes are usually multiples of 7 MHz, ranging up to 112 MHz, supporting modulations up to 1,024 QAM, which renders (given appropriate coding schemes) 9 bps/Hz [18]. Based on the above, the link capacity is then $C_L = 112 \times 9 = 1,008$ Mbps. Even higher capacity can be achieved when combined with robust signal processing and multiple antennas as reported in [18].

Microwave links support point-to-point as well as point-to-multipoint communication. In essence, the air interfaces are the same as well as the spectral efficiency [18]. In point-to-point the spectrum is leased on a link-basis (see Table 12.1), and, as observed in [39], the available bands are becoming congested due to increasing demand, thus a need for new frequency bands. Point-to-point often uses highly directional antennas with narrow beams and high gains. In multipoint, a single hub is able to cover a sector that serves multiple users and operates under an area licensed spectrum, which allows the bandwidth to be shared within the sector [38]. Besides, multipoint is up to 40% more spectral efficient than is the single link counterpart. Moreover, users can be added to a sector on-the-fly (no installation or additional spectrum needed), thus becoming more effective for dense networks [38].

Microwave solutions offer 1 ms round trip latency per hop [18], which complies with current requirements of real-time video and gaming applications [26]. The coverage usually ranges from 2 to 4 km with high availability, longer ranges can also be reached, though the availability reduces [18]. According to [39] prices vary with respect to the carrier frequency, distance, bandwidth, availability, and location. Lower frequency bands are more expensive, since longer links can be supported and those are often more robust to weather impairments (e.g., rain and fog [39,40]). Moreover, multipoint solutions may offer lower operational cost [38]. As pointed out in [34], higher frequencies are subject to stronger attenuation, therefore short-range communications are perhaps a more suitable application for these technologies.

12.2.2.3   Sub-6 GHz

Currently most access services resort to sub-6 GHz frequency bands, which include a wide range of wireless networks: cellular networks, WiFi, and Bluetooth for instance. A key characteristic of this frequency band is propagation under nLOS conditions, which is also desirable for some backhaul applications.

One drawback of this frequency band is its fragmentation in terms of licensing, whose regulations vary from country to country. In general, frequencies from 1 to 3 GHz are mainly used for mobile access [18]. Another point to be considered is the interference management and congestion, besides the high licensing costs [34].

Similarly to the microwave case, the link capacity is determined by the bandwidth and spectral efficiency; however, an additional step is needed in sub-6 GHz frequency bands to determine the system capacity, which is then related to its ability to perform frequency re-use. In order to increase spectral efficiency, MIMO techniques for spatial-multiplexing are employed jointly with interference management and coordination, which also allows the use of high-order modulations [18].

Example: System Capacity at Sub-6 GHz

The link capacity (C_L) is calculated in Mbps just as in the case of microwave links [18]. In order to define the system capacity herein we need to consider frequency reuse, so that the system capacity is $C = \delta \times C_L$, where δ is the reuse factor. In general, the channel size ranges from 5 to 20 MHz, and 256 QAM in a 2×2

MIMO renders a peak spectral efficiency of 14 bps/Hz according to [18]. Bearing this in mind, a link capacity of $C_L = 20 \times 14 = 280$ Mbps, with a frequency reuse of $\delta = 1/2$ renders $C = 140$ Mbps.

In order to attain the capacity calculated in the above example, robust interference management protocols and algorithms are needed, besides scheduling, coordination, and multiple access protocols, which influence the latency of the communication link, being in the order of tens of milliseconds [18]. Coverage varies with the environment, for instance, in a dense urban scenario the coverage may reach few kilometers, while in sub-urban areas (where there are few high buildings, for example) it reaches up to 5 km according to [18].

Compared with microwave, sub-6 GHz has larger coverage, since it is possible to communicate in harsh nLOS environments, facilitating fast deployments; however, it has much lower link capacity and sophisticated interference mitigation schemes are needed. Nonetheless, such solutions are also applicable to mobile backhaul nodes, such as trains and road units [18], besides spatial-multiplexing, directional antennas, as well as sophisticated modulation and coding schemes may be used to improve the link capacity.

TV white spaces
Are mostly used to mobile access and are unlicensed spectrum, possessing another advantage in terms of larger coverage (up to tens of kilometers) attained in lower frequency bands, which turns this spectrum region a suitable option for remote and rural areas [18]. One more advantage is the reduced cost compared to the licensed options; on the other hand, interference and co-existence are issues that require robust solutions such as spectrum sensing and interference management protocols. As pointed out in [18], the capacity is limited, currently up to 18 Mbps per channel, with latency of a few milliseconds, but coverage is large, tens of kilometers in LOS conditions, and costs are smaller than licensed solutions.

12.2.2.4   mmWave

In the recent years, mmWave has gained considerable attention from both academia and industry [2,37,40–43], because of the massive amount of spectrum available and the possibility to achieve gigabit per second throughput. There are several bands available, but the most promising ones, due to availability and industry interest [2,42,43] are: 28 and 38 GHz, E-band (71–76 GHz, 81–86 GHz) [37], and 92–95 GHz. Several research institutes around the world have started campaign measurements in order to characterize fading and path loss at those frequencies for indoor and outdoor scenarios. For instance, [43] reports propagation measurements for an office environment at frequencies of 28 and 73 GHz, path loss exponents are characterized in the order of 1.1 and 1.3, respectively, under LOS conditions with 1 m reference. Note that the exponents in these cases are lower than in free-space (which has exponent 2) due to the constructive interference from multipath. Also, in [43] nLOS measurements are performed and the path loss exponents are 2.7 and 3.2, and it is pointed out that the importance of beamforming at both terminals in order to enhance the signal-to-noise ratio as well as coverage.

Overall, mmWave suffers from weather impairments which affect propagation due to severe attenuation. In lower frequencies (below 30 GHz) and at E-band, water molecules in the air cause most of the attenuation, while around 60 GHz attenuation is caused by oxygen [18]. In general, the coverage is quite limited even under LOS conditions, reaching up to a few hundred meters and greatly affected by the weather. We remark that the coverage might be greater depending on the frequency band, for instance, in E-band the distances may reach a few kilometers as indicated in [44]. On the other hand, the throughput achieves gigabit data rates, and is increasing steadily with new beamforming techniques associated with directional antennas and sophisticated signal processing [44,45]. As pointed out in [44], a large number of antennas can be deployed to a mmWave transceiver due to its small form-factor, which helps to reduce interference due to antenna directive and beamforming, and may as well enhance reliability by improving signal-to-noise ratio, which was also indicated in [43]. Additionally, round trip delay latency in this regimes are expected to be lower than 200 μs [18], since less processing is needed under LOS conditions. However, sophisticated beamforming and massive number of antennas as envisioned in [43] may increase round trip latency.

Some of the mmWave frequency band regions are license exempt or have light licensing, which considerably reduces the operational and management costs. In the region around 60 GHz there are large chunks of available spectrum, going up to 9 GHz, which allow for wide bandwidths with limited interference, besides the lighter regulation [18]. Also, due to the small form-factor of the transceiver, production and dissemination are facilitated [18]. On the other hand, the complexity of the transceiver is greater than conventional microwave communications, especially for the baseband unit components, requiring new designs as proposed in [44].

Solutions based on mmWave are seen as a key technology not only for backhauling, but also for broadband applications where extremely high data rates are needed to cope with the crescent traffic demand [2].

12.2.2.5 Final remarks on wireless solutions

Table 12.1 summarizes the key characteristics of the wireless solutions discussed above. Other alternatives for wireless backhaul are summarized in [8,18,44–46].[1] Moreover, [8] provides an overview of the challenges and research directions for backhaul in future wireless networks. Herein, though, we focus on the most promising wireless technologies that could potentially be employed in the next generation of wireless networks.

---

[1]Although this chapter focuses on wired and RF wireless solutions, we would like to point out that Free Space Optics (FSO), which uses light to transmit data through free space, has gained considerable attention as candidate solution for backhauling [46–48]. FSO is highly dependent of LOS conditions in order to achieve high data rates and low latency, and thus hybrid RF/FSO alternatives have been proposed [48] in order to reduce such dependency and increase reliability via NLOS links. An overview on related research challenges and applications is found in [47].

Figure 12.6   Wireless backhaul solutions: spectrum availability, bandwidth, cell size, coverage, and throughput comparison

Table 12.1   Key characteristics of wireless solutions

	Technology	Carrier (GHz)	Licensing	Connection	Throughput (Mbps)[†]
LOS*	Satellite	4–6, 10–12, 20–30	Licensed	Point-to-point	15–50
	Microwave	6–56	Link licensed	Point-to-point	1,000
	Microwave	6–56	Area licensed	Point-to-multipoint	1,000
	mmWave	56–64	Unlicensed	Point-to-point	1,000
	mmWave	28, 70–80	Area/light licensed	Point-to-point	10,000
nLOS*	White spaces	0.6–0.8	Dynamic	Point-to-multipoint	18
	Sub-6 GHz	0.8–6	Licensed	Point-to-multipoint	170
	Sub-6 GHz	2.4, 3.5, 5	Unlicensed	Point-to-multipoint	150–450

*Refer to [18].
[†]Refer to [8].

Figure 12.6 compares some key characteristics of the different wireless technologies discussed herein, including coverage, capacity, and bandwidth. Notice that, in general, communication under nLOS conditions is feasible specially in lower frequencies, which also allows for mobility and wide coverage. On the other hand, high frequencies usually require at least some LOS (though some recent results show the feasibility of nLOS communication in mmWave [43]) which implies in static applications with limited coverage, which is suitable for small cell deployments.

Figure 12.7 further compares the wireless technologies discussed herein. Notice that, as in the wired case, there is no single solution able to cope with all requirements. For instance, satellite communication have ubiquitous coverage, but limited capacity and low throughput, besides high operational costs. Meanwhile, mmWave solutions offer high throughput and low latency, as the cost of shorter links and LOS. All in all, these technologies complement each other, thus a robust backhaul network need to encompass these different solutions—wireless and wired—in order to meet the stringent requirements of the upcoming wireless networks.

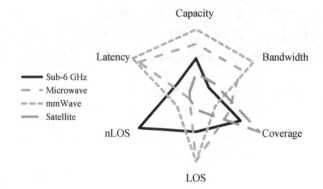

Figure 12.7   Comparative analysis of wireless technologies

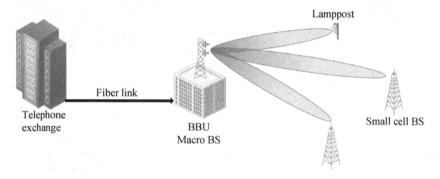

Figure 12.8   FTTX connection associated with wireless solution for backhaul

Figure 12.8 illustrates the backhaul deployment of FTTX associated with wireless solutions. Notice that fiber link is established from the telephone exchange (which may be seen as the core network) up to the macro base station (or to a common BBU). From the base station, the backhaul connection is wireless, through one or even a combination of the solutions discussed in this section. A microwave solution may be adopted between the base station and the lamppost, which are perhaps also interconnected via microwave once in street level. On the other hand, the small cell base station is perhaps indoors or blocked by some obstacle, therefore nLOS solutions (such as sub-6 GHz) are more appropriate in this case.

12.3   Conclusion

Recently wireless networks have become ubiquitous and an indispensable part of our life due to a broad range of applications. An exponential growth in the number of connected devices and a 1,000-fold increase in traffic volume is expected by 2020. In order to cope with these demands, future wireless networks will become more

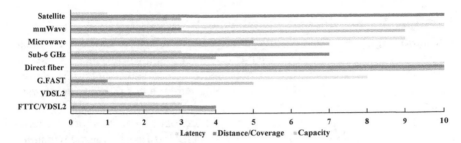

*Figure 12.9   Qualitative comparison of different wired and wireless technologies,
where the worst relative performance is denoted as 0 and the best
as 10*

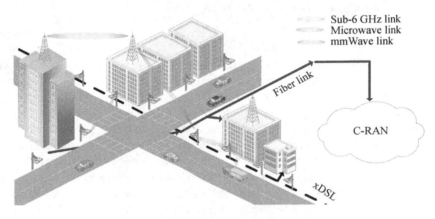

*Figure 12.10   Future hybrid backhaul deployment. Wired and wireless solutions are
combined to ensure best coverage and QoS. Notice that large
bandwidths are available at mmWave frequencies*

dense and heterogeneous, where devices will have many RATs. Then, in order to
enable these advances, it is necessary to guarantee connectivity and capacity in a
cost-effective and sustainable way to all kinds of applications [7]. A robust backhaul
network is mandatory. In this chapter, we addressed some of the most promising
technologies that may be part of the solution for such endeavors. Figure 12.9 depicts
a qualitative comparison (though, the quantitative values can be found in the previous
sections as well as in [8,18,22]) of different wired and wireless solutions discussed in
this chapter. Notice that the worst relative performance is denoted as 0 while the best
relative performance is denoted as 10. In the previous sections we have shown that
a few solutions are able to achieve gigabit throughput and low latency (in the order
to micro seconds), as direct fiber, which would be an excellent solution, if only the
deployment cost were not exorbitant. Next, we provide an example and an illustration
of the envisioned future deployment in Figure 12.10.

Example: Future Backhaul Deployment

A final example of a hybrid wired and wireless backhaul solution is provided. Figure 12.10 envision a future backhaul network deployment where wired and wireless solutions are combined to offer the best performance in terms of coverage, capacity, reliability, and availability. Notice that in this example, microwave solutions are used to connect base stations, while mmWave is perhaps used not only for data traffic but as well for in-band backhaul. High mobility use cases are covered by sub-6 GHz solutions, while xDSL and fiber cover the remaining cases. Each of these applications has its own set of requirements and constraints, and some nodes may have multiple radio and wired interfaces. One major research challenge is then to design efficient solutions that account for all these heterogeneous requirements and characteristics of each wired and wireless technologies, while performing seamlessly [8,14].

Some challenges ahead

An important fact is that the backhaul technologies may complement each other, and as discussed before, there is no "one-size-fits-all" type of solution. The backhaul network is becoming even more complex and composed of a combination of diverse technologies, which brings a great challenge for designing such a dynamic, flexible and heterogeneous network. One promising alternative to tackle such diversity issue is via distributed, self-optimized networks where user association and backhauling are dynamically integrated as in [14]. Another key challenge is the backhaul latency which is required to be in the order of few hundreds of micro seconds to enable all the envisioned functionalities of the C-RAN (see [8, Table 4]). Fiber and mmWave solutions can cope with that requirement. However, fiber incurs in high installation costs and limited availability, while most of mmWave solutions rely on LOS and are high susceptible to environmental and propagation conditions. Nonetheless, much research effort is being put in these directions [42,44,45,49,50]. Some challenges related to mmWave solutions are beamsteering and beam alignment, which require advanced signal processing and synchronization, but are fundamental to achieve high data rates and low latency [51,52].

All in all, there are several possible solutions, open challenges and opportunities ahead which show that backhauling is an exciting and dynamic research area, of fundamental importance for the development of future wireless networks.

References

[1] Nokia, "White paper: future works – 5G use cases and requirements," *Nokia White Papers*, 2014.
[2] Ericsson, "5G radio access: technology and capabilities," *Ericsson White Paper*, 2015. [Online]. Available: https://www.ericsson.com/res/docs/white papers/wp-5G.pdf

[3] J. Manyika, M. Chui, P. Bisson, *et al.*, "Unlocking the potential of the internet of things," *McKinsey Global Institute*, 2015.

[4] C. Perera, C. H. Liu, S. Jayawardena, and M. Chen, "A survey on internet of things from industrial market perspective," *IEEE Access*, vol. 2, pp. 1660–1679, 2014.

[5] F. Boccardi, R. W. Heath, A. Lozano, T. L. Marzetta, and P. Popovski, "Five disruptive technology directions for 5G," *IEEE Communications Magazine*, vol. 52, no. 2, pp. 74–80, Feb. 2014.

[6] A. Osseiran, F. Boccardi, V. Braun, *et al.*, "Scenarios for 5G mobile and wireless communications: the vision of the METIS project," *IEEE Communications Magazine*, vol. 52, no. 5, pp. 26–35, May 2014.

[7] X. Ge, H. Cheng, M. Guizani, and T. Han, "5G wireless backhaul networks: challenges and research advances," *IEEE Network*, vol. 28, no. 6, pp. 6–11, Nov. 2014.

[8] M. Jaber, M. A. Imran, R. Tafazolli, and A. Tukmanov, "5G backhaul challenges and emerging research directions: a survey," *IEEE Access*, vol. 4, pp. 1743–1766, 2016.

[9] J. G. Andrews, S. Buzzi, W. Choi, *et al.*, "What will 5G be?" *IEEE Journal on Selected Areas in Communications*, vol. 32, no. 6, pp. 1065–1082, Jun. 2014.

[10] D. C. Chen, T. Q. S. Quek, and M. Kountouris, "Backhauling in heterogeneous cellular networks: modeling and tradeoffs," *IEEE Transactions on Wireless Communications*, vol. 14, no. 6, pp. 3194–3206, Jun. 2015.

[11] O. Galinina, A. Pyattaev, S. Andreev, M. Dohler, and Y. Koucheryavy, "5G multi-RAT LTE-WiFi ultra-dense small cells: performance dynamics, architecture, and trends," *IEEE Journal on Selected Areas in Communications*, vol. 33, no. 6, pp. 1224–1240, Jun. 2015.

[12] A. Orsino, G. Araniti, A. Molinaro, and A. Iera, "Effective RAT selection approach for 5G dense wireless networks," in *2015 IEEE 81st Vehicular Technology Conference (VTC Spring)*, May 2015, pp. 1–5.

[13] R. Wang, H. Hu, and X. Yang, "Potentials and challenges of C-RAN supporting multi-RATs toward 5G mobile networks," *IEEE Access*, vol. 2, pp. 1187–1195, 2014.

[14] M. Jaber, M. A. Imran, R. Tafazolli, and A. Tukmanov, "A distributed SON-based user-centric backhaul provisioning scheme," *IEEE Access*, vol. 4, pp. 2314–2330, 2016.

[15] O. Tipmongkolsilp, S. Zaghloul, and A. Jukan, "The evolution of cellular backhaul technologies: current issues and future trends," *IEEE Communications Surveys Tutorials*, vol. 13, no. 1, pp. 97–113, First 2011.

[16] R. Taori and A. Sridharan, "Point-to-multipoint in-band mmwave backhaul for 5G networks," *IEEE Communications Magazine*, vol. 53, no. 1, pp. 195–201, Jan. 2015.

[17] H. Lundqvist, "Death by starvation? backhaul and 5G," *IEEE CTN*, Accessed on 30 Aug. 2016, 2015. [Online]. Available: http://www.comsoc.org/ctn/death-starvation-backhaul-and-5G

[18] Small Cell Forum, "Backhaul technologies for small cells: use cases, requirements and solutions," *Small Cell Forum*, Feb. 2013.

[19] A. Pizzinat, P. Chanclou, F. Saliou, and T. Diallo, "Things you should know about fronthaul," *Journal of Lightwave Technology*, vol. 33, no. 5, pp. 1077–1083, Mar. 2015.

[20] A. D. L. Oliva, X. C. Perez, A. Azcorra, *et al.*, "Xhaul: toward an integrated fronthaul/backhaul architecture in 5G networks," *IEEE Wireless Communications*, vol. 22, no. 5, pp. 32–40, Oct. 2015.

[21] T. Pfeiffer, "Next generation mobile fronthaul and midhaul architectures [invited]," *IEEE/OSA Journal of Optical Communications and Networking*, vol. 7, no. 11, pp. B38–B45, Nov. 2015.

[22] NGMN, "Deliverable: RAN evolution project backhaul and fronthaul evolution," *NGMN Alliance*, vol. 1.01, Feb. 2015.

[23] J. Allen, F. Chevalier, and B. Bora, "Report for vodafone – mobile backhaul market: phase 1 report," *Vodafone*, Feb. 2014.

[24] J. Allen and F. Chevalier, "Report for vodafone – mobile backhaul market: phase 2 report," *Vodafone*, Feb. 2014.

[25] NGMN, "White paper: small cell backhaul requirements," *NGMN Alliance*, vol. 1.0, Jun. 2012.

[26] G. P. Fettweis, "The tactile internet: applications and challenges," *IEEE Vehicular Technology Magazine*, vol. 9, no. 1, pp. 64–70, Mar. 2014.

[27] K. Moskvitch, "Tactile internet: 5G and the cloud on steroids," *Engineering Technology*, vol. 10, no. 4, pp. 48–53, May 2015.

[28] J. Y. Sung, C. W. Chow, C. H. Yeh, Y. Liu, and G. K. Chang, "Cost-effective mobile backhaul network using existing ODN of PONs for the 5G wireless systems," *IEEE Photonics Journal*, vol. 7, no. 6, pp. 1–6, Dec. 2015.

[29] OECD, "Fixed and wireless broadband subscriptions by technology," *OECD Broadband Portal*, Dec. 2015. [Online]. Available: http://www.oecd.org/sti/broadband/oecdbroadbandportal.htm

[30] V. Oksman, R. Strobel, X. Wang, *et al.*, "The ITU-T's new G.fast standard brings DSL into the gigabit era," *IEEE Communications Magazine*, vol. 54, no. 3, pp. 118–126, Mar. 2016.

[31] J. R. Schneir and Y. Xiong, "Cost assessment of FTTdp networks with G.fast," *IEEE Communications Magazine*, vol. 54, no. 8, pp. 144–152, Aug. 2016.

[32] Y. Huang, E. Medeiros, N. Fonseca, *et al.*, "LTE over copper – potential and limitations," in *2015 IEEE 26th Annual International Symposium on Personal, Indoor, and Mobile Radio Communications (PIMRC)*, Aug. 2015, pp. 1339–1343.

[33] X. Ge, S. Tu, G. Mao, C. X. Wang, and T. Han, "5G ultra-dense cellular networks," *IEEE Wireless Communications*, vol. 23, no. 1, pp. 72–79, Feb. 2016.

[34] U. Siddique, H. Tabassum, E. Hossain, and D. I. Kim, "Wireless backhauling of 5G small cells: challenges and solution approaches," *IEEE Wireless Communications*, vol. 22, no. 5, pp. 22–31, Oct. 2015.

[35] M. Mueck, W. Jiang, G. Sun, H. Cao, E. Dutkiewicz, and S. Choi, "White paper novel spectrum usage paradigms for 5G," *IEEE Cognitive Networks Technical Committee, Special Interest Group Cognitive Radio in 5G*, Nov. 2014.

[36] J. Wells, "Licensing and license fee considerations for E-band 71-76 GHz and 81-86 GHz wireless systems," *E-Band Communications Corp.*, 2010.

[37] M. G. L. Frecassetti, "E-band and V-band – survey on status of worldwide regulation," *ETSI White Paper*, vol. 9, 2015.

[38] C. B. Networks, "White paper – spectrum and technology issues for microwave backhaul in Europe," *Cambridge Broadband Networks*, Feb. 2014. [Online]. Available: http://cbnl.com/resources/white-papers

[39] Innovation Observatory, "White paper – spectrum and technology issues for microwave backhaul in Europe," *Cambridge Broadband Networks*, Nov. 2010. [Online]. Available: http://cbnl.com/resources/white-papers

[40] M. V. Perič, D. B. Perič, B. M. Todorovič, and M. V. Popovič, "Dynamic rain attenuation model for millimeter wave network analysis," *IEEE Transactions on Wireless Communications*, vol. 16, no. 1, pp. 441–450, Jan. 2017.

[41] Z. Pi, J. Choi, and R. Heath, "Millimeter-wave gigabit broadband evolution toward 5G: fixed access and backhaul," *IEEE Communications Magazine*, vol. 54, no. 4, pp. 138–144, Apr. 2016.

[42] T. S. Rappaport, S. Sun, R. Mayzus, *et al.*, "Millimeter wave mobile communications for 5G cellular: it will work!" *IEEE Access*, vol. 1, pp. 335–349, 2013.

[43] G. R. Maccartney, T. S. Rappaport, S. Sun, and S. Deng, "Indoor office wideband millimeter-wave propagation measurements and channel models at 28 and 73 GHz for ultra-dense 5G wireless networks," *IEEE Access*, vol. 3, pp. 2388–2424, 2015.

[44] Z. Gao, L. Dai, D. Mi, Z. Wang, M. A. Imran, and M. Z. Shakir, "mmwave massive-MIMO-based wireless backhaul for the 5G ultra-dense network," *IEEE Wireless Communications*, vol. 22, no. 5, pp. 13–21, Oct. 2015.

[45] S. Hur, T. Kim, D. J. Love, J. V. Krogmeier, T. A. Thomas, and A. Ghosh, "Millimeter wave beamforming for wireless backhaul and access in small cell networks," *IEEE Transactions on Communications*, vol. 61, no. 10, pp. 4391–4403, Oct. 2013.

[46] Y. Li, N. Pappas, V. Angelakis, M. Pióro, and D. Yuan, "Optimization of free space optical wireless network for cellular backhauling," *IEEE Journal on Selected Areas in Communications*, vol. 33, no. 9, pp. 1841–1854, Sep. 2015.

[47] Z. Ghassemlooy, S. Arnon, M. Uysal, Z. Xu, and J. Cheng, "Emerging optical wireless communications-advances and challenges," *IEEE Journal on Selected Areas in Communications*, vol. 33, no. 9, pp. 1738–1749, Sep. 2015.

[48] H. Dahrouj, A. Douik, F. Rayal, T. Y. Al-Naffouri, and M. S. Alouini, "Cost-effective hybrid RF/FSO backhaul solution for next generation wireless systems," *IEEE Wireless Communications*, vol. 22, no. 5, pp. 98–104, Oct. 2015.

[49] N. Benzaoui, Y. Pointurier, B. Uscumlic, T. Bonald, Q. Wei, and S. Bigo, "Transport mechanisms for mobility support in optical slot switching-based

next-generation mobile backhaul networks," *Journal of Lightwave Technology*, vol. 34, no. 8, pp. 1946–1955, Apr. 2016.

[50]   L. Simic, N. Perpinias, and M. Petrova, "60 GHz outdoor urban measurement study of the feasibility of multi-Gbps mm-wave cellular networks," in *2016 IEEE Conference on Computer Communications Workshops (INFOCOM WKSHPS)*, pp. 554–559, Apr. 2016.

[51]   M. Tercero, P. von Wrycza, A. Amah, *et al.*, "5G systems: the mmMAGIC project perspective on use cases and challenges between 6–100 GHz," in *2016 IEEE Wireless Communications and Networking Conference*, pp. 1–6, Apr. 2016.

[52]   R. W. Heath, N. González-Prelcic, S. Rangan, W. Roh, and A. M. Sayeed, "An overview of signal processing techniques for millimeter wave MIMO systems," *IEEE Journal of Selected Topics in Signal Processing*, vol. 10, no. 3, pp. 436–453, Apr. 2016.

Chapter 13

Spectral coexistence for next generation wireless backhaul networks

Shree Krishna Sharma[1], Eva Lagunas[2], Christos Tsinos[2],
Sina Maleki[2], Symeon Chatzinotas[2],
and Björn Ottersten[2]

One of the main bottlenecks for terrestrial wireless operators to meet the capacity demands of future wireless networks is the limitation in the backhaul capacity. On the other hand, satellite community is putting significant efforts in implementing high throughput multibeam satellites in order to achieve the aggregated capacity up to Terabit/s. In this context, one of the emerging solutions to address the backhaul capacity demand is to enable Hybrid Satellite-Terrestrial Backhauling (HSTB) by combining terrestrial backhaul networks with the satellite backhaul in a seamless manner. The incorporation of a satellite into the wireless backhaul network can bring several advantages such as cost-effective coverage extension to the remote areas, quick deployment, bandwidth extension and the backup route for delivering wireless traffic. However, due to the limitation in the available radio spectrum, the main research challenge is how to enable the sharing of radio spectrum among different segments of these HSTB networks in order to enhance the overall spectral efficiency.

In this chapter, starting with the recent trend in terrestrial and satellite backhaul technologies, we provide possible use cases for HSTB networks and their potential benefits and challenges. Subsequently, we focus on the spectrum sharing aspects of wireless backhaul networks considering the following two categories of enabling techniques: (i) spectral awareness techniques and (ii) spectral exploitation techniques. The first category mainly comprises radio environment awareness techniques such as spectrum sensing and databases while the second category includes interference mitigation and resource allocation techniques. Furthermore, we present three case studies along with the numerical results considering the coexistence of satellite and terrestrial systems in the Ka-band. Finally, this chapter provides some interesting recommendations for future research directions.

[1] Department of Electrical and Computer Engineering, Western University, Canada
[2] Interdisciplinary Centre for Security, Reliability and Trust (SnT), University of Luxembourg, Luxembourg

13.1   Introduction

The term "Backhaul" in telecommunication networks generally refers to the intermediate connections between the backbone (core) network and the local sub-networks
which provide connectivity to the last-mile users. Besides, the term "Mobile Backhaul" is commonly used to denote the connectivity between Base Stations (BSs) and
radio controllers in cellular systems with the help of various transport media [1]. As
an example, the Abis interface between a BS Controller (BSC) and a Base Transceiver
Station (BTS) may be of the following types: (i) Digital Subscriber Line (DSL)/copper
technology, (ii) microwave, (iii) optical fiber network, and (iv) satellite technology.
On the basis of the above transmission technology/medium, backhaul technologies
can be widely categorized into [1]: (i) wired and (ii) wireless. The wired backhaul
may comprise of copper or optical fiber and mainly used for residential or enterprise
purposes. On the other hand, wireless backhaul medium may comprise microwave or
satellite, and is mainly utilized for public access scenarios.

Over the recent years, the demand for backhaul capacity has tremendously
increased due to exponentially increasing number of high data-rate services and
broadband mobile users. With the current technologies and architectures, it seems
infeasible to support Gigabit-level data traffic in low cost and resource (power, frequency) efficient manner. Besides, in the terrestrial wireless context, there is an
increasing trend of adopting new wireless solutions such as small/femto/pico cells,
large-scale antenna array-based systems such as massive Multiple Input Multiple Output (MIMO) systems and three-dimensional (3D) beamforming, Content Delivery
Network (CDN), Internet of Things (IoT), millimeter wave (mmWave) technology, Cognitive Radio (CR), self-backhauling mechanisms, and Cloud-Radio Access
Network (C-RAN) [2]. On the other hand, in the satellite domain, the recent trend
is toward exploring multibeam satellites, onboard processing, full frequency reuse,
advanced Non-Geostationary (NGEO) satellite constellations such as large Low Earth
Orbit (LEO) or Medium Earth Orbit (MEO) constellations, nano satellites [3], and
the role of satellite in the fifth generation (5G) of wireless communications [4].
As the number of wireless systems in the access side increases, the design of cost-
effective integrated backhaul solutions under practical spectrum and energy constraint
scenarios becomes challenging.

In 5G and beyond wireless networks, Satellite Communication (SatCom) solutions can complement terrestrial solutions in all geographical regions including rural/
inaccessible places and urban/suburban areas in terms of providing telecommunication services to the end users. Moreover, in the remote and rural areas, it is difficult to
deploy wired backhaul solutions due to cost and implementation issues. In this regard,
satellite backhaul becomes an ideal solution to deliver telecommunication services
to these geographically challenging areas. As compared to the terrestrial backhaul,
satellite backhaul can significantly reduce the infrastructure cost and also helps to
deploy the services quickly over a wide area [5]. Besides, satellite backhaul networks
can relay the backhaul traffic to the core network with the help of one or a very few
hops and can also provide backup solution to the terrestrial backhaul links in case of

failure or for load balancing in the places/events with high traffic demand. Therefore, SatCom can be an important means to support the expanding of the 5G ecosystem toward highly reliable and secure global networks.

In the above context, Hybrid Satellite-Terrestrial Backhauling (HSTB), which can grasp the benefits from both satellite and terrestrial technologies, is considered as a promising approach in order to meet the backhaul capacity requirements of future wireless networks. This novel solution can provide the following benefits [6]: (i) enhancing mobile backhaul network capacity, (ii) improving the resilience of wireless networks, (iii) extension of the coverage area, (iv) enhancing the spectral efficiency, and (v) improving energy efficiency. However, there are several issues in realizing HSTB networks in practice. The main issues are (i) optimal placement of the HSTB node, (iii) designing low cost and power efficient transceiver for the HSTB node, and (iv) spectrum sharing issues and the resulting co-channel inter-system/intra-system interferences. Out of these issues, this chapter mainly focuses on the spectrum sharing issues and investigates suitable techniques to enable the spectrum sharing among different segments in HSTB networks.

In this chapter, first, we highlight the recent trends in wireless backhaul considering both terrestrial and satellite technologies. Subsequently, we introduce the concept of HSTB and present various possible scenarios and their associated benefits and challenges. Next, we present various enabling techniques for spectrum sharing in wireless backhaul networks. In addition, we present three case studies considering the coexistence of satellite and terrestrial network segments as well as HSTB networks in the Ka-band. Finally, we provide some interesting recommendations for future research directions.

13.2   Research trends in wireless backhaul

In this section, we describe the recent research trends in terrestrial and satellite backhaul technologies from the existing state of the art.

1. *Terrestrial backhaul technologies*: The upcoming 5G deployment has posed numerous challenges, mainly in terms of supporting very high data rates with low end-to-end delays [7]. One of the key technologies for meeting the 5G target rates is the massive MIMO technology [8], which considers the BSs equipped with antenna arrays with a few hundred antennas that can simultaneously serve the tens of users using the same frequency resource. In addition, one of the potential approaches to improve the spectral efficiency in 5G networks is millimeter wave (mmWave) communication technology [9].

Massive MIMO and mmWave are considered as key technologies to address the requirements for 5G systems, but their benefits cannot be fully achieved without substantially improving the underlying backhaul infrastructure. The expected traffic growth will cause significant burdens on the capacity of the wireless backhaul network. In fact, backhauling has been identified as the key challenge for the 5G deployment [7,10].

Current wireless backhaul networks mostly operate at the frequencies above 6 GHz and at these high frequencies, signal attenuation is quite high resulting in short-range links (few tens of kilometers). Therefore, either using mmWave or using microwave bands, multiple hops are required to overcome long distances and/or obstacles in the propagation path [11,12]. Summarizing, and following the indications in [13], the technical solutions that could be considered by market players to facilitate the roll out, to reduce the backhaul cost, and to meet the traffic needs are the following: (i) using spectrum efficient techniques such as adaptive modulation techniques, bandwidth adaptive systems, polarization multiplexing, massive MIMO, full duplex radios and asymmetrical point-to-point links, (ii) adaptation of network topology to cope with network failures, (iii) cell densification by incorporating more smaller cells, (iv) aggressive frequency reuse in backhaul links, and (v) self-backhauling, i.e., using the same set of frequencies in the access side and the backhaul side.

2. *Satellite backhaul technologies*: As mentioned earlier in Section 13.1, satellite backhaul is one of the emerging wireless backhaul solutions and this refers to providing backhaul capacity/links via satellite. The satellite acts as a repeater between the teleport (which is connected to the core mobile network) and the remote sites connected to BTS/eNodeB. The backhaul links to the earth sites can be provided with the help of either Geostationary (GEO) satellite or NGEO satellite. These two approaches have their own advantages and disadvantages. In GEO satellite backhaul networks, one GEO satellite can provide backhauling service over a wide area and there is no need of tracking at the mobile site. However, it suffers from the problem of latency. On the other hand, backhauling with NGEO satellites such as MEO and LEO satellites improve the latency for delay-sensitive applications. This solution requires the user site and the teleport to be equipped with tracking antennas in order to track the NGEO satellites, and also needs to deal with the handover operation [14].

Satellite backhaul is suitable not only for rural/remote cases but also for urban/sub-urban areas. As illustrated in Figure 13.1, the main application areas of satellite backhauling are [14]: (i) mobile sites, (ii) rural small cells, and (iii) backup solution. The main access techniques used for satellite backhauling are Time Division Multiplexing/Time Division Multiple Access (TDM/TDMA) and Single Carrier Per Channel (SCPC) [14]. In general, TDM technique is used in the forward link, i.e., from the gateway to the remote site, and TDMA or SCPC technique is employed in the return link, i.e., from the remote sites to the gateway. Due to the emerging trend of migrating satellite traffic to the IP-based platform, TDM/TDMA has got significant importance over the conventional SCPC approach. Regardless of the traffic types, many TDM/TDMA networks are designed to handle all types of IP traffic with better Quality of Service (QoS) [15]. Although many conventional SCPC based networks have already been converted to TDM/TDMA platform, there exist hybrid TDM/SCPC networks, with the SCPC technology being used in the return link and a Digital Video Broadcasting-Satellite-Second Generation (DVB-S2) TDM carrier in the forward link, due to its several benefits such as better model efficiency, more resilient to rain fades/interference, higher speed, and lower latency.

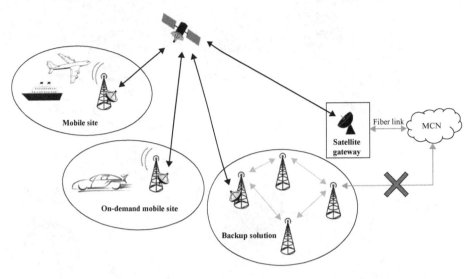

Figure 13.1   Illustrations of various applications of satellite backhaul (MCN stands for Mobile Core Network) [14]

13.3   Hybrid satellite-terrestrial backhaul

The main objectives behind the considered HSTB network are [5]: (i) incorporation of satellite segment in the terrestrial backhaul networks in order to take advantage from specific properties of satellite systems, (ii) possible spectrum sharing among various segments of the network in order to enhance overall spectral efficiency, and (iii) offloading traffic to the satellite depending on the nature of traffic and latency requirements.

13.3.1   Scenarios

A well-defined scenario is a combination of various elements as depicted in Figure 13.2. An HSTB scenario mainly depends on deployment types, satellite link characteristics, terrestrial link characteristics, used frequency bands (shared or exclusive) and new network features such as content delivery. Regarding possible shared frequency bands, the Ka-band segments, i.e., 17.7–19.7 and 27.5–29.5 GHz are primarily allocated to Fixed Service (FS) microwave systems and can be utilized by the Fixed Satellite System (FSS) in a secondary way provided that they can protect themselves in the forward link case and they can guarantee the protection of the primary FS receivers against the harmful interference in the return link case. Besides, as highlighted in Figure 13.2, various characteristics of terrestrial links such as channel models, antenna characteristics, carrier frequency and operating bandwidth can be considered in characterizing an HSTB network. In addition, satellite link characteristics such as antenna structures, carrier frequency, operating bandwidth, and channel

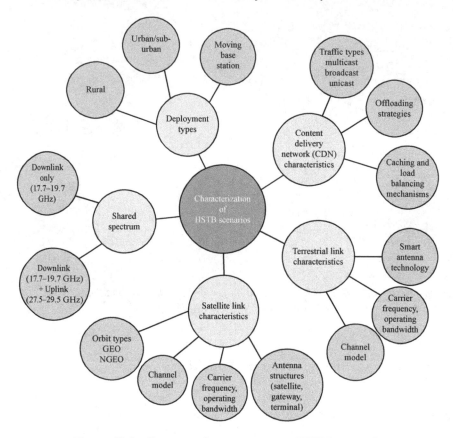

Figure 13.2   Scenario characterization of HSTB networks

models are also significantly important to be considered while defining an HSTB scenario.

CDN is an emerging concept to serve the contents to the end-users with better availability and higher performance. This is done by caching highly demanded content in the edge caches, thereby bringing resources closer to users and reducing the round trip time. A simplified architecture of a hybrid satellite-terrestrial CDN is provided in Figure 13.3. Several characteristics of CDN networks such as traffic types, offloading strategies, caching and load balancing strategies need to be considered while investigating hybrid satellite-terrestrial CDN networks.

From the deployment perspective, HSTB scenarios can be broadly divided into the following categories [16]: rural, urban, and moving BS. Besides, a suitable hybrid satellite-terrestrial CDN architecture can be investigated in all these scenarios. These scenarios are briefly described below.

1. Rural scenario: In this scenario, suitable solutions can be developed in order to support the existing backhauling networks with the objective of providing faster

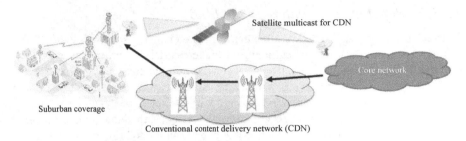

Figure 13.3   *Illustration of a hybrid satellite-terrestrial content delivery network*
(CDN) scenario

and more reliable mobile networks. One of the potential solutions is to use CDNs to bring the contents closer to the users. Besides, resilience and high speed satellite connections to offload the terrestrial traffic are the desired characteristics of this scenario. To achieve these objectives, future HSTB networks should consider satellite segment as an important component as well as other innovative key enabling components such as smart antennas, intelligent backhaul nodes and a hybrid network manager.[1]

2. Urban scenario: The important aspects to be considered in urban environment are end-users' expectation on high QoS from any place/platform, gathering of people forming a dynamic crowd, and significant traffic growth during particular social events/concerts/sports. In order to support these cases, network densification with a suitable network topology is essential. In this regard, HSTB networks can be used to complement the traffic demand by enabling a subset of the backhaul nodes to operate with both the satellite and terrestrial systems. Such a hybrid solution can potentially bridge the gap for the coexistence of terrestrial and satellite backhaul technologies, and will also reduce the costs involved in running two separate wireless backhaul solutions.

3. Moving BS scenario: HSTB solutions can be useful for various mobile platforms such as trains, cruise ships, and airplanes. By deploying the hybrid network capability in the mobile BSs, high speed broadband connectivity can be provided to the end-users at the cheaper rate. For example, in cruise ships, one or more BSs/access points can be installed to provide Internet access to the passengers and the traffic can be backhauled with the help of a satellite link. Such a setup can largely benefit from the capabilities of an HSTB network since it provides the flexibility of using both terrestrial and satellite links for mobile backhauling. Depending on the application cases, satellite only or both backhauling can be used, and also satellite backhaul can be also used as a backup solution to the case where terrestrial only backhauling is used.

---

[1] For further details on these components, authors may refer to [16].

13.3.2   Benefits and challenges of HSTB

In this subsection, we discuss the benefits and the challenges caused due to collaboration between the satellite segment and the terrestrial system in future backhaul networks. In particular, it is expected that satellite component will provide many possibilities and opportunities in terms of supported capacity, resilience and in extending backhaul services to anywhere and anytime. While wireless backhaul terrestrial systems lack the ability to leap across continents and oceans, the inherent large coverage footprint of satellites make them the most suitable solution to bring backhauling solutions over the world. Furthermore, the transparent integration of satellite and terrestrial systems should be done in such a way that spectral and energy efficiencies are maximized. The main benefits of such a hybrid network are the following: (i) increase in the capacity of mobile backhaul networks, (ii) improvement in resilience, (iii) extension of coverage area, (iv) improvement in spectrum efficiency, and (v) enhancement in spectrum efficiency. However, the following challenges need to be successfully tackled while designing and deploying such a hybrid network.

   1. Deployment of satellite terminals: One of the main challenges is to decide on where to deploy satellite terminals within the satellite's coverage. A deployment decision is usually based on technological versus performance factors, and also cost and regulatory aspects. Also, the deployment decision is scenario-based. For example, in the rural case, satellite terminals should be installed at the nodes that serve as a unique link to the core network while in the urban scenario, link failure and congestion issues become critical and satellite link can help with offloading and resilience.

   In CDN networks depicted in Figure 13.3, edge caches are good places to incorporate satellite terminals, which can serve as complementary components of the system in order to boost the network capacity while ensuring the uninterruptable delivery of services. In the case of urban deployments, hybrid satellite-terrestrial nodes can be connected to the core through a high speed connection (optical fiber). In the case of the moving BS scenario, all nodes are expected to be equipped with the satellite terminals since these platforms need to spend long periods of time without having access to the terrestrial networks.

   2. Interference: Considering a hybrid backhaul network, we can distinguish the following two types of interference: (1) intra-system interference and (2) inter-system interference. Intra-system interference is caused by the transmitters that belong to the same hybrid network, whereas inter-system interference is caused from the network's nodes belonging to external systems, which use the same frequencies. Focusing on the intra-system interference, we can distinguish the following sources of interference: (i) interference caused by terrestrial nodes to the satellite terminals when these links operate in the downlink mode, (ii) interference caused by terrestrial nodes to the satellite in the uplink mode, (iii) interference caused by the satellite terminals to the terrestrial nodes in the uplink mode, and (iv) interference caused by the satellite to the terrestrial nodes when the satellite terminals operate in the downlink mode. Due to long propagation distance, directional characteristics of satellite transmission and different receiver sensitivity levels of satellite and terrestrial terminals, the second and the fourth interference cases can be neglected in many cases.

13.4 Spectrum sharing in wireless backhaul networks

In spectrum sharing scenarios, where two or more networks share the same set of radio frequencies, Cognitive Radio (CR) principles can be used in order to devise opportunistic schemes as well as to mitigate the harmful interference. Various possible spectral coexistence scenarios in the considered HSTB network are listed below: (i) Case 1: Coexistence of satellite segment (uplink) with the terrestrial segment in one HSTB network, (ii) Case 2: Coexistence of satellite segment (downlink) with the terrestrial segment in one HSTB network, (iii) Case 3: Coexistence of one HSTB network with another HSTB network both operating under Case 1, and (iv) Case 4: Coexistence of one HSTB network with another HSTB network both operating under Case 2. Spectrum sharing techniques, which can enable the aforementioned coexistence scenarios, can be broadly categorized into: (i) spectrum awareness techniques and (ii) spectrum exploitation/utilization techniques [17]. In the following, we provide the main principles and the issues related to different techniques under these categories.

13.4.1 Spectrum awareness techniques

Radio environment awareness is a key requirement in wireless communication systems that share the same set of radio frequencies. Various levels of awareness can be obtained depending on the employed signal processing technique. The acquired information can be in various forms such as spectrum occupancy over the available bands, Signal-to-Noise Ratio (SNR) of the primary signal, channel toward the Primary (licensed/incumbent) Users (PUs), modulation and coding used by the PUs, pilot/header information in the PU transmit frame, etc. [17,18]. The higher the amount of information a node can gather from the environment, the better becomes the utilization of the available radio resources. For acquiring information about the spectrum occupancy, several techniques such as Spectrum Sensing (SS), spectrum cartography, and geolocation databases can be utilized, which are briefly described below along with their associated challenges.

13.4.1.1 Spectrum sensing

There are several SS techniques depending on the employed signal processing mechanisms. In this regard, the main techniques discussed in the literature are [17]: (i) energy detection, (ii) matched filter-based detection, (iii) cyclostationary feature detection, (iv) covariance-based detection, (v) eigenvalue-based detection, and (vi) waveform-based detection. These techniques have their own characteristics and complexity requirements from the deployment perspective. Some of the important research challenges of SS techniques are provided below [19].

(i) Restricted sensing ability: Most devices do not have specific Radio Frequency (RF) transceiver to sense the environment, and therefore the common assumption is that they typically access the spectrum following a two-stage "listen-before-talk" protocol in which the device performs sensing and transmission independently in two different time intervals. There exists an evident trade-off between the sensing

capabilities and the throughput that can be achieved by these devices [20,21]. Recently, there is an increasing research interest in exploring "Listen-and-Talk" protocol [22], however, the close proximity of transmit and receive antennas might pose severe self-interference levels.

(ii) Wideband sensing: The traditional way for sensing a wideband spectrum is channel-by-channel sequential scanning [23], which introduces a large latency. One way to solve the latency problem at the cost of implementation complexity is to use an RF front-end with a bank of narrow band-pass filters. Besides, one of the emerging trends in the research community is to directly sense a wide frequency range at the same time [24], however, this needs high sampling rates and costly Analog-to-Digital Converters (ADCs). To address this drawback, Compressive Sensing (CS) [25] based wideband sensing has attracted important attention in the literature [26,27].

(iii) Practical imperfections: In practice, there may occur several imperfections such as noise uncertainty, channel/interference uncertainty, and transceiver hardware imperfections like amplifier non-linearity, quantization errors, and calibration issues [17]. These imperfections may severely deteriorate the performance of the employed spectrum awareness mechanisms.

13.4.1.2   Spectrum cartography

While individual SS techniques provide valuable information about spectrum activity at a specific location, frequency and time, they are not suitable to acquire the network-wide spectrum usage knowledge . This network-level spectrum knowledge can be obtained using a spectrum map, commonly known as Radio Environment Map (REM) [28,29]. The main research challenges related to this case are listed below.

(i) Decision fusion: Assuming that sensing information is well received by all the devices, one important issue is how to combine all the acquired information gathered by heterogeneous sensing nodes to produce an accurate map. In general, there is a central node which collects information from multiple devices about the spectrum occupancy of the whole area of interest. Based on this information, spatial interpolation [30] can be used to estimate missing information at the unobserved locations.

(ii) Learning: The REM information can be updated with observations from the cognitive nodes. Therefore, learning mechanisms and knowledge management techniques such as in [31] can be applied in this context.

(iii) Localization of emitters: The knowledge of the emitters' location can improve the accuracy of the final estimate of the REM. The process of REM construction becomes easier if information about the incumbent users' locations is known beforehand by using some estimation methods [32].

(iv) Sparsity-based cartography: In most of the spectrum cartography literature, the spatial sparsity of multiple sensing devices over a geographical area as well as the sparsity on the frequency domain of the spectrum measurements have not been taken into account. In this regard, CS enabled spectrum cartography can be a promising solution to improve the REM construction process [33].

13.4.1.3   Database

The process of REM construction can be improved with the help of information from the national and international spectrum regulators, since they usually register the licensed systems into their databases. The main advantage of the availability of regulatory databases is that they can provide information about the radio spectrum environment, and hence can reduce the required sensing burden. Based on the information from the database, suitable prediction methods can be utilized to estimate the REM. Some challenges related to use of regulatory databases are listed below.

(*i*) *Accuracy of parameters*: In practice, the actual values such as power levels, antenna patterns, etc. may vary from that of the specified values in the database. In such cases, databases may be considered as an initial platform and it can be updated based on the outcomes of dynamic spectrum awareness mechanisms.

(*ii*) *Temporal-variations*: Databases usually cannot capture the temporal variations on the spectrum opportunity. The information provided by the database can be considered as side-information to devise sensing mechanisms, which can find the temporal spectral opportunities.

(*iii*) *Out-of-date*: Completeness of available databases determines the performance of the database approach. Unfortunately, databases may not fully reflect the reality since they may be out-of-date. In this regard, maintaining and updating of databases in time is important for the operators. However, in HSTB networks, the involved terrestrial links and the satellite network are not dynamic as in a mobile system and, thus, it may not be necessary to update the database so frequently. In any case, verification of available database via measurements is an essential step for the formation of an accurate REM.

13.4.2   Spectrum exploitation techniques

13.4.2.1   Interference mitigation techniques

Existing interference mitigation techniques in spectral coexistence scenarios can be broadly categorized into: cognitive beamforming (BF), cognitive Interference Alignment (IA), and cognitive zone. A short description about these techniques and the associated research challenges are provided in the following.

1. Cognitive beamforming: BF can be applied as an underlay spectrum sharing technique in order to maximize the Signal to Interference plus Noise Ratio (SINR) toward the desired user while guaranteeing the protection of the PUs against harmful interference. The main difference between cognitive BF and the conventional BF problem arises due to the introduction of interference constraints in order to restrict the interference toward/from the victim/interfering stations. In this context, cognitive BF approaches have been widely studied with different objectives such as SINR/rate balancing, sum rate maximization, and power minimization with QoS constraints [34]. The main research challenges associated with the cognitive BF in the considered HSTB scenario are summarized below [34].

(*i*) *Channel State Information (CSI)*: The cognitive transmitter, i.e., microwave tower equipped with the multiple antennas, requires the knowledge of CSI toward other HSBT nodes and FSS terminals, which is difficult to obtain in practice. In this context, how to obtain the perfect CSI at the HSBT node in order to design a beamformer is a crucial challenge.

(ii) Uncertainties in the array response vector: Besides CSI robustness, BF solutions should also be robust to the uncertainties in the array response vector, inaccurate Direction of Arrival (DoA) information, and transceiver hardware imperfections such as phase noise, quantization errors, etc.

(iii) Interference threshold: The additional constraint, i.e., tolerable interference threshold at the unintended (may be primary or secondary depending on the scenarios) receivers to be incorporated in the cognitive BF design makes the problem more complex than conventional BF problems. In this regard, computationally efficient solutions need to be investigated in order to solve cognitive BF problems in the considered scenarios.

(iv) DoA of devices: The acquisition of accurate DoA information of the desired and interfering sources/receivers is crucial for implementing cognitive BF in the considered scenario. In practice, this information can be obtained either from the databases or by employing a suitable DoA estimation algorithm.

(v) Calibration: Accurate calibration of the antenna array equipped at the FS station is crucial, especially while creating nulls to the directions of the interfering/victim sources. In the case of calibration errors, suitable compensation mechanisms need to be investigated.

2. Cognitive interference alignment: In spectrum sharing context, IA can be used as an interference mitigation tool which aligns interference in space, time or frequency domain using suitable precoding techniques. In this approach, signals transmitted by all users can be designed in such a way that the interfering signals fall into a reduced dimensional subspace at each receiver. Each receiver can then apply an interference removal filter in order to project the desired signal onto the interference free subspace [35].

In the context of spectrum sharing networks, IA techniques can be broadly classified into non-cooperative and cooperative [36] and they can be either CSI-aware or blind. The main research challenges involved with cognitive IA technique are specified below [37,38].

(i) CSI acquisition and uncertainty: In many cases except in the distributed IA, the global channel knowledge is required to carry out IA operation. Further, in distributed IA techniques, local CSI knowledge is needed. In this context, it is a crucial aspect to investigate suitable blind and semi-blind IA techniques in order to reduce the overhead for acquiring the sufficient channel knowledge. In addition, the penalty of the residual channel uncertainty at the transmitters and the impact of channel correlations are other aspects to be explored for IA implementation in practice. Moreover, tracking the IA solution under time varying channels is another research issue.

(ii) Synchronization: IA techniques require strict synchronization in order to avoid any timing and carrier frequency offsets between cooperating nodes. If not

synchronized sufficiently, additional interference terms may be introduced to the signal model, making the IA solution ineffective.

(iii) Low SNR scenarios: IA algorithms are mostly investigated for high SNR regime where they can achieve the channel's maximum multiplexing gain. Investigation of new IA algorithms which can provide better sum capacity in moderate or low SNR region is another interesting research challenge.

(iv) Dimensionality of interference networks: Another main limitation for the IA technique is the requirement of the large dimensionality of interference networks. The practical achievable scheme which requires finite dimensions for the case of multiple non-intended receivers is still an open research problem.

3. Cognitive zone: In spectrum sharing scenarios, a Cognitive Zone (CZ) is usually designed around the primary receiver based on its interference threshold, within which secondary users are not allowed to reuse the frequencies used by the primary user [39]. However, since the interference may occur in both directions in practice, CZs can be created around one of these receivers or both depending on the considered scenarios and the level of interference between these two systems. The size of the CZ has a significant impact on the QoS of the primary system since it affects the level of the secondary interference, and also on the secondary system's capacity since it affects the available amount of primary spectrum at a given location [40]. The main research challenges involved with the CZ method are mentioned below.

(i) Unpredictable propagation conditions and inaccuracy of the database: Static CZ method may not always guarantee the perfect avoidance of the co-channel interference. The received interference level may vary based on different factors such as terrain variations, environmental conditions, and antenna gain patterns. Since the CZ method is mostly based on database created using some propagation models, the accuracy of the database is crucial for the interference protection guarantee.

(ii) Terrain-enabled propagation models: Most of the CZ/protection zone related works consider the worst case scenario considering only the path loss case without considering the effect of the actual terrain. In practice, terrain variations often induce more path loss [41], and hence, resulting in the reduction the size of CZ than that generated by the worst case model.

(iii) Multi-tiered CZs: In order to enhance the spectrum utilization, a CZ can be divided into multi-tiered CZs, where the boundary between two tiers is depicted by the tolerable interference threshold of the primary receivers [42]. Defining the proper boundaries considering realistic propagation environment for the implementation of multi-tiered CZs is another important research challenge.

(iv) Combination of exclusion/protection zones, spectrum awareness mechanisms, and resource allocation mechanisms: The combination of suitable dynamic spectrum awareness mechanisms with the CZ method can enable a spectrum sharing system to enhance the spectrum utilization by exploiting both spatial and temporal spectral opportunities. Also, CZ method can be further combined with the power

control approach in order to exploit the unused spectral resources in an efficient manner. How to combine these different interdisciplinary schemes effectively is another interesting aspect to be considered.

13.4.2.2 Resource allocation techniques

Resource allocation plays an important role in both the wireless radio access and the backhaul networks to enhance the overall system performance. On the top of the interference mitigation schemes, resource allocation in dynamic spectrum sharing networks is responsible for distributing/allocating the available spectrum in combination with efficient power control strategies [43,44,45]. In this regard, the scarce wireless backhaul frequency spectrum should be allocated wisely in such a way that the demands of the network are satisfied while reserving enough resources for possible future demands. In addition to the smart spectrum allocation, power control mechanisms can further improve spectral efficiency to enable more interference avoidance, and further enable the spatial reuse of resources in addition to its important role in the energy saving [46]. Some of the research challenges related to the resource allocation in the spectrum sharing context include the following.

(i) Imperfect SS information: Designing the resource allocation algorithms while considering that sensing information is perfectly known may lead to inefficient distribution of the resources and cause harmful interference to the nodes that share the spectrum. Any allocation process should consider the effect of sensing errors such as false alarm and miss detection probabilities and the amount of the resulted interference should be adapted accordingly.

(ii) Imperfect CSI: In practice, there always arises some uncertainty in the CSI due to feedback channels errors, sampling time limitation, etc. Accordingly, resource allocation at the microwave backhaul nodes should be developed in order to reduce the negative impact of the lack of this information [47]. Significant improvements can be obtained by applying the resource allocation schemes which are able to adapt the measured/reported CSI by incorporating the possible estimation errors.

(iii) Cross-layer resource allocation: The performance of a wireless backhauling network depends on all the layers of the network. Although, independent optimization per layer is common and can lead to system performance improvement but exploiting several layers may result in better enhancement and allow more adaptability to different variations in the system [48,49]. Accordingly, backhauling nodes can take into account several information from the access as well as network layers in addition to possible environmental and traffic changes.

(iv) Inter-network interference consideration: Generally, resource allocation schemes in the spectrum sharing scenarios make a high consideration for the interference induced to the neighboring networks and take into account the received interference as well [44]. Depending on the sharing schemes and the approach of distributing the spectrum, i.e., centralized, distributed, or cluster-wise, interference between different nodes should be included considering possible interferences caused by the communication between the terrestrial nodes and/or the communication between the satellite and terrestrial nodes.

13.5 Case studies

In this section, we provide three different case studies for the coexistence scenarios in the Ka-band. The first case considers the coexistence of FSS downlink with the FS microwave links in 17.7–19.7 GHz while the second case deals with the coexistence of FSS uplink with the FS microwave links in 27.5–29.5 GHz. In the third case, we provide the interference analysis of hybrid satellite-terrestrial backhaul scenario considering their coexistence in 28 GHz.

13.5.1 FSS-FS coexistence in the forward Link (17.7–19.7 GHz)

For this case study, we consider the spectral coexistence of GEO FSS downlink with FS microwave links in 17.7–19.7 GHz as depicted in Figure 13.4. In this scenario, FS microwave links and FSS downlink are incumbent and cognitive links, respectively. The downlink interference from the cognitive FSS satellite to the cochannel FS receivers is taken into account during system planning and can be kept below the defined regulatory limitations in terms of the maximum power flux-density at the earth's surface [50]. However, the interference from the FS transmitter to the cognitive FSS terminal should be effectively mitigated in order to guarantee the QoS of FSS downlink. In this scenario, an FSS terminal may receive aggregate interference from several cochannel FS transmitters via their side-lobes even if FS stations use highly directive antennas.

The harmful interference at the FSS terminal can be mitigated using different approaches such as by employing interference detection [51] and/or receiver BF at

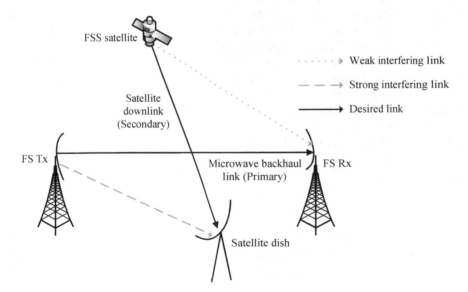

Figure 13.4 Spectral coexistence of FSS downlink with FS link in the Ka-band (17.7–19.7 GHz)

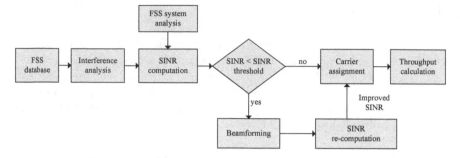

Figure 13.5   Spectrum exploitation framework for employing joint BF and carrier allocation in the considered scenario

the terminal-side and/or the carrier allocation at the system level [52,53]. In the following, we provide a case study of joint Carrier Allocation (CA) and receive BF in order to mitigate harmful interference in the downlink coexistence scenario presented in [52,53].

The generalized framework to implement this joint approach is illustrated in Figure 13.5 and the methodology followed in [52,53] is described below. First, based on the FS database, which can be available from regulators/operators or can be extracted from REMs created with the help of sensing, the interference analysis between FS and FSS systems is carried out. Subsequently, based on the desired FSS signal level received from the FSS satellite and on the calculated interference level received from the FS transmitters, SINR value is computed for all the FSS terminals considering all the carrier frequencies. Then, BF is applied to only the FSS terminals which receive interference levels above a certain threshold. In the next step, the improved SINR obtained from the BF is fed to the CA module in order to make an effective allocation of the available shared and exclusive carriers.

In our analysis, the parameters related to FS links are obtained from ITU-R BR International Frequency Information Circular (BR IFIC) database. The desired signal levels at the FSS terminals are calculated using the realistic beam patterns of a multibeam FSS satellite considered during CoRaSat project [54]. In order to carry out interference analysis, a free space path loss model is used to consider the worst case interference level at the FSS terminal.

The CA problem reduces to find an assignment matrix in a way that the carriers are optimally allocated to the FSS users with a certain objective, for example, to maximize the total throughput. In the considered FSS downlink-FS coexistence scenario, this CA problem with the objective of maximizing the total throughput has been solved in [44,53] by employing the widely used Hungarian algorithm.

For the system level evaluation, first, we evaluate the performance results in terms of the per beam throughput considering five representative beams having different FS densities over France as depicted in Table 13.1.[2] Then, the total system

---

[2]The geographical coordinates (latitude and longitude) denote the location of the beam center.

Table 13.1 Selected beams over France

Beam number	Number of FS links	Weight	Latitude °N	Longitude °E
1	1,681	0.077	50.0072	2.0397
2	1,522	0.039	48.3364	3.5137
3	635	0.5	46.4896	1.5707
4	906	0.269	43.5876	3.9157
5	1,220	0.115	43.7635	5.3898

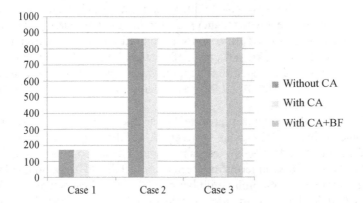

Figure 13.6 Total system throughput (Gbps) of FSS satellite over Europe for different cases

level throughput over whole Europe is calculated by extrapolating the total system throughput over 250 beams. For the comparison of the considered framework with the benchmark case, we consider the following cases: (i) **Case 1-Exclusive only**: This conventional approach considers that FSS system only uses the exclusive carriers in 19.7–20.2 GHz band, (ii) **Case 2-Shared plus exclusive without FS interference**: This case considers that the FSS satellite uses all the carriers in (17.7–20.2 GHz band) exclusively, and assumes that there is no presence of co-channel FS interference, and (iii) **Case 3-Shared plus exclusive with FS interference**: This case represents the scenario where FSS system utilizes the non-exclusive Ka-band (17.7–19.7 GHz) in the shared basis with the FS system and the band (19.7–20.2 GHz) exclusively.

From the detailed analysis provided in [44], the values of the total FSS throughput over Europe (250 beams), obtained by employing the exploitation framework in the above three cases, are presented in the form of a bar chart in Figure 13.6. From the chart, it can be observed that the system throughput obtained by employing the considered exploitation techniques (CA and BF) in the presence of FS interference is closer (even slightly higher) to that can be achieved in the absence of co-channel FS interference. More importantly, with respect to the conventional exclusive only case, the performance improvement is around 405% while employing the joint CA and BF approach. This performance gain is mainly achieved due to the flexibility of the using additional 2 GHz non-exclusive spectrum with the help of the employed CA and BF schemes.

13.5.2   FSS-FS coexistence in the return link (27.5–29.5 GHz)

This section explores the scenario where the return link of the FSS system reuses the frequency bands of FS terrestrial microwave links (incumbent systems) in the Ka 27.5–29.5 GHz band, as depicted in Figure 13.7. The considered scenario entails the FSS terminals operating in the uplink and thus is a potential interferer to the incumbent terrestrial links. To enable this scenario, it should be guaranteed that FSS terminals will not interfere with the incumbent FS links. Consequently, satellite uplink communication in the shared band is not performed unless the interference caused at the incumbent system is below a pre-defined threshold. In this section, we review the techniques adopted for spectrum exploitation for the FSS-FS coexistence in the return link (27.5–29.5 GHz). A detailed description of the work summarized here can be found in [44,45,55,56].

In [55], a simple and conservative technique that assumes perfect knowledge of cross-channel gains, i.e., the gains between the satellite terminals and the different terrestrial antennas, at the FSS system is proposed. The proposed approach identifies the worst FS link per user in terms of interference and divides the amount of tolerable interference among the maximum number of FSS terminal users that can potentially interfere with it. With this information, the power and carrier frequencies are allocated in a way that the prescribed interference threshold is never exceeded. This technique was evaluated in [44] by considering realistic multiple beams provided by a satellite manufacturer and assuming a precise propagation model.

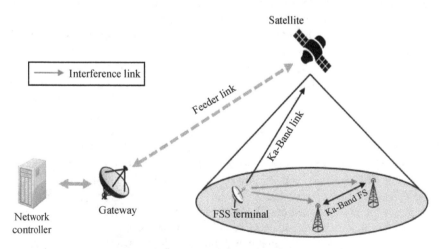

Figure 13.7   Spectral coexistence of FSS return link with FS terrestrial link in the Ka-band (27.5–29.5 GHz)

The country chosen in [44] for evaluation was Finland. The reason behind this choice was that the FS database of Finland obtained from the FInnish COmmunications Regulatory Authority (FICORA) was one of the most complete. From all the satellite beams covering Finland, a selection of the most representative ones was performed based on FS link deployment density. The selected beams are depicted in Figure 13.8, and Table 13.2 provides the detailed locations and weighting factors of each selected beams. Next, throughput evaluation was carried out separately for each of them and the results were lastly combined with a proper weighting factor to give the final throughput result.

In our analysis, we evaluated three different cases highlighted in the previous section considering the exclusive band of 29.5–30 GHz and the non-exclusive band of 27.5–30 GHz. Beam throughput results are shown in Table 13.3, where "w/ CA" stands for "with carrier allocation," which is always better or as good as the random CA indicated with "w/o CA." Clearly, we observe that the additional spectrum together with the optimal power and CA technique provides 398.12% improvement over the

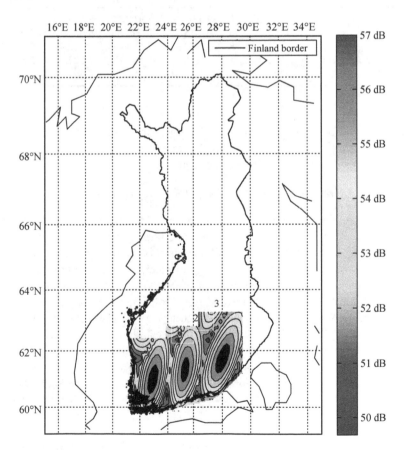

Figure 13.8   Selected beams for return link evaluation

Table 13.2   Detailed information about the selected beams

Beam number	FS links	Weight	Latitude* (°N)	Longitude* (°E)
1	32	0.222	60.8801	22.8749
2	902	0.111	61.2905	25.4073
3	6	0.667	61.7522	27.878

*Location of the beam center.

Table 13.3   Throughput per beam (Mbps) – FSS-FS Coexistence in the Return Link

Beam number	Weight	Case 1		Case 2		Case 3 w/ Power & Carrier allocation
		w/o CA	w/ CA	w/o CA	w/ CA	
1	0.222	1,098.79	1,098.79	5,467.67	5,474.58	5,474.58
2	0.111	1,015.55	1,015.55	5,077.65	5,081.19	5,081.19
3	0.667	1,127.58	1,127.58	5,607.49	5,612.45	5,612.45
Per beam average throughput	1	1,108.75	1,108.75	5,517.64	5,522.87	5,522.87

conventional exclusive band (29.5–30 GHz) case. From Table 13.3, we can conclude that using the proposed joint power and CA, we can achieve the same throughput as if there were no FS system, while ensuring protection to the FS receivers.

13.5.3   Satellite-terrestrial backhaul coexistence

As mentioned earlier, the coexistence of satellite and terrestrial backhaul networks in the Ka-band leads to potential intra-system interference among the terrestrial interfaces as well as among the satellite and terrestrial ones. In this section, we develop a toy example based on a specific topology extracted from the Finnish database for Ka-band terrestrial networks. The idea of this section is to provide an insight into the requirements of a network consisting of multiple antenna terrestrial networks and satellite terminals working in the same frequency band. These requirements can in turn be used by PHY and MAC layers for resource allocation, BF, etc. However, in this section, we only stick to the interference analysis example.

The selected topology is depicted in Figure 13.9. As we can see, this topology consists of a number of interconnected star topologies. This topology based on a database is of 28 links and 15 actual locations (nodes).[3] It should be noted that all the

---

[3] Readers may refer to [16] for the underlying parameters defining each link and the specific locations of all the nodes.

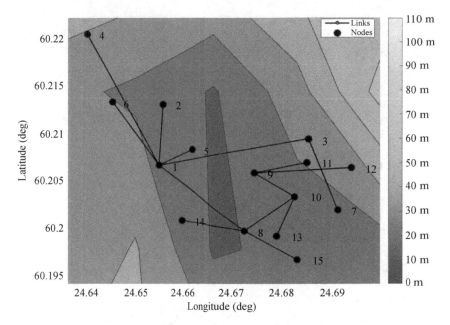

Figure 13.9   Example backhaul topology obtained from the Finnish 28 GHz database, Helsinki

depicted links are bidirectional. In the following, we present the interference analysis/ SINR computation in different contexts of the considered topology.

1. Benchmark SINR distribution: For the benchmark SINR calculation, we employ the ITU-R 452-16 interference modelling, including the free space loss as well as the diffraction loss based on the Bullington model to derive the SINR of each receiver based on the coordinated frequency plan in Figure 13.9. This result which can be considered as the benchmark SINR distribution is presented in Figure 13.10(a). We can note that all the receivers experience the SINR value of > 42 dB, while a significant number of them experience SINR > 60 dB. This is as expected since the benchmark topology is the outcome of careful network planning through link registration by the national regulator.

2. SINR distribution with aggressive frequency reuse: Herein, we move toward the concept of HSTB network enabled by smart antennas, and analyze the performance of each link, when all the links employ the same frequency band. It should be noted that this is a worst case scenario and less aggressive frequency reuse could be used in practice. We further would like to estimate the number of required nulls to be produced by each smart antenna to tackle the strong interferers. Here, we define the strong interferers based on the ITU-R recommendations [57]. An interferer is considered to be harmful if the level of the received interference increases the noise floor by 10%.

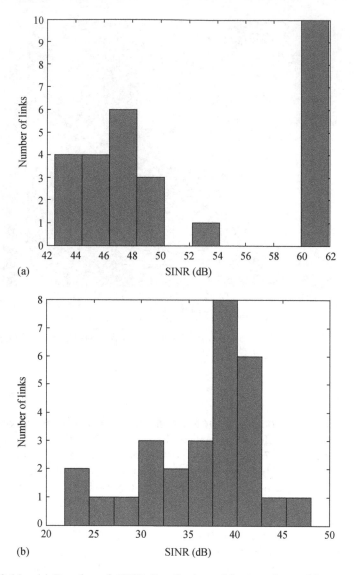

Figure 13.10   *(a) Benchmark SINR distribution of the coordinated frequency plan,*
(b) SINR distribution of the links with aggressive frequency reuse

After calculations, we can note in Figure 13.10(b) that in this case, the lower value of SINR is reduced to around 22 from 42 dB in the benchmark model. Further, none of the links experiences SINR > 48 dB. This is explained by the increased internal interference among the HSTB links. After evaluation of the number of strong interferers in each link and thus the required number of nulls in each smart antenna, we noted that each node should be able to produce 7 nulls in average. It is expected that if less aggressive frequency is used in combination with suitable carrier allocation schemes, a smaller number of nulls will be required. Thus, this number can be considered as an upper bound requirement while using smart antenna-based techniques.

3. Interference from terrestrial nodes to satellite terminals: We also studied the impact of employing satellite links for backhauling for the example topology of Figure 13.9. To this end, we replaced the terrestrial terminals of nodes 1, 8, and 10 of the existing topology with the satellite ones and calculated the interference from the remaining terrestrial nodes. It is assumed that satellite terminals on these three nodes are pointing to the HYLAS 2, 31.7° East. Based on the node's position, satellite antennas should have elevation and azimuth 21.7° and 176°, respectively, in order to point to this satellite.

Next, we carried out interference analysis to calculate interference generated to the satellite terminals by the terrestrial nodes. To this end, for the considered topology and elevation angles of the satellite terminals, the interference levels were found to be very low and there is no need of placing nulls toward the interfering directions. Each node experiences interference that decreases its SINR less than 10% of the SNR floor, so based on the ITU-R recommendations, this interference can be considered as non-harmful [57]. Note that in cases where the satellite nodes experience harmful interference from the terrestrial ones, the latter may apply transmit BF techniques in order to null out the interference to the satellite ones.

4. Interference from satellite terminals to terrestrial nodes: For this case, we considered the uplink scenario where interference occurs from the satellite terminals to the terrestrial nodes. It is assumed that satellite terminals have the same azimuth and elevation parameters as in the previous case. The satellite link characteristics and simulation parameters used for this evaluation are listed in Table 1 of [55]. Our calculations showed that there is a requirement for at most 1 null out of 9 links. The latter result is very promising since the required number of nulls can be easily handled by the antenna infrastructures of the typical backhaul nodes by the application of receive BF. For illustrations of this result, one may refer to [16].

13.6   Future recommendations

In Sections 13.3 and 13.4, we have highlighted the challenges related to HSTB network, and various spectrum sharing techniques. Following these challenges, we provide some recommendations for future research directions below.

1. Optimal strategies for the placement of the HSTB node: In HSTB networks, several aspects such as connectivity, resilience, and coverage as highlighted in Section 13.3 need to be considered while selecting the nodes with the hybrid backhauling capability. Besides, it is not possible to replace all the existing nodes with the HSTB node at the same time due to cost and implementation issues. Also, this full replacement may also be not needed in order to meet the desired set of objectives. In this regard, it is an important future aspect to investigate the optimal placing of the HSTB nodes in the existing backhaul networks.

2. Interference analysis and modeling: After placing the HSTB nodes in the desired points in the network, another important step is to carry out interference analysis using realistic channel models for different satellite and terrestrial segments involved in the HSTB network. For this, suitable channel models and antenna patterns need to be investigated in order to reflect the practical scenario.

3. Frequency reuse schemes for HSTB networks: As in the terrestrial cellular, the aspect of full frequency reuse is getting attention in the satellite community. In order to maximize the utilization of the available radio resources, it is important to devise suitable frequency reuse strategies between different segments of the HSTB networks in order to fulfill the desired performance objectives.

4. Choice of suitable spectrum sharing techniques: As highlighted in Section 13.4, there are several techniques to enable the spectral coexistence of different segments in HSTB networks. However, the choice of a particular technique depends on several factors such as the desired performance target, performance achieved by employing a particular method, complexity of implementation, interference scenario as well as traffic variations. In this regard, future research works need to focus on the selection of suitable techniques based on the desired requirements.

5. Hybrid analog-digital BF at the HSTB node: In order to reduce the cost of RF circuitry as well as to take the advantage of digital BF technologies, the concept of hybrid analog–digital BF technique is receiving important attention in the literature. In this direction, developing suitable hybrid BF solutions in the HSTB nodes equipped with smart antennas is another important future direction.

6. Adaptive resource allocation schemes: Resource allocation problem in HSTB network becomes different than that of the ones in individual satellite and terrestrial networks since the constraints arising from both sides should be considered. Besides, depending on the nature of the traffic, a hybrid manager may decide whether the traffic will be forwarded via satellite segment or via the terrestrial segment. In this regard, it is crucial to investigate suitable resource allocation schemes to be applicable in HSTB networks by taking all the involved constraints.

7. Self-backhauling in 5G networks: One possible way to address the scarcity of available spectrum in 5G wireless networks is to enable the access and backhaul transmissions in the same radio spectrum using either in-band full duplex or out of band full duplex at the edge nodes of the network. For this purpose, backhaul capacity constraints as well as self-interference constraints (in case of in-band full duplex) need to be considered.

8. Cross-layer approach for resource allocation: The interactions of physical layer parameters with the link layer, network layer as well as the transport layer parameters are important in order to design a reliable end-to-end HSTB system. In this direction, investigating novel cross-layer solutions for resource allocation is an important research direction.

13.7 Conclusions

In order to meet the ever-increasing demand for wireless broadband and multimedia applications/services, one of the main challenges for telecommunication operators is how to enhance the capacity and spectral efficiency of the existing wireless backhaul networks. In this regard, researchers are investigating the cost-effective ways to enable the sharing of available radio resources among heterogeneous wireless networks.

One of the promising solutions in this context is an HSTB network, which can provide various benefits in terms of meeting upcoming 5G requirements such as extension of coverage area, increasing the resilience of the networks, enhancing the total backhaul capacity, and improving spectral efficiency and energy efficiency. Starting with the detailed review on the terrestrial and satellite backhaul technologies, this chapter has discussed various use cases for HSTB networks. Subsequently, various spectrum sharing techniques such as spectrum awareness and spectrum exploitation techniques have been described along with their associated challenges in practical coexistence scenarios. Furthermore, three case studies have been provided considering the spectral coexistence of FSS system with the terrestrial backhaul links and HSTB networks. The first case considered the coexistence of FSS system with the incumbent terrestrial microwave backhaul (FS) links in the forward link (17.7–113.7 GHz) by employing joint BF and CA mechanisms while the second case considered the coexistence of FSS system with the incumbent FS links in the return link (27.5–29.5 GHz) by employing joint power and carrier allocation schemes. It has been shown that the considered techniques can enable the sharing of non-exclusive Ka-band spectrum between FSS and FS systems in both the forward and return links while providing significant capacity gains over the conventional systems. Similarly, the third case provided the preliminary analysis for the coexistence of satellite and terrestrial segments in an HSTB network in terms of SINR distribution.

Finally, some important future research directions for enabling spectrum sharing among different segments of HSTB networks include the investigation of techniques for optimal placing of satellite terminals, interference analysis and modeling, frequency reuse schemes, hybrid analog–digital beamforming, adaptive cross-layer resource allocation schemes, and self-backhauling mechanisms.

Acknowledgments

This work was supported by EU H2020 project SANSA (Grant agreement no. 645047), and also by FNR, Luxembourg under the CORE projects "SeMIGod"and "SATSENT".

References

[1] Tipmongkolsilp O, Zaghloul S, and Jukan A. "The evolution of cellular backhaul technologies: current issues and future trends," *IEEE Communications Surveys Tutorials*. 2011 First;13(1):97–113.

[2] Agiwal M, Roy A, and Saxena N. "Next generation 5G wireless networks: a comprehensive survey," *IEEE Communications Surveys Tutorials*. 2016 thirdquarter;18(3):1617–1655.

[3]　Simons RN and Goverdhanam K. "Applications of nano-satellites and cube-satellites in microwave and RF domain," in: 2015 IEEE MTT-S International Microwave Symposium; 2015. p. 1–4.

[4]　Evans B, Onireti O, Spathopoulos T, and Imran MA. "The role of satellites in 5G," in: Signal Processing Conference (EUSIPCO), 2015 23rd European; 2015. p. 2756–2760.

[5]　Cell-Sat. 3G/4G mobile backhaul via satellite. Available: http://www.cell-sat.com/en/solutions/3G-4G-mobile-backhaul-via-satellite.html, Access date: 31/07/2016. Online.

[6]　SANSA – Shared Access Terrestrial-Satellite Backhaul Network enabled by Smart Antennas. European Union's Horizon 2020 research and innovation programme. Feb. 2015.

[7]　Ge X, Cheng H, Guizani M, and Han T. "5G wireless backhaul networks: challenges and research advances," *IEEE Network*. Dec. 2014;28(6):6–11.

[8]　Larsson EG, Tufvesson F, and Marzetta TL. "Massive MIMO for next generation wireless systems," *IEEE Communications Magazine*. Feb. 2014;52(2):186–195.

[9]　Gao Z, Dai L, Mi D, Wang Z, Imran MA, and Shakir MZ. "mmWave massive-MIMO-based wireless backhaul for the 5G ultra-dense network," *IEEE Wireless Communications*. Oct. 2015;22(5):13–21.

[10]　Hossain E and Hasan M. "5G cellular: key enabling technologies and research challenges," *IEEE Instrumentation and Measurement Magazine*. Jun. 2015;18(3):11–21.

[11]　Cao M, Wang X, Kim SJ, and Madihian M. "Multi-hop wireless backhaul networks: a cross-layer design paradigm," *IEEE Journal of Selected Areas Communication*. May 2007;25(4):738–748.

[12]　Anastasopoulos MP, Arapoglou PDM, Kannan R, and Cottis PG. "Adaptive routing strategies in IEEE 802.16 multi-hop wireless backhaul networks based on evolutionary game theory," *IEEE Journal of Selected Areas Communication*. Sep. 2008;26(7):1218–1225.

[13]　EC RADIO SPECTRUM POLICY GROUP. RSPG Report on Spectrum issues on Wireless Backhaul. RSPG 15-607. Jun. 2015.

[14]　SANSA. State-of-the-art of cellular backhauling technologies. Horizon 2020 project; 2015. Deliverable 2.2.

[15]　STM. TDMA vs. SCPC SatLink System Technical Notes; 2013. Online: ftp://nic2b.whoi.edu/pub/.

[16]　SANSA. Definition of reference scenarios, overall system architectures, research challenges, requirements and KPIs. Horizon 2020 project; 2015. Deliverable 2.3.

[17]　Sharma SK, Bogale TE, Chatzinotas S, Ottersten B, Le LB, and Wang X. "Cognitive radio techniques under practical imperfections: a survey," *IEEE Communications Surveys Tutorials*. 2015 fourthquarter;17(4):1858–1884.

[18]　Lagunas E and Najar M. "Sparse correlation matching-based spectrum sensing for open spectrum communications," *EURASIP Journal on Advances in Signal Processing*. 2012:31. Feb. 2012.

[19]   Höyhtyä M, Hekkala A, Katz M, and Mämmelä A. "Spectrum Awareness: Techniques and Challenges for Active Spectrum Sensing," in: Fitzek HP, Katz MD, editors. Cognitive Wireless Networks. The Netherlands: Springer; 2007

[20]   Lee WY and Akyldiz IF. "Optimal spectrum sensing framework for cognitive radio networks," *IEEE Transactions on Wireless Communications*. Oct. 2008;7(10):3845–3857.

[21]   Hamdi K and Ben Letaief K. "Power, sensing time and throughout tradeoff in cognitive radio systems: a cross-layer approach," *Wireless Communications and Networking Conference (WCNC)*, Budapest, Hungary. Apr. 2009.

[22]   Liao Y, Wang T, Song L and Han Z. "Listen-and-talk: full-duplex cognitive radio networks," *IEEE Global Telecommun Conf (GLOBECOM)*, Austin, TX, USA. Dec. 2014.

[23]   Kim M and Takada J. "Efficient multichannel wideband spectrum sensing technique using filter bank," *IEEE International Symposium on Personal Indoor Mobile Radio Communications (PIMRC)*, Tokyo, Japan. Sep. 2009.

[24]   Sun A H Nallanathan, Wang CX, and Chen Y. "Wideband spectrum sensing for cognitive radio networks: a survey," *IEEE Wireless Communications*. Aug. 2013;20(2):74–81.

[25]   Candes E and Wakin M. "An introduction to compressive sampling," *IEEE Signal Process Magazine*. 2008;25(2):21–30.

[26]   Lagunas E and Najar M. "'Spectral feature detection with sub-Nyquist sampling for wideband spectrum sensing," *IEEE Trans Wireless Communications*. Mar. 2015;14(7):3978–3990.

[27]   Sharma SK, Lagunas E, Chatzinotas S, and Ottersten B. "Application of compressive sensing in cognitive radio communications: a survey," *IEEE Communications Surveys Tutorials*. 2016 thirdquarter;18(3):1838–1860.

[28]   Yilmaz HB, Tugcu T, Alagoz F, and Bayhan S. "Radio environment map as enabler for practical cognitive radio networks," *IEEE Communications Magazine*. Dec. 2013;51(12):162–169.

[29]   Wei Z, Zhang Q, Feng Z, Li W, and Gulliver TA. "On the construction of radio environment maps for cognitive radio networks," *Wireless Communications and Networking Conference (WCNC)*, Shanghai, China. Apr. 2013.

[30]   Alaya-Feki A, Ben-Jemaa S, Sayrac B, Houze P, and Moulines E. "Informed spectrum usage in cognitive radio networks: interference cartography," *IEEE International Symposium on Pers Indoor Mobile Radio Commun (PIMRC)*, Cannes, France. Sep. 2008.

[31]   Atanasovski V, van de Beek A, Denkovski D, *et al.* "Constructing radio environment maps with heterogeneous spectrum sensors," in: *IEEE Symposium on New Frontiers in Dynamic Spectrum Access Networks (DySPAN)*; May 2011. p. 660–661.

[32]   Bolea L, Perez-Romero J, Agusti R, and Sallent O. "Context discovery mechanisms for cognitive radio," *IEEE Vehicular Technology Conference (VTC-Spring)*, Budapest, Hungary. May 2011.

[33]   Jayawickrama BA, Dutkiewicz E, Oppermann I, Fang G, and Ding J. "Improved performance of spectrum cartography based on compressive sensing in

cognitive radio networks," *IEEE International Conference on Communications (ICC)*, Budapest, Hungary. Jun. 2013.

[34]   Sharma SK, Chatzinotas S, and Ottersten B. "Cognitive beamforming for spectral coexistence of hybrid satellite systems," in: Chatzinotas S, Ottersten B, Gaudenzi R, editors. Cooperative and Cognitive Satellite Systems. London, Elseiver; 2015.

[35]   Jafar SA. "Interference alignment – a new look at signal dimensions in a communication network," *Foundations and Trends in Communication and Information Theory*. 2010;7(1):1–134.

[36]   Pantisano F, Bennis M, Saad W, Debbah M, and Latva-aho M. "Interference alignment for cooperative femtocell networks: a game-theoretic approach," *IEEE Transactions on Mobile Computing*. Nov. 2013;12(11):2233–2246.

[37]   Ayach OE, Peters SW, and Heath RW. "The practical challenges of interference alignment," *IEEE Wireless Communications*. Feb. 2013;20(1):35–42.

[38]   Sharma SK, Chatzinotas S, and Ottersten B. "Cognitive interference alignment for spectral coexistence," in: Di Benedetto, M-G, Fabio Cattoni, A, Fiorina, J, Bader, F, De Nardis, L, editors. Cognitive Radio and Networking for Heterogeneous Wireless Networks. Cham, Springer; 2014.

[39]   Maleki S, Chatzinotas S, Krause J, Liolis K, and Ottersten B. "Cognitive zone for broadband satellite communications in 17.3–17.7 GHz Band," *IEEE Wireless Communications Letters*. Jun. 2015;4(3):305–308.

[40]   Sharma SK, Chatzinotas S, and Ottersten B. "Cognitive beamhopping for spectral coexistence of multibeam satellites," *International Journal of Satellite Communications and Networking*. Mar. 2014;33(1):69–91.

[41]   Lagunas E, Sharma SK, Maleki S, Chatzinotas S, and Ottersten B. "Impact of terrain aware interference modeling on the throughput of cognitive Ka-band satellite system," in: *21st Ka-band and Broadband Communications Conference*; 2015.

[42]   Ullah A. "Multi-tier exclusion zones for dynamic spectrum sharing," Available online: http://www.arias.ece.vt.edu/pdfs/multitierexclusions.pdf, Access date: 28/07/2016. Online.

[43]   Maleki S, Chatzinotas S, Evans B, *et al.* "Cognitive spectrum utilization in Ka band multibeam satellite communications," *IEEE Communications Magazine*. Mar. 2015; 53(3):24–29.

[44]   Lagunas E, Sharma SK, Maleki S, Chatzinotas S, and Ottersten B. "Resource allocation for cognitive satellite communications with incumbent terrestrial networks," *IEEE Transactions on Cognitive Communications and Networking*. Sep. 2015;1(3):305–317.

[45]   Lagunas E, Maleki S, Chatzinotas S, Soltanalian M, Pérez-Neira AI, and Ottersten B. "Power and rate allocation in cognitive satellite uplink networks," in: *2016 IEEE International Conference on Communications (ICC)*; 2016. p. 1–6.

[46]   Chatzikokolakis K, Spapis P, Kaloxylos A, and Alonistioti N. "Toward spectrum sharing: opportunities and technical enablers," *IEEE Communications Magazine*. Jul. 2015;53(7):26–33.

[47] Al-Khasib T, Shenouda MB, and Lampe L. "Dynamic spectrum management for multiple-antenna cognitive radio systems: designs with imperfect CSI," *IEEE Transactions on Wireless Communications*. Sep. 2011;10(9):2850–2859.

[48] Berry RA and Yeh EM. "Cross-layer wireless resource allocation," *IEEE Signal Processing Magazine*. Sep. 2004;21(5):59–68.

[49] Zhang Y and Leung C. "Cross-layer resource allocation for mixed services in multiuser OFDM-based cognitive radio systems," *IEEE Transactions on Vehicular Technology*. Oct. 2009;58(8):4605–4619.

[50] ITU. Radio Regulations; 2004. ITU-R, Article 21.

[51] Sharma SK, Maleki S, Chatzinotas S, Grotz J, and Ottersten B. "Implementation Issues of Cognitive Radio techniques for Ka-band (17.7–19.7 GHz) SatComs," in: *2014 7th Advanced Satellite Multimedia Systems Conference and the 13th Signal Processing for Space Communications Workshop (ASMS/SPSC)*; 2014. p. 241–248.

[52] Sharma SK, Maleki S, Chatzinotas S, Grotz J, Krause J, and Ottersten B. "Joint carrier allocation and beamforming for cognitive SatComs in Ka-band (17.3–18.1 GHz)," in: *2015 IEEE International Conference on Communications (ICC)*; 2015. p. 873–878.

[53] Sharma SK, Lagunas E, Maleki S, *et al.* "Resource allocation for cognitive satellite communications in Ka-band (17.7–19.7 GHz)," in: *2015 IEEE International Conference on Communication Workshop (ICCW)*; 2015. p. 1646–1651.

[54] COgnitive RAdio for SATellite Communications – CoRaSat. European Commission FP7. Oct. 2012.

[55] Lagunas E, Sharma SK, Maleki S, *et al.* "Resource allocation for cognitive satellite uplink and fixed-service terrestrial coexistence in Ka-band," *International Conference on Cognitive Radio Oriented Wireless Networks (CROWNCOM)*, Doha, Qatar. Apr. 2015.

[56] Lagunas E, Sharma SK, Maleki S, Chatzinotas S, and Ottersten B. "Power control for satellite uplink and terrestrial fixed-service co-existence in Ka-band," *IEEE Vehicular Technology Conference* (VTC-Fall), Boston, USA. Sep. 2015.

[57] ITU-R. System parameters and considerations in the development of criteria for sharing or compatibility between digital fixed wireless systems in the fixed service and systems in other services and other sources of interference; 2015. Recommendation ITU-R F.758-6. Online.

Chapter 14

Control data separation and its implications on backhaul networks

Abdelrahim Mohamed[1], Oluwakayode Onireti[2], and Muhammad Imran[2]

Abstract

As soon as 2020, network densification will be the dominant theme to support enormous capacity and massive connectivity. However, such deployment scenarios raise several challenges and they impose new constraints. In particular, signalling load, mobility management and energy efficiency will become critical considerations in the fifth generation (5G) era. These aspects suggest a paradigm shift towards a signalling and energy conscious radio access network (RAN) architecture with intelligent mobility management. In this direction, the conventional RAN design imposes several constraints due to the tight coupling between the control plane (CP) and the data plane (DP). Recently, a futuristic RAN architecture with CP/DP separation has been proposed to overcome these constraints. In this chapter, we discuss limitations of the conventional RAN architecture and present the control/data separation architecture (CDSA) as a promising solution. In addition, we identify the impact of the CDSA on the backhaul network. An analytical framework is developed to model the backhaul latency of the CDSA, and a densification limit under latency constraints is derived. Furthermore, the impact of the backhaul technology on the CDSA energy saving gains is discussed and an advanced non-direct backhaul mechanism is presented.

14.1 Introduction

Future cellular systems need to cope with a huge amount of data and diverse service requirements in a flexible, sustainable, green and efficient way with minimal signalling overhead. This calls for network densification, a short length wireless

[1] 5G Innovation Centre, University of Surrey, UK
[2] School of Engineering, University of Glasgow, UK

link, efficient and proactive control signalling and the ability to switch off the power consuming devices when they are not in use. In this direction, the conventional always-on service and worst-case design approach has been identified as the main source of inefficiency, and a paradigm shift towards adaptive and on-demand systems is seen as a promising solution. However, the conventional radio access network (RAN) architecture limits the achievable gains due to the tight coupling between network and data access points, which in turn imposes strict coverage and signalling requirements irrespective of the spatio-temporal service demand, channel conditions or mobility profiles.

At the signalling dimension, conventional RANs adopt a worst-case design approach to ensure acceptable performance for all users including those in severe conditions. However, such an approach may lead to a dramatic increase in signalling overhead under futuristic deployment scenarios. From an energy point of view, there are concerns about coverage holes and thus an always-on service concept was considered in the conventional RANs to ensure ubiquitous connectivity. Moreover, mobility management is a major concern in dense deployment scenarios, since it has a direct impact on both the signalling overhead and the quality of service (QoS). These limitations can be traced to the conventional RAN design that couples the control plane (CP) and the data plane (DP). This suggests a new clean slate RAN architecture with a logical separation between the ability to establish availability of the network and the ability to provide functionality or service. This separation of CP and DP provides a framework where limitations and constraints of the conventional RAN can be overcome.

In this chapter, we set out to discuss limitations of the conventional RAN design and the main sources of inefficiency. Then, the futuristic control/data separation architecture (CDSA) is presented as a candidate solution that relaxes the RAN design constraints. Next, the implications of the CDSA on the backhaul network are discussed. In particular, an analytical framework is proposed to analyse the impact of the CDSA separation schemes on the backhaul latency. In addition, the impact of the backhaul technology on energy saving in the CDSA is discussed and alternative backhauling mechanisms are presented. Finally, concluding remarks are drawn.

14.2 RAN design in legacy standards

The end-to-end network chain of wireless systems in general, and of cellular systems in particular, consists of three main elements: mobile devices known as user equipment (UE), a RAN and a core-network (CN). Typically, the UE communicate with the RAN over the air-interface through wireless links, while the RAN is connected to the CN through backhaul networks formed of wired links or a combination of wired and wireless links. A simplified hierarchy of these elements and their connections is shown conceptually in Figure 14.1. Legacy cellular standards define the RAN

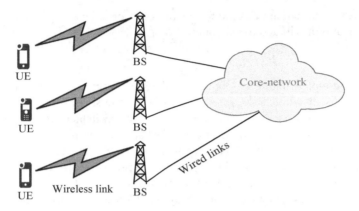

Figure 14.1   Main elements of cellular networks

components along with the protocols, procedures and interfaces. A brief description of each standard's RAN is provided below:

- Global System for Mobile communications (GSM): it describes the second generation (2G) cellular systems, where the RAN is known as GRAN[1] (i.e., GSM RAN). The latter consists of base stations (BSs) and base station controllers (BSCs), and it adopts time division multiple access (TDMA) as the air-interface multiplexing scheme.
- Universal Mobile Telecommunications System (UMTS): it describes the third generation (3G) cellular systems, where the RAN is known as UTRAN (i.e., Universal Terrestrial RAN). The latter consists of BSs known as Node-Bs in addition to radio network controllers (RNCs). The air-interface multiplexing scheme in this RAN adopts the code division multiple access (CDMA).
- Long-Term Evolution (LTE): it describes the fourth generation (4G) cellular systems. The RAN is known as EUTRAN (i.e., Evolved Universal Terrestrial RAN) which consists of BSs known as eNode-Bs (i.e., evolved Node-Bs). Orthogonal frequency division multiple access (OFDMA) is adopted here as the air-interface multiplexing scheme.

Although the RANs of these standards have diverse requirements and they adopt different protocols and multiplexing techniques, they share the same concept of using a single node to provide coverage and data services. In other words, the CP and the DP are coupled and transmitted by the same RAN node. Such an approach raises several challenges and offers limited adaptation opportunities. In particular, the coupled

[1]The RAN of the GSM extension that includes Enhanced Data rates for GSM Evolution (EDGE) services, is known as GERAN (i.e., GSM EDGE RAN) which has the same basic concept as the GRAN.

CP/DP RAN approach implies the following design criteria: always-on RAN, worst-case design and distributed management.

14.2.1   Always-on design

The CP/DP coupled nature of conventional RANs requires a system design with an always present wireless channel irrespective of the spatio-temporal demand of service. The RAN nodes, i.e., the BSs,[2] cannot be switched off even if they are not needed for data transmission. This can be traced to the fact that a switched off BS creates a coverage hole when there is no other BS that can serve the affected area. Such coverage holes break the anywhere/anytime service paradigm because idle UE may not be able to issue service requests and detached UE may not be able to camp on the network. As a result, an always-on RAN design was adopted in legacy standards.

This approach can be justified when the BSs are heavily utilised for most of the time. However, BSs of current cellular systems are significantly underutilised [1]. For 45% of the time, only 10% of the BS resources are utilised [2]. In addition, only a small percentage of the deployed BSs carry most of the network traffic at a given time [3]. Furthermore, the BS traffic profile fluctuates over both time and space dimensions. In the former, daily traffic profiles have a sinusoidal shape rather than a zero slope line, see, for example, the traffic profiles in [4, Figure 4.c], [5, Figure 1] and [6, Figure 3]. Under such utilisation status and traffic profiles, the always-on design approach becomes inefficient from several perspectives. At the interference dimension, an always-on BS acts as an always-interferer to neighbouring BSs. This can be linked to the fact that each BS continuously transmits pilot and reference signals, system information, etc. to preserve the coverage irrespective of the UE activity state. As a result, the UE always suffer from downlink inter-cell interference (ICI) even if the interfering BSs are not serving any user.

At the energy/power dimension, the always-on service concept results into load-independent power consumption profiles as shown in Figure 14.2. This in turns yields negative economic and environmental impacts in the form of increased energy bill and CO_2 emissions, respectively. Several techniques have been proposed to reduce the RAN energy consumption and to improve the load/energy profile. These include: cell zooming [7] along with cell wilting and blossoming [8] where BS coverage is reduced in low load periods to reduce the energy loss, time domain sleep modes with duty cycles and discontinuous transmission where the traffic is scheduled in few subframes to enable subframe-level sleep opportunities [9] and space domain sleep modes where some of the BS antenna ports are switched off [10]. The reader is referred to [10,11] for a comprehensive survey of energy saving techniques in cellular RANs. Despite their potential gains, most (if not all) of these energy saving mechanisms are constrained by the requirement to maintain an ubiquitous coverage, resulting into marginal energy saving gains in practical scenarios.

---

[2]In the rest of this chapter, the term "BS" will be used to denote the RAN transmission/reception nodes irrespective of the cellular standard/generation.

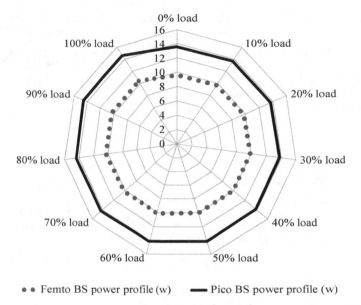

Figure 14.2   Power consumption profile of LTE femto and pico BSs. Based on the model of [4], with a discretised load step of 10%

14.2.2   Worst-case design

In legacy RAN standards, different BS types (e.g., macro, micro, pico, etc.) use a unified radio frame structure dimensioned based on the worst-case scenario. For instance, dimensions and allocations of the LTE radio frame are based on a macro cell environment with 991 ns rms delay spread, and a very high mobility assumption of 500 km/h. The rationale behind this approach originates from the UE-BS association criteria adopted in conventional RANs, which does not differentiate between idle and active UE. This state-independent RAN node selection means that, regardless of the UE activity state, the same BS serves all UE and the same radio frame is used by all UE. This leads to two main restrictions in the RAN radio frame and resource grid design:

1. All UE, including idle UE in severe conditions such as at cell edge or moving at high speed, should be able to detect and decode the frame contents correctly.
2. Control signals (e.g., pilots, synchronisation, etc.) are cell-specific rather than UE-specific resources.

The worst-case design approach guarantees acceptable performance for all UE within the considered RAN standard/generation limits. However, it over-provisions the radio frame in moderate and good channel environments [12]. As an illustrative example, Figure 14.3 shows the downlink radio frame structure and allocations of the LTE in frequency division duplex (FDD) mode, while Figure 14.4 shows the

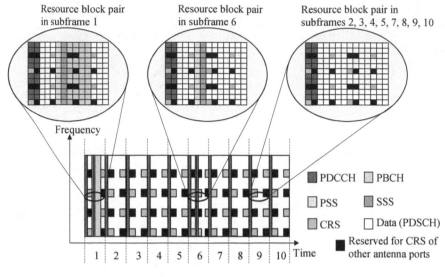

Figure 14.3   LTE FDD downlink radio frame structure and allocations

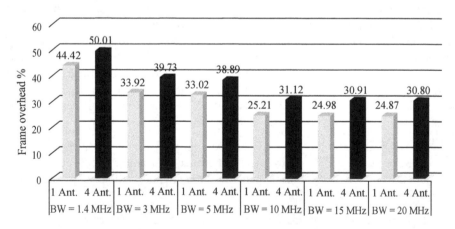

Figure 14.4   LTE FDD downlink frame overhead, based on 3GPP allocation guidelines in [13], Acronyms: Ant.: Antenna, BW: Bandwidth

frame overhead. As can be seen in the latter, 25%–50% of the LTE RAN resources are reserved for signalling. Thus the peak data rate of the RAN is at least 25%–50% less than the over-the-air rate. An interesting conclusion that can be drawn from Figures 14.3 and 14.4 is that the signalling allocations and overhead in the RAN frame do not depend on the UE channel conditions and data allocations, but rather they depend on the network configuration in terms of bandwidth, number of antennas and duplex mode.

14.2.3 Distributed management design

Mobility support is the fundamental requirement of any cellular system. When users move from one location to another, their serving BS is changed via a BS handover (HO) process for users in active state or via a BS reselection process for users in idle state. To enable these processes, each UE monitors signal strength and/or quality of the serving and the neighbouring BSs, and reports the monitored values to the serving BS either periodically or on an event basis. Depending on the adopted HO decision criteria, the serving BS issues a HO command to the UE when a better serving candidate is found. The HO procedure consists of preparation, execution and completion phases during which the BSs reserve resources and exchange the necessary parameters, the UE detaches from the old BS and accesses the candidate BS and the DP path is switched from the old to the candidate BS, respectively [14].

It can be noticed that the active state mobility management has a distributed nature. HO decisions are taken by the serving BS based on information provided by the UE.[3] In addition, each HO generates signalling towards the CN to switch the DP path towards the candidate BS. This may not be a challenge in macro BS deployment scenarios, where HO rates are typically low-moderate. In the dominant fifth generation (5G) theme of dense small cell (SC) deployments [15,16], however, the HO rates are expected to be significantly high which in turn increases the HO signalling load and failure rates [17–19]. In such scenarios, advanced mechanisms such as predictive mobility management, BS clustering and coordinated multipoint (CoMP) could provide significant gains in terms of reduction in HO rate and signalling load. These mechanisms require centralised decision entities with a wider view of the network rather than fully distributed decisions based on the local scope of a single (or a few) BS(s).

Mobility management for idle UE in conventional RANs can be described as an extreme distributed approach, where the BS reselection decision is taken by the UE.[4] This approach can be justified for idle UE since the BS reselection does not require resource assignment. Nevertheless, when these UE become active, they are served by the same BS which was initially selected by the UE. Such a UE-driven

---

[3] In some standards, the HO decision can also be taken by the UE with network assistance.
[4] The network can provide offset parameters to be added to the measured signal strength/quality to privilege some cells.

approach puts constraints on the resource assignment, management and optimisation process.

14.3   5G RAN with control/data separation

A futuristic RAN architecture with separation of CP and DP has been recently proposed in research community to overcome limitations of the conventional RAN design. The basic idea of this architecture depends on a dual connection approach to enable providing data services under the umbrella of a coverage layer. The latter is supported at low frequency bands by macro BSs known as control base stations (CBSs) where the large service area ensures robust connectivity and mobility. Within the CBS footprint, dedicated small BSs known as data base stations (DBSs) provide high data rate transmission. As shown in Figure 14.5, idle UE are associated with the CBS only for basic connectivity services, while active UE maintain a dual connection with both the CBS and the DBS [11,19].

The separation property of this RAN architecture can be viewed from three closely related perspectives: plane, state and signalling. A plane separation is achieved by separating the CP from the DP while a state separation is achieved by dedicating RAN nodes (i.e., DBSs) for active UE. On the other hand, the signalling separation property originates from the functionality/service/signalling mapping [20] of this architecture, where the broadcast-type/cell-specific signals are provided by the CBS, while the unicast-type/UE-specific signals are provided by the DBS. In the rest of this chapter, this architecture will be referred to as CDSA, however its worth emphasising that the CDSA supports all of the aforementioned separation types.

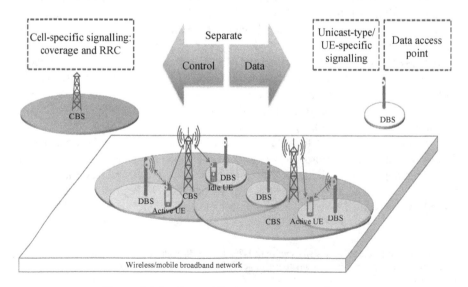

Figure 14.5   Control/Data separation architecture

The CDSA offers a wide range of benefits. In particular, it overcomes the main design limitations of the conventional RAN discussed in Section 14.2 by allowing a paradigm shift

- from always-on RAN to on-demand always-available RAN.
- from worst-case static design to adaptive dynamic design.
- from distributed management to almost centralised (or hybrid centralised-distributed) management.

14.3.1 On-demand always-available design

Unlike the conventional RAN where an always-on design was adopted because of the coverage constraints, the CDSA allows switching off the DBSs when they are not needed for data transmission. Since the coverage is maintained by the underlay CBSs, a switched off DBS does not create coverage holes. Typically, idle UE are associated with the CBSs only, which are kept on regardless of the traffic load. Notice that the CBS density is much lower than the DBS density. When a UE becomes active, e.g., starting a session or receiving a call, service requests are handled by the CBS which associates the requesting UE with the best serving DBS. The latter can be switched on, if necessary, before establishing the high rate DBS-UE connection. Expressed differently, the CDSA enables an on-demand RAN design in the DBS layer without breaking the anywhere/anytime service concept.

The main benefit of the on-demand RAN design can be observed in the energy dimension. It allows the CDSA energy consumption profile to scale with the traffic load as opposed to the load-independent energy profile of conventional RANs, see for example [20, Figure 4]. The results reported in [21–23] show a significant improvement in the energy efficiency when the CDSA with on/off DBSs is adopted instead of the conventional RAN. Additional energy saving gains can be achieved when the DBSs are serving active UE. These can be realised by transmitting the data towards the active UE only, e.g., by using cell structuring and high gain beamforming techniques which can be applied in the DBS layer with relaxed coverage constraints [24].

14.3.2 Adaptive design

As discussed in Section 14.2.2, the inefficiency of the conventional RAN architecture in terms of radio frame overhead and in-band signalling is linked to the worst-case design approach. The latter can be traced to the fact that all users of the conventional architecture are connected to the same BS irrespective of their activity state. In the CDSA, however, only the active UE are connected to the DBSs which do not transmit broadcast-type signals [12,20]. Thus the one-to-one nature of the UE-DBS link motivates replacing the cell-specific signals (whose allocations are based on the worst-case scenario) with adaptive and dynamic UE-specific signals, since the DBS frame is not constrained by the unknown channel conditions of the idle/detached UE.

Considering this feature, an adaptive DBS pilot signalling scheme was proposed in [25] to minimise the CDSA in-band signalling used for data detection and channel estimation in the DBS layer. This scheme considers the downlink of a multi-carrier air-interface and it takes into account both frequency and time domain variations. It depends on estimating the actual channel correlation function under realistic conditions rather than considering the worst-case correlation function. A theoretical comparison between the adaptive scheme and the conventional worst-case design indicated promising gains of more than 90% reduction in the pilot overhead in local area and low mobility scenarios. Simulation results show that the adaptive design reduces the pilot overhead by up to 78% w.r.t. the LTE cell-specific reference signal (CRS)[5] pattern with marginal performance penalty.

14.3.3 Almost centralised management design

The CDSA system model, where active UE maintains a dual connection with both the CBS and the DBS, enables centralised (or hybrid centralised-distributed) mobility and resource management procedures. The dual connectivity feature allows exploiting the CBS (which has a large footprint) as a radio resource control (RRC) anchor point in order to reduce the HO rate and signalling overhead [26]. Thus the RRC anchor point remains the same as long as the UE mobility is within the same umbrella (i.e., CBS). Although DBS-to-DBS HOs will be required, such HOs can be performed with lightweight and fast procedures [27]. In addition, the serving node (i.e., the serving DBS) selection is delegated to the CBS. This network-driven approach enables optimised resource management with a wide view of network status and parameters.

14.4 Main challenge: backhaul networks

The CDSA concept opens a wide range of benefits and offers relaxed constraints in the RAN design as discussed in Section 14.3. Energy efficiency, signalling overhead, interference management and mobility management are identified as promising dimensions that can be substantially improved under CDSA configuration, especially in dense deployment scenarios. These benefits are enabled in the CDSA by the separation and the dual connection features. Nonetheless, the latter have a direct impact on design, requirements and constraints of the backhaul networks that connect the CBSs with the DBSs, as well as the CBSs/DBSs with the CN.

As opposed to the conventional RAN, active UE in the CDSA maintains a dual connection with the CBS and the DBS. Consequently, an additional CBS-DBS backhaul link may be required to enable signalling and/or data flow between the CBS and the DBS [28]. The requirements of this link in terms of bandwidth, latency, reliability, etc. depend on the adopted separation scheme, the functionalities supported by each plane, the DP route from the CN to the DBS and the level of the CBS-DBS coordination. A separation scheme with centralised scheduling functionalities at the CBS requires a low latency CBS-DBS link to avoid delays in the downlink scheduling. Similarly, a tight CBS/DBS coordination generates excessive signalling exchange which suggests robust and low latency backhaul links [29,30]. On the other hand,

---

[5]The CRS is used in the LTE as a pilot for channel estimation and for channel quality measurements.

a mobility-efficient RRC signalling with a centralised DP anchor point at the CBS requires a high bandwidth CBS-DBS link to allow (downlink) data forwarding from the CBS to the DBS. In addition, superiority of the CDSA over the conventional RAN depends on the employed backhaul technology. For instance, the overall energy saving gains in the CDSA might be marginal when the backhaul network consumes high power. These aspects are discussed in detail in the following subsections.

14.4.1 Impact of separation schemes on data plane backhaul latency

Mobility robustness is considered as one of the main benefits of the CDSA. The dual connection feature allows exploiting the CBS as a mobility anchor point for the UE, and as a DP anchor point for the DBSs. Thus the CP link and the DP path from the CN to the CBS remain unchanged as long as the UE mobility is within the same CBS umbrella. This in turns improves mobility performance and alleviates the associated HO signalling overhead [11,19,20,26].

Such an architecture allows the intra-CBS HOs (i.e., between DBSs within the footprint of the same CBS) to be transparent to the CN. This can be traced to the CDSA model where the DP path is switched locally at the CBS without informing the CN, as shown in Figure 14.6(a). This approach has been found to be very efficient in reducing the HO-related signalling overhead, especially in dense deployment scenarios [19,26,31], as compared with the conventional CN-visible HO approach. In the latter, the DP path is routed directly to the DBS as in Figure 14.6(b). Hence each HO triggers a path switch request towards the CN. Despite the potential gains, the CDSA with DP routing through both the CBS and the DBS (to enable the CN-transparent HO) induces an additional DP backhaul latency (DP-BL) due to the additional hop in this case.

To investigate this penalty, we propose a stochastic geometry model to derive closed-form expressions for the DP-BL under both HO/separation schemes in the CDSA. This model is used to analyse the impact of several parameters, such as DBS and CBS densities, data load and processing capabilities, on the DP-BL. In addition, we derive a DBS densification limit under DP-BL constraints. This ensures a fair comparison by setting a DBS density limit, below which the peak CDSA HO signalling gains can be achieved/assessed without violating the backhaul latency constraints, and above which these gains degrade the performance.

The tractability, lack of edge effects and natural inclusion of different cell sizes have made the Poisson point process (PPP) a popular modelling approach in cellular research community. It has been adopted in [32] to derive coverage probability and downlink throughput of homogeneous cellular networks, and in [33] to model the inter-cell signalling overhead in heterogeneous networks (HetNets). The authors of [34] use the PPP model to analyse the impact of user density on the signal-to-interference-plus-noise ratio (SINR) distribution. Similarly, [35] optimises the UE-BS association in HetNets by using the PPP to model both user and network nodes. Following these mature studies, the proposed framework adopts the PPP to model the CDSA nodes.

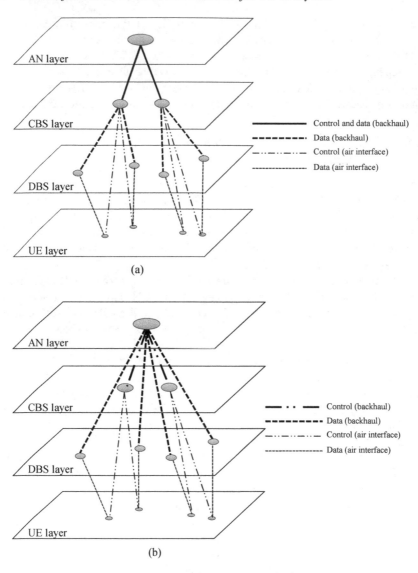

Figure 14.6   Voronoi spanning trees between CDSA nodes: (a) separation at CBS to enable CN-transparent HO and (b) separation at CN to enable CN-visible HO.

The system model consists of DBSs, CBSs and CN aggregation nodes (ANs) that are modelled as three independent PPPs in $\mathbb{R}^2$: Φ_d with density λ_d for the DBSs, Φ_c with density λ_c for the CBSs and Φ_a with density λ_a for the ANs, where $\lambda_d > \lambda_c > \lambda_a$. d, c and a are arbitrary points of Φ_d, Φ_c and Φ_a, respectively. The total DP latency consists of backhaul latency and air-interface latency. The former is the delay experienced by the packets from the CN to the DBS, while the latter represents the DBS-UE delay. As can be seen in Figure 14.6, the air interface latency is the same under both CDSA HO/separation schemes and it does not depend on the separation scheme but rather it depends on frame structure and access procedure. Thus, we focus only on the backhaul latency by considering the backhaul and the RAN nodes in the analytical model. This results in voronoi spanning trees, where any realisation of Φ_d, Φ_c and Φ_a yields random trees. However, these trees depend on the adopted CDSA separation and HO schemes:

- **Separation at CBS to enable CN-transparent HO:** the RRC connection is maintained at the CBS where the CP/DP separation is performed. Thus the data path switching is performed locally at the CBS for intra-CBS HOs rather than triggering a switch request towards the CN. Each DBS is connected to a single CBS, where the CBS-DBS association is based on the least distance rule, i.e., DBS_i is connected to CBS_j if and only if DBS_i is in the voronoi cell of CBS_j. In addition, each CBS is connected to a single AN based on the least distance rule, i.e., CBS_j is connected to AN_k if and only if CBS_j is in the voronoi cell of AN_k. Figure 14.6(a) shows the CDSA voronoi spanning trees of this scheme.
- **Separation at CN to enable CN-visible HO:** each DBS acts as a DP anchor point and has a separate RRC entity. Consequently, the CP/DP separation is performed at the CN. Thus all HOs (i.e., intra- and inter-CBS HOs) are visible to the CN since they require data path switching operation at the CN. Each DBS is connected directly to a single AN, where DBS_i is connected to AN_k if and only if DBS_i is in the voronoi cell of AN_k. In addition, each CBS is connected to a single AN based also on the least distance rule. The CBS-AN link is used for control signalling and it is not involved in the data path. A backhaul link (not shown in Figure 14.6) may be required between the CBS and the DBS for signalling and coordination purposes. As with the CBS-AN link, the DBS-CBS link is not involved in data transmission. Figure 14.6(b) shows the CDSA voronoi spanning trees of this scheme.

The DP-BL t consists of three components: propagation delay t_p, transmission delay t_t and processing delay t_s. For the CN-transparent HO scenario, the DP backhaul path traverses two links AN → CBS → DBS. On the other hand, the DP backhaul path of the CN-visible HO case traverses a single link AN → DBS. Assuming error free transmission across the backhaul network, each latency component is modelled as in the following.

14.4.1.1 Data plane backhaul propagation delay

It can be seen in Figure 14.6(a) that the DP backhaul path of the CDSA with CN-transparent HO is longer than the DP backhaul path of the network configuration that enables the conventional CN-visible HO. Notice that PPPs of all network nodes are distributed in the same two-dimensional plane, however they are shown in Figure 14.6 as a layered three-dimensional architecture for readability. For instance, the AN-DBS link in Figure 14.6(b) is not necessarily longer than the AN-CBS link. In other words, the length and the propagation delay of the direct link between any two nodes of two different layers (i.e., PPPs) depend on the distribution and the density of these two layers only. In the CN-visible HO network, the propagation delay can be calculated by the following lemma.

Lemma 14.1. *With a propagation speed $s_{a,d}$, the propagation delay between an arbitrary AN in Φ_a and an arbitrary DBS in Φ_d, is a Rayleigh distributed random variable with scale parameter $\sigma = \frac{1}{s_{a,d}\sqrt{2\pi\lambda_a}}$, expected value $\mathbb{E}\left(t_p\right) = \frac{1}{2s_{a,d}\sqrt{\lambda_a}}$, and probability density function (PDF) for $t_p \geq 0$:*

$$f\left(t_p\right) = \frac{t_p}{\sigma^2}\, e^{\frac{-t_p^2}{2\sigma^2}} = 2\,\pi\,\lambda_a\, s_{a,d}^2\, t_p\, e^{-\pi\,\lambda_a\, s_{a,d}^2\, t_p^2}. \tag{14.1}$$

Proof. See Appendix A. □

Lemma 14.1 can also be used to calculate the propagation delay across a single link in the CN-transparent HO network. However, the DP path of the latter traverses two links. Thus the total propagation delay in this case is a summation of two Rayleigh variables. The method in [36] can be used to calculate the PDF of this summation. The expected value of the total propagation delay can be written as:

$$\mathbb{E}\left(t_p\right) = \int_0^{\infty} t_p f(t_p)\, \mathrm{d}t_p = \sum_{\forall x \in \Lambda \setminus \{d\}} \frac{1}{2\, s_{x,y}\, \sqrt{\lambda_x}}, \tag{14.2}$$

where Λ is the set of all nodes in the DP backhaul path, i.e., $\Lambda = \{a, d\}$ for the CN-visible HO network and $\Lambda = \{a, c, d\}$ for the CN-transparent HO network and y is the node in the layer immediately below the layer of node x. It can be noticed in (14.2) that $\mathbb{E}(t_p)$ is inversely proportional to λ_x. Thus the DBS density does not affect the DP backhaul propagation delay of both HO/separation scheme. In the CN-transparent HO network, the additional propagation delay depends on the CBS density. Hence increasing the latter reduces the impact of the two hop DP path on the overall propagation delay.

14.4.1.2   Data plane backhaul transmission delay

The transmission delay can be modelled as the ratio between the data load (D) and the backhaul capacity. Depending on the backhaul technology (e.g., copper, fibre, wireless, etc.), repeater(s) may be needed if the link transmission range is smaller than the nodes' distance. These repeaters result in additional transmission delays. The expected number of repeaters between x in Φ_x and y in Φ_y, i.e., $\mathbb{E}(h_{x,y})$, is:

$$\mathbb{E}(h_{x,y}) = \left\lceil \frac{\mathbb{E}(r_{x,y})}{R_{x,y}} - 1 \right\rceil = \left\lceil \frac{1}{2\, R_{x,y}\, \sqrt{\lambda_x}} - 1 \right\rceil, \tag{14.3}$$

where $\mathbb{E}(r_{x,y}) = \int_0^{\infty} r f(r)\, \mathrm{d}r$ is the expected value of the distance between nodes x and y. $R_{x,y}$ is the transmission range of the backhaul link between x and y, and $\lceil m \rceil$ means the smallest integer $\geq m$. The transmission delay between x and y is $\left(\frac{D}{L_{x,y}}\right)(\mathbb{E}(h_{x,y}) + 1)$, which consists of delays at x and at each repeater, with $L_{x,y}$

being the capacity of the link connecting x with y through the repeaters. The expected value of the DP backhaul transmission delay can be written as:

$$\mathbb{E}(t_t) = \sum_{\forall x \in \Lambda \setminus \{d\}} \frac{D}{L_{x,y}} \left[\frac{1}{2 R_{x,y} \sqrt{\lambda_x}} \right], \tag{14.4}$$

where x and y have the same definitions as in (14.2). As with the latter, (14.4) indicates that the DBS density does not affect $\mathbb{E}(t_t)$, while a higher CBS density reduces the impact of the second hop on the overall DP backhaul transmission delay in the CN-transparent HO network.

14.4.1.3   Data plane backhaul processing delay

We follow [35,37] by assuming that t_s in a single node has gamma distribution, since it has been found to provide a good fit for real measurements [37,38]. The processing delay in a single node x depends on: processing capabilities of the node, DBS data load and the number of DBSs connected to the node. The latter can be captured by the shape parameter k of the gamma distribution, i.e., $k_x \propto \frac{\lambda_d}{\lambda_x}$, while the scale parameter θ_x captures processing capabilities and DBS data load. Based on the classical definition of gamma distribution [39], the PDF of node x processing delay is:

$$f(t_s) = \frac{t_s^{k_x - 1} e^{-\frac{t_s}{\theta_x}}}{\theta_x^{k_x} \Gamma(k_x)} = \frac{t_s^{\frac{\mu_x \lambda_d}{\lambda_x} - 1} e^{-\frac{t_s}{\alpha_x + \beta_x D}}}{(\alpha_x + \beta_x D)^{\frac{\mu_x \lambda_d}{\lambda_x}} \Gamma\left(\frac{\mu_x \lambda_d}{\lambda_x}\right)}, \tag{14.5}$$

with $\mathbb{E}(t_s) = k_x \theta_x$, $k_x = \frac{\mu_x \lambda_d}{\lambda_x}$, $\theta_x = \alpha_x + \beta_x D$, where μ_x is the constant of proportionality, α_x is the static processing delay, β_x is the processing delay per bit and $\Gamma(z)$ is the gamma function evaluated at z. The processing delay at the AN, the CBS or the DBS can be modelled by (14.5). For a tandem topology based repeaters, the amount of DP traffic traversing a repeater between x and y is the same amount of DP traffic traversing y because there could be more than a node of Φ_y connected to x. Thus, the scale parameter $k_{x,y}$ of the processing delay of all repeaters between x and y is proportional to $\frac{\lambda_d}{\lambda_y} \mathbb{E}(h_{x,y})$.

We assume that the processing capabilities of the nodes in the same tier is the same (e.g., all ANs have the same processing capabilities). However, nodes in different tiers may have different processing capabilities (e.g., the nodes towards the CN may have higher processing capabilities than the nodes towards the edge). As a result, the distribution of the total processing delay of all nodes and repeaters is not a simple gamma distribution because the scale parameter θ of different nodes is not constant. In other words, the total DP backhaul processing delay is a summation of n independent gamma variables $g_1 \sim \Gamma(k_1, \theta_1)$, $g_2 \sim \Gamma(k_2, \theta_2)$, $\ldots, g_n \sim \Gamma(k_n, \theta_n)$. Such a summation can be represented by a single gamma-series with coefficients that are computed by recursive relations. The reader is referred to [39] for the PDF of this

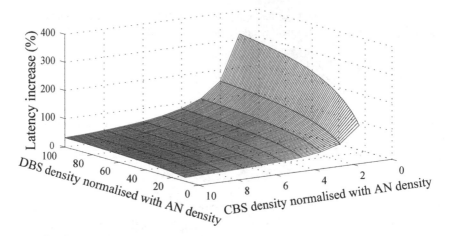

Figure 14.7   Effect of DBS and CBS densities on the DP-BL, D = 1,500 bytes

gamma-series. The expected value can be directly obtained as the summation of $\mathbb{E}(t_s)$ at each node and repeater, i.e.,

$$\mathbb{E}(t_s) = \sum_{l=1}^{n} k_l\, \theta_l$$

$$= \sum_{\forall x \in \Lambda \setminus \{d\}} \frac{\mu_{x,y}\, \lambda_d}{\lambda_y} \left[\frac{1}{2\, R_{x,y}\, \sqrt{\lambda_x}} - 1 \right] (\alpha_{x,y} + \beta_{x,y} D) + \sum_{\forall x \in \Lambda} \frac{\mu_x\, \lambda_d}{\lambda_x} (\alpha_x + \beta_x D)$$

$$(14.6)$$

where $\mu_{x,y}$, $\alpha_{x,y}$ and $\beta_{x,y}$ are the processing parameters of the repeaters between nodes x and y. Unlike the propagation and the processing delays, (14.6) indicates that the DP backhaul processing delay depends on the DBS density.

Figure 14.7 compares the effect of the DBS and the CBS densities on the DP-BL, i.e., the summation of (14.2), (14.4) and (14.6), of the two CDSA HO/separation schemes while Figure 14.8 shows the effect of the data load. The comparison is based on the increase of the DP-BL in the CN-transparent HO scheme w.r.t. the CN-visible HO model by considering the backhaul parameters in Table 14.1. As can be seen in Figure 14.7, the rate of increase in the DP-BL with the increase of the DBS density is higher in the CN-transparent HO network as compared with the CN-visible HO network. Expressed differently, although the DP-BL of both networks is proportional to the DBS density, the latter has a higher impact on the DP-BL of the CN-transparent HO network. However, the increase in the DP-BL depends on the CBS density as can be noticed in Figure 14.7.

For the CN-transparent HO network, the proportional DBS-density/backhaul-latency relationship has a larger (smaller) slope with low (high) CBS densities. This

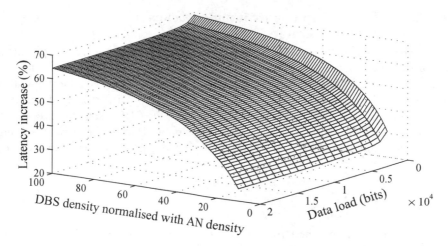

Figure 14.8   Effect of data load on the DP-BL, with $\frac{\lambda_c}{\lambda_a} = 5$

Table 14.1   Backhaul link and processing parameters

Link parameters		Processing capabilities	
Parameter	**Value**	**Parameter**	**Value**
$s_{a,c}$	2.1×10^8 m/s	μ_a	1
$R_{a,c}$	20 km	α_a	1 μs
$L_{a,c}$	1 Gb/s	β_a	1 ns/bit
$s_{c,d}$	2.1×10^8 m/s	$\mu_c, \mu_{a,c}$	2
$R_{c,d}$	20 km	$\alpha_c, \alpha_{a,c}$	2 μs
$L_{c,d}$	0.1 Gb/s	$\beta_c, \beta_{a,c}$	2 ns/bit
$s_{a,d}$	2.1×10^8 m/s	$\mu_d, \mu_{c,d}, \mu_{a,d}$	4
$R_{a,d}$	20 km	$\alpha_d, \alpha_{c,d}, \alpha_{a,d}$	4 μs
$L_{a,d}$	0.1 Gb/s	$\beta_d, \beta_{c,d}, \beta_{a,d}$	4 ns/bit

can be linked to the fact that increasing the CBS density distributes the processing load (i.e., smaller number of DBSs associated with each CBS), which reduces the DP-BL. Moreover, a higher CBS density results in a smaller CBS-DBS distance which reduces the propagation delay and alleviates the additional delay at intermediate repeaters. Thus it can be said that the CBS density is an important design parameter that can be controlled to reduce the DP-BL of the CN-transparent HO network. On the other hand, Figure 14.8 shows that the amount of increase in the DP-BL of the CN-transparent HO network w.r.t. the CN-visible HO network remains roughly constant irrespective of the data load when $D_l > 2{,}000$ bits (i.e., 250 bytes). Thus, the latter has a negligible impact on the rate of increase in the DP-BL as compared with the CBS and the DBS densities.

14.4.1.4   DBS densification limit under backhaul latency constraints

It has been shown in [19,26,31] that the CDSA is more beneficial in dense deployment scenarios, where the peak gains (in terms of capacity expansion and HO signalling reduction) increase with the DBS density. However, the DP-BL, i.e., the summation of (14.2), (14.4) and (14.6), is also proportional to the DBS density. Expressed differently, a dense DBS deployment increases both the CDSA signalling reduction gains and the DP-BL. This suggests a latency constrained DBS deployment. Assuming a maximum tolerable DP-BL of $\mathbb{E}(t_{thr})$, then a latency protection condition can be formulated as:

$$\mathbb{E}(t) \leq \mathbb{E}(t_{thr}). \tag{14.7}$$

Based on this condition, the DBS densification limit λ_d^* without violating (14.7) can be obtained by substituting the summation of (14.2), (14.4) and (14.6) in the left-hand side of (14.7) and solving the resultant equation for λ_d. The exact closed form expression of λ_d^* is given by

$$\lambda_d^* \leq \frac{\mathbb{E}[t_{thr}] - B - \mu_{i,d} \left\lceil \frac{1}{2 R_{i,d} \sqrt{\lambda_i}} - 1 \right\rceil (\alpha_{i,d} + \beta_{i,d} D) - \mu_d (\alpha_d + \beta_d D)}{\frac{\mu_a}{\lambda_a} (\alpha_a + \beta_a D) + j \left(\frac{\mu_c}{\lambda_c} (\alpha_c + \beta_c D) + \frac{\mu_{a,c}}{\lambda_c} \left\lceil \frac{1}{2 R_{a,c} \sqrt{\lambda_a}} - 1 \right\rceil (\alpha_{a,c} + \beta_{a,c} D) \right)}, \tag{14.8}$$

with

$$B = \sum_{\forall x \in \Lambda \setminus \{d\}} \frac{1}{2 s_{x,y} \sqrt{\lambda_x}} + \frac{D}{L_{x,y}} \left\lceil \frac{1}{2 R_{x,y} \sqrt{\lambda_x}} \right\rceil, \tag{14.9}$$

where $i = c$ and $j = 1$ in the CN-transparent HO case, while $i = a$ and $j = 0$ in the CN-visible HO scenario.

Figure 14.9 provides the DBS densification limit under latency constraints. It can be noticed that relaxing $\mathbb{E}(t_{thr})$ increases the DBS densification limit. This can be traced to the proportional DBS-density/DP-BL relationship. With a conservative latency threshold of $\mathbb{E}(t_{thr}) < 1$ ms, both HO/separation schemes can operate with roughly the same DBS density without violating the latency constraint. Although increasing $\mathbb{E}(t_{thr})$ increases the allowed DBS density for both HO/separation schemes, the rate of increase in the allowed density is higher in the CN-visible HO network as compared with the CN-transparent HO network. In other words, the former can operate with a higher DBS density than the later. As with Figure 14.7, Figure 14.9 indicates that the CBS density has a significant impact on the allowed DBS for the CN-transparent HO network. Thus it can be concluded that the latency threshold and the CBS density are important design parameters that have a significant impact on the DP-BL and the allowed DBS density.

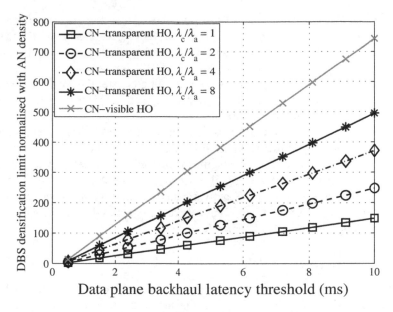

Figure 14.9   DBS densification limit vs DP-BL threshold, $D_l = 1,500$ bytes

14.4.2   Impact of backhaul technology on energy efficiency

Traditionally, cellular users camp on the network by selecting the BS that offers the strongest signal strength. Measuring the latter requires the BS to periodically transmit pilot signals regardless of the users' activity state. Once a candidate BS is found, the UE establishes a random access procedure via uplink signalling with the selected BS. Irrespective of the employed energy saving technique, the users need to be able to camp on the network and issue service requests whenever needed. Expressed differently, the downlink pilot signal and the ability to receive uplink signalling from the UE are two basic requirements that have to be met by all BSs. Such an ubiquitous connectivity requirement can be satisfied in the CDSA even when the DBSs are switched off for energy saving. This can be achieved by maintaining:

1.   An always-on CBS layer.
2.   A CBS-controlled (or CBS-assisted) serving DBS selection.
3.   A CBS-DBS backhaul link to transmit DBS activation messages.

Since the energy consumption increases linearly with the number of active BSs [40–42], the requirement to keep an always-on CBS layer to allow switching off the DBSs can be justified when the DBS density is higher than the CBS density. The CBS-controlled (or CBS-assisted) DBS-UE association allows the UE to issue service requests whenever needed even when the DBSs are switched off. In the latter case, the CBS-DBS backhaul link can be used by the CBS to send wakeup signals to the selected DBS. In addition, this link can be used to exchange load information, network status, service type, etc. to optimise the DBS selection and on/off decisions.

Several backhaul options such as ideal fibre optic, non-ideal fibre optic, wireless links and wired links can be used for the CBS-DBS link. A low latency and a reliable backhaul link is of great importance in enabling a faster DBS wakeup to deliver the required QoS. This suggests using the ideal fibre optic for the CBS-DBS link due to the low latency and the high reliability of this technology. Nonetheless, the fibre optic link and the associated interfaces consume the highest power among the aforementioned options, when the backhaul links are in active state [43,44]. On the other hand, the wireless backhaul link consumes the lowest active state power [43,44], at the cost of a lower reliability and a higher latency due to retransmissions. Expressed differently, enabling a faster DBS activation/deactivation could reduce the energy consumption at the RAN side, however this could come at the cost of an increased energy consumption at the backhaul side.

14.4.3 Alternative backhaul mechanisms

Recently, advanced techniques have been proposed to replace the conventional direct backhaul networks in the CDSA. These techniques exploit the broadcast nature of the CBS transmission under CDSA network configuration where the DBSs are deployed within the CBS footprint. In other words, the DBSs can be viewed as users that are able to detect and decode the CBS transmission to the UE, and then act accordingly. As an illustrative example, consider a separation scheme that enables centralised scheduling functionalities at the CBS, i.e., the DBS scheduling functionalities and decisions are controlled by the CBS. The latter issues scheduling grants to the UE which can be overheard by the DBSs [45]. Such an approach is known as over-the-air signalling [46] and it can significantly reduce the CAPEX of the backhaul network. However, it is worth emphasising that this mechanism requires a robust signalling design with optimum transmission gaps [28].

14.5 Conclusion

Network densification is being considered as one of the main themes of 5G cellular systems. Despite the potential gains at the capacity dimension, such deployment scenarios raise several challenges and may require redesigning the RAN architecture. From an energy perspective, the always-on service approach has been identified as the main source of inefficiency in the traditional system. In addition, the conventional worst-case design approach may not be suitable because it wastes the transmission resources by over-provisioning the radio frame. In addition, the expected high HO rates in dense deployment scenarios could lead to a dramatic increase in mobility-related signalling overhead. In this direction, the CDSA has been proposed as a candidate architecture that enables a paradigm shift towards an on-demand RAN with adaptive allocations and efficient mobility management.

The main concept of the CDSA is the separation of the coverage signals from those needed to support high data rate transmission. This allows data services to be provided by efficient DBSs under the umbrella of a coverage layer supported by CBSs. Such an approach opens a wide range of benefits with relaxed constraints. The latter have been discussed in this chapter and compared with the conventional approach. In addition, the new backhaul challenges and constraints imposed by the CDSA are presented. The impact of the CDSA on the backhaul latency is discussed and modelled by using stochastic geometry. Moreover, a DBS densification limit under DP-BL constraints has been derived. This limit ensures a fair comparison by setting a DBS density limit, below which the CDSA gains can be achieved/assessed without violating the latency constraints, and above which these gains degrade the performance. Furthermore, the impact of the backhaul technology on the CDSA energy efficiency has been discussed and the concept of over-the-air signalling has been presented as an alternative to using the conventional direct backhaul networks. To summarise, exploiting benefits of the CDSA requires a careful and backhaul-aware CBS/DBS deployment to ensure that the gains at the RAN side do not come at the expense of performance degradation at the backhaul side.

Appendix A: Proof of lemma 14.1

The probability that the distance r between a in Φ_a and d in Φ_d is greater than a distance r_0 is equivalent to the null probability in an area with radius r_0 centred at d, i.e.,

$$\mathbb{P}[r > r_0] = e^{-\pi \lambda_a r_0^2}, \tag{A.1}$$

where $\mathbb{P}[B]$ is the probability of event B. The cumulative distribution function of r is $F(r) = \mathbb{P}[r \le r_0] = 1 - e^{-\pi \lambda_a r_0^2}$ and the PDF of r can be obtained as:

$$f(r) = \frac{d}{dr}(1 - e^{-\pi \lambda_a r^2}) = 2 \pi \lambda_a r \, e^{-\lambda_a \pi r^2}. \tag{A.2}$$

t_p is the ratio between r and $s_{a,d}$. Since t_p is a monotonic increasing function of r, then $F(r)$ can be written as:

$$F(r) = \int_{m_1}^{m_2} f(r) \, dr = \int_{t_p(m_1)}^{t_p(m_2)} f(r(t_p)) \frac{dr}{dt_p} \, dt_p. \tag{A.3}$$

Expressed differently, $f(t_p)$ can be derived from $f(r)$ as: $f(t_p) = f(r) \left| \frac{dr}{dt_p} \right|$. Substituting (A.2) and $r = t_p \, s_{a,d}$ in this equation and solving it yields (14.1). The expected value of the propagation delay can be directly obtained by substituting (14.1) in $\mathbb{E}(t_p) = \int_0^\infty t_p f(t_p) \, dt_p$.

References

[1] Auer G, Blume O, Giannini V, *et al*. EARTH project deliverable D2.3: Energy efficiency analysis of the reference systems, areas of improvements and target breakdown; 2010. p. 1–69.

[2] Son K, Kim H, Yi Y, and Krishnamachari B. "Base station operation and user association mechanisms for energy-delay tradeoffs in green cellular networks," *IEEE Journal on Selected Areas in Communications*. 2011 September;29(8): 1525–1536.

[3] Guan H, Kolding T, and Merz P. "Discovery of Cloud-RAN," in: *Proceedings of Cloud-RAN Workshop*; 2010. p. 1–13.

[4] Auer G, Giannini V, Desset C, *et al*. "How much energy is needed to run a wireless network?," *IEEE Wireless Communications*. 2011 October;18(5): 40–49.

[5] Oh E and Krishnamachari B. "Energy savings through dynamic base station switching in cellular wireless access networks," in: *Proceedings of IEEE Global Telecommunications Conference (GLOBECOM)*. Miami, FL, IEEE; 2010. p. 1–5.

[6] Alcatel-Lucent. 900 Wireless Network Guardian; 2008. White Paper.

[7] Niu Z, Wu Y, Gong J, and Yang Z. "Cell zooming for cost-efficient green cellular networks," *IEEE Communications Magazine*. 2010 November;48(11):74–79.

[8] Conte A, Feki A, Chiaraviglio L, Ciullo D, Meo M, and Marsan MA. "Cell wilting and blossoming for energy efficiency," *IEEE Wireless Communications*. 2011 October;18(5):50–57.

[9] Mclaughlin S, Grant P, Thompson J, *et al*. "Techniques for improving cellular radio base station energy efficiency," *IEEE Wireless Communications*. 2011 October;18(5):10–17.

[10] Domenico AD, Strinati EC, and Capone A. "Enabling green cellular networks: a survey and outlook," *Computer Communications Journal*. 2014 January; 37:5–24. Available from: http://www.sciencedirect.com/science/article/pii/S0140366413002168.

[11] Mohamed A, Onireti O, Imran M, Imran A, and Tafazolli R. "Control-data separation architecture for cellular radio access networks: a survey and outlook," *IEEE Communications Surveys and Tutorials*. 2016 Firstquarter;18(1): 446–465.

[12] Mohamed A, Onireti O, Qi Y, Imran A, Imran M, and Tafazolli R. "Physical layer frame in signalling-data separation architecture: overhead and performance evaluation," in: *Proceedings of the 20th European Wireless Conference*; 2014. p. 820–825.

[13] 3GPP. User equipment (UE) radio transmission and reception; 2012. 3GPP TS 36.101 version 10.0.0 Release 10. Technical Specification. Available from: http://www.3gpp.org/dynareport/36101.htm.

[14] 3GPP. Feasibility study for evolved Universal Terrestrial Radio Access (UTRA) and Universal Terrestrial Radio Access Network (UTRAN); 2012.

3GPP TR 25.912 version 11.0.0 Release 11. Technical Report. Available from: http://www.3gpp.org/DynaReport/25912.htm.

[15] Andrews JG, Buzzi S, Choi W, *et al*. "What will 5G be," *IEEE Journal on Selected Areas in Communications*. 2014 June;32(6):1065–1082.

[16] Hoydis J, Kobayashi M, and Debbah M. "Green small-cell networks," *IEEE Vehicular Technology Magazine*. 2011 March;6(1):37–43.

[17] 3GPP. Study on small cell enhancements for E-UTRA and E-UTRAN: higher layer aspects; 2013. 3GPP TR 36.842 version 12.0.0 Release 12. Technical Report. Available from: http://www.3gpp.org/DynaReport/36842.htm.

[18] 3GPP. Evolved Universal Terrestrial Radio Access (E-UTRA); mobility enhancements in heterogeneous networks; 2012. 3GPP TR 36.839 version 11.1.0 Release 11. Technical Report. Available from: http://www.3gpp.org/DynaReport/36839.htm.

[19] Ishii H, Kishiyama Y, and Takahashi H. "A novel architecture for LTE-B: C-plane/U-plane split and phantom cell concept," in: *Proceedings of IEEE Globecom Workshops*; 2012. p. 624–630.

[20] Xu X, He G, Zhang S, Chen Y, and Xu S. "On functionality separation for green mobile networks: concept study over LTE," *IEEE Communications Magazine*. 2013 May;51(5):82–90.

[21] Capone A, Filippini I, Gloss B, and Barth U. "Rethinking cellular system architecture for breaking current energy efficiency limits," in: *Proceedings of Sustainable Internet and ICT for Sustainability (SustainIT)*; 2012. p. 1–5.

[22] Ternon E, Bharucha Z, and Taoka H. "A feasibility study for the detached cell concept," in: *Proceedings of 9th International ITG Conference on Systems, Communication and Coding*; 2013. p. 1–5.

[23] Romanous B, Bitar N, Zaidi SAR, Imran A, Ghogho M, and Refai HH. "A game theoretic approach for optimizing density of remote radio heads in user centric cloud-based radio access network," in: *Proceedings of IEEE Global Communications Conference (GLOBECOM)*; 2015. p. 1–6.

[24] Olsson M, Cavdar C, Frenger P, Tombaz S, Sabella D, and Jantti R. "5GrEEn: Towards green 5G mobile networks," in: *Proceedings of IEEE 9th International Conference on Wireless and Mobile Computing, Networking and Communications (WiMob)*; 2013. p. 212–216.

[25] Mohamed A, Onireti O, Imran MA, Imran A, and Tafazolli R. "Correlation-based adaptive pilot pattern in control/data separation architecture," in: *Proceedings of IEEE International Conference on Communications (ICC)*; 2015. p. 2233–2238.

[26] Mohamed A, Onireti O, Imran M, Imran A, and Tafazolli R. "Predictive and core-network efficient RRC signalling for active state handover in RANs with control/data separation," *IEEE Transactions on Wireless Communications*. 2016;PP(99):1–14.

[27] Mohamed A, Onireti O, Hoseinitabatabae S, Imran M, Imran A, and Tafazolli R. "Mobility prediction for handover management in cellular networks with

control/data separation," in: *Proceedings of IEEE International Conference on Communications (ICC)*; 2015. p. 3939–3944.

[28]   Nakamura T, Nagata S, Benjebbour A, *et al.* "Trends in small cell enhancements in LTE advanced," *IEEE Communications Magazine*. 2013 February;51(2): 98–105.

[29]   Zhao T, Yang P, Pan H, Deng R, Zhou S, and Niu Z. "Software defined radio implementation of signalling splitting in hyper-cellular network," in: *Proceedings of ACM 2nd Workshop on Software Radio Implementation Forum*; 2013. p. 81–84.

[30]   Aydin O, Valentin S, Ren Z, *et al.* METIS project deliverable D4.1: Summary on preliminary trade-off investigations and first set of potential network-level solutions; 2013. p. 1–97.

[31]   3GPP, Nokia Siemens Networks. Mobility statistics for macro and small cell dual-connectivity cases. 3GPP TSG-RAN WG2 Meeting, Chicago, USA, 15–19 April 2013. Technical Report. Available from: http://www.3gpp.org/DynaReport/TDocExMtg–R2-81b–30048.htm.

[32]   Andrews JG, Baccelli F, and Ganti RK. "A tractable approach to coverage and rate in cellular networks," *IEEE Transactions on Communications*. 2011 November;59(11):3122–3134.

[33]   Xia P, Jo HS, and Andrews JG. "Fundamentals of inter-cell overhead signalling in heterogeneous cellular networks," *IEEE Journal of Selected Topics in Signal Processing*. 2012 June;6(3):257–269.

[34]   Martin-Vega FJ, Renzo MD, Aguayo-Torres MC, Gomez G, and Duong TQ. "Stochastic geometry modelling and analysis of backhaul-constrained hyper-dense heterogeneous cellular networks," in: *Proceedings of 17th International Conference on Transparent Optical Networks (ICTON)*; 2015. p. 1–4.

[35]   Zhang G, Quek TQS, Kountouris M, Huang A, and Shan H. "Fundamentals of heterogeneous backhaul design, analysis and optimization," *IEEE Transactions on Communications*, 2016 February;64(2):876–889.

[36]   Archer C. Some properties of Rayleigh distributed random variables and of their sums and products; 1967. DTIC Document. Technical Report.

[37]   Bovy C, Mertodimedjo H, Hooghiemstra G, Uijterwaal H, and Mieghem P. "Analysis of end-to-end delay measurements in Internet," in: *Proceedings of Passive and Active Measurement Workshop-PAM*; 2002. p. 1–8.

[38]   Hooghiemstra G and Van Mieghem P. Delay distributions on fixed internet paths; 2001. Delft University of Technology. Technical Report.

[39]   Moschopoulos PG. "The distribution of the sum of independent gamma random variables," *Annals of the Institute of Statistical Mathematics* 1985;37(1): 541–544.

[40]   Ashraf I, Boccardi F, and Ho L. "SLEEP mode techniques for small cell deployments," *IEEE Communications Magazine*. 2011 August;49(8):72–79.

[41]   Ternon E, Agyapong PK, and Dekorsy A. "Impact of varying traffic profile on phantom cell concept energy savings schemes," in: *2015 IEEE 81st Vehicular Technology Conference (VTC Spring)*; 2015. p. 1–6.

[42] Ashraf I, Boccardi F, and Ho L. "Power savings in small cell deployments via sleep mode techniques," in: *2010 IEEE 21st International Symposium on Personal, Indoor and Mobile Radio Communications Workshops (PIMRC Workshops)*, 2010. p. 307–311.

[43] Prasad A and Maeder A. "Backhaul-aware energy efficient heterogeneous networks with dual connectivity," *Telecommunication Systems*. 2015;59(1): 25–41. Available from: http://dx.doi.org/10.1007/s11235-014-9893-4.

[44] Prasad A and Maeder A. "Energy saving enhancement for LTE-advanced heterogeneous networks with dual connectivity," in: *Proceedings of IEEE 80th Vehicular Technology Conference VTC*, Fall; 2014. p. 1–6.

[45] Aydin O, Valentin S, Ren Z, *et al.* METIS project deliverable D4.3: Final Report on Network-Level Solutions; 2015. p. 1–148.

[46] Yang C, Wang J, Wang M, Zou KJ, Yang KW, and Hua M. "Over-the-air signalling in cellular communication systems," *IEEE Wireless Communications*, 2014 August;21(4):120–129.

Chapter 15

Backhaul relaxation through caching

Keivan Bahmani[1], Antonios Argyriou[2],
and Melike Erol-Kantarci[1,3]

The advent of smartphones and portable devices have led to exponential growth in the demand for multimedia contents, especially video over wireless networks. According to Cisco's data traffic forecast [1], the global mobile traffic will increase 8-fold between 2015 and 2020, reaching around 30 exabytes per month. At the same time, the effect of mobile video traffic is expected to be even dominant in this traffic expansion, reaching about 75% of mobile data traffic by 2020. Historically the Radio Access Networks (RANs) have been the bottleneck of the mobile wireless networks and even the orders of magnitude increase in the capacity of wireless networks pledged by the upcoming 5G standard may not be able to fully address the expected growth in the mobile traffic. This drastic increase is expected to disturb the more stable parts of the mobile wireless networks such as mobile backhaul and the core network [2].

Mobile Network Operators (MNOs) traditionally relied on the combination of three main approaches, namely, allocating more spectrum to the system, utilizing spatial reuse and increasing the spectral efficiency to address the traffic expansion problem. The physical layer of the Long-Term Evolution (LTE) standard has near-optimal performance by adapting multiple antennas and Orthogonal Frequency Division Multiplexing (OFDM). This fact means that any considerable increase in the spectral efficiency of the physical layer would require novel approaches like interference alignment [3] and cooperative multipoint [4]. Allocating more spectrum bands to wireless systems is a tedious and time-consuming task. The fact that spectrum is a finite resource makes this approach unsustainable. MNOs have lately shown considerable interest in utilizing the spatial domain by integrating femtocells in their networks and working towards a large-scale heterogeneous network. As we will discuss in Section 15.2, Small Cell Base Stations (SCBSs) traditionally require a high speed backhaul in order to provide a considerable increase in data rate of the users. In this chapter, we introduce the studies that bring the content closer to the user,

[1]Department of Electrical and Computer Engineering, Clarkson University, USA
[2]Department of Electrical and Computer Engineering, University of Thessaly, Greece
[3]School of Electrical Engineering and Computer Science, University of Ottawa, Canada

i.e., caching at Macrocell Base Stations (MBSs), SCBSs or even devices. Those caching strategies allow backhaul relaxation through responding to content request before the request is forwarded to the content providers residing at core networks. Bringing content closer to the users also aids in minimizing the content delivery delay and enhancing the Quality of Service (QoS) of the users.

The rest of the chapter is organized as follows. In Section 15.1, we discuss the idea of Information Centric Networking (ICN) and the motivation behind adopting such schemes in the next generation of wireless systems. Content Centric Networking (CCN) as an example of ICN networks, will be discussed in detail to explain the adaptation of in-network caching and named data objects in wireless networks. Caching strategies and the metrics used to quantify their performance are presented next. We also explore the idea of placing contents closer to the edge of the network and its associated benefits for users and MNOs. In Section 15.2, the current research carried out for improving caching schemes in 5G wireless systems, especially for the case of video content delivery, are presented. The improvements expected in 5G systems are analysed and presented with respect to three main categories, namely, QoS, backhaul relaxation through caching and energy consumption. The relation between backhaul links and content placement in the Heterogeneous Networks (HetNets) is also investigated. Lastly in Section 15.3 we present an elaborate analysis of some of the prosperous works on the energy consumption optimization in caching schemes proposed for 5G systems.

15.1 Background on content caching

Mobile traffic is undergoing a transition from being "connection-centric", like text and calls, to a more content-centric traffic such as video [1]. Despite this trend and the tremendous increase in the total volume of the mobile traffic, recent research results indicate that the number of distinct content items do not increase in a similar rate [5]. This observation makes mobile wireless networks an ideal candidate for adopting a subset of the schemes from the traditional Content Delivery Networks (CDNs) in order to improve the overall efficiency of the network. Unfortunately current wireless networks are neglectful to these changes and traditionally have been optimized for host-centric communication where unique information is being transferred from a particular source to a destination. This design renders such networks inadequate to the task of delivering myriad of multimedia contents requested by today's mobile users.

Addressing the aforementioned traffic explosion challenge and the changes in the characteristics of mobile traffic motivated researchers to come up with a novel networking paradigm to alleviate these problems. ICN is a promising architecture for the next generation networks [6]. Contrary to the host-centric networks, ICN is based on the Named Data Object (NDO) which allows the information/content to be decoupled from the source. ICN enables MNOs to name and store each content parts know as Content Objects (COs) anywhere in the network, while simultaneously granting them the flexibility of delivering COs through different paths and assemble them for a personalized viewing experience which has been tailored for each particular user. Researchers in [7,8] have proposed several ICN architectures, however Content Centric Network (CCN) [7] and its implementation [9] have been the most bolstered architectures. CCNs rely heavily on NDOs and in-network caching. Each CO can be stored and retrieved from the Caching Store (CS) of any device. In CCN networks each router will generate a Pending Interest Table (PIT) and store the information regarding each previously forwarded interest packet so it would be used in order to push the cached or fetched content toward the requesters. Due to the cache dependent nature of CCNs, their performance greatly depends on the caching mechanism adapted in their architecture. This fact makes the study on caching algorithms even more significant. The default caching policy proposed for the CCN architecture is Leave Copy Everywhere (LCE) [10]. We will discuss some of the most crucial caching policies for CCNs in this chapter.

Traditionally practical caching systems adopt a two-phase caching scheme, namely, placement and delivery phases [11]. In the placement phase the COs will be delivered to the nearby caching nodes. This phase mainly aids with reducing the content delivery delay. The second phase called delivery phase, attempts to minimize the over the air wireless traffic by leveraging the temporal variability. These schemes mainly work by placing content on the nodes during off-peak hour to drastically reduce the peak rate and smooth out the peak-time traffic spikes [11]. Neisen *et al.* in [12] addressed the problem of content caching in wireless networks from the information theoretic perspective. The authors defined the performance of a caching network as a capacity region and present the inner and outer bound of performance of a $\mathcal{N}$ node caching wireless network. The other significant outcome presented in [12] is that the problem of cache selection and load balancing should be performed jointly.

In [13], Liu *et al.* summarized the benefits of caching in wireless networks as follows. The first benefit of caching in wireless networks would be the minimization of the time for content delivery. Placing content closer to users can drastically improve the content delivery time and improve the QoS. Reducing the core and backhaul traffic would be the second perk of caching in wireless networks. Caching in the radio access networks can curb down the backhaul's and core network's traffic by delivering the content from cache and eliminating the superfluous requests for COs to travel into the core network. Minimizing the amount of over the air wireless traffic to redress the already strained wireless access networks is an another benefit of employing caching schemes in wireless networks.

Performance improvement promised by ICN architecture motivated research into adapting caching and leveraging computing capability of nodes in the next generation wireless systems. In this chapter, we will discuss pros and cons of some of the most promising caching mechanisms proposed for the upcoming 5G wireless systems. An illustration of caching over small cells is given in Figure 15.1.

The main metric for defining the cacheability of a content in both wired and wireless networks is revisit rate, which has been given in (15.1). The revisit rate is used to quantify the gain one can expect from caching such content. Equation (15.2) proposed in [14] represents the same notion of revisit rate while also considering the cacheable data volume. k_i and s_i, respectively, represent the total number of downloads

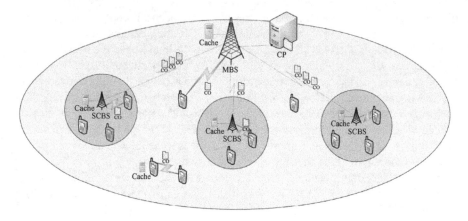

Figure 15.1   System model view of a cache enabled HetNet

and the size of the content i and n denotes the total number of contents also known as catalogue size of the caching scheme:

$$\text{Revisit } \% = \frac{100 * \Sigma_{i=1}^{n} (k_i - 1)}{\Sigma_{i=1}^{n} k_i} \tag{15.1}$$

$$\text{Revisit } DV\% = \frac{100 * \Sigma_{i=1}^{n} (k_i - 1) s_i}{\Sigma_{i=1}^{n} k_i s_i} \tag{15.2}$$

Authors in [15] monitored a live LTE network in a dense urban metropolitan area for 24 h using the Bell Lab LTE Xplorer network monitoring tool. The results from monitoring suggested that a central caching mechanism (Network Proxy) placed at the Serving Gate Way (SGW) can reduce the backhaul traffic by 6.6% and can improve the latency of content delivery by 16% . It is worth mentioning that such network proxy would require 9 TB of storage capacity to be able to achieve the aforementioned improvements. According to the collected information authors suggested that if the MNO is aiming to reduce the content delivery time and user's perceived delay, it is more fitting to cache image, text and applications and caching video content will aid in conserving the bandwidth and backhaul relaxation. As we discussed in Section 15.1 a considerable part of the mobile traffic consists of videos. This fact motivated research to design video transmission schemes to facilitate video content delivery over content delivery networks. The MPEG Media Transport (MMT) [16] developed by Moving Picture Expert Group (MPEG) and the Adaptive mobile video streaming with offloading and sharing in wireless named data networking (AMVS-NDN) [17] are the two of the most bolstered schemes. Video content delivery in the heterogeneous 5G systems is discussed in detail in Section 15.2.2 of this chapter. Cache replacement policies are another essential aspect of the caching mechanism. These policies determine evacuation time of contents from the cache storage. These policies are specially important in the wireless systems where the storage is limited. The default cache replacement policy proposed for the CCNs is the Least Recently Used (LRU) [18]. The LRU is considered the simplest cache replacement policy, however, as indicated in [19], LRU has difficulty coping with the bursty access patterns. There are algorithms designed to work adequately in real systems such as Least Frequently Used (LFU) [20] and First In First Out (FIFO). The complete survey of these algorithms is out of the scope of this chapter. Detailed research can be found in [21,22].

Traditionally LTE networks employ caching at two places in the network, namely, at a central caching mechanism in the Evolved Packet Core (EPC) or at the Radio Access network (RAN). Caching at the EPC using conventional caching methods has been addressed in the [23,24]. RAN caching in LTE will take place in the Evolved NodeBs (eNBs). Authors in [25,26] discussed RAN caching in detail. The ever increasing adoption of SCBSs in the RAN section of current 4G LTE networks and the upcoming 5G standard opens up the door to employ caching at SCBSs [5,26–28] are examples of research carried out on employing caching at SCBSs. The elaborate explanation of these works is presented in Sections 15.2 and 15.3. Another promising area of improvement is emerging from caching and sharing content among mobile users adopting Device-2-Device (D2D) communication and exploiting user social relationship [29–32].

Caching in EPC and RAN can be adopted to achieve varying improvements, Wang *et al.* pointed in [33] that both caching in EPC and RAN can drastically improve the content delivery delay and hence user-perceived latency. Authors in [33] also formulated the total delay of users, given in (15.3). Equation (15.3) emphasizes the important effect of proper content placement and exploiting content popularity on QoS of users, where o_i represents the content, p_i denotes the associated popularity of content i and t_{eNB}, t_{RAN}, t_{EPC}, t_{CP} represent the transmission delays of eNB, RAN, EPC and the content providers, respectively:

$$\text{Minimize} \sum_{o_i \in Locale\_eNB} p_i . t_{eNB}$$

$$+ \sum_{o_i \in Neighbor\_eNB} p_i . (t_{eNB} + t_{RAN})$$

$$+ \sum_{o_i \in EPC} p_i . (t_{eNB} + t_{RAN} + t_{EPC})$$

$$+ \sum_{o_i \in CP} p_i . (t_{eNB} + t_{RAN} + t_{EPC} + t_{CP})$$

subject to : for any node $n_j(eNB, RAN, EPC)$,

$$\sum_{o_i \in cache \ of \ n_j} file\_size(o_i) \leqslant cache\_storage\_size(n_j) \tag{15.3}$$

MNOs not only adopt caching at the EPC and RAN to improve the QoS of the users by bringing content closer to users, but also can benefit from drastic reduction in the inter and intra Internet Service Provider (ISP) traffic and Operational Expense (OPEX). MNOs can minimize their inter-ISP traffic also known as the outbound traffic between the ISP and the Over The Top Content Providers (OTTCP) such as Google or Netflix, by ensuring that the content in all storages (EPC- eNodBs or SCBS) has the highest content diversity possible. This would ensure that most of the requests will be answered using the content already cached in the infrastructure of the MNO. This is especially a daunting task in the case of caching in eNodBs and SCBSs as the storage space is the limiting factor in these caching nodes. On the other hand, MNOs are also interested in reducing the intra-ISP traffic. Intra-ISP traffic is the amount of internal traffic that travels inside the ISP to satisfy the requests of users. This is a crucial aspect for MNOs. The radical increase in the over-the-air wireless traffic already over-strains the backhaul network. Hence, ISPs should try to ensure that the most popular contents are cached as close as possible to the users. This would lead to a significant decrease in the backhaul traffic as most of the requests will be satisfied locally without the need to fetch content from the core network. The decrease in the inter-ISP traffic comes hand in hand with a reduction in cost of bandwidth payment and cost of infrastructure's maintenance which aids in diminishing the OPEX of the MNOs [34]. Results presented in [34] indicate a 35% reduction in the ten-year network OPEX by employing caching at EPC and RAN. Cache hit is the main metric for evaluating the performance of caching schemes. Cache hit for a network of caches such as CCN can be evaluated using (15.4) [35]. An interest message answered by the cached content at the node i will a hit_i and in the same way the unanswered interest message will trigger a $miss_i$. N represents the number of nodes in the network [35]:

$$CacheHit = \frac{\Sigma_{i=1}^{|N|} hits_i}{\left(\Sigma_{i=1}^{|N|} hits_i\right) + \Sigma_{i=1}^{|N|} miss_i} \tag{15.4}$$

Bernardini *et al.* in [35] compared and evaluated the performance of the most promising caching algorithm for the CCN networks under realistic scenarios. LCE [10], LCD [10], ProbCache [36], Cache "Less for more" [37] and MAGIC [38] has been evaluated. It is worth mentioning that all of the evaluated algorithms other than MAGIC have a complexity of $O(m)$ and only MAGIC has the higher complexity of $(O(m \times n))$. The results suggested that MAGIC has the best performance in terms of cache hit, however its higher complexity would limit its performance in a real implementation. The authors also noted that based on their results, LCD would be a more suitable choice as a default caching policy for CCN networks [35]. Authors of [39] proposed a proactive caching scheme to leverage the recent advances in machine learning in order to determine the contents to be cached at different locations in the network. Even though this scheme would require a lot of computation, the outcome could drastically improve the performance of the system in terms of storage and backhaul usage.

15.2 Caching in 5G wireless systems

There has been a substantial amount of research on 5G wireless systems. The upcoming 5G standard will be based on heterogeneous networks [40] and even in today's 4G Long-Term Evolution (LTE) networks, MNOs are increasingly adopting femtocells, picocells and even WiFi hotspots. Employing these schemes gives the MNOs the ability to grant users higher speeds by effectively reusing the spatial dimension of the communication system as well as providing better coverage for users at the edge of the cell. In this section we explore caching in HetNets, a fundamental aspect of 5G systems.

15.2.1 Caching in 5G HetNets

As we discussed in Section 15.1 video content currently constitutes a considerable fraction of the mobile wireless traffic, and it will be even more significant in the future. The work in [41] addressed the video content delivery problem over a heterogeneous cellular network. The authors investigated balancing the performance of the users (average video delivery delay) and the service cost. To do that MNOs have to jointly optimize video encoding, caching and routing policies simultaneously. After the video delivery problem is formulated the authors proved that the problem is NP-hard. For its solution, the authors employed partial relaxation for the SCBS service rate constraints and then used a primal–dual Lagrangian approximation algorithm for solving the problem in an iterative manner. The relaxed primal problem obtained after the aforementioned steps was later decomposed into two sub-problems, namely, *caching* and *bandwidth allocation* sub-problems. The authors argued that the first subproblem can be rewritten as a series of one-dimensional knapsack problems and hence the sub-problem can be solved in pseudo-polynomial time. Authors also showed that the second subproblem is convex and can be solved using traditional standard convex optimization techniques and hence their proposed algorithm for the complete problem is easily proven to converge asymptotically to the optimal solution.

Another significant contribution of the research in [41] is that the work also address the issue of delivering the same video content in different qualities and in different encoded forms. The authors investigated both cases of encoding the content in different qualities with the Scalable Video Coding (SVC) extension of the H.264/MPEG4-AVC standard [42] where the quality of the decoded video can be dynamically determined by the number of *layers* delivered to the user, and also using different quality versions of the same content as existing systems do. The results presented in [41] indicated that the decision of adopting a caching approach based on layers or versions is related to the level of homogeneity of user's requests. In cases where requests are homogenous, the version caching would be the desirable solution as this approach eliminates the overhead associated with the SVC. In scenarios where videos with different quality are requested, adopting the caching of layers is preferable as this approach will lead to conserving the storage space of the caching nodes.

Authors of [41] also investigated the relation between the user's delay and servicing cost of the MNO. The simulation results indicated that 10% decrease in the average delay of users (achieved by tuning the objective's balancing parameter) comes hand in hand with 10%–30% increase in the servicing cost of the MNO.

Authors of [43] presented a generalized information centric model of the two tier HetNet cache enabled network. They introduced the notion of information centric local coverage probability in order to represent the probability that users would retrieve the content from the neighbouring MBS or SCBSs under the LRU content eviction policy. Small cells play a crucial role in the heterogeneous 5G systems, however, providing high speed backhaul for these SCBSs is the bottleneck that hinders their progressive deployment [44]. Authors of [5] proposed adopting SCBSs with large storage capacity and a distributed caching infrastructure for video traffic to eliminate the requirement for a high speed backhaul. Authors envisioned a heterogeneous system with low complexity base stations called "helpers". These helpers have large storage and weak backhaul. One can see that the ever decreasing trend in the cost of storage is making such scheme even more practical. The aforementioned scheme named femtocaching, promises order of magnitude increase in the video delivery throughput by caching video files in the helpers, and delivering them from MBS or the closest helper with a very high efficiency. As we discussed in Section 15.1 with the help of (15.3), bringing content closer to the edge of the network can be very beneficial to both users and MNOs. Authors have shown that as the majority of a mobile traffic is due to very few popular (viral) videos, an efficient in-network caching scheme with appropriate amount of storage in the helper can eliminate the high speed backhaul (fibre) requirement for SCBSs. By adopting such scheme, slower backhaul links such as DSL lines can be employed to refresh the cached content in an off-peak cycles. The core idea stems from the fact that if there would be considerable content reuse, the cached content at the helper can mask the effect of the weak backhaul.

Golrezaei *et al.* also addressed the problem of cache placement in such heterogeneous system with helpers for cases of coded and uncoded video content delivery. The coded file assignment problem has been addressed using rateless MDS codes such as fountain codes [45]. These codes produce an arbitrary long coded sequence from a file with size B while the original file can be decoded using any B bits of the coded sequence. The B bits can be acquired through either a BS or a helper hence the video can be decoded using the combination of the data received from different sources. The simulation results indicated that the performance of the scheme for coded and uncoded cases were very close and in both cases a drastic increase in the video delivery throughput is expected. The problem can be described with two main scenarios. If the number of helpers are small each user generally will be in the range of only one helper. In this case the problem is trivial as each helper should be instructed to cache the files with highest popularity in the decreasing order, however, if the helper deployment will be dense enough similar to practical deployments in the crowded urban areas, each user will be in the range of more than one helper and at this point users will see a distributed cache containing the union of the caches of all the helpers covering the user. It is worth mentioning that in practical wireless system due to the interference, users will simultaneously be covered by only few helpers [26].

Shanmugam *et al.* showed that the uncoded case is NP-complete. Authors formulated the file placement problem with the objective of minimizing the average delay of all users as a maximization of a submodular function subject to matroid constraints. They proposed a greedy algorithm that populates the cache of each helper by including the file which maximizes the marginal benefit. This greedy algorithm provides $1 - (1 - \frac{1}{d})^d$ approximation to the problem where d represents the number of helpers. It is noteworthy that this greedy algorithm provides 50% of the optimum solution in worst case. Shanmugam *et al.* also addressed the coded file placement problem by reducing the problem to a Linear Program (LP). The aforementioned LP has the worst-case complexity of $O((U + H)^{3.5} F^{3.5})$ where U, H and F denote the number of users, helpers and files in the catalogue. Authors of [26] also considered mobility of the users and simulation results carried out in their work indicated that the effect of the mobility of users on the average download rate of users is negligible. Golrezaei *et al.* carried out some numerical simulations in [5]. They considered a wireless system with an LTE macro-base station (MBS) and helpers operating in the short range WiFi-like communication with users. Each helper is equipped with a 60 GB of hard drive. The simulation results show that on some important aspects of the file placement problem. Authors showed that for limited number of helpers which would likely be the case for practical systems, the simple file placement scheme which populates the helper caches with the most popular files is almost as effective as the greedy algorithm. Another crucial outcome of the result is that even with a small number of helpers, which is a more pragmatic scenario, we can expect almost 500% increase in the number of satisfied users. It is worth mentioning that in both [5,26], the effect of interference between the helpers with each other and the MBS has been omitted.

The work presented in [46] also addressed the content placement problem in a generic wireless network with multiple single antenna base stations while also considering the effect of the backhaul link between the base stations and a central controller containing all the COs. This topology can closely resemble a cache enabled HetNet. Authors defined the content placement problem as a minimization of the average content delivery delay. The authors formulated the average download delay as a Mixed Integer Nonlinear Programming (MINLP), however, they argued that due to the high complexity, such scheme would not be suitable for practical systems. They proposed a novel scheme by introducing relaxations to the problem and finding the suboptimal solution using the Successive Convex Approximation (SCA) algorithm. The numerical results presented in the work indicated that, there is a direct relation between backhaul delivery delay and the performance of the content placement schemes from a Quality of Experience (QoE) perspective. Authors compared their proposed backhaul aware scheme with the LCD [47] and MPC [48] algorithms where their backhaul-aware scheme outperforms both.

15.2.2 *QoE-aware caching & 5G HetNet configuration*

Even though multi-tier HetNets are expected to be an important architectural component of 5G systems, they introduce a new problem that of intra-cell interference.

A macrocell BS (MBS) and the SCBSs must coordinate to minimize it since they are deployed in very close proximity (see Figure 15.1). However, when the small cells in the network coordinate their transmissions with the macrocell to minimize interference, this affects the source location of the user data (MBS or SCBS) and consequently data caching decisions. The authors in [49] investigated a first version of this problem by studying the impact of decisions like interference control on the QoE of multiple users that receive video streams from either the MBS or an SCBS. The problem is far more challenging than it might seem initially when low-level details of the wireless PHY are considered.

The reason is a technical detail at the PHY that LTE-based HetNets have introduced for a new intuitive way for handling interference. The scheme, that will also be adopted in 5G systems is the time domain resource partitioning (TDRP) mechanism [50], where the MBS shuts off its transmissions for a time fraction equal to η during which the small cells can achieve a higher data rate. During the fraction $1 - \eta$, there is *intra-cell* interference since the MBS transmits simultaneously with the small cells. This technique was standardized through the concept of almost blank subframes (ABS) and regular subframes (RS) in 3GPP LTE-A under the more general name of enhanced inter-cell interference coordination (eICIC) [50]. One important detail is that the LTE-A standard currently allows the dynamic adaptation of η but it does not specify how it should be configured. Given the increasing number of video streaming users in cellular networks, and the necessity of TDRP, it is of outmost importance to configure the network optimally. The specific questions that should be answered in the case of 5G HetNets that support caching are the following: What is the optimal η when we have video traffic? What is the best video quality that each user should receive? From which BS should a user receive the video and where data should be cached?

15.2.2.1   Problem formulation and solution

Let us consider more details for the HetNet model we introduced in this section. In this case we consider that a set $\mathcal{N}$ of the HetNet users receive video. We desire to configure the HetNet that consists of the set of $\mathcal{J}$ of BSs in such a way that the users receive the highest video QoE, i.e., low number of video freezes and the best video quality/resolution (or representation in video terminology). For each user i associated to a HetNet BS j we must select the video representation with the highest quality, and the rate allocated to it. For the complete HetNet the globally optimal TDRP must be calculated. First we define the optimization variables. Let $x_{ir}^{ABS}, x_{ir}^{RS} \in \{0, 1\}$ indicate whether user $i \in \mathcal{N}_j$ is served with video representation r in an ABS and RS, respectively. Let also $z_{ir}^{ABS} \in [0, 1]$ denote the fraction of the ABS resources that the SCBS allocates to $i \in \mathcal{N}_j$ for streaming the video representation r. Similarly for the RS, we define $z_{ir}^{RS} \in [0, 1]$. Hence, the decisions of each BS j are: (a) the *video quality selection (VQS)* vector for all the associated users, i.e., $x_j = (x_{ir}^{ABS} \geq 0 : i \in \mathcal{N}_j, r \in \mathcal{R}_i)$, and (b) the *rate allocation (RA)* vector for all users, i.e., $z_j^{ABS} = (z_{ir}^{ABS} \geq 0 : i \in \mathcal{N}_j, r \in \mathcal{R}_i)$. Similarly the VQS and RA vectors for the regular slots. Also the global resource partitioning decision η.

The objective for the HetNet operator is to maximize the average aggregate delivered video quality captured by:

$$\sum_{j \in \mathscr{J} \setminus \{0\}} \sum_{i \in \mathscr{N}_j} \sum_{r \in \mathscr{R}_i} (x_{ir}^{ABS} + x_{ir}^{RS}) Q_{ir0} + \sum_{i \in \mathscr{N}_0} \sum_{r \in \mathscr{R}_i} x_{ir}^{RS} Q_{ir0} \tag{15.5}$$

In the above Q_{ir0} is the average quality of representation r for video of user i. Thus, the objective expresses the video quality delivered to the complete HetNet. In the second term we have the quality for the users associated to the MBS since they cannot transmit during an ABS.

For the first set of constraints we have to recall that the fraction of the blank ABS resources available for the SCBSs (there is resource re-use across the SCBSs) is η. This leads to:

$$\sum_{i \in \mathscr{N}_j} \sum_{r \in \mathscr{R}_i} z_{ir}^{ABS} \leq \eta, \quad \forall j \in \mathscr{J} \setminus \{0\} \tag{15.6}$$

In the above we excluded again the MBS since it cannot transmit during an ABS. During the RS all the BSs transmit:

$$\sum_{i \in \mathscr{N}_j} \sum_{r \in \mathscr{R}_i} z_{ir}^{RS} \leq 1 - \eta, \quad \forall j \in \mathscr{J} \tag{15.7}$$

When a particular representation r is selected, the average rate R_{ir} in bits/s that must be sustained by a user i is less than the rate that can be achieved during both the ABS and RS. Also the resources allocated during ABS and RS will determine the average rate. The above can be formally written as:

$$x_{ir}^{ABS} R_{ir} \leq (z_{ir}^{ABS} C_i^{ABS} + z_{ir}^{RS} C_i^{RS}), \quad \forall r \in \mathscr{R}_i, \, i \in \mathscr{N}_j, \, j \in \mathscr{J} \tag{15.8}$$

C_i^{ABS} and C_i^{RS} are the capacities of the user i's links, from its respective BS, according to Shannon formula and during the ABS and RS, respectively. This constraint ensures that the average number of rebuffering time over the complete video playback is zero since the streaming rate R_{ir} is lower than the average throughput on the RHS of this expression.

We also have that resources cannot be allocated to a video representation r if it is not actually selected:

$$z_{ir}^{ABS} \leq x_{ir}^{ABS}, \quad \forall r \in \mathscr{R}_i, \, i \in \mathscr{N}_j, \, j \in \mathscr{J} \tag{15.9}$$

$$z_{ir}^{RS} \leq x_{ir}^{RS}, \quad \forall r \in \mathscr{R}_i, \, \forall i \in \mathscr{N}_j, \, j \in \mathscr{J} \tag{15.10}$$

We also need the integer constraints according to which only one video representation r can be used for each user. Thus:

$$\sum_{r \in \mathscr{R}_i} x_{ir}^{ABS} \leq 1, \quad \forall i \in \mathscr{N}_j, \, j \in \mathscr{J} \tag{15.11}$$

$$\sum_{r \in \mathscr{R}_i} x_{ir}^{RS} \leq 1, \quad \forall i \in \mathscr{N}_j, \, j \in \mathscr{J} \tag{15.12}$$

During the regular slots the SCBSs can also transmit together with the MBS, albeit with lower spectral efficiency. In this case the rate will be lower.

Solving this problem can be accomplished in different ways. A primal–dual decomposition method has been proposed in [49] and allows for localized video quality decisions at each BS and a centralized network configuration for configuring η.

15.2.3 Caching in D2D 5G systems

As we discussed in Section 15.1, D2D communication is another paradigm which is expected to play a crucial rule in the 5G wireless systems [51]. Authors of [5] also proposed to adapt D2D communication and using users' devices as helpers. Golrezaei *et al.* noted the importance of power control in this scheme and proposed a cooperation between the MBS and users to deliver content based on a clustering approach to minimize the interference. In [32] Lan *et al.* proposed to use D2D communication coupled with proactive caching in a heterogeneous mobile cellular network to maximize the amount of offloaded data. In their system model, authors assume users have different caching capacity and unlike systems with a specific helper nodes, each user act both as a data requester and server. The problem has been a maximization of the expected total size of offloaded data subject to the cache capacity of users. Authors corroborated that the problem is NP-Hard and the solution would require an exponential computational complexity $O(2^{FN})$, where F and N are the numbers of files and nodes, respectively. As in a practical system M and N tends to be larger solving the problem would be trivial. Authors proposed two novel schemes namely, Infrastructure Assisted Data Offloading Algorithm (IADOA) and Fully Distributed Data Offloading Algorithm (FDDOA), to tackle the problem by leveraging the D2D communication between the users and solve the problem in a distributed manner. In IADOA, the MBS as a central node, collects all the necessary information from nodes and provides them with the information required to make a local decision, while, FDDOA works in a fully autonomous and distributed manner adopting the information estimated in each node to make a decision. Authors compared the performance of aforementioned algorithms with traditional LFU [20] and Randomize Caching Strategy (RCS) [52]. Both IADOA and FDDOA outperform the LFU and RCS, while IADOA has a slightly better performance due to the greater accuracy of the acquired information. Authors of [13] proposed to adapt a content-centric approach to wireless content delivery, utilizing the computation and caching capability of wireless networks to leverage both content and network diversities. Authors coined the term content rate (R_c) represented in (15.13) and formulate the content delivery task over wireless network into a content rate maximization problem. Equation (15.13) denotes the content rate of a wireless system, where B represents the bandwidth of wireless channel and T states the latest time granted for content to be delivered. $y_k (t_k)$ represents the request from the kth user at time t_k:

$$R_c = \frac{\sum_{k=1}^{K} |y_k (t_k)|}{B \times T} \tag{15.13}$$

As can be seen from (15.13) the content rate is defined as the total amount of content successfully delivered to users and as content rate is inversely proportional to the bandwidth B, the maximum content rate is achieved when all of the content will be delivered with the minimum amount of B. The authors then demonstrated the upper and lower bounds of the content rate. Authors pointed out the upper bound of content rate corresponds to the case of infinite caching storage where all the content are cached at the user ends and all the COs has been delivered using broadcast. These aforementioned results are consistent with the results presented in [53,54] which indicated that the optimality of the so-called converged networks for wireless content delivery. On the other hand, the lower bound of content rate is associated with the case of content delivery using only unicast and zero caching. It is worth mentioning that in this case the content rate will decrease to the bit rate of the wireless system. Authors also proposed to use the Priority Encoding Transmission (PET) scheme proposed in [55] in order to harness the content diversity by making the priority associated with each content, inversely proportional to its popularity. This approach will increase the content rate while it will simultaneously minimize the average caching size requirements [13].

15.3　Energy efficiency and caching in 5G systems

Achieving better energy efficiency is also an important objective in the 5G systems. In [27], Erol-Kantarci considered the content placement problem from an energy consumption perspective. Energy efficiency and the effect of caching in relays of the cellular networks has been addressed in [56] however, in [27] the author introduced the notion of solving the cache placement problem with the objective of prolonging the life time of User Equipments (UEs).

Extending life time of UEs is a very sought after feature in today's heterogeneous wireless networks [57]. The main contribution of [27] is to propose a novel scheme called Cache-at-Relay (CAR). CAR incorporates three Integer Linear Programming (ILP) models, namely, Select Relay (SER), Place Content (PCONT) and Place Relay (PREL). In this section we first discuss the system model used in CAR and then take an elaborate look at the inner workings of each these algorithms. The 5G HetNet has been modelled as a connected network graph $G = (V, E)$, T and A, respectively represent the sets of UEs and relays. N_{UE} and N_R denote the number of UEs and relays. In [27], for simplicity each SCBS (relay) has been equipped with an identical storage capacity, represented by St_R. Users denoted by n have been randomly placed in the area. The content popularity has been modelled according to the findings in [58,59] to reflect the effect of social networks on the traffic characteristics. The system has been considered to have a limited catalogue, where the number of files are limited to C. The author adopts a content centric approach for content delivery. In this case each video content is partitioned in COs with size S_K. As we discussed in Section 15.1, each user can acquire COs from different sources (MBS or SCBS) in order to construct the original content in its entirety. Other parameters involved in the work has been listed

Table 15.1   Frequently used notations

$\alpha_{xy,i}^k$	Binary variable that is 1 if UE i is downloading content k from the relay at (x,y)
$\beta_{i,BS}^k$	Binary variable that is 1 if UE i is downloading content k from the BS
$\Delta_{xy,i}^k$	Binary variable that is 1 if the k_{th} content is cached at the relay (x,y) and UE i requests it from that particular relay
$\Delta_{i,BS}^k$	Binary variable that is 1 if the k_{th} content is cached at the BS and UE i requests it from the BS
R_{xy}	Binary variable that is 1 if there is a relay at (x,y)
η_{xy}^k	Binary variable that is 1 if there is content k at the relay at (x,y)
$\Psi_{xy}^{i,k}$	Binary variable 1 if there is a relay at (x,y) and UE i is receiving k_{th} content from (x,y)
D_R	Range of a relay
D_{BS}	Range of the BS
H_i^k	Binary value indicating whether UE i demands content k or not
$d_{i,BS}$	Distance between UE i and the BS
$d_{xy,i}$	Distance between UE i and relay at (x,y)
$P_{xy,i,k}^{tr,relay}$	Uplink power of UE i to download k_{th} content from the relay at (x,y)
$P_{i,k}^{tr,BS}$	Uplink power of UE i to download k_{th} content from the BS
$P_{xy}^{cache,k}$	Caching power for the relay at (x,y) for the k_{th} content
$P_{BS}^{cache,k}$	Caching power for the BS for the k_{th} content
p_R^{eff}	Power efficiency values of the storage employed at the relay
p_{BS}^{eff}	Power efficiency values of the storage employed at the BS
St_R	Storage size for a relay
ρ	Weight factor

in Table 15.1. Authors assume that the MBS has enough storage capacity and all the content in the catalogue has been previously cached in the MBS.

15.3.1   SER

Select Relay (SER) [27] is the most rudimentary model proposed in the CAR scheme. The main objective of SER is to minimize the uplink power of the UEs. SER assumes that the SCBSs have been previously deployed and the information about the distribution of cached contents among them and their locations are available. Energy budget of users has been defined by B_n which represents the battery capacity of user n. B_n will be consumed by uplink transmission, downlink refection and so on and so forth. As we noted in Section 15.1 in CCN networks each node will transmit an interest packet to demand each COs. Based on the given assumption above the problem of minimizing the uplink power of UEs can be formulated as choosing the closest SCBS or MBS based on the uplink power budget of the UE. Equation (15.14) represents the objective function of the SER model. This objective function has been solved subject to the constraints given in (15.15)–(15.20):

$$\min \quad \sum_i \sum_k \sum_x \sum_y P_{xy,i,k}^{tr,relay} R_{xy} \eta_{xy}^k \alpha_{xy,i}^k + \sum_i \sum_k \sum_x \sum_y P_{i,k}^{tr,BS} \beta_{i,BS}^k \quad (15.14)$$

$$\alpha_{xy,i}^k d_{xy,i} \leq D_R, \qquad \forall x, y, i, k \tag{15.15}$$

$$\alpha_{xy,i}^k = 0, \qquad \forall d_{xy,i} > D_R \tag{15.16}$$

$$\alpha_{xy,i}^k \leq R_{xy} \eta_{xy}^k H_i^k, \qquad \forall x, y, i, k \tag{15.17}$$

$$\alpha_{xy,i}^k d_{xy,i} \leq d_{i,BS}, \qquad \forall x, y, i, k \tag{15.18}$$

$$\beta_{i,BS}^k + \sum_x \sum_y \alpha_{xy,i}^k = H_i^k, \qquad \forall x, y, i, k \tag{15.19}$$

$$\beta_{i,BS}^k d_{i,BS} < d_{xy,i}, \qquad \forall x, y, i, k | d_{xy,i} \leq D_R \wedge \eta_{xy}^k = 1 \tag{15.20}$$

Equations (15.15) and (15.16) guarantee that each UE only download contents only from the SCBSs which cover its location. Equation (15.17) enforces the existence of an SCBS at location (x,y) and the availability of the demanded content in the cache of SCBS. Equation (15.18) deals with the fact that if the MBS is closer to the user than any of the SCBS, as the MBS has all the catalogue cached the user must demand the COs from the MBS rather than any of the SCBSs. Equation (15.19) represents the constraint for limiting the maximum number of server nodes. Each UE can connect to either the MBS or one of the SCBS. Equation (15.20) denotes the condition on downloading the content from the MBS. The UE download COs from the MBS if the MBS is closer to the UE than all other relays that have the demanded content in cache.

15.3.2 PCONT

The Place Content model [27] is designed with the objective of placing content in the SCBs in order to jointly minimize the uplink power of users and caching power of MBS and all the SCBSs. The objective function of the PCONT model is described in (15.21) and it has been solved subject to constraints given in (15.22)–(15.28). It is worth mentioning that due to adapting different technologies in the MBS and SCBSs there are different power efficiencies for the storages in the MBS and SCBSs represented by p_R^{eff} and p_{BS}^{eff}, respectively:

$$\min \quad \sum_i \sum_k \sum_x \sum_y P_{xy,i,k}^{tr,relay} R_{xy} \Delta_{xy,i}^k + \sum_i \sum_k \sum_x \sum_y P_{i,k}^{tr,BS} \Delta_{i,BS}^k$$

$$+ \sum_k \sum_x \sum_y P_{xy}^{cache,k} + \sum_k P_{BS}^{cache,k} \tag{15.21}$$

$$S_k \sum_k \eta_{xy}^k \le St_R, \qquad \forall x, y \tag{15.22}$$

$$\beta_{i,BS}^k \le \eta_{BS}^k H_i^k, \qquad \forall i, k \tag{15.23}$$

$$\alpha_{xy,i}^k \le \eta_{xy}^k H_i^k R_{xy}, \qquad \forall x, y, i, k \tag{15.24}$$

$$\eta_{xy}^k \le R_{xy}, \qquad \forall x, y, k \tag{15.25}$$

$$\eta_{xy}^k \le \sum_i H_i^k R_{xy}, \qquad \forall x, y, i, k \tag{15.26}$$

$$\sum_x \sum_y \eta_{xy}^k + \eta_{BS}^k \ge 1, \qquad \forall x, y, k \tag{15.27}$$

$$\sum_x \sum_y \eta_{xy}^k sgn(D_R + \epsilon - d_{xy,i}) + \eta_{BS}^k \ge H_i^k, \qquad \forall x, y, k \tag{15.28}$$

Equation (15.22) enforces the restriction on the caching storage of the relays. Equations (15.23) and (15.24) represent the condition on the existence and availability of the content in the storage of SCBSs for downloading COs. Equation (15.25) guarantees that a CO can be stored in the SCBS at (x,y) if there exist and SCBS at the (x,y). Equations (15.26) and (15.27), respectively, impose conditions on COs as follows. If any COs is cached at an SCBS at least one UE should have been downloaded it and each CO should be at least stored in one of the SCBS in the system. Equation (15.28) imposes a constraint that a demanded CO should be cached either at the SCBSs available to the demanding UE or at the BS. The rest of the model linearizes $\Delta_{xy,i}^k$ and $\Delta_{i,BS}^k$ and can be found in [27].

15.3.3   PREL

Place Relay model [27] jointly aims to minimize the capital expenditure of the MNO by reducing the number of required relays and the uplink power of UE. One can intuitively understand that placing more SCBSs with the cached content closer to user will drastically reduce UE's uplink power, however, as we discussed before placing SCBSs is a costly process [34]. In this case, the author relaxed the constraint on the storage capacity of SCBSs and assumed that each SCBS have the ability of caching all the COs. Author adopted a normalized weight factor, ρ, to indicate the trade of between cost of deploying SCBSs and uplink power consumption of UEs. Equation (15.29) represents the objective function of the PREL model. This objective function has been solved subject to (15.31)–(15.36):

$$\min \quad \rho \sum_x \sum_y R_{xy} P^{coff} + (1 - \rho) \sum_i \sum_k \sum_x \sum_y P_{xy,i,k}^{tr,relay} \Psi_{xy}^{i,k} \tag{15.29}$$

$$\Psi_{xy}^{i,k} = R_{xy} \alpha_{xy,i}^k \tag{15.30}$$

The objective function of PREL is solved subject to the following constraint set:

$$\alpha_{xy,i}^k \leq R_{xy}H_i^k, \qquad \forall x,y,i,k \tag{15.31}$$

$$\sum_x \sum_y \alpha_{xy,i}^k = H_i^k, \qquad \forall i,k \tag{15.32}$$

$$\sum_i \sum_k \alpha_{xy,i}^k \geq R_{xy}, \qquad \forall x,y \tag{15.33}$$

$$\Psi_{xy}^{ik} \leq R_{xy}, \qquad \forall x,y,i,k \tag{15.34}$$

$$\Psi_{xy}^{ik} \leq \alpha_{xy,i}^k, \qquad \forall x,y,i,k \tag{15.35}$$

$$\Psi_{xy}^{ik} - R_{xy} - \alpha_{xy,i}^k \geq -1, \qquad \forall x,y,i,k \tag{15.36}$$

Equation (15.31) enforces the existence of SCBS if any UE would request to download COs from an SCBS located at (x,y). Equation (15.32) constrain UEs to download content from only one SCBS. Equation (15.33) ensures that at least one UE download a content from each SCBS while other constraints have been added for the purpose of linearization of binary parameters.

Figures 15.2 and 15.3 illustrate the total uplink energy consumption of the UE and the caching energy at the SCBSs (relays) in the SER, PCONT and PREL models,

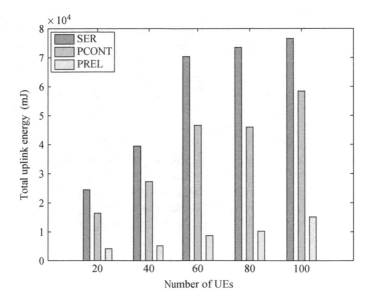

Figure 15.2   Total uplink energy for SER, PCONT and PREL under varying number of UEs [27]

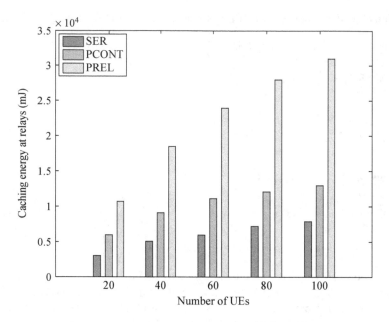

Figure 15.3   Total caching energy for SER, PCONT and PREL under varying number of UEs [27]

respectively. It can be clearly observed that the PREL algorithm has the lowest total uplink energy consumption between the models. This was expected as in the PREL model all the chunks are cached in each SCBSs and hence UEs will be served locally instead of demanding contents from the MBS. However results presented in Figure 15.3 indicates the PREL has the highest caching energy consumed at the SCBSs while SER has the lowest. It is worth mentioning that even though balancing the energy consumed in the UE and SCBSs is an important factor, conserving the energy consumed in the UEs is more desired than that of the SCBSs. Author compared the runtime of the three algorithms under the same condition and noted that the runtime of SER is considerably lower than that of the PCONT and PREL. The author also investigates the cache utilization of SER and PCONT under different number of SCBSs (relays) and catalogue size of ten contents. Authors noted that the downward trend in the cache utilization of the PCONT indicates that the content is being placed to better extend as the number of SCBSs increases, however, increasing number of SCBSs wont assist the SER due to the fact that the model is placing content uniformly among SCBSs.

15.4   Summary

In this chapter the idea of content centric mobile wireless networks introduced as a solution to the traffic expansion problem faced by MNOs. We denoted the essential role that in-network caching plays in the transition from our current host centric to the content centric networking paradigm and we presented the metrics required to define the performance of such caching mechanisms. A comprehensive study on the content placement problem in the heterogeneous cellular networks has been presented while some of the most promising methods has been described in detail. The content placement problem is investigated from different objectives. QoS, backhaul relaxation through caching and energy consumption were the main aspects that we aimed to address in this chapter. These principles has been mainly investigated for the case of video content delivery due to its prominent role in the future mobile wireless traffic.

References

[1] "Cisco Visual Networking Index: Global Mobile Data Traffic Forecast Update, 2015–2020 White Paper" – Cisco.

[2] X. Wang, A. V. Vasilakos, M. Chen, Y. Liu, and T. T. Kwon, "A survey of green mobile networks: opportunities and challenges," *Mobile Network Applications*, vol. 17, pp. 4–20, Jun. 2011.

[3] H. Ning, C. Ling, and K. K. Leung, "Feasibility condition for interference alignment with diversity," *IEEE Transactions on Information Theory*, vol. 57, pp. 2902–2912, May 2011.

[4] A. Liu and V. K. N. Lau, "Exploiting base station caching in MIMO cellular networks: opportunistic cooperation for video streaming," *IEEE Transactions on Signal Processing*, vol. 63, pp. 57–69, Jan. 2015.

[5] N. Golrezaei, A. F. Molisch, A. G. Dimakis, and G. Caire, "Femtocaching and device-to-device collaboration: a new architecture for wireless video distribution," *IEEE Communications Magazine*, vol. 51, pp. 142–149, Apr. 2013.

[6] G. Xylomenos, C. N. Ververidis, V. A. Siris, *et al.* "A survey of information-centric networking research," *IEEE Communications Surveys Tutorials*, vol. 16, no. 2, pp. 1024–1049, 2014.

[7] V. Jacobson, D. K. Smetters, J. D. Thornton, M. F. Plass, N. H. Briggs, and R. L. Braynard, "Networking named content," in *Proceedings of the 5th international conference on Emerging networking experiments and technologies*, pp. 1–12, ACM, 2009.

[8] SAIL NetInf [online]. Available from http://www.netinf.org/.

[9] CCNx, PARC's implementation of content-centric networking [online]. Available from http://blogs.parc.com/ccnx/.

[10] G. Zhang, Y. Li, and T. Lin, "Caching in information centric networking: a survey," *Computer Networks*, vol. 57, pp. 3128–3141, Nov. 2013.

[11] M. A. Maddah-Ali and U. Niesen, "Fundamental limits of caching," in *2013 IEEE International Symposium on Information Theory Proceedings (ISIT)*, pp. 1077–1081, Jul. 2013.

[12] U. Niesen, D. Shah, and G. W. Wornell, "Caching in wireless networks," *IEEE Transactions on Information Theory*, vol. 58, pp. 6524–6540, Oct. 2012.

[13] H. Liu, Z. Chen, X. Tian, X. Wang, and M. Tao, "On content-centric wireless delivery networks," *IEEE Wireless Communications*, vol. 21, pp. 118–125, Dec. 2014.

[14] B. Ager, F. Schneider, J. Kim, and A. Feldmann, "Revisiting cacheability in times of user generated content," in *INFOCOM IEEE Conference on Computer Communications Workshops , 2010*, pp. 1–6, Mar. 2010.

[15] B. A. Ramanan, L. M. Drabeck, M. Haner, N. Nithi, T. E. Klein, and C. Sawkar, "Cacheability analysis of HTTP traffic in an operational LTE network," in *Wireless Telecommunications Symposium (WTS), 2013*, pp. 1–8, Apr. 2013.

[16] Y. Lim, K. Park, J. Y. Lee, S. Aoki, and G. Fernando, "MMT: an emerging MPEG standard for multimedia delivery over the internet," *IEEE MultiMedia*, vol. 20, no. 1, pp. 80–85, 2013.

[17] B. Han, X. Wang, N. Choi, T. Kwon, and Y. Choi, "AMVS-NDN: adaptive mobile video streaming and sharing in wireless named data networking," in *2013 IEEE Conference on Computer Communications Workshops (INFOCOM WKSHPS)*, Turin, Italy, pp. 375–380, IEEE, 2013.

[18] E. J. Rosensweig, D. S. Menasche, and J. Kurose, "On the steady-state of cache networks," in *2013 Proceedings of the IEEE INFOCOM*, pp. 863–871, Apr. 2013.

[19] M. N. Nelson, B. B. Welch, and J. K. Ousterhout, "Caching in the Sprite Network File System," *ACM Transactions on Computing Systems*, vol. 6, pp. 134–154, Feb. 1988.

[20] J. T. Robinson and M. V. Devarakonda, "Data cache management using frequency-based replacement," in *Proceedings of the 1990 ACM SIGMET-RICS Conference on Measurement and Modelling of Computer Systems*, SIGMETRICS '90, Boulder, CO, pp. 134–142, ACM, 1990.

[21] H. Chen and Y. Xiao, "On-bound selection cache replacement policy for wireless data access," *IEEE Transactions on Computers*, vol. 56, pp. 1597–1611, Dec. 2007.

[22] J. Xu, Q. Hu, W.-C. Lee, and D. L. Lee, "Performance evaluation of an optimal cache replacement policy for wireless data dissemination," *IEEE Transactions on Knowledge and Data Engineering*, vol. 16, pp. 125–139, Jan. 2004.

[23] J. Erman, A. Gerber, M. Hajiaghayi, D. Pei, S. Sen, and O. Spatscheck, "To cache or not to cache: the 3g case," *IEEE Internet Computing*, vol. 15, no. 2, pp. 27–34, 2011.

[24] S. Woo, E. Jeong, S. Park, J. Lee, S. Ihm, and K. Park, "Comparison of caching strategies in modern cellular backhaul networks," in *Proceedings of the 11th Annual International Conference on Mobile Systems, Applications, and Services*, Taipei, Taiwan, pp. 319–332, ACM, 2013.

[25] H. Ahlehagh and S. Dey, "Video caching in radio access network: impact on delay and capacity," in *2012 IEEE Wireless Communications and Networking Conference (WCNC)*, pp. 2276–2281, Apr. 2012.

[26] K. Shanmugam, N. Golrezaei, A. G. Dimakis, A. F. Molisch, and G. Caire, "FemtoCaching: wireless content delivery through distributed caching

helpers," *IEEE Transactions on Information Theory*, vol. 59, pp. 8402–8413, Dec. 2013.

[27] M. Erol-Kantarci, "Cache-at-relay: energy-efficient content placement for next-generation wireless relays," *International Journal of Network Management*, vol. 25, pp. 454–470, Nov. 2015.

[28] A. A. AlMomani, A. Argyriou, and M. Erol-Kantarci, "A heuristic approach for overlay content-caching network design in 5G wireless networks," in *2016 IEEE Symposium on Computers and Communication (ISCC)*, Messina, Italy, pp. 2276–2281, Apr. 2012.

[29] X. Wang, M. Chen, Z. Han, D. O. Wu, and T. T. Kwon, "TOSS: traffic offloading by social network service-based opportunistic sharing in mobile social networks," in *IEEE INFOCOM 2014 – IEEE Conference on Computer Communications*, pp. 2346–2354, Apr. 2014.

[30] K. Doppler, M. Rinne, C. Wijting, C. B. Ribeiro, and K. Hugl, "Device-to-device communication as an underlay to LTE-advanced networks," *IEEE Communications Magazine*, vol. 47, pp. 42–49, Dec. 2009.

[31] M. Ji, G. Caire, and A. F. Molisch, "Wireless device-to-device caching networks: basic principles and system performance," *IEEE Journal on Selected Areas in Communications*, vol. 34, pp. 176–189, Jan. 2016.

[32] R. Lan, W. Wang, A. Huang, and H. Shan, "Device-to-device offloading with proactive caching in mobile cellular networks," in *2015 IEEE Global Communications Conference (GLOBECOM)*, pp. 1–6, Dec. 2015.

[33] X. Wang, M. Chen, T. Taleb, A. Ksentini, and V. C. Leung, "Cache in the air: exploiting content caching and delivery techniques for 5G systems," *IEEE Communications Magazine*, vol. 52, no. 2, pp. 131–139, 2014.

[34] H. Sarkissian, "The business case for caching in 4G LTE networks," *Wireless 20*, vol. 20, 2012.

[35] C. Bernardini, T. Silverston, and O. Festor, "A comparison of caching strategies for content centric networking," in *2015 IEEE Global Communications Conference (GLOBECOM)*, San Diego, CA, pp. 1–6, IEEE, 2015.

[36] I. Psaras, W. K. Chai, and G. Pavlou, "Probabilistic in-network caching for information-centric networks," in *Proceedings of the Second Edition of the ICN Workshop on Information-centric Networking*, Helsinki, Finland, pp. 55–60, ACM, 2012.

[37] W. K. Chai, D. He, I. Psaras, and G. Pavlou, "Cache 'less for more' in information-centric networks (extended version)," *Computer Communications*, vol. 36, pp. 758–770, Apr. 2013.

[38] J. Ren, W. Qi, C. Westphal, *et al.*, "MAGIC: a distributed MAx-Gain In-network Caching strategy in information-centric networks," in *2014 IEEE Conference on Computer Communications Workshops (INFOCOM WKSHPS)*, pp. 470–475, Apr. 2014.

[39] E. Baştuğ, M. Bennis, E. Zeydan, *et al.*,"Big data meets telcos: a proactive caching perspective," *Journal of Communications and Networks*, vol. 17, no. 6, pp. 549–557, 2015.

[40]   P. Demestichas, A. Georgakopoulos, D. Karvounas, *et al.* "5G on the horizon: key challenges for the radio-access network," *IEEE Vehicular Technology Magazine*, vol. 8, pp. 47–53, Sep. 2013.

[41]   K. Poularakis, G. Iosifidis, A. Argyriou, and L. Tassiulas, "Video delivery over heterogeneous cellular networks: optimizing cost and performance," in *2014 Proceedings IEEE INFOCOM*, Toronto, ON, pp. 1078–1086, IEEE, Apr. 2014.

[42]   H. Schwarz, D. Marpe, and T. Wiegand, "Overview of the scalable video coding extension of the H.264/AVC standard," *IEEE Transactions on Circuits and Systems for Video Technology*, vol. 17, pp. 1103–1120, Sep. 2007.

[43]   S. A. R. Zaidi, M. Ghogho, and D. C. McLernon, "Information centric modelling for two-tier cache enabled cellular networks," in *2015 IEEE International Conference on Communication Workshop (ICCW)*, 2015, pp. 80–86.

[44]   V. Chandrasekhar, J. G. Andrews, and A. Gatherer, "Femtocell networks: a survey," *IEEE Communications Magazine*, vol. 46, pp. 59–67, Sep. 2008.

[45]   A. Shokrollahi, "Raptor codes," *IEEE Transactions on Information Theory*, vol. 52, pp. 2551–2567, Jun. 2006.

[46]   X. Peng, J. C. Shen, J. Zhang, and K. B. Letaief, "Backhaul-aware caching placement for wireless networks," in *2015 IEEE Global Communications Conference (GLOBECOM)*, pp. 1–6, Dec. 2015.

[47]   N. Golrezaei, P. Mansourifard, A. F. Molisch, and A. G. Dimakis, "Base-station assisted device-to-device communications for high-throughput wireless video networks," *IEEE Transactions on Wireless Communications*, vol. 13, pp. 3665–3676, Jul. 2014.

[48]   H. Ahlehagh and S. Dey, "Video-aware scheduling and caching in the radio access network," *IEEE/ACM Transactions on Networking*, vol. 22, pp. 1444–1462, Oct. 2014.

[49]   A. Argyriou, D. Kosmanos, and L. Tassiulas, "Joint time-domain resource partitioning, rate allocation, and video quality adaptation in heterogeneous cellular networks," *IEEE Transactions on Multimedia*, vol. 17, pp. 736–745, May 2015.

[50]   3GPP, "LTE-Advanced." [online]. Available from http://www.3gpp.org/specifications/releases/68-release-12, 2013.

[51]   L. Lei, Z. Zhong, C. Lin, and X. Shen, "Operator controlled device-to-device communications in LTE-advanced networks," *IEEE Wireless Communications*, vol. 19, pp. 96–104, Jun. 2012.

[52]   K. Psounis and B. Prabhakar, "A randomized web-cache replacement scheme," in *INFOCOM 2001. Twentieth Annual Joint Conference of the IEEE Computer and Communications Societies. Proceedings of the IEEE*, vol. 3, pp. 1407–1415, IEEE, 2001.

[53]   S. Y. Baek, Y.-J. Hong, and D. K. Sung, "Adaptive transmission scheme for mixed multicast and unicast traffic in cellular systems," *IEEE Transactions on Vehicular Technology*, vol. 58, no. 6, pp. 2899–2908, 2009.

[54]   A. Bria and A. G. Font, "Cost-based resource management for filecasting services in hybrid DVB-H and 3G systems," in *2008 IEEE International Symposium on Wireless Communication Systems*, pp. 159–163, Oct. 2008.

[55] A. Albanese, J. Blomer, J. Edmonds, M. Luby, and M. Sudan, "Priority encoding transmission," *IEEE Transactions on Information Theory*, vol. 42, pp. 1737–1744, Nov. 1996.

[56] X. Wang, Y. Bao, X. Liu, and Z. Niu, "On the design of relay caching in cellular networks for energy efficiency," in *2011 IEEE Conference on Computer Communications Workshops (INFOCOM WKSHPS)*, pp. 259–264, Apr. 2011.

[57] Y. Jiang, G. Yu, J. Wu, and R. Yin, "Energy consumption tradeoff between network and user equipment in small cell networks," in *2013 IEEE International Conference on Communications Workshops (ICC)*, pp. 396–401, Jun. 2013.

[58] D. Romani, "A popularity-based approach for the design of mobile content delivery networks," Oct. 2013.

[59] C. Bernardini, T. Silverston, and O. Festor, "SONETOR: A social network traffic generator," in *2014 IEEE International Conference on Communications (ICC)*, pp. 3734–3739, Jun. 2014.

Part IV

System Integration and Case Studies

Chapter 16

SDN and edge computing: key enablers toward the 5G evolution

Ali Hussein[1], Ola Salman[1], Sarah Abdallah[1], Imad Elhajj[1], Ali Chehab[1], and Ayman Kayssi[1]

Abstract

"2020 and Beyond" is announced by the International Telecommunication Union to be the era of the next mobile network generation. After Long-Term Evolution (LTE)/fourth generation, fifth generation (5G) is promising to be a major evolution in the communication domain, not simply due to the acceleration of the data rate, but rather due to the new applications. The challenging objectives such as minimum user-plane latency, uninterruptable connectivity, high quality of service, high data rate communications, and network capacity, while dealing with ubiquitous and heterogeneous network access, call for a major overhaul of the whole mobile network architecture. The limitations of today's mobile systems, derived from their dependency on hardware-based designs, led to inflexible and limited architectures. It is essential to have dynamic and flexible management systems at several levels, starting from the radio access network (RAN), passing by the Evolved Packet Core (EPC), up to the application interfaces. These future demands and the requirement for a self-adaptive system can be realized by adopting the software-defined-networking (SDN) paradigm, which leads to the integration of SDN in the network components of the upcoming 5G technology. Benefiting from software flexibility on one hand and control and management centralization on the other hand, SDN has its positive impact in the communication world from several aspects. SDN will provide 5G with a smooth transition and unified management among various wireless standards and among different RANs and wired core networks. Furthermore, SDN can optimally orchestrate the interference between cells, handovers, roaming process, routing, and signaling between access and core networks, management of the gateways, and even the management of user data. Furthermore, the new bandwidth and latency requirements, and the ability to support the innovative 5G applications, cannot be satisfied by centralizing the data in the cloud. Pushing the data to the user's proximity will be

[1] Electrical and Computer Engineering (ECE) Department, American University of Beirut (AUB), Lebanon

vital for some time-critical applications, which is a requirement supported by edge computing. Therefore, the geo-distribution of data requires an optimal networking design for these edges. On top of this distributed data layer, the network function virtualization coupled with SDN promises easier management of such an infrastructure. In this chapter, we investigate how SDN can be integrated into the 5G network architecture at different levels (RAN, EPC, security, etc.) and highlight the solutions, challenges, and benefits resulting from such integration. Also, we present the mobile edge computing concept, its integration with SDN, and its implications on 5G. We review the designs and architectures that have been proposed in this area and others that are under development, including the innovative applications and use cases that will be enabled by the SDN–5G combination. Our work will be concluded by proposing an architecture for SDN–5G for telecom operators.

16.1   Introduction (mobile network evolution toward 5G)

From a voice call dedicated mobile network with the first analog mobile network generation (1G) with 2.4 kbps, passing by the text messaging added service (SMS) in the second digital mobile network generation (2G) with 64 kbps data rate, toward the multimedia data traffic support in the packet switched third mobile generation (3G) with 2 Mbps data rate, up to the all IP network in the fourth generation (4G) with a data rate reaching 100 Mbps, the mobile network is in continuous and aggressive evolution. However, the future generation [fifth generation (5G)] is not expected to be a simple upgrade of the older generations. Connected cars, remote healthcare operations and surveillance, smart cities, and smart homes are some of the future applications foreseen to be part of the 5G era (Figure 16.1). Being the enabler of the Internet of Things (IoT)-based applications, the 5G infrastructure is depicted to be the future global network infrastructure [1].

As Cisco estimates in its VNI global mobile data traffic forecast [2], by 2020, there will be more than 5.5 billion global mobile users (up from 4.8 billion in 2015) and about 11.6 billion mobile-ready devices and connections (nearly 4 billion more than in 2015). In addition, the global mobile IP traffic will reach 367 exabytes per year (up from 44 exabytes in 2015) having the video traffic at 75% of the total mobile traffic (up from 55% in 2015) with an average of a 3.3 GB generated mobile traffic per mobile-connected user per month (up from 495 MB in 2015). The explosion in the number of connected mobile devices along with the generated mobile data traffic calls for new architectural considerations. In this context, there are essential agreed upon requirements that shape the 5G era: high capacity (10,000× more traffic and [10–100]× more connected devices), low latency ($\leq$1 ms), high data rate (10 Gbps peak data rate with 100 Mbps wherever needed i.e., [10–100]× data rate), high reliability and availability, high security, and low energy consumption (100 years M2M battery life) [3].

In the light of these revolutionary requirements, there is a need for new disruptive technologies enabling the 5G realization. At radio access level, the use of unlicensed spectrum (e.g., millimeter wave), the cell densification (Femto-, Pico-, Micro-, and Macrocells), and employing massive MIMO with the use of multiple antennas are

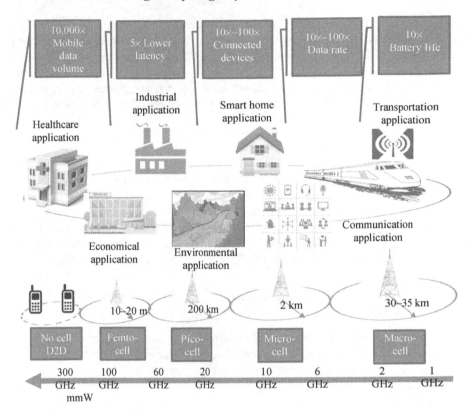

Figure 16.1 5G characteristics, applications, and requirements

the main proposed solutions [4]. However, the 5G requirements cannot be met by a simple upgrade of the access network technologies. A more sophisticated high-level network orchestration architecture is needed. The centralized hardware-based network functions in the old generations are no more suitable for this new "network of networks." Added to high scalability, big data management, security and privacy concerns, and high mobility, the future heterogeneity calls for a new control plane design allowing for more flexibility, and dynamic adaptability, while meeting the stringent above mentioned requirements [5].

In this context, "softwarization" and "cloudification" are two paradigms promising to overcome these challenges. Software-defined networking (SDN) aims at providing network programmability and dynamicity by detaching the network functions from the hardware boxes. In addition, the "virtualization" allows for infrastructure sharing between different operators reducing the CapEx and OpEx in the mobile network domain. To handle the 5G big data, the integration of the cloud is essential in any future mobile network solution. However, to meet the "zero latency" requirement, new cloud-based solutions (e.g., mobile edge computing) pushing the data to the edge network must be considered [6].

This chapter is organized as follows: in Section 16.2, we introduce the SDN and network function virtualization (NFV), and cloud/edge computing as 5G-enabling technologies. In Section 16.3, we list the main 5G challenges with their proposed solutions in the literature. Section 16.4 reviews the SDN and cloud/edge-based 5G proposed architectures. In Section 16.5, we propose a global 5G architecture integrating SDN and NFV and cloud/edge computing. Finally, we conclude in Section 16.6.

16.2   5G enabling technologies

5G represents the next major phase of mobile telecommunications standards beyond the current mobile technologies. For this technology to meet the requirements of the future users, operators, and network demands, the integration with some evolutionary network techniques is essential. It is for this purpose that SDN, NFV, and cloud and edge-fog computing are converging together into a single systemic transformation termed "softwarization" [7] that will build a concrete foundation in 5G systems.

16.2.1   Software-defined networking and network function virtualization

SDN is the new promise toward an easily configured and remotely controlled network. Based on centralized control, SDN technology has proven its positive impact in the world of network communications from different aspects. SDN has been extended beyond the network world and recently has been proposed to satisfy upcoming telecom requirements.

SDN is based on physically decoupling the intelligence of a network (i.e., control plane) from network forwarding devices (i.e., data plane) such as switches and routers (Figure 16.2).

The control plane provides a bidirectional communication tunnel for rule insertion and packet exchange, whereas the data plane represents the packet forwarding medium

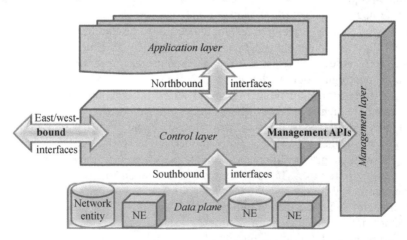

Figure 16.2   SDN architecture

between the switches. This separation allows the centralization of network logic and topology in the network operating system or the control plane, hence reducing network forwarding devices to simple network switches. SDN provides operators with more flexibility to configure their networks while having complete visibility and granularity about their networks.

In SDN-based networks, the controller [8] is responsible for the whole network manageability. Consequently, the controller keeps a global view of the network and dictates the entire network behavior. The data plane consisting of network switches is responsible for traffic forwarding based on installed rules in their flow tables. If there is no corresponding rule for a certain packet, the packet is forwarded to the controller. Forwarding rules are defined by applications running on top of the controller. Thus, SDN offers an application-programming interface (API) where the data plane of the network can be modified by external applications.

There could be more than one controller, especially in large networks. These controllers synchronize their tasks via east/westbound APIs. The data plane interconnects to the control plane via the southbound API. The controller connects to the applications (e.g., SDN management applications) via a northbound API. OpenFlow [9], which is standardized by the Open Networking Foundation [10], is the leading southbound API. The controller uses OpenFlow to push rules to switches that enable administrators to partition traffic, enhance performance, and execute applications and configurations.

SDN has been essentially applied to wired networks and datacenters. However, there are active efforts to apply it to the wireless and mobile domains. SoftCell [11] is one of the first attempts to use SDN in the cellular domain by inserting managed switches at base stations and in the core network to perform fine-grained traffic classification and target good Quality of Experience (QoE) for mobile users. OpenRAN [12] and OpenRadio [13] were depicted to manage the radio access network (RAN) through the SDN OpenFlow paradigm. MobileFlow [14] introduces a new mobile network paradigm, the software-defined mobile network (SDMN).

NFV is a technology that complements SDN. It aims at detaching the network functions from the network hardware devices and transforming them into logical entities that can be deployed anywhere (essentially in cloud-based environments). This virtualization permits multiple operators to use the same network in an isolated manner (isolated slices of the same network), which essentially results in decreasing the functional cost. The virtualization already exists in the computing domain and was widely employed in the cloud, but pushing it to the network has a great role in supporting innovative network applications and opening the network and telecommunication domains to the "pay-per-use" service paradigm. In addition, virtualization overcomes the complexity associated with multiple middle boxes that are used for different functions by allowing a more flexible and simpler software-based service chaining.

16.2.2 Cloud and edge computing

The cloud was a revolutionary technology in the last decade, aiming at delivering on-demand computing services. Four main services are provided by cloud computing: Software as a Service, Platform as a Service (PaaS), Infrastructure as a Service, and more recently Network as a Service. The "pay-as-you-go" paradigm has contributed to the decrease of both CapEx and OpEx in the IT domain. Also, the cloud has been widely integrated into the mobile domain. Mobile cloud computing (MCC) is one of the direct applications of cloud computing in the wireless mobile domain [15]. MCC aims at merging mobile and cloud computing giving the mobile users the capability to access a plethora of applications and data. Cloud-RAN (C-RAN) is another cloud application in the mobile domain at the access network level [16]. C-RAN aims at centralizing the radio resources management for a set of remote radio heads at the base band unit (BBU) level. C-RAN provides efficient radio access management and permits the reuse of the available spectrum frequencies.

However, acquiring the cloud-based services through the core network presents challenges in terms of overhead and latency, which are unsuitable for the 5G mobile network given that it requires low latency, real-time processing, and immediate responses in some cases (e.g., healthcare urgency, traffic collision, etc.). Accordingly, pushing data to the edge [17] is vital for 5G. Mobile edge computing (MEC) is an emerging technology initiated by European Telecommunications Standards Institute (ETSI) through the Industry Specification Group that was set up by Huawei, IBM, Intel, Nokia Networks, NTT DOCOMO, and Vodafone. MEC aims at pushing data processing and computation resources to the mobile user's proximity. It provides better user experience with lower latency, more effective location awareness, better mobility management, and higher data rate [18]. MEC is a concept related to fog computing and sometimes, they are used interchangeably in the mobile domain. Thus, fog extended the cloud to the network edge. As we see in Figure 16.3, the edge nodes are

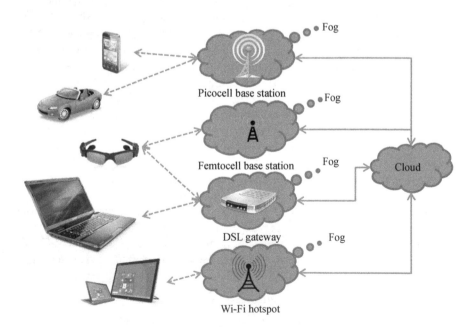

Figure 16.3   Fog-tiered architecture

deployed in a tiered architecture where the devices are directly connected to the edge nodes and these nodes are connected to the cloud. Edge computing associated with SDN and NFV can deliver a new service at the network access level: PaaS opening the RAN to third-party innovative applications.

16.3   5G characteristics, challenges, and solutions

As is the case with the emergence of any new technology, new requirements are brought about and 5G is no exception. In fact, given the drastic changes in human lifestyle nowadays, the world as we know it is shifting toward a more connected and virtualized world. This poses many challenges that 5G needs to overcome to deliver the required user experience.

16.3.1   5G requirements/challenges

Agyapong *et al.* thoroughly discussed six challenges that 5G should promise to solve [19]. Higher capacity and higher data rate, which are crucial due to the expected traffic increase, can be achieved by shifting to new communication and antenna techniques. It is also favorable to separate the user data channel from the control channel, in order to allow the system to scale data channels without having to increase control channels and reduce cost. Lower end-to-end latency is another design objective to be achieved, to enable the network to support real-time and critical applications. Requirements as such specify a delay of 1 ms E2E. Furthermore, as more devices are expected to be connected, new concerns arise due to the heterogeneity of devices and the amount of signaling incurred. Reducing capital and operational costs are also challenges that are easily solved by employing techniques such as SDN, NFV, and energy management at the RAN. Finally, improved quality of experience is an important requirement in future networks, and it can be achieved through all the above enhancement techniques.

The authors of [20,21] also focused on the 5G network challenges, pointing out two main approaches for addressing the increase in traffic, the needed low latency and spectral efficiency. The first approach relies on dense deployment of small cells, which leads to bigger data rates because of the shorter distance between the devices and the access points, and more efficient use of the spectrum, because the latter can be subject to time–frequency reuse across many cells. The second approach relies on centralized processing: it provides the needed flexibility and ease of management to be able to efficiently manage the dense deployment.

In [20], Tikhvinskiy *et al.* tackled 5G requirements from a different angle: trust in 5G networks. They argued that users' trust is a major factor that should be achieved in 5G, and this trust is correlated to two factors: security and Quality of Service (QoS). They elaborated on the importance of QoS in establishing trust since 5G will be used in critical applications such as medical or business apps, and any degradation in the performance can lead to a human death or financial theft. Virtualization plays an important role in achieving QoS objectives; the authors discuss the high-level features of a 5G QoS architecture, which are centered around NFV. This architecture consists of a QoS manager that performs traffic flow control and network monitoring, and a Spectrum Manager performing channel aggregation when needed. Both communicate together to guarantee efficient allocation of resources.

Panwar *et al.* intended to answer the question on "what will be done by 5G and how?" [21]. They presented a comprehensive study on the proposed architectures of 5G that are categorized based on energy efficiency, network hierarchy, and network types. They argued that the issues of interference management, economic impacts, QoS guarantee, handoff management, security and privacy protection, channel access management, and load balancing hugely affect the realization of 5G networks. The authors further argued that the requirements of each terminal (users, operators, and network) should be independently studied and satisfied. The authors also shed light on some requirements that are essential to 5G, such as zero latency, self-interference cancellation, and fast caching; and some methodologies such as dramatic upsurge in device scalability, downlink and uplink decoupling, and spectrum utilization. Moreover, they discussed some required technologies of the 5G network that include SDN, NFV, millimeter wave, M2M communications, massive MIMO, and visual light communications. These technologies and others such as C-RANs and D2D communications were also detailed in [22,23] as solution for the different 5G challenges.

The authors of [24] discussed how a 5G network would need a multiservice and context-aware adaptation mechanism, mobile network multitenancy, and network slicing, in order to allow the same network infrastructure to be used by multiple operators, implementing their own logical networks. Consequently, each network slice can fulfill different requirements and serve different purposes.

The main security challenges/requirements for the 5G future network are tackled in [25]. First, not all applications require the same level of security. In addition, adopting the same security scheme to the tremendous number of connected things leads to unnecessary overhead, especially if the security level is not needed for most connections. The main requirements are user identity protection, mutual authentication between user and network, data and signaling confidentiality and integrity, security configurability and visibility, and the flexibility of any security solution.

16.3.2　Proposed solutions

Researchers did not only identify the requirements of 5G, but also many works proposed solutions to the cited challenges from different aspects. In [26], Kukliński *et al.* explored handover management in 5G and analyzed the benefits of using SDN for the connected state (where the mobile node is in live session when moving from cell to cell). They proposed two approaches to handle it: the cut-establish method and the establish-cut method. The drawbacks of the current handover management are due to the distributed management aspect and the associated exchange of messages. SDN overcomes the cited drawbacks because it relays the control to a high-level entity having a relative global view of the network. Three cases were presented: centralized, semicentralized, and hierarchical SDN-based control. While the centralized solution suffers from scalability and reliability issues, the semicentralized one relaxes the global view by relying on domain-based cooperation to handle the handover process. However, this method presents overhead due to the exchange of state information between neighbor domains. The hierarchical approach consists of multiple domains with a master controller, which has an abstracted global view of the whole network giving it the ability to manage the handover between domains. It was found to be the most suitable approach.

The mobility management issue was tackled also in [27]. Network Address Translation (NAT) is a strongly related concept to the mobility management with the heterogeneous addressing scheme in today's networks. The vertical integration of different networks in the future 5G network will make this issue harder and more complex to manage. Consequently, a horizontal consolidation for the control task is required to support future innovative applications. The proposed scheme relies on an overlay network controller (ONC)-centric model of combined NAT and mobility management operations. This model uses SDN control to establish a connection between two mobile NATed devices. The ONC establishes two Transmission Control Protocol (TCP) connections with the NATed devices and acts as relay for any connection between these devices. After establishing the connection, the communication becomes direct between the devices.

In [28], Chen *et al.* surveyed the most promising technologies enabling the SDMN. The principal concern when applying SDN to mobile network is to retrieve an efficient spectrum management method. Three main problems hinder this attempt: the dispersion of network state information over heterogeneous networks making the coordination between nodes a critical task, the network configuration at different level network nodes (where details are left to lower level entities and global configuration to high level nodes), and fine-grained cooperation between different network entities. The SDMN application in 5G will enable convergence in the vertical and horizontal dimensions, as also discussed in [29]. In the vertical dimension, cooperation is attainable between different cells and networks. In the horizontal one, a cooperation between network entities is ensured.

Security solutions were also proposed in [25,30,31]. The authors of [30] presented a multitier approach to secure SDMNs by handling security at different levels to protect the network and its user entities (UEs). They started by securing the communication channels between network elements by enforcing Host Identity Protocol and IPSec tunneling. Then, they used policy-based communications to restrict any unwanted access to the mobile backhaul network. While [31] worked on implementing an adaptive trust evaluation framework to achieve software-defined 5G network security, the proposed solution in [30] uses phantom cell separating the user and control planes; small cells handle the user plane for higher data rate and more flexible network efficiency, and macrocells handle the control plane to maintain mobility and connectivity.

Table 16.1 summarizes the above-mentioned challenges and solutions.

Table 16.1   5G challenges and proposed solutions

Challenges	Solutions
Higher capacity and data rate	New communication and antenna techniques: • Carrier aggregation • Antenna arrays • Network densification • Separation data channel and user channel
Higher capacity	• Improvement in architecture, channel receiver, and design • Use of optical wave technology • Moving intelligence to the edge
Heterogeneity/convergence	• SDN and NFV-based gateways and traffic clustering
Reduce CapEx and OpE	• SDN and NFV-based energy management at RAN
Traffic increase	• Cloud-based solution • Cell densification • Processing centralization
Better spectral and signaling efficiency	• SDN and NFV, and cloud-based solution at the RAN level
Handoff/mobility management	• SDN-based NA • Centralized, semicentralized, or hierarchical SDN-based control architecture
Security and privacy	• SDN and NFV-based security architecture providing the essential security services (authentication, confidentiality, and integrity) and more important flexibility, configurability, and visibility

16.4   Literature review

Most of the literature related to architecture design emanate from the challenges and solutions discussed in Section 16.2. These designs basically rely on network virtualization, SDN, edge computing, or combinations of these technologies. Different 5G trials were carried out to test the effectiveness of these technologies. The author of [32] announced that in a joint 5G technologies trial, the companies achieved a cumulative 20 Gb/s of data throughput in an outdoor environment, using the 15-GHz frequency band with two simultaneously connected mobile devices of a downlink bit rate of more than 10 Gb/s each. Separate trials were also done to test the variation of the throughput as a function of distance from the base stations.

16.4.1   SDN and NFV based 5G architectures

As previously mentioned, virtualization and softwarization of the network are major drivers for future 5G networks. In fact, legacy vertical architectures failed to answer the needs of 5G (low latency and flexibility) because they were designed for individual services with predicted traffic. SDN and NFV-based clouds allow horizontal architectures that are more flexible in terms of scalability, response speed, and adaptability. In addition, an SDN infrastructure provides a better environment for virtualization because of the ease of management and abstractions that it introduces.

In fact, the work in [33] proposed integrating SDN and NFV within a 5G network architecture. It described the requirements that SDN and NFV need to provide for the integration to work and discussed if these requirements can be met. First, the technology should support the migration from old mobile systems to 5G and provide backward compatibility. SDN and NFV meet this requirement as they provide easy and seamless network management, which makes it possible for new and legacy technologies to coexist. The second requirement is security and availability, which SDN should critically be able to provide, especially that a single-centralized controller could pose a security threat. Network monitoring should also be supported to facilitate the deployment of SLA and QoS/QoE solutions. Both requirements along with network monitoring are ensured with SDN, which exposes an application layer that allows direct supervising and control of the network via specific applications. Finally, cost is expected to be reduced; however, this is yet to be verified.

16.4.1.1 Software-defined radio access network

SDN was extensively considered as a technology for the RAN. One proposal is SoftAir [34]. It is composed of two layers, data and control, where the data layer encompasses both the RAN and core networks. The RAN, called SD-RAN, is composed of software-defined base stations, while the core network, called SD-CN, is composed of software-defined switches. The RAN is deployed in a distributed manner, where hardware radio heads are connected to software-based base band units via fiber or microwave fronthaul. Functions are divided between the two components to enhance scalability and reduce data traffic on the fronthaul; and the radio heads are allowed to perform some of the baseband processing functions such as the modem. The base stations, basically Open vSwitch, have OpenFlow-enabled interfaces which allow them to communicate to the core switches through the OpenFlow protocol. The control layer runs all the network applications and the management tools. The mobility-aware traffic balancer application, the resource-efficient network virtualization application, and the traffic classifier are of particular interest.

Yazıcı *et al.* also consider extending SDN to the RAN network [35]. The authors argued that the centralized virtualization of the network functions at the cloud will be critical in terms of latency and overload on the core network. Thus, we need to recur to the SDN central control while pushing some functions to the user proximity, balancing between response time and global network view. Consequently, the authors proposed to divide the network into two parts: the RAN and the CN. The all-SDN proposed architecture does not rely on a central control but on a hierarchical set of controllers allowing for local decision optimization and an opportunity to provide fine-grained service differentiation. Cloudification and softwarization are mentioned as two key paradigms needed in the 5G network to overcome the incapability of the 4G network in term of flexibility and dynamicity and especially in the

D2D domain. The proposed architecture consists of the UE controller (RAT selection, D2D discovery, and communication operation), the BS controller (wireless resource management and scheduling, D2D resource management and synchronization), RAN controller (C-RMM control with possibility to be done at the BS controller as the case in LTE), and the network controller (end-to-end QoS provisioning, application-aware route establishment, service chaining, mobility management, and policy and charging management). A controller of higher hierarchy injects rules (⟨Match, action⟩) into a lower level controller, which sends feedback and periodical reports to the higher level controller.

In [36], Cho *et al.* proposed a crosslayer architecture combining SDN and software-defined radio (SDR) characteristics in a 5G network. They show the relevance between frequency spectrum and network information, and hence propose to consider spectrum utilization and channel flow interaction. They also discussed how an effective processing of flows can avoid spectrum competitions and bandwidth overloading. The architecture presented the SDR layer, which contains all devices that can access the radio. This layer monitors all frequencies and avoids many cases of illegal use and interference. Units that can send and receive packets (such as switches and routers) are stored in the SDN layer which enables access to different heterogeneous networks (wireless 5G environment, WiMAX, LTE, and 802.11). A crosslayer controller handles administration over both layers. All spectrum resource usage requests should be sent to the controller, which checks the flow traffic information and allows access or suggests a better band.

Sun *et al.* revisited the integration of SDR and SDN in 5G [37]. They also included NFV in their proposed generic high-level architecture. An SDN controller sits on the head of the cloud and connects to a set of physical switches, below which virtual controllers and gateways are deployed to connect to the RAN. This latter spans different technologies that are promising for future 5G deployments, such as C-RAN, HetNets, millimeter wave, and massive MIMO.

Moreover, [19] proposed a design that implements the improvements discussed in Section 16.2; the architecture is separated into two logical planes, a limited L1/L2 function Radio Network and an SDN/NFV-based cloud that deploys L3 and higher functionalities. This latter is composed of separate user and control planes, which allow the use of different frequency bands for data and control. In the radio network, small cell base stations are deployed at high frequency bands, whereas macrocell base stations are deployed at lower frequency bands, in order to increase capacity and data rates. Moreover, some capable devices are used as relays to other devices, in order to aggregate traffic and allow device heterogeneity as well as scalability. A proof of concept implemented on a real-time simulator showed improvement in throughput gains up to 1,000 times as compared to a macro-only 3GPP Release 8 LTE deployment.

In [38], Raza *et al.* extended the SDN-based designs to span the RAN and the devices in heterogeneous radio networks. Their architecture, SDN-5G, uses bandwidth from different available networks to achieve high data rate and QoS requirements. It is constructed as a three-level scheme: device, RAN, and Core network. A Mobile Controller (M-OFC) and a Mobile Gateway (M-OFG) are implemented on the device. The M-OFG sits below the application layer; it oversees multiplexing packets from the application onto the different available RANs. The M-OFC monitors the bandwidth requirements of the different applications and specifies the RANs that will be used for the application. At the RAN level, the radio access interfaces (N-RAI) are OpenFlow enabled. They are managed by the radio access controller (R-OFC), which also monitors the load and the bandwidth available on each N-RAI. This information is constantly shared with the core controller N-OFC. The core also contains core gateways (N-OFG); whenever a service connects to a N-RAI, the N-OFC automatically creates a tunnel between the N-RAI and the N-OFG for bidirectional communication.

16.4.1.2   Software-defined core network

Many schemes were introduced where SDN was deployed in the core network. In [36], network functions such as MME, HSS, load balancing, QoE, and so on are all mapped to SDN and deployed in the mobile network cloud, which corresponds to the application layer of SDN. The network elements expose a 3GPP interface to support the coexistence with legacy mobile and the eventual migration from it. The authors developed a test bed composed of off-the-shelve equipment from Nokia, Coriant Oy, EXFO, and nwEFC. Results were favorable of the integration of SDN with 5G except for some properties like robustness and reliability, where the test bed showed an increased latency when migrating VMs with network functions.

In [38], Sun *et al.* employed heterogeneous networks (HetNets) to solve the 5G capacity challenge. They proposed the use of SDN as an intelligent management plane, for the HetNets that are composed of Low Power Nodes (LPN). Their architecture builds on the traditional Core Network/RAN architecture. The SDN controller is deployed on a central server and runs intelligent software to efficiently operate and manage the network. SDN is used for dynamic allocation of resources in the core network; at the E-UTRAN level, SDN allows the virtualization and management of the eNBs. Different SDN-based solutions were proposed such as creating a virtualized "node" that analyzes and stores the resource parameters of each node in the 5G network in the context of a vector (CPU usage/available bandwidth/storage/uplink and downlink dataflows/number of users attached). The authors proposed using back-propagation and support vector machine for traffic behavior prediction as an intelligent function in the SDN/5G network. In addition, they discussed using the classical single-server processor sharing queue that can provide a feasible and dynamic solution to inter-cell load balancing. Another research topic discussed by the authors is the importance of user density prediction. With accurate user density prediction, the network can allocate resources more "wisely" and purposefully. Finally, they proposed to make the SDN controller a global optimizer to jointly manage resources among the networks for network utility maximization and for the UEs with multiple interfaces to have the flexibility to select the cell to attach to in order to improve local capacity.

Moreover, SoftNet is a software-defined 5G network aiming to hide the LTE challenges as result of new communication paradigm integration (Mobile Cloud, Social Mobile Network, IoT) [39]. To overcome the signaling overhead and system incapability to provide low latency due to centralized control in LTE, the proposed architecture adopts a distributed mobility management scheme over a set of access servers at the edge of a software-defined core network. The two main components of this architecture are: the core network and the unified access network. The main function at the core network is the communication-control function responsible of the mobility management, network management function, network policy management, and VNF orchestration. The unified RAN is responsible of the multi-RATS coordination function, the distributed control function, and the gateway control function.

In [40], the authors proposed the use of SDN for network management in the backhaul. They modeled this part of the network with SDN switches: the first level of OpenFlow (OF) switches is called the eOF and is connected directly to the base stations. On the other side of the backhaul, another level of OF switches called mOF is in charge of connecting the access network to the Internet. The eOFs encapsulate all traffic coming from users toward the network in 802.1ah (Provider Backbone Bridges) to isolate its own traffic from the rest of the nodes. The mOFs also tag traffic coming back from the mobile access to the appropriate eNB. Other types of switches are deployed throughout the network: notably, carrier grade switches and OF switches are deployed between eOFs and mOFs to aggregate traffic coming from eNBs, and Internet access OF switches (iOFs) that connect the mOFs to the Internet. The deployment of OF switches at the different levels in the backhaul network allows SDN applications to manage the backhaul network. In fact, the Mobility Management App keeps track of devices within the wireless area and modifies rules in eOFs and mOFs, whenever a device moves from the area of one eNB to another. It also assumes the role of an MME. The access app controls the iOFs; it assigns virtual IP addresses to mobile devices to connect them to the Internet. Moreover, it implements load balancing, firewalling, and NAT services.

In [41], the authors proposed an SDN-based backhaul architecture for 5G, which relies on the LTE architecture with added virtualization of the core network elements in the cloud and the introduction of an SDN controller in the backhaul network. Each access domain is managed by an SDN controller coupled with an MME running at the cloud. The introduction of rule-based switches allows the replacement of the GTP scheme (used in the LTE network to support IP mobility) by VLAN and MPLS or GRE tunnels providing tunneling flexibility and reducing the network overhead presented by the GTP protocol.

A method for spectrum sharing in 5G heterogeneous networks using SDN called harmonized SDN-enabled approach (HSA) was presented in [42]. The proposed architecture combines a two-level centralized control with distributed spectrum monitoring. It consists of macrocells whose base stations are directly connected to the SDN controller through fiber optic and use the OpenFlow protocol to control the data plane. The architecture also supports multilayered cells. For example, a macrocell can contain many microcells that in turn contain many picocells. Base stations of the small cells are connected to the bigger cells through wireless or fiber optic backhaul to inform them of the state of their cell. In this architecture, the network control is hierarchical: local control is carried out by base stations, whereas network decision-making is relayed to the SDN controller. At the macrocell level, the BS stores a local database that contains information about primary and secondary users within the cell and the different load conditions. The SDN controller stores a global database, which is updated by the LDPs. The BS relies on the data received from secondary users instead of performing spectrum sensing itself. If the cell is overloaded, a request is sent to the SDN controller, which checks for underutilized cells in the GDB, acquires bandwidth from them, and grants it to the overloaded cell based on operator-defined policies.

The authors of [43] also considered HetNets and focused on an architecture for handling plane failure in 5G heterogeneous networks. The authors proposed the small cell controller (SCC) scheme that takes control of small cell BSs in a clustered fashion in case of the failure of a controlling macrocell BS. For this reason, they proposed a new signaling mechanism based on central control in the core network (Evolved Packet Core [EPC]) to avoid frequent handover signaling in a failure situation.

16.4.1.3   Software-defined 5G architectures

In [44], SDN-based solutions for fronthaul and backhaul 5G architecture were proposed. The fronthaul solution resides on the common public radio interface over Ethernet providing a unified transport format. Thus, the sharing of the infrastructure between multiple mobile operators exposes the network to security breaches. In this context, a software-defined distributed security solution is proposed for the backhaul network. Adopting a central scheme, where the private data is copied over X2 using IPSec, is costly from a latency perspective. An alternate solution, consisting of choosing the aggregation point with operator-specific end-to-end encryption is proposed to overcome the latency issue.

Other propositions addressed architectures with hierarchical deployment of SDN control plane. In [45], Giraldo *et al.* presented a two-level software-defined approach for the 5G architecture at the network and radio access layers. At each level, a dedicated centralized controller and northbound/southbound API are defined. Two crosslayer schemes are also introduced for interaction between the radio access and the network controllers, which act as typical SDN controllers with network functions such as load balancing and firewalling deployed as software-based application modules. The radio access controller centralizes the radio access functions of the base stations within a region. Resources are no longer links or switch ports but frequencies, channels, or antenna ports. The paper investigates two levels of crosslayer optimization to establish coordination between the radio and network controllers. At the controller level, optimization is done by introducing a new API, between the network controller and the radio access controller, the west/eastbound API. At the application level, a more advanced crosslayer mechanism is used to implement a software-defined application on top of both controllers, and each of these would be accessed through the corresponding northbound API.

A "plastic" architecture consisting of a hierarchical set of controllers was proposed in [46–48]. This architecture consists of three levels of controllers: device controller (DC), edge controller, and network controller. The DC has the role of access selection and network selection, the edge controller is responsible for network access control, packet routing and transfer, mobility management, security control, radio resource management, addressing functions, proximity service, and relaying and mutual authentication. The network controller is responsible for network management (control plane overload control, data plane load control/balancing, 5G C-plane instantiation, 5G C-plane maintenance). The main goal of this architecture, in addition to allow for device heterogeneity, is to reduce the latency. An analysis of end-to-end FTP file sharing use case shows an enhancement of 75% in latency in comparison to LTE.

Other works built on the generic architecture and extended it to address specific aspects and challenges of 5G. For example, in [49], Bradai *et al.* proposed the Cellular SDN architecture (CSDN), which also employs SDN and NFV in the aim of resolving the challenges of 5G. The proposed architecture is composed of the forwarding plane, which resides under the controller plane. The control plane supports an application plane, which runs various applications such as routing or resource management. However, the authors augmented the SDN layers with a knowledge plane, which gathers data and policies from the network and from an operator's portal, as well as data about customer preferences. The availability of this information in one plane allows the mobile service provider to make intelligent decisions concerning resource allocation and delivering services to users with the appropriate quality. Once data is analyzed and decisions are made, they are translated into flow rules, which are then easily deployed in the network.

Kaleem *et al.* proposed an SDN-based extension of the LTE-A architecture using 5G mPCs (mobile personal cells) [50]. The architecture introduces an additional personal cell domain (PCD) supporting ubiquitous connectivity with high data rate, high mobility, and low transmission delay. In this architecture, the network components are moved to the cloud on top of an SDN controller connected to the EPC domain by custom APIs. The architecture is divided into three domains: the network domain consisting of the EPC and the SDN controller having the role of RAT management, the PCD consisting of a personal cloud hub for relaying and content caching functions, and a terminal domain. The PCD is connected to the network domain by means of high capacity wireless backhaul links. The mPC users can also communicate with each other by using the wireless sidehaul links. This architecture presents many benefits in terms of high speed and high data rate, but it also presents challenges in terms of mobility and interference management.

Moreover, Inam *et al.* proposed a service-oriented architecture for the 5G network as the enabler of the IoT [51]. The authors discussed three main challenges: QoS differentiation, mobile network dynamicity, and the lifecycle of service management. The proposed architecture is composed of five layers: the user layer (different users and API to connect user and network), the 5G service lifecycle management layer (SLA management), the orchestration and control layer (the translation of SLAs to physical control), the radio access and core network layer (physical radio access management), and the device layer (all physical devices).

Figure 16.4 presents the different contributions that SDN and NFV have to offer for 5G.

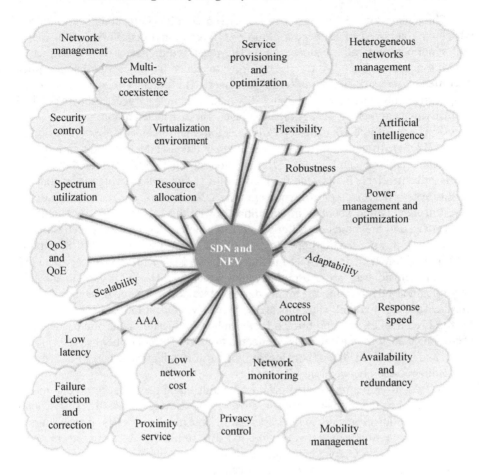

Figure 16.4   SDN and NFV contributions in 5G

16.4.2   Edge/cloud-based 5G architecture

Some work extended the previous work done at the core and backhaul to the edges of the network, at the fronthaul and base stations level.

16.4.2.1   Edge/cloud-based access network

Checko *et al.* investigated the benefits of using cloud-based RAN (C-RAN) for 5G networks to answer requirements such as low delay or high capacity [52]. However, they emphasized the importance of virtualization of the baseband resources for the C-RAN implementation and discussed technologies such as SDN and NFV for virtualization purposes. In the proposal, the base station functions are divided between the resources pool and the cell.

The authors of [53] proposed an Ethernet over radio (EoR) architecture for 5G, which is composed of an access cloud (AC), a regional distributive cloud (DC), and a national centralized cloud (NC). The tough requirements of 5G are met by pushing access to the network edge in order to reduce the processing latency when it comes to network decisions (e.g., handover). The mobility management in the proposed architecture relies on L2 mobility management at the access network and L3 mobility management at the aggregation layer. The distributed mobility management functions are enforced by rules on Policy Enforcement Points by Policy Distributed Points (PDPs). In addition to latency reduction, this architecture helps in reducing CapEx and OpEx by employing SDN. The control function is distributed over three levels of controllers (PDPs): the local controller (L-controller), which is responsible of the access network control (manages the L2-switches), the regional controller (R-controller) which operates in coordination with the L-controllers to manage the L3-routers, and the main controller which is the highest level network orchestrator.

In [54], a more detailed architecture of the AC part was presented. It employs Ethernet with MIPv6 in an SDN-managed manner. SDN and NFV, hierarchical distribution, subsidiarity are the main design principles. The GTP used in the LTE architecture is no more suitable for 5G due to the scalability and tough latency requirements. The edge network elements are connected via OpenFlow switches controlled by the L-controller. To keep track of user mobility, a location table and cache tables are updated each time a device is connected to an AP. The controller has an entry for each, and it is responsible of configuring the corresponding paths end-to-end in case of communication between mobile entities.

In [55], Roozbeh *et al.* also proposed a distributed control scheme that moves some control to the base stations and closer to the devices drastically. This scheme reduces signaling especially in networks with a large number of users. It also achieves the scalability required in networks like IoT. The author defines two types of nodes in the core network. On one hand, gateways are responsible for the data plane traffic, on the other hand, virtual control nodes oversee all functions related to signaling and handling traffic control. These control functions are further divided into parent and child. Child functions are control functions that utilize the network frequently and are deployed closer to the access points. They update the corresponding parent functions deployed farther in the core, when necessary. In this architecture, SDN is used for the communication between child and parent modules. This hybrid control, distributed and centralized, reduces latency of control requests. It also reduces the load on the core network because of the signaling at the UE level, as these will be handled at the edge (child functions).

Some researchers went even further by considering the mobile devices. For example, in [56], Moser *et al.* proposed to deploy an OpenFlow virtual switch within the operating system of the mobiles devices (Android). In this case, the communication layer of the device would be forwarded to the OF virtual switch, without affecting the normal behavior of the legacy network features and protocols. The SDN controller can also be implemented on the device or externally. The authors also elaborated on the case of many SDN controllers, internal and external, and proposed using virtual networks and slicing in order to separate the traffic control of different controllers.

If the different control apps affect the same network parameter, then an SDN proxy controller such as FortNOX can be used to specify permissions based on roles.

Moreover, proposals like the one in [55] took advantage of user location. In fact, Wang *et al.* presented an architecture which differentiates between indoor and outdoor users and employs the most appropriate and efficient access technology in each case. For outdoor communications, the authors refer to massive MIMO and distributed antenna techniques. As for indoor communications, the authors use high-frequency technologies like Wi-Fi, mm-wave, or Li-Fi because they deliver high data rates, and they are not degraded in indoor environments. Finally, the core network is deployed with virtualization technologies such as SDN or NFV, or both, to ensure flexibility. However, the authors claimed that SDN and NFV do not have to be restricted to the core and can be deployed gradually all the way till the edge.

16.4.2.2   Edge/cloud-based 5G architecture

In [47], Guerzoni *et al.* introduced edge computing with SDN and NFV in an architecture for 5G networks. The architecture features three types of control that are implemented at three different levels: the device level, the edge level, and the cloud level. The DC is responsible for the functions related to the access stratum and to the connectivity to the wireless network. At the edge, the controller (EC) is in charge of managing the 5G network. It implements functions such as routing, security, mobility and resource management, and all the AS/NAS functions. At this level, two types of controllers are implemented to account for out-of-coverage connection: the EC(i) is located at the edge, whereas the EC(ii) is located on the device itself. EC(ii) can be permanent or temporary; it allows the device itself to perform AS and NAS functions when it is not able to communicate with the network in order to support the mission-critical communication requirements of 5G. And finally, the orchestration controller (OC) manages the cloud and the allocation of its resources. It runs two modules, topology management and resource orchestration, which are in charge of physical resource management and instantiation of virtual EC control applications given these resources, respectively. As for the data plane, the architecture does not use anchor points nor gateways. Alternatively, the device is given an address and automatically linked to a last hop routing element (LHRE). This latter sets up the connection between the AP and the backhaul, and a forwarding path is created to transport packet from the device to the network and vice-versa. The architecture also supports device-to-device communication via a wireless connection managed by the EC.

Many works addressed specifically the heterogeneity of networks in their designs for edge-based architectures. Iovanna and Ubaldi presented an architecture using SDN for heterogeneous multidomain transport, where each domain runs a certain technology [57]. A radio orchestrator sits on top of radio domains and processes all functions between the remote radio unit (RRU) and the baseband unit (BBU). They perform operations that are slow and depend mostly on the technology in the domain. On the other hand, an E2E multidomain orchestrator layer implements all connectivity services related to the radio domains. It is connected to the radio network layer via local virtualizers, which provide it with virtual abstractions of their domains that will be then translated to service-specific parameters at the E2E level such as speed,

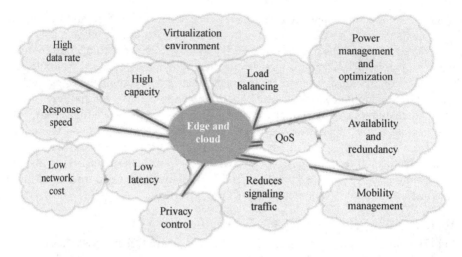

Figure 16.5   Edge and cloud contributions in 5G

delay, bandwidth, and so on. This solution proved optimal resource utilization where 90% to 98% of requests were served. A similar HetNet cloud consisting of a layered architecture of SDN and radio access services was proposed in [58].

Furthermore, Yang *et al.* proposed an architecture based on a three-level cloud design: access, control, and forwarding [59]. The control cloud, implemented in NFV, is responsible for all network management functions and creates a control plane, which is completely separated from the data plane. In fact, data is never forwarded to the control cloud; it only traverses the access and forwarding cloud. The access cloud contains the RAN and supports all technologies. It features centralization of control over the clustered underlying cells. Finally, the forwarding cloud thrives to ensure fast forwarding and services such as NAT, firewalling, etc. The authors implemented the proposed architecture with a virtualization hypervisor and proposed different cases where the architecture can be deployed: wide area coverage, high-capacity hot zone, ultrareliable and low-latency communications, and massive machine-type communications scenarios.

Figure 16.5 presents all the contributions that edge and cloud computing offer for 5G.

Each of the proposed SDN, NFV, or Edge-based architectures above thrives to address one or many challenges of 5G networks. They also target different levels of the mobile network; for example, some propose solutions are for the core network, whereas others consider the RAN, the backhaul, or the fronthaul network.

16.5 Proposed architecture

After reviewing the different proposed 5G architectures, we see that there is no global architecture integrating SDN, NFV, edge and cloud computing at different levels while taking into consideration the most important 5G requirements/challenges of low latency, heterogeneity, high scalability, big data management, and network security management.

Thus, we propose a global 5G architecture integrating the most promising technologies at the access network level as well as the core network level. Although many previous works considered the centralization of control at the radio access and core networks, we argue that the large number of connected devices call for a hierarchical control architecture. In addition, to handle the 5G big data, a distributed data management platform at the access network is a key to overcoming the resulting challenges from the data centralization paradigm (core network overload and high latency). Our proposed architecture consists of five layers (shown in Figure 16.6):

The access layer: This layer consists of multiple cells with different sizes (Femto-, Pico-, Micro-, and Macrocells) each having special characteristics in terms of the used frequencies, power, and coverage area. Using different cell types conforming to the densification paradigm aims at providing more frequencies, allowing multiple frequency reuse, and consequently providing high data rates. Each cell has a software-defined access point (SD-AP) to which different user devices will be connected with different access technologies (Wi-Fi, Li-Fi, ZigBee, 5G, etc.). The SD-APs are equipped with OF-like switches and are fog enabled with data handling and analysis capabilities. It is worth noting that cells in 5G can be created and deleted as needed and a cell might be mobile (e.g., cell tower at a train head wagon). Moreover, sometimes no cell technology is used (i.e., D2D communication). Thus, a cell discovery is mandatory in this case. In the context of big data management, employing distributed edge nodes keeps part of the data at the user proximity, which is beneficial in terms of latency, location awareness, augmented reality and at the same time decreases the amount of data to be transferred to the cloud through the core network.

The access control layer: This layer resides on top of the Marco cells access points. The SD-APs are equipped with SDN controllers and are called eNBC (evolved Node Base Controller) as an extension to the control functions of the eNodeB element of the LTE architecture. eNBCs are responsible of controlling and managing the underlying access layer. The main control functions handled by this layer are: interference management between the managed cells, mobility and handover management when a UE moves from a cell to another (intra or inter-macrocell domains), UE to SD-AP authentication, authorization, accounting (AAA), QoS management, spectrum access management, and cell discovery. In addition, eNBCs are the gateways between the access network or user plane and the core network or control plane. They are connected via wired connections to the core network using three separate planes: the control, security, and verification planes. The control plane is responsible for transmitting the control packets for rule and policy insertion, along with network management. The security plane [60] is designed to exchange security-related information (certificates, authentication keys, security analysis data, etc.) to enforce different levels of real-time security with low overhead and minimal configuration. Finally, the verification plane [61]. handles consistency and error management. Our proposal is based on physically separating the main functionality of the core network, which strikes a good balance between network performance, security, and consistency features. These channels are

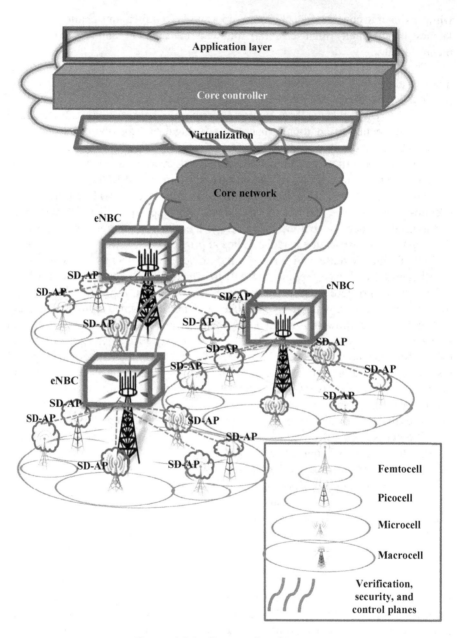

Figure 16.6   Proposed architecture

also designed to be separately secured and authenticated using different keys. This technique mitigates the effect of an attack over one channel on the others. Thus, compromising a single channel will not compromise all system information. In addition, from a performance perspective, this separation minimizes the overloading effect of continuous monitoring operation (collection of security analysis data for verification tests) and does not allow slowing down the control connection and degrading the network performance.

The core network layer: This layer consists of the 5G backbone network devices, which are managed by the above layer and might be SDN or non-SDN based network elements. They are connected to the eNBC through specific northbound interfaces and to the above control layer via specific southbound interfaces. The components of this layer act per the installed policies by the core controller. This core network might rely on the existing Internet backbone with the ability of SDN controller to communicate with legacy network elements via traditional protocols via southbound interfaces. The hybrid nature of this network calls for SDN controllers to support both legacy and SDN protocols (e.g., OpenDaylight). Furthermore, the core elements might have a hybrid nature, which means that they can act as SDN elements in some cases, and alternatively, they might switch to the distributed routing behavior at certain points following certain defined algorithms depending on the network state (failure, load, etc.). Intermingling centralized and distributed networking behaviors in a reactive controlled manner results in better core network design, while profiting from these two paradigms advantages and overcoming their disadvantages.

The core control layer: This layer consists of the main SDN controller responsible of the whole network orchestration. Three sublayers are encompassed in this layer: virtualization layer, verification layer, and the security layer. The core controller is responsible of QoS management (load balancing, traffic classification), network slicing, AAA function (for the eNBCs, SD-APs, and UEs), network virtualization, security analysis (e.g., blacklist, traffic analysis, and attack predictions), network verification (e.g., routing flow consistency, error handling), and spectrum access management. As mentioned above, the core controller is an SDN controller having an essential role of managing the routing and the corresponding QoS guarantees at the core network level. In addition, being the highest authority in this hierarchical network, it has the role of managing the AAA operations. It is also designed to be aware of network changes (e.g., mobility and handover) through control periodic updates from the eNBCs and, thus, it has a continuously updated database of the whole network. The core network could be handled by several SDN controllers, synchronized through west/eastbound interfaces. Moreover, the network's geographical size could be extended to an international domain and, thus, we could have different underlying SDN controllers handling national regions. Consequently, a roaming operation should be performed when moving from one core controller domain to another. Having the ability to communicate via east/westbound interfaces, the horizontal set of controllers can share the necessary information to perform the roaming task (similarly to the existing HLR and VLR concepts in today's mobile networks).

The application layer: This layer is the most distinctive layer added by the SDN and NFV integration in the telecom domain. Residing on top of the control layer, this layer consists of different applications profiting from the collected data at the control layer via the northbound interfaces (e.g., REST). Similarly, on top of the eNBCs, we can have an application layer but for preventing unauthorized use of the access network, it is preferable to keep the application layer on top of the core controller. So, any application should be authenticated by the core control entity before being deployed over a slice of the underlying network. We take advantage of the network slicing concept provided by NFV technology to allow the deployment of an application per network slice. Examples of applications that can be deployed over the 5G network are: network monitoring, voice call applications, social networking applications, remote home security management, etc.

The core SDN controller is equipped with security and verification modules, each connected to the eNBCs over the security and verification planes, respectively. The security module is responsible for collecting and analyzing all data traffic coming from all eNBCs to detect abnormal events (security breaches, UE spoofing, access control manipulation, and authentication attempts) and to trigger alarms to the controller to take necessary preventive actions. The verification module is responsible for analyzing all cell topology changes, benefiting from the controller's view of the network and verifying them against routing and management policies, including spectrum utilization policies. Other functions of the verification module include flow rules consistency check in inter and intracell domains and error handling such as destination unreachable or connection loss events.

16.6 Discussion

The different features of our proposed architecture aim to overcome some of the challenges of the future 5G network. First, the architecture is hierarchical, which makes it very scalable. Second, it illustrates the use of cell densification (small cells), which supports higher network capacity. The control at the access level in such an architecture provides easier radio management, better bandwidth allocation, and interference management. In addition, the control at the core level provides different features including verification, virtualization, and security. Other features such as QoS management, network slicing, and spectrum access management are also satisfied through the hierarchical scheme.

In our architecture, security is an important feature that can be provided with the integration of the security and verification planes, through the separation between the control, data, and security channels. In addition, trust is an important feature in any hierarchical based network. In this context, we propose a hybrid authentication scheme. This scheme is centralized at the initial authentication step, where each newly connected device should be registered and authenticated by the core controller. The authentication request, in this case, is transmitted by the intermediate levels. Similarly, each device at each level should be authenticated by the core controller. Thus, the device certification included in the core controller reply should be stored at each level. This will eliminate the need to communicate with the controller each time the device wants to connect to the network. In this highly dynamic network, connections and disconnections are very frequent and, thus, the handover process must be as fast as possible to meet the low latency requirement. So, when the device moves to another cell within the same macrocell (same eNBC), the device authentication will be accomplished by the new SD-AP. Initially, when the handover signaling occurs, the device certificate and keys are sent by the old SD-AP to the new one. In this case, we recur to a distributed authentication scheme. This can reduce real-time signaling traffic sent to the core network, thus lowering latency and core network overhead. If the device moves to another cell attached to another macrocell, the exchange of the device certificate and keys is done by the eNBCs. Afterward, the new eNBC communicates the device credentials to the new SD-AP.

Thus, in the proposed scheme, the core controller is assumed to be a trusted party, and the SD-APs and eNBCs are allowed to perform authentication functions. Nevertheless, security is still an open issue in SDN due to the centralization of the controller, which makes the security of the whole network related to the security of the control plane. Consequently, securing the controller is a priority of any future SDN scheme. In addition, authorization and authentication of the edge nodes should always be enforced by the controlling entities up to the controller, to overcome security threats that are tied to dynamic and mobile nodes because of the potentially malicious UEs.

16.7   Conclusion

5G came to express the human need to be connected to everyone, to everything, anywhere, and at any time. This poses many challenges in terms of the required capacity, low latency for time-critical applications, management of the huge number of connected devices, the heterogeneity associated with the integration of multiple vertical silos, the required high data rates for augmented reality and better QoE, the management of the big data, energy efficiency, in addition to the security and privacy concerns. Therefore, new architectural designs have to be investigated, as the current mobile and Internet infrastructures are incapable of handling these tight requirements because of their inflexibility, hardware dependency, and static configurability. The upcoming challenges call for new revolutionary technologies to enable network flexibility, dynamicity, and configurability. The theme of this chapter is about the main 5G enabling technologies. We argue that SDN coupled with NFV have the power to overcome the added complexity of the future networks. In addition, the extended cloud technology to the edge network is also promising in context of reducing latency, providing location-based services and augmenting QoS. We have reviewed most of the previously proposed architectures integrating these technologies. Furthermore, we proposed a new 5G architecture intermingling the centralizing of control inherited from the SDN paradigm and the distributed scheme inherited from the edge computing paradigm. However, the scalability of such a network calls for hierarchical centralized control as discussed in the proposal.

A main drawback in the 5G previous research works is the lack of a unified vision on the future 5G challenges, requirements, and characteristics. Also missing are real-case scenario implementations, simulations and emulations; thus, the correctness of the proposed solutions has not been completely verified. Regulations for data use rights are mandatory to prevent any misuse of private user data. Furthermore, the mobile technology leaders should coordinate with the mobile device manufacturers, as the new network infrastructure will impose new technological requirements (mobile devices integrating millimeter wave receivers, massive MIMO antennas, etc.) in addition to the new energy efficiency requirements. Therefore, applying SDN in the mobile domain calls for further consideration in terms of new mobile controller design. Although SDN was successfully used in the networking domain, the telecommunication domain requires different control mechanisms (charging, availability, user registration, etc.) On the other hand, the edge/fog computing is in its early stage, and further work and standardization efforts are under way.

Finally, the D2D communication paradigm involved in the 5G mobile network imposes new challenges in terms of charging process, security and privacy inquiries, etc. in absence of any authorized control party in this case. These and other challenges have to be considered in future 5G research work to ensure a safe, efficient, and real evolutionary upgrade of the current mobile infrastructure to the "5G dream world."

Acknowledgment

This research is funded and supported by TELUS Corp., Canada, the American University of Beirut (AUB) Research Board, Lebanon, and the National Council for Scientific Research (NCSR), Lebanon.

References

[1]   P. Hu, "A system architecture for software-defined industrial internet of things," in *2015 IEEE International Conference on Ubiquitous Wireless Broadband (ICUWB)*, 2015, pp. 1–5.

[2]   Cisco [Online]. Available: http://www.cisco.com/c/en/us/solutions/service-provider/visual-networking-index-vni/index.html#~mobile-forecast.

[3]   I. F. Akyildiz, S. Nie, S. Lin and M. Chandrasekaran, "5G roadmap: 10 key enabling technologies," *Computer Networks*, vol. 106, pp. 17–48, 2016.

[4]   R. Wang, J. Cai, X. Yu and S. Duan, "Disruptive technologies and potential cellular architecture for 5G," *Open Electrical & Electronic Engineering Journal*, vol. 9, pp. 512–517, 2015.

[5]   P. Demestichas, A. Georgakopoulos, K. Tsagkaris and S. Kotrotsos, "Intelligent 5G networks: managing 5G wireless/mobile broadband," *IEEE Vehicular Technology Magazine*, vol. 10, pp. 41–50, 2015.

[6]   T. Maksymyuk, M. Klymash and M. Jo, "Deployment strategies and standardization perspectives for 5G mobile networks," in *2016 13th International Conference on Modern Problems of Radio Engineering, Telecommunications and Computer Science (TCSET)*, 2016, pp. 953–956.

[7]   P. Rost, A. Banchs, I. Berberana, *et al.*, "Mobile network architecture evolution toward 5G," *IEEE Communications Magazine*, vol. 54, pp. 84–91, 2016.

[8]   O. Salman, I. H. Elhajj, A. Kayssi and A. Chehab, "SDN controllers: a comparative study," in *2016 18th Mediterranean Electrotechnical Conference (MELECON)*, 2016, pp. 1–6.

[9]   N. McKeown, T. Anderson, H. Balakrishnan, *et al.*, "OpenFlow: enabling innovation in campus networks," *ACM SIGCOMM Computer Communication Review*, vol. 38, pp. 69–74, 2008.

[10] OpenNetwotking [Online]. Available: https://www.opennetworking.org/.

[11] X. Jin, L. Li, L. Vanbever and J. Rexford, "Softcell: scalable and flexible cellular core network architecture," in *Proceedings of the Ninth ACM Conference on Emerging Networking Experiments and Technologies*, 2013, pp. 163–174.

[12] M. Yang, Y. Li, D. Jin, L. Su, S. Ma and L. Zeng, "OpenRAN: a software-defined RAN architecture via virtualization," in *ACM SIGCOMM Computer Communication Review*, 2013, pp. 549–550.

[13] M. Bansal, J. Mehlman, S. Katti and P. Levis, "OpenRadio: a programmable wireless dataplane," in *Proceedings of the First Workshop on Hot Topics in Software Defined Networks*, 2012, pp. 109–114.

[14] K. Pentikousis, Y. Wang and W. Hu, "Mobileflow: toward software-defined mobile networks," *IEEE Communications Magazine*, vol. 51, pp. 44–53, 2013.

[15] J. Samad, S. W. Loke and K. Reed, "Mobile cloud computing," in *Cloud Services, Networking, and Management*, John Wiley & Sons, Inc., 2015, pp. 153–190.

[16] A. Checko, H. L. Christiansen, Y. Yan, *et al.*, "Cloud RAN for mobile networks—a technology overview," *IEEE Communications Surveys & Tutorials*, vol. 17, pp. 405–426, 2015.

[17] O. Salman, I. Elhajj, A. Kayssi and A. Chehab, "Edge computing enabling the internet of things," in *Internet of Things (WF-IoT), 2015 IEEE Second World Forum On*, 2015, pp. 603–608.

[18] M. Patel, B. Naughton, C. Chan, N. Sprecher, S. Abeta and A. Neal, "Mobile-edge computing introductory technical white paper," *White Paper, Mobile-Edge Computing (MEC) Industry Initiative*, 2014.

[19] P. K. Agyapong, M. Iwamura, D. Staehle, W. Kiess and A. Benjebbour, "Design considerations for a 5G network architecture," *IEEE Communications Magazine*, vol. 52, pp. 65–75, 2014.

[20] V. Tikhvinskiy, G. Bochechka and A. Gryazev, "QoS requirements as factor of trust to 5G network," *Journal of Telecommunications and Information Technology*, vol. 1, pp. 3–8, 2016.

[21] N. Panwar, S. Sharma and A. K. Singh, "A survey on 5G: the next generation of mobile communication," *Physical Communication*, vol. 18, part 2, pp. 64–84, 2015.

[22] N. Nikaein and R. Knopp, "Software-defined 5G networks: challenges, and technologies," *ACM Sigmetrics*, France, 2016.

[23] L. Wang and H. Tang, *Device-to-Device Communications in Cellular Networks*. Berlin: Springer, 2016.

[24] E. Ayanoglu, "5G today: modulation technique alternatives," in *2016 International Conference on Computing, Networking and Communications (ICNC)*, 2016, pp. 1–5.

[25]   P. Schneider and G. Horn, "Towards 5G security," in *Trustcom/BigDataSE/ISPA, 2015 IEEE*, 2015, pp. 1165–1170.

[26]   S. Kukliński, Y. Li and K. T. Dinh, "Handover management in SDN-based mobile networks," in *2014 IEEE Globecom Workshops (GC Wkshps)*, 2014, pp. 194–200.

[27]   M. Peradilla and Y. Jung, "Combined operations of mobility and NAT management on the horizontal model of software-defined networking," in *Proceedings of the International Conference on Internet of Things and Cloud Computing*, 2016, pp. 31–40.

[28]   T. Chen, M. Matinmikko, X. Chen, X. Zhou and P. Ahokangas, "Software defined mobile networks: concept, survey, and research directions," *IEEE Communications Magazine*, vol. 53, pp. 126–133, 2015.

[29]   H. J. Einsiedler, A. Gavras, P. Sellstedt, R. Aguiar, R. Trivisonno and D. Lavaux, "System design for 5G converged networks," in *2015 European Conference on Networks and Communications (EuCNC)*, 2015, pp. 391–396.

[30]   M. Liyanage, I. Ahmed, M. Ylianttila, *et al.*, "Security for future software defined mobile networks," in *2015 Ninth International Conference on Next Generation Mobile Applications, Services and Technologies*, 2015, pp. 256–264.

[31]   Z. Yan, P. Zhang and A. V. Vasilakos, "A security and trust framework for virtualized networks and software-defined networking," *Security and Communication Networks*, 2015.

[32]   J. Gozalves, "Fifth-generation technologies trials [mobile radio]," *IEEE Vehicular Technology Magazine*, vol. 11, pp. 5–13, 2016.

[33]   J. Costa-Requena, J. L. Santos, V. F. Guasch, *et al.*, "SDN and NFV integration in generalized mobile network architecture," in *2015 European Conference on Networks and Communications (EuCNC)*, 2015, pp. 154–158.

[34]   I. F. Akyildiz, S. Lin and P. Wang, "Wireless software-defined networks (W-SDNs) and network function virtualization (NFV) for 5G cellular systems: an overview and qualitative evaluation," *Computer Networks*, vol. 93, pp. 66–79, 2015.

[35]   V. Yazıcı, U. C. Kozat and M. O. Sunay, "A new control plane for 5G network architecture with a case study on unified handoff, mobility, and routing management," *IEEE Communications Magazine*, vol. 52, pp. 76–85, 2014.

[36]   H. H. Cho, C. F. Lai, T. K. Shih and H. C. Chao, "Integration of SDR and SDN for 5G," *Access, IEEE*, vol. 2, pp. 1196–1204, 2014.

[37]   S. Sun, L. Gong, B. Rong and K. Lu, "An intelligent SDN framework for 5G heterogeneous networks," *Communications Magazine, IEEE*, vol. 53, pp. 142–147, 2015.

[38]   S. M. Raza, D. S. Kim and H. Choo, "The proposal for SDN supported future 5G networks," in *Proceedings of the 2014 Conference on Research in Adaptive and Convergent Systems*, 2014, pp. 180–185.

[39]   H. Wang, S. Chen, H. Xu, M. Ai and Y. Shi, "SoftNet: a software defined decentralized mobile network architecture toward 5G," *IEEE Network*, vol. 29, pp. 16–22, 2015.

[40] J. Costa-Requena, R. Kantola, J. Llorente, "Software defined 5G mobile backhaul," in *2014 First International Conference on 5G for Ubiquitous Connectivity (5GU)*, 2014, pp. 258–263.

[41] J. Costa-Requena, V. F. Guasch and J. L. Santos, "Software defined networks based 5G backhaul architecture," in *Proceedings of the Ninth International Conference on Ubiquitous Information Management and Communication*, 2015, Article No. 35.

[42] A. M. Akhtar, X. Wang and L. Hanzo, "Synergistic spectrum sharing in 5G HetNets: a harmonized SDN-enabled approach," *IEEE Communications Magazine*, vol. 54, pp. 40–47, 2016.

[43] J. S. Thainesh, N. Wang and R. Tafazolli, "A scalable architecture for handling control plane failures in heterogeneous networks," *IEEE Communications Magazine*, vol. 54, pp. 145–151, 2016.

[44] V. Jungnickel, K. Habel, M. Parker, *et al.*, "Software-defined open architecture for front-and backhaul in 5G mobile networks," in *2014 16th International Conference on Transparent Optical Networks (ICTON)*, 2014, pp. 1–4.

[45] C. Giraldo, F. Gil-Castiñeira, C. López-Bravo and F. J. González-Castaño, "A software-defined mobile network architecture," in *2014 IEEE 10th International Conference on Wireless and Mobile Computing, Networking and Communications (WiMob)*, 2014.

[46] D. Soldani and A. Manzalini, "A 5G Infrastructure for 'Anything-as-a-Service'," *Journal of Telecommunications System & Management*, vol. 3, no. 2, pp. 1–10, 2014.

[47] R. Guerzoni, R. Trivisonno and D. Soldani, "SDN-based architecture and procedures for 5G networks," in *2014 First International Conference on 5G for Ubiquitous Connectivity (5GU)*, 2014, pp. 209–214.

[48] R. Trivisonno, R. Guerzoni, I. Vaishnavi and D. Soldani, "Towards zero latency software defined 5G networks," in *2015 IEEE International Conference on Communication Workshop (ICCW)*, 2015, pp. 2566–2571.

[49] A. Bradai, K. Singh, T. Ahmed and T. Rasheed, "Cellular software defined networking: a framework," *IEEE Communications Magazine*, vol. 53, pp. 36–43, 2015.

[50] Z. Kaleem, Y. Li and K. Chang, "Architecture and features for 5G mobile personal cell," in *Information and Communication Technology Convergence (ICTC), 2015 International Conference On*, 2015, pp. 164–166.

[51] R. Inam, A. Karapantelakis, K. Vandikas, L. Mokrushin, A. Vulgarakis Feljan and E. Fersman, "Towards automated service-oriented lifecycle management for 5G networks," in *2015 IEEE 20th Conference on Emerging Technologies & Factory Automation (ETFA)*, 2015, pp. 1–8.

[52] A. Checko, M. Berger, G. Kardaras, L. Dittmann and H. L. Christiansen, "Cloud radio access network architecture: towards 5G mobile networks," *PhD Thesis, Technical University of Denmark*, pp. 1–170, June 2016.

[53] A. F. Cattoni, P. E. Mogensen, S. Vesterinen, *et al.*, "Ethernet-based mobility architecture for 5G," in *2014 IEEE Third International Conference on Cloud Networking (CloudNet)*, 2014, pp. 449–454.

[54] P. Ameigeiras, J. J. Ramos-munoz, L. Schumacher, J. Prados-Garzon, J. Navarro-Ortiz and J. M. Lopez-soler, "Link-level access cloud architecture design based on SDN for 5G networks," *IEEE Network*, vol. 29, pp. 24–31, 2015.

[55] A. Roozbeh, "Distributed cloud and de-centralized control plane: a proposal for scalable control plane for 5G," in *2015 IEEE/ACM Eighth International Conference on Utility and Cloud Computing (UCC)*, 201.

[56] M. Moser and F. Jaramillo, "Extending software defined networking to end user devices," in *2015 24th International Conference on Computer Communication and Networks (ICCCN)*, 2015, pp. 1–8.

[57] P. Iovanna and F. Ubaldi, "SDN solutions for 5G transport networks," in *2015 International Conference on Photonics in Switching (PS)*, 2015, pp. 297–299.

[58] M. Rahman, C. Despins and S. Affes, "HetNet cloud: leveraging SDN and cloud computing for wireless access virtualization," in *2015 IEEE International Conference on Ubiquitous Wireless Broadband (ICUWB)*, 2015, pp. 1–5.

[59] F. Yang, H. Wang, C. Mei, J. Zhang and M. Wang, "A flexible three clouds 5G mobile network architecture based on NFV and SDN," *China Communications*, vol. 12, pp. 121–131, 2015.

[60] A. Hussein, I. H. Elhajj, A. Chehab and A. Kayssi, "SDN security plane: an architecture for resilient security services," in *2016 IEEE International Conference on Cloud Engineering Workshop (IC2EW)*, Berlin, Germany, 2016.

[61] A. Hussein, I. H. Elhajj, A. Chehab and A. Kayssi, "SDN verification plane for consistency establishment," in *The Twenty-First IEEE Symposium on Computers and Communications*, Messina, Italy, 2016.

Chapter 17

Low latency optical back- and front-hauling for 5G

Pandelis Kourtessis[1], Milos Milosavljevic[1], and Matthew Robinson[1]

Abstract

Looking forward to the not-so-far future, wireless networks will comprise centralised processing, mixed macro and small cells deployments as well as new radio technologies in order to support very high data rates and traffic types that are characterised by connectionless and sporadic transmission of short packets such as those employed by machine to machine communication devices. In current radio access networks, there is a very clear difference between the fronthaul and backhaul. In comparison with the advent of centralised radio access networks, the difference between the fronthaul and backhaul networks has been shifted further away from the user. Subsequently, in the latest propositions for 5G architectures, they are being merged into an 'xhaul' (crosshaul) network where the fronthaul and backhaul concepts no longer exist. In particular instead of using a dedicated centralised processing pool for a set of cellular access points, the idea of using centralised processing within the core network has emerged. With the use of network function virtualisation and centralised processing in data centres, cloud-RANs can be used to provide access to the distribution antenna, removing the need for backhauling and fronthauling. The cloud-RAN can perform the duties of the Mobility Management Entity and Serving Gateway and at the same time can also process each cellular access point's analogue processing using a flexible virtualised software environment. Assuming this is used along with split processing, what needs also to be addressed is the means of communication between the cloud-RAN and the distribution antenna. Traditional solutions such as the common public radio interface might not be sufficient to meet all requirements. Largely, Ethernet is being proposed. When Ethernet is used, software-defined networking (SDN) can also be used to dynamically control data flows to and from the cloud-RAN, as well as providing additional benefits, such as network slicing, allowing multiple cellular operators to use the same xhaul infrastructure. This chapter, therefore, largely elaborates on xhaul networks by investigating the potential of SDN to provide an effective

[1]Engineering and Technology, University of Hertfordshire, UK

user experience for the services provided. The control of specific services such as billing, roaming and registration could then be sent via alternative links such as satellite links, as latency for these packets are not critical, resulting in reduced packet delay on the data plane. It is apparent that for Gbps wireless connectivity, targeted by 5G, the data rate requirements on the centralised cloud xhaul link will be in the range of several Gbps with a latency requirement close to 1 ms.

17.1   Introduction

The currently installed access networks have been designed and built to support a set of services that are now outdated and soon to be superseded. Replacement services are growing at an exponential rate, and the fundamental underlying networks that support accessing these services will soon no longer be able to support them. This is most easily seen in the traffic forecasts from the core network providers. In Cisco's annual global mobile data traffic forecast update, the expected increase in content demand from users can be seen in four different areas [1].

First, within cellular communications, the amount of traffic requested by users in 2015 was 74% higher than in 2014; an increase of 1.6 EB per month from 2.1 EB per month at the end of 2014 to 3.7 EB per month at the end of 2015. It is forecast that by 2020, global mobile data traffic will increase by 700% compared to 2015, with data throughput exceeding 30.6 EB per month. In addition, the number of mobile devices per capita is expected to increase by 46% from 2015 to 2020; totalling 11.6 billion mobile-connected devices by the end of 2020 [1].

Second, the fixed access networks (femtocell and WiFi) are expected to play a much larger role in the future. In 2015, 51% of the total mobile data traffic was offloaded onto the fixed access network either through femtocell or through WiFi technology. This accounted for an additional 3.9 EB per month of data traffic in 2015 and is set to account for 38.1 EB per month of traffic by 2020. Without WiFi and femtocell technology, it is predicted that the total mobile data traffic would grow at a compound annual growth rate (CAGR) of 67% instead of 57% between 2015 and 2020. This can be seen in Figure 17.1. Furthermore, the number of public WiFi hotspots and femtocells (globally) is predicted to increase 7× between 2015 and 2020 from 64.2 million to 432.5 million, and the number of home WiFi hotspots and femtocells (globally) is expected to increase from 56.6 million to 423.2 million by 2020 [1].

Third, the type of content being requested by users is also changing to primarily video-centric services. It is forecast that mobile video will grow at a CAGR of 62% between 2015 and 2020. This CAGR is higher than the overall average mobile traffic CAGR of 53%, meaning the video services will slowly demand a greater ratio of the overall bandwidth. Video has already become the primary requested service in 2012, by representing more than half of the global mobile traffic, and this increase is set to continue. Out of the 30.6 EB total crossing the network per month, 23.0 EB is expected to be for video services by 2020, which is approximately 75% of the total [1]. This can be seen in Figure 17.2.

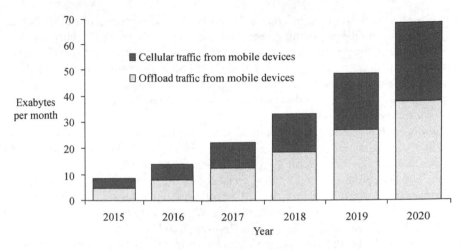

Figure 17.1   Offload and cellular traffic forecast [1]

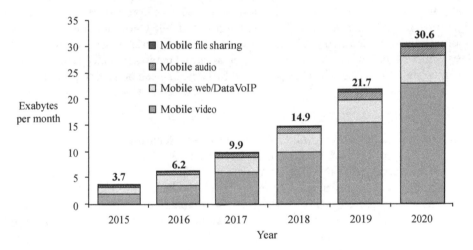

Figure 17.2   Mobile video traffic forecast [1]

Finally, although video services already existed, and are simply growing in popularity and therefore consumption, there is also a new type of service being introduced as people's day-to-day lives change to be more technology oriented. These services fall under the category of machine-to-machine (M2M) communication. M2M is growing very quickly because a multitude of devices are becoming more autonomous and cloud oriented. Everyday appliances such as fridges and washing machines are becoming more and more connected to the cloud by offering intelligence not seen before to users that allow the devices to autonomously order new components for themselves when they are needed, or order food for their owners when they're running low [1].

Equally, on a more extravagant side, higher risk applications of M2M such as autonomous vehicles are being researched and will likely be slowly introduced to everyday life in the near future. These high risk applications require a set of network capabilities that are currently not possible. M2M communication typically needs an end-to-end latency of less than 1 ms, but conversely only requires low data rates. This is because they typically send very small amounts of data that are only useful when they are received immediately.

In 2015, 604 million M2M connections were made; this is set to increase to 3.1 billion by 2020, which is a 38% CAGR and a 5× increase. Wearable devices with either embedded cellular technology or connections to the network using partnered devices by Bluetooth are forecast to account for a large percentage of this increase. It is estimated that by 2020, there will be 601 million wearable devices globally, growing from 97 million in 2016 at a CAGR of 44%. Wearables are expected to account for 1.1% of the total global mobile data traffic from 2020 [1].

To conclude, for these new demands to be met by the access network operators, a shift in infrastructure needs to be undertaken. The next generation of mobile communications, 5G, is currently being designed to fulfil these service requirements based on a new set of targets. The target latency has been set at 1 ms end-to-end delay, and the target throughput has been set at 10 Gbps. These targets fulfil the requirements of M2M communications and also fulfil the requirements of video streaming at ever increasing resolutions and immersive techniques such as virtual reality, 3D video, and ultrahigh frame rates [2]. These targets as produced by IMT-2020 can be seen in Figure 17.3.

17.1.1   Key-enabling technologies and services

This chapter elaborates on the back-/front-haul network infrastructure able to meet the requirements of an increasingly cloud-based, centrally oriented 5G network. In current radio-access networks, there is a very clear difference between the fronthaul and backhaul. The fronthaul section connects the user to their local cellular access point, for example the E-UTRAN Node B (eNodeB), whereas the backhaul section connects the cellular access point to the cellular infrastructure, such as the Serving Gateway (SGW) and Mobility Management Entity (MME). In this context, the eNodeB is performing all of the functions (including analogue waveform composition) associated with the fronthaul and is using the backhaul network only to connect to the rest of the cellular network.

With the advent of centralised radio-access networks (CRANs), the difference between the fronthaul and backhaul networks is shifted further away from the user. In a CRAN, the processing for multiple cellular access points is performed in one location. In this context, the fronthaul can be viewed as the connection between the centralised processing pool and the user. The fronthaul, therefore, includes the cellular access section between the distribution antenna and the user and also the connection between the distribution antenna and the centralised processing pool. The backhaul is still used to connect the central processing pool to the MME or SGW.

In these scenarios, there is a clear difference between the fronthaul and backhaul, whereas in the proposed 5G architectures of the future, the difference between the

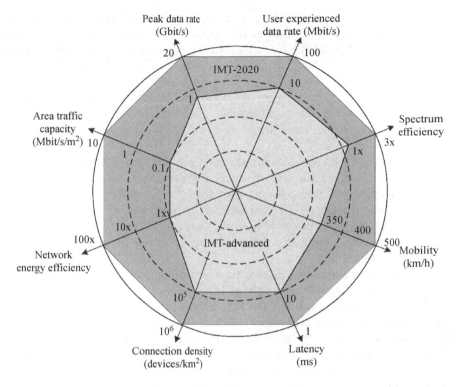

Figure 17.3   5G targets [2]

fronthaul and the backhaul is becoming increasingly blurred. One of the proposed techniques for 5G features Split Processing, where the waveform is no longer all created in the centralised processing pool. Instead, some of the latter stages of the waveform creation are moved to the distribution antenna to be completed before the analogue waveform is broadcasted to the user. In this context, the fronthaul is now part of the central processing unit, extending all the way to the distribution antenna and the user, where the fronthaul is actually preforming some of the processing of the analogue waveform (but in a digital format). The backhaul network is still connecting the centralised processing pool to the rest of the cellular network resources.

In 5G, the fronthaul and backhaul networks are being merged into a 'xhaul' (crosshaul) network where instead of using a dedicated centralised processing pool for a set of cellular access points, centralised processing is performed within the core network. Cloud-RANs can be used to provide the access to the distribution antenna and therefore software-defined networking (SDN) if applied could provide intelligent networking allowing very high data-rate services and very low latency services to co-exist in the same architecture with assured user experience. It is worth noting that from this point of this chapter, the terms xhaul (xhauling) and fronthaul (fronthauling) are used interchangeably.

The research community has been paying a lot of attention to SDN for access communications in a myriad of areas. Key aspects of common public radio interface over Ethernet (CPRIoE) [3] have been investigated in the last year, providing key results that show how CPRIoE fronthauling can be used for SDN-enabled baseband unit (BBU) pooling [4–9].

SDN-enabled quality-of-experience (QoE) and quality-of-service (QoS) assurance for end users using video services has also been looked at, where sweet-point bandwidth allocation has been introduced to video service bandwidth allocation research [10–16].

SDN-enabled Time-Wavelength Division Multiplexing–Passive Optical Networks (TWDM-PONs) have been researched by the optical community to provide fast reconfigurability at the optical network units (ONUs) and optical line terminals (OLTs) by using the OpenFlow [17] protocol and explicitly developed SDN controllers. These TWDM-PONs have been shown to provide legacy support for xPON and GPON1 services, while also potentially providing 100-GbE connectivity using next-generation techniques [18–25].

SDN-enabled content distribution network (CDN) and Internet service provider (ISP) collaboration has also been very recently explored [26]. It has been suggested that SDN technology could allow CDNs to influence the ISP architecture and topology to provide their users with exemplary video services, while not negatively impacting ISP workloads. Equally important SDN-enabled caching has only been discussed where it was suggested that intelligent caching at the access network can rapidly reduce the load on the backhaul network and also decrease user latency experience in high use cases [27–30].

To provision the forecasted increase in video consumption, there are two main areas of relevance: live video and on-demand video. An example of live video is DVB broadcast services such as DVB-T2 BBC One, whereas on-demand services are available for streaming at any time, there is no broadcast time. To ease the access network of live video delivery, live video offloading can be used to move the video load onto another source. Instead of streaming the live video from the Internet via the backhaul network, the video can be sourced closer to the user, in the access network, by capturing the live video from another feed, for example satellite, cable, or terrestrial services. This captured live video can then be sent to the user as a replacement service, reducing the load on the backhaul network, reducing the latency experienced by the user, and providing a higher quality video than is typically available via Internet-streaming services.

An example of on-demand video is BBC iPlayer, where pre-recorded programmes can be streamed by the user from CDNs at any time of day at their convenience. To ease the access network of on-demand video playback, intelligent caching can be used within the access network to reduce the bandwidth needed on the backhaul link. This requires a caching infrastructure to be installed within the access networks that can intelligently cache content for users based on the most likely requested content. Intelligent caching in turn can be improved by allowing CDN and ISP cross-platform cooperation.

The large difference in characteristics between M2M and video services can be helped by splitting the network into different virtual networks that are designed for difference services. For example, video and M2M communications could run in different 'splits' that are tailors towards either high data rates or low latencies. Splits can also be used to create separate networks for difference network operators using the same underlying hardware.

All of these opportunities discussed above are achievable when using an SDN access network. Live and on-demand video feeds can be automatically re-routed to local alternatives within the access network by using intelligent network controllers. CDNs and ISPs can collaborate by using SDN controllers to provide instant network reconfigurability based on current network demands and topology constraints. Network splitting can be achieved by using network slicing in an SDN network [31]. This allows several individual network controllers to run at the same time on their own virtual subset of a real physical network. Each slice can be managed by individual network operators in complete anonymity of each other and can therefore be tailored by those network operators to provide different services such as low latency or high bandwidth specific slices.

Following a review of the most up to date solutions presented in literature to enable intelligently programmable networks and services, this chapter will present, describe, and critically evaluate the operation of an SDN network featuring replacement SAT>IP and cached video services, an SDN controllable physical layer, a heterogeneous SDN-enabled access infrastructure that allows cellular, legacy PON, and fixed wireless networks to run in isolation at the same time all using their own network controllers with the use of network slicing. The SDN-based SAT>IP delivery access network is being researched with the aim of increasing the user QoE [32]. This is being done by developing network controller applications that adapt the network to let the user achieve the optimum QoE based on live feedback from the user's video client to the network application.

17.2   CPRI over Ethernet mobile fronthauling

CPRIoE is the concept of packetizing CPRI data into Ethernet frames for transportation over an Ethernet network. A brief overview of CPRI [1] and its integration to CPRIoE is shown in Figures 17.4 and 17.5, respectively.

Within the topic of mobile fronthauling, CPRIoE has widely been accepted as one of the probable techniques for next generation systems. In [4], an overview of CPRI for centralised radio access network (CRAN)-based long-term evolution (LTE) [5] scenarios was given, wherein it was stated that plain Ethernet is asynchronous and best effort, therefore making it unsuitable for the transport of CPRI traffic. Concurrently, the Time Sensitive Networking Task Group (TSNTG) [6] was formed within IEEE 802.1 to develop new extensions to support the forwarding of highly time sensitive Ethernet traffic with particular jitter and delay guarantees [4].

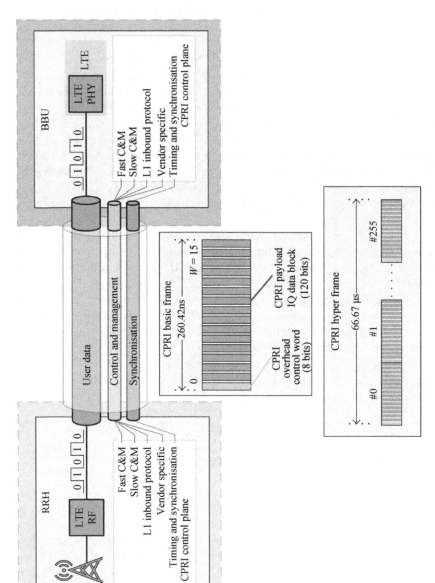

Figure 17.4 Standard CPRI

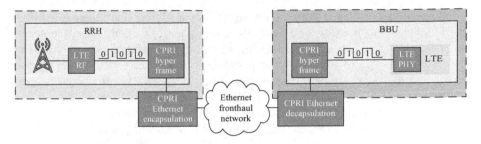

Figure 17.5   CPRI over Ethernet implementation

In [7], insight is given to how CPRI data would be affected by the two Ethernet enhancements that have been suggested by the TSNTG. The two propositions are frame pre-emption (802.1Qbu) and scheduled traffic (802.1Qbv). In [7], a CPRIoE network was simulated using NS3 [33] with these two additional techniques applied. It was found that frame pre-emption in both dedicated CPRI and shared channel scenarios could not meet the CPRI jitter requirement of 8.138 ns [3,7]. However, when scheduled traffic was included in the simulation with a well-defined scheduling algorithm, the jitter could be significantly lowered or removed completely, thus allowing the CPRI jitter requirement to be met [7]. In the experiment, a schedule was computed prior to simulation and then checked by the switching nodes to see which queue was allowed to transmit at a given time. It was noted by the authors that their implementation is a proof of concept and not 100% accurate to the 802.1Qbv standard that is currently in the process of being drafted; however, it is also stated that in the case of CPRIoE, their proof of concept is completely valid for demonstrating the Ethernet enhancement [7].

Following the theme of enabling CPRIoE, authors in [8] used a novel CPRI compression technique to reach the 8.138-ns jitter threshold. The author encapsulated CPRI data that had been compressed by up to 45% within 1,064-byte Ethernet frames for transportation over a PON to multiple ONUs. The CPRIoE convertors were equipped with jitter absorbers to reduce the jitter at the remote radio head (RRH). The jitter absorber buffered the recovered CPRI frames and transmitted them at a constant latency time, therefore offsetting any positive or negative latency to a universal latency time, which had very little jitter. The CPRI reference clock was also extracted from the CPRI dataflow at the OLT for PON synchronisation so the ONU could be provided with a synchronised clock and time. A fixed bandwidth allocation algorithm was also used instead of conventional dynamic bandwidth algorithms to eliminate round trip delay and achieve low latency [8].

As can be seen from the above references, the methods of attaining CPRIoE are still being researched. Authors in [8] are opting for in-phase and quadrature-phase (IQ) compression and fixed bandwidth allocation algorithms, and authors in [4] are not using IQ compression but are using a similar type of technique for bandwidth allocation by using the scheduled traffic draft standard set out in 802.1Qbv.

Authors in [9] agreed with the sentiments above stating that Ethernet for fronthaul is appealing because of Ethernet's maturity and adoption within the access network currently. It is noted that the operations, administration, and maintenance (OAM) capabilities of Ethernet offer a standardised way of fault finding, support performance monitoring, and management of the network. By using Ethernet, key advantages would be seen such as lowering the cost of hardware by using industry standard equipment, enabling convergence by sharing equipment in fixed access networks, allowing the move to network function virtualisation and overall network orchestration [9]. Authors in [9] also suggested that the delay and jitter requirements of CPRI [3] could potentially be met by allowing Ethernet switches to operate in two modes: store-and-forward, where the whole Ethernet frame is received at the switch, and then checked before it is forwarded on to the output port; and cut through, where the Ethernet frame is only buffered until the destination data is received, at which time the Ethernet frame is forwarded on immediately to the output port, therefore lowering delay time. It was noted by the authors that the errors associated with not error checking at switches would not be significant due to the usually very low error rate. Authors in [9] then suggested that SDN and self-optimisation networks (SONs) could be used for path management through the fronthaul network, and that a network control function could set up paths in the network between the BBU pool and the RRHs with knowledge that there will be no contention.

The above literature explains how research is currently providing solutions that will allow CPRIoE to be become a reality. Furthermore, it shows how SDN-enabled networks can provide the management needed in the next generation of mobile fronthaul networks.

17.3   QoE for video services

At present, there is key research being made into SDN-enabled QoE improvement for video services [10–16]. The ideas of sweet-point bandwidth for users, buffer aware applications, dynamic route and video quality to client, and CDN integration to the SDN stack have been presented recently.

In [11], a QoE aware video-streaming application using SDN was introduced. It was summarised that over-the-top video delivery is challenging because clients, content service providers, and ISPs do not have a global view of the end-to-end networking conditions which means there is no method for video service providers to track video streams in real time. In current systems, the video server is also rarely changed during video playback even when playback is poor at the client side, leading to bad QoE for the users. It was noted that this was combated by client-side video rate selection algorithms so the users experience less stutter when a lower resolution is selected for poor network conditions. But, this is not a solution to the network congestion; it is an attempt to patch a poorly designed system. Therefore, authors in [11] proposed a system wherein the network congestion can be mitigated by using an SDN enabled network with the use of intelligent adaptive routing. In [11], an HTML5 plugin was created to monitor QoE factors at the client side, where buffering status and video resolution were used to analyse the user's perceived experience while a video was playing. This information was periodically reported to the video distribution node and was made available to the SDN controller so northbound applications could make changes to the network based on current user QoE statistics. The experimental setup consisted of 8 real Junos Space SDN switches using Juniper for SDN control and management. The network reconfiguration application was created to communicate with the controller using RESTful application programming interfaces (APIs), and the HTML5 application used HTTP POST messages to periodically deliver the QoE information to the network controller. Overall, this technique provided an improvement of 55.9% compared to a non-QoE aware reference technique [11].

Similarly to authors in [11], authors in [10] also use client side applications to change the network configuration based on the users' perceived QoE. Authors in [10] used a buffer aware technique to produce the QoE metric to the SDN controller, and also used the SDN controller's knowledge of the current network capability to choose between real-time live streaming and client-side buffering modes, and also change to the most appropriate transmission paths based on the utilisation and stability of the network switches. The two buffering modes aim to preload video data so even when network bandwidth is poor, the user can still view high-quality video for a period of buffered time, t. When the video buffer status is higher than $2t$, the quality of the video is increased, therefore requiring more bandwidth. When the buffer status is less than $2t - 1$, the video quality is decreased so less bandwidth is required. This effectively means that when the network condition is good, the download speed increases, and the video segments accumulate rapidly in the buffer so future network bandwidth low points can be suppressed. By monitoring the change in buffer status, a trend can be derived which is used to forecast the network behaviour and is therefore used to trigger video scale up or down events [10]. In poor networking conditions, when the buffering status has reached 0, the SDN controller switches the client to real-time video mode. The authors experimentally proved that their mechanism could maintain a certain level of streaming quality for SDN-enabled 5G wireless networks and ensure smooth and complete streaming services [10].

In [13], essentially, a caching technique is applied to the end users point of contact in the network to increase the QoE for video delivery. Authors stated that SDN centralised control and programmable node storage features are used to optimise HTTP video quality. The programmable storage node is a router with storage functionality. The node periodically requests and accepts new video content from source servers under the control of an SDN controller that can load balance, redirect videos, and manage content. When the user requests HTTP video services, the request is forwarded to the SDN controller by the programmable storage router. According to the utilisation of the programmable storage routers in the network, the SDN controller specifies the relevant programmable storage router for video requests from that particular user. The technique was experimentally evaluated via simulation, and the authors stated that the SDN scheme greatly improved the HTTP video quality and user experience.

In [14], bandwidth was allocated in a fibre-to-the-home (FTTH) network based on video QoE using SDN. The idea of 'sweet points' was introduced. This is the bandwidth x at which the perceived QoE of a user will not be increased once it is reached. For example, if a video stream requires 10 Mbps to achieve 100% QoE from the user, there is no reason to allocate the service with more than 10 Mbps. On the other hand, a bandwidth lower than 10 Mbps rapidly decreases the QoE at the client side; therefore, 10 Mbps is the sweet point for the service, and slightly larger than 10 Mbps should be allocated to it. The authors stated that the QinQ technology that is currently used in Broadband Remote Access Servers mainly divides bandwidth per service as QinQ technology does not allow the differentiation of different applications within each service [14]. A system is proposed wherein an SDN application running on a controller on the OLT side of the FTTH network hosts a socket server that can receive application bandwidth requests from users based on their QoE sweet spots. When the user requests a bandwidth allocation for an application, the user provides a value for the sweet-point bandwidth, the source IP and port, and the destination IP and port. The port information is needed because the broadband remote access server (BRAS) has no concept of the application layer. This effectively allows OpenFlow flows to be set up based on application as well as device. The SDN application then either grants or restricts the bandwidth being requested based on the ISPs maximum agreed bandwidth and network parameters such as congestion. The SDN controller can then be used by the application to set up new flows in the relevant switches to allow the minimum bandwidth required for the application. The concept was experimentally trialled using an OpenVSwitch based network running OpenFlow on a Linux server. Linux hierarchical token buckets (HTB) were used to ensure the minimum bandwidth to each queue, and a Floodlight [34] controller was used for SDN control. In the experiment, it was shown that the QoE seen by the user was met due to the sweet-point allocation, and the authors stated 'all in all, the experimental results show the feasibility of QoE-based bandwidth allocation method' [14].

In [15], local caching on client machines is used to create a CDN-using SDN. The main concept was to enable peer-assisted content delivery. The SDN controller became the point of contact for clients to reach external network servers. It kept a list of files that were currently cached in the network on client devices and could redirect end users to these locations within the local network to retrieve the content instead of using external connections to the original server. If the file was not present on the local network, the end user was redirected to the original server like in a normal network implementation. This technique effectively reduced congestion on the link to the external networks by moving network congestion to the local network, which therefore could increase the QoE to users both requesting data from local clients due to the higher internal data rates and to the users requesting data from external clients due to the reduced congestion on the external link.

As can be seen from the literature above, there are many techniques for improving the QoE to end users when an SDN enabled network is used. The techniques in [10,11] used QoE feedback to the SDN controller from the end user to enable network changes to be made, whereas [14] required the user application to proactively request a bandwidth allocation to be made. All three of these papers however used bandwidth allocation to ensure the QoE to users. On the other hand, [13,15] both used a type of local network caching to achieve higher QoE where [13] used a dedicated storage node and [15] used caches within the clients machines. There was not enough data available to determine which technique was the best at increasing QoE, but clear advantages can be seen when combining the techniques. It is evidential that a proactive and reactive technique for determining and setting bandwidth allocation could be combined with local caching in dedicated and/or client distributed form in the correct networking environment with SDN capability.

17.4   CDN and local caching

In relation to improving the QoE for users, the integration of CDNs into the ISP SDN-controlled network is also being proposed. In [26], the collaboration of CDNs and ISPs was proposed and demonstrated using SDN switches in the ISP network to provide the CDNs with a better integration with the ISPs traffic management system. The authors provide background on the current CDN ISP situation. CDN providers cannot rely on accurate link-layer information to achieve an optimal assignment of end users to CDN servers. It is also not possible to react to ISP network congestion in a timely manner, which is disadvantageous for both the CDN and the ISP. This means the CDN cannot provide optimal performance to its end users, and the ISP has to transport data on a non-optimal path across its network. In the worst case scenario, the user could be assigned to a CDN distribution node that is not connected to the users ISP network directly, so the traffic would need to transverse a peered link, which in turn would decrease the QoE for the end user and cost more for the ISP. Therefore, it can be deduced that cooperation between the CDN and ISP when making traffic engineering decisions is worthwhile both financially and in terms of QoE assurance to users.

A system is proposed in [26] that allows the CDN and ISP to communicate with each other for traffic engineering purposes via a novel redirection centre that resides within the ISP network. When the client requests content from the CDN, the uniform resource locator (URL) is resolved by a standard authoritative domain-name-system server which points to the ISPs local redirection centre. From this point on the request is handled by the ISP network only. The redirection centre then acts like the CDN primary cache by terminating the user's transfer control protocol (TCP) session as if the redirection centre would deliver the actual content. In parallel, the CDN local cache capable of delivering the content is calculated and selected by the redirection centre. The complete HTTP and TCP session is then migrated to the suitable CDN cache. The redirection centre takes care of all signalling at the BRAS close to the client. OpenFlow switches are used to redirect the flows to and from the client to ensure a working TCP flow. Finally, the content is delivered to the client from the selected CDN cache.

This architecture is dependent on the use of OpenFlow hardware in the ISP network and an OpenFlow capable SDN controller being located within the redirection centre. The proposed solution to CDN ISP integration was evaluated using four virtual machines (VMs). One VM ran the novel redirection centre, 2 VMs acted as CDN caches, and the last VM acted as the end user. An OpenFlow-enabled Mininet [35] setup was used to imitate the ISPs network connecting all of the VMs together. Through experimentation, it was concluded that using the novel approach led to a highly decreased number of video playback stalls in flash crowd scenarios [26].

Although CDNs are caches of contents that are set up in advance with pre-synchronised content for users to consume, local caches are also available, which use an on-demand strategy to help relieve network congestion, reduce latency, and speed up end user download speeds in a similar way to CDNs but by using a different mechanism. CDN caches reside within the core network at statically selected locations, whereas recently, caches located within the access network have been proposed and demonstrated [27–30].

In [27], a pushing and caching technique for wireless heterogeneous networks was analysed. The pushing and caching technique uses the wealth of processing power located in the access network within the base stations, relays, and end users' equipment to provide caching functionality within the access network itself. During low traffic load periods, such as during the night, the most popular multimedia content can be pushed and cached at the relays and cache enabled user devices via broadcasting. All of the user equipment and relays with caching enabled store the same content in their local cache. When multimedia content is requested by a user, the caching space on the local device is first checked. If the content is stored within the cache, the content can be obtained immediately; otherwise, the user is directed to the closest user or relay with a copy or simply to the original source of the content using normal streaming methods if no nearly caches are available. Numerical analysis was performed in [27] based on this technique. It was found that the caching system was suitable for high density networks, and the performance gain could reach $9.2\times$ a baseline no-caching example when only 30% of the local users have the ability to cache [27]. This effectively increased the QoE to end users by reducing latency for users accessing cached content and by reducing network congestion on shared links to the original content source.

In [28], a joint-wireless-and-backhaul-load-balancing (JWBLB) framework was proposed to minimise the system content transmission delay, wherein an SDN-inspired load balancing control architecture consisting of a controller and distributed and cooperative caches was introduced. In the architecture, the distributed caches can intercept end user application payloads and extract content URLs or identification. The small-cell base stations can be connected together through local links such as X2 or over digital subscriber line (xDSL) Ethernet and are connected to access routers before converging to the mobile core network via backhaul links. An orchestrator was introduced to coordinate all of the distributed caches, to load balance, and to schedule end user content requests to distributed caches or the mobile core network. The framework operates as follows: If the requested content is cached in the end users local base station, the content is served locally. If not, the request is forwarded to the orchestrator. If the orchestrator does not have the content location information stored in its local content table, the request is forwarded to the CDN. If the orchestrator does have a listing for the content in the local caches, a JWBLB strategy is used to schedule the end users content connection concurrently with the aim of minimising the system content transmission time. The authors used a statistical simulation framework developed in MATLAB® to analyse the content delivery time based on different orchestrator algorithms. It was found that the JWBLB technique they developed achieved a great reduction in the average transmission delay and outgoing backhaul traffic load [28].

In [29], the use of caching helpers is introduced. The helpers are femto base stations that are assumed to have high storage capabilities and low-bandwidth backhaul links; they form a wireless distributed caching infrastructure. The helpers are located in static positions within a larger macro-cell and have high-bandwidth connectivity with frequency reuse capabilities. The concept is for the cache to store the most popular multimedia files and serve requests from mobile end users by enabling localised communication and hence high frequency reuse. It is proposed that if enough users request the same multimedia content, caching via helpers can effectively replace backhaul communication. When users request content that is not cached by helpers, the regular base station fulfils the request using backhaul communication. To analyse the proposed technique, a university campus scenario was simulated using a single 3rd Generation Partnership Project (3GPP) LTE R8 cell for standard mobile backhaul, and several helpers were simulated using a simplified 802.11n protocol. Real video request data for YouTube was used in the simulation to analyse different algorithms for caching control. It was found that there were significant gains in user QoE even when very simple algorithms were used. Overall, analysis proved that 400%–500% more users were able to connect to the network and consume video content with acceptable QoE that without the caching helpers [28].

References [27–29] show a progression into advanced caching techniques for access networks due to the clear improvements that can be seen from implementing caches closer to the end users, essentially decentralising network functions. The next step to including caching into an SDN capable network is to make the caching mechanism software definable, and therefore intelligently remotely controllable. Reference [30] introduces software called OpenCache [36] which is aimed at doing exactly this.

Authors in [30] introduce a software defined content caching platform called OpenCache. It is similar to SDN in that the control plane is separated from the underlying technology of the cache and APIs are used for configuration and control of the processing technology. The authors state that OpenCache is an API capable of defining fine-granular behaviour and functions of a cache deployment, but that it is not itself SDN compatible, and needs its own novel controller to run in a network. This is because there is no provision for extra communication within the current OpenFlow protocol, but there is no reason that OpenCache could not run over an OpenFlow protocol version designed for the application.

In addition, in such a scenario, the OpenCache controller could be programmed to run within an SDN controller application instead of running in its on controller. The OpenCache API provides remote control of the following functions: The start function is used to start the cache service on a particular node. Once the OpenCache controller has started a cache service, it automatically informs the network controller (which could be an SDN controller) so the forwarding layer can be modified intelligently for user to cache communication. The stop function removes cache services from a node and also informs the network controller so forwarding rules can be removed from the network. There is also a pause function to allow caching services to be stopped and started without the need for forwarding layer changes. There is also a move function that allows caches to be dynamically repositioned in the network. The controller can also use a fetch function to request that a cache preloads a specific file before it is requested by the end user. The seed function is used to reduce instances of duplication in the cache object store. Using these functions, the authors developed applications for load balancing and failover monitoring that were evaluated using the SDN capable Fed4FIRE [37] testbed with OpenCache prototypes deployed on the network.

17.5 Software-enabled passive optical networks

Until about a year ago, there were very few research results published in software enabled optical access and aggregation networks. Reference [18], for example, simply applied the OpenFlow protocol to the control of optical network elements. More recently, there has been an increased interest in how SDN can be utilised in TWDM-PONs to create a more efficient network that can: allocate bandwidth to users, support legacy PON architectures transparently, reduce the end-to-end delay in virtual setup times, and provide software reconfigurability using novel digital filter multiple access and orthogonal frequency division multiplexing (OFDM) techniques.

In [19], SDN-enabled tuneable laser ONUs and optical service units (OSUs) are used to provide reconfigurability to a TWDM-PON with energy saving features that allow the TWDM to move into time division multiplexing (TDM) mode when optical units on different wavelengths are not functioning. In [20], the same authors explored the same setup in more depth and concluded that the impact of the reconfiguration time on the average frame delay using their SDN approach is negligible.

In [21], digital filter multiple access techniques were used to produce a software reconfigurable PON where all ONUs and OLTs in the network could dynamically share the transmission medium under the control of a centralised SDN controller. In the setup each ONU and OLT used a digital-signal-processing (DSP) controller that communicated to the SDN controller to compute a set of shaping filter coefficients to perform the filtering process required by the ONU or OLT. The data to be transmitted was intensity-modulation direct detection (IMDD) based, which is digitally filtered based on the intelligently selected coefficients by the DSP controller before it is converted to analogue and then optically intensity modulated for transmission over fibre. At the receiver, after photo detection and analogue to digital conversion the relevant matching filter selected by the DSP controller is applied to the digital signal and the original data is recovered. The DSP controllers are all connected to the SDN controller by extended OpenFlow. The SDN controller is able to see all of the frequencies in use and dynamically allocate resources based on bandwidth allocation algorithms. This means the bandwidth of the links are elastic, so they can grow or shrink with demand and utilisation [21].

In [22], a 40-Gbps TWDM-PON access architecture was proposed and demonstrated. OFDM downstream traffic was launched from the OLT side using 4 SDN controllable 2.5 GHz directly modulated lasers at 4×10 Gbps. Four upstream lasers were also generated in the OLT for seeding upstream modulation in the ONUs each running at 2.5 Gbps using on–off-keying (OOK). As the OLT provides both laser sources for upstream and downstream, the wavelengths can be dynamically provisioned using SDN with OpenFlow connections at only one location, which negates the need to also make the ONU SDN enabled. This is in contrast to the majority of other works, where the ONU normally needs to be SDN enabled to function in a dynamically controlled environment.

In [23], it is stated that previous SDN based control algorithms targeted fixed PON architectures only, and that in order to enable SDN-based control that is malleable across different topologies a novel network virtualisation technique is needed. To do this the authors created an SDN controller that first abstracts the central nodes into physical resource blocks that can be shared among remote nodes, for example bandwidth on one wavelength or a fibre optic link. These available bandwidth metrics along with start and end points are stored within a central table without saving information about the medium over which they are transported. This allows completely disparate network topologies such as wavelength division multiplexed (WDM)—PON and fixed wavelength point to point networks to be abstracted for use by the central controller at the same time. A fully physical layer agnostic interface is therefore achieved so SDN-traffic based prioritisation and resource algorithms can run on a multitude of incompatible platforms at the same time, in the same network. The authors conclude that the proposed approach enabled 30%–50% performance gains in terms of request serving ratio, priorities and revenue; therefore making it attractive for future optical access and mobile fronthaul and backhaul networks in an Internet of things (IoT) era [23].

Similarly to [23], [24] also looks into abstracting underlying networks so SDN control is more efficient, but instead is focussed on CRANs which also encompasses wireless technologies to the abstraction model. The abstraction is presented to an orchestrator that provisions the services on the network based on the policies set by the network management administrators beforehand. The first abstraction model presented in [24] is called big switch (BS). The transport network is presented to the orchestrator as a single node with no information about the internal transport interconnections. This is a very easily maintained model because it does not need any updates from the transport controller. The second abstraction model presented is called virtual link with constant weights (VLCW), wherein the orchestrator is presented with a set of virtual links interconnecting the transport input and outputs (IOs). Each virtual link is weighted according to the physical shortest path distance between the IO ports. Using VLCW, each newly active remote radio unit (RRU) is connected to the evolved packet core (EPC) using the closest BBU Pool with available resources, so is more intelligent at allocation than the BS approach. The last abstraction model is called virtual link with variable weights (VLVW). This is similar to the VLCW abstraction but also presents the number of available light paths on each link the orchestrator enabling more sophisticated traffic engineering solutions such as load balancing. VLVW is also more complicated that BS and VLCW because every time a new RRU is successfully activated, the transport controller is required to update the orchestrator with a recalculated set of weights for all the virtual links in the topology. Reference [24] used a reference topology coded in C++ consisting of 38 nodes, 59 bidirectional fibre-links, and 256 wavelengths per fibre to evaluate the scalability limitations of each abstraction technique. It was found that BS did not require any updates to the orchestrator and therefore did not introduce any scalability limitations. VLCW and VLVW required non-negligible updates to maintain their abstracted topology. Through simulation it was found that VLCW needed between 15% and 35% less updates than VLVW, which demonstrates the high complexity. It was also found that when resources for BBU were scarce, BS was the most effective solution. On the other hand, when wavelength resources become the bottle neck of the system, VLVW provided the best blocking performance at the cost of relatively high complexity. VLCW sat in between at a happy medium of blocking performance and complexity [24].

As can been seen from the above, abstraction methods such as ones discussed in [23,24] are very necessary in optical and wireless networks if dependable and cost efficient dynamic allocation for non-vendor specific topologies is to become a reality in the future. The next step for SDN enabled TWDM-PONs is to create hardware that is fully OpenFlow capable. In [25], a gigabit-PON (GPON)-based virtual OpenFlow-enabled SDN switch was designed and implemented. The authors explained that the GPON is well known for its network management capabilities and its ability to accurately control bandwidth for each subscriber. However, it was also explained that the GPON's configuration and service provisioning is basically static, and therefore, the bandwidth assigned to each ONU cannot be changed dynamically on demand. Because the GPON is not a switch, it was not included in the OpenFlow specifications. The authors of [25] abstracted the entire underlying GPON into an OpenFlow enabled switch by implementing an embedded GPON agent that resides in the OLT. The agent was able to communicate with the OpenFlow controller and the ONU management and control interface (OMCI) module for GPON control. From the OpenFlow controller's perspective, there was no difference in controlling the GPON in comparison to a normal OpenFlow switch. All of the GPON functionality is available to the OpenFlow controller. For example, bandwidth metering is performed by invoking the standard GPON bandwidth allocation scheme via the OMCI module.

17.6 Enabling SDN-based high performance heterogeneous access networks

Starting with this section, a unique network design is presented that can fulfil the requirements presented in this chapter so far, based on components that are fully SDN controllable, and are modularised to work both independently of each other and with each other with minimal changes. Figure 17.6 presents the network breakdown into subsystems, comprising the SAT>IP [32] subsystem, CPRIoE subsystem, fixed wireless network subsystem, an intelligent caching subsystem, and finally a TWDM-PON subsystem for transportation. In addition, this modularisation means that each service subsystem can run within its own virtual network to provide numerous benefits that will be discussed in detail later. In this scape, SAT>IP, CPRIoE, and fixed wireless

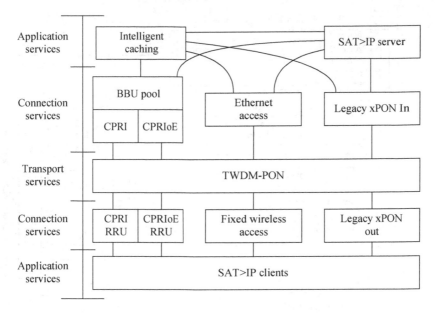

Figure 17.6   System breakdown

network access subsystems are regarded as services, and the TWDM-PON subsystem is regarded as the means of transportation for these services.

In the first subsystem, satellite TV is distributed to consumers by using SAT>IP in an Ethernet access network utilising SDN controllable switches, an intelligent controller, and accompanying tailored network applications. To enable intelligent network changes to be made by the network applications, QoE information from each user is fed back regularly using a custom made SAT>IP client.

The SAT>IP client application allows the user to view SAT>IP content served from a SAT>IP server on the same network. The application uses the real time protocol to receive real time video and audio data from the SAT>IP server in user datagram protocol (UDP) frames, and uses the real-time-streaming-protocol control protocol to set up, close down, and configure connections with the server. The fundamental novelty with this unique SAT>IP video client is its ability to calculate a QoE metric based on the decoded video feed that is then sent to the SDN controller. The SDN controller and SAT>IP network applications can then use these QoE metrics from each user to make positive changes to the network based on the current network configuration and demand.

Figure 17.7 shows the structure of the proposed SAT>IP subsystem utilising an SDN.

In the second subsystem, 5G mobile operator's data is front hauled using CPRIoE in an Ethernet access network utilising SDN controllable switches and an intelligent controller. The system is designed to be most intelligent in a CRAN topology, where a BBU pool can process multiple mobile fronthaul connections simultaneously.

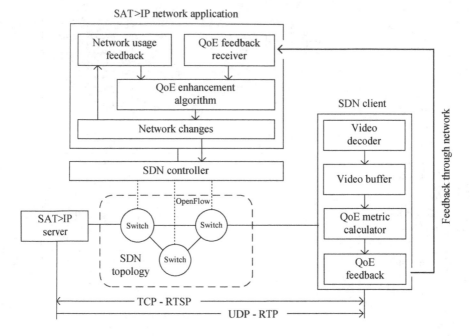

Figure 17.7   The SAT>IP subsystem

Mobile fronthaul information including link latency and jitter can then be made available to the access network's SDN controller and mobile access network's network applications so intelligent network changes can be made. In addition to this, the new IEEE 802.1Qbu- and IEEE 802.1Qbv-proposed enhancements can be incorporated into the current SDN switches so centralised changes to scheduled traffic and traffic pre-emption strategies and algorithms can be made using an evolution of the Open-Flow protocol. This subsystem is also designed to be capable of using CPRI without Ethernet conversion for transport, therefore allowing legacy support for CPRI systems. This is achieved by running CPRI and CPRIoE on different wavelengths within a TWDM-PON. Figure 17.8 shows the structure of the proposed CPRIoE subsystem utilising an SDN.

In the third subsystem, a fixed wireless access network is introduced to provide support for WiFi and femtocells. The WiFi and femtocells only need an Ethernet connection, and don't need any centralised control or administration to work unlike cellular networks so therefore can run in an Ethernet based TWDM-PON natively. The fixed wireless access network subsystem can run within its own network slice in the SDN network and can also use new techniques to broadcast SAT>IP to multiple users with the introduction of WiFi packet forward error correction. Figure 17.9 shows the structure of the proposed fixed wireless access network subsystem utilising SDN.

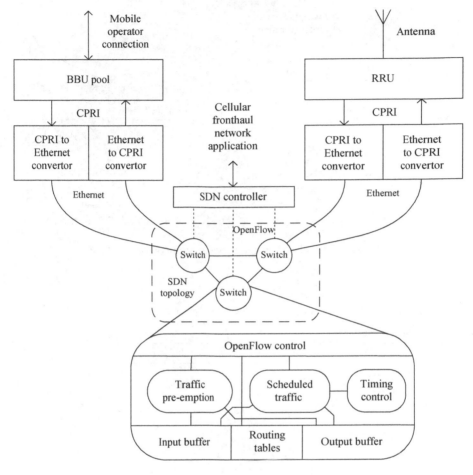

Figure 17.8   Cellular fronthauling using SDN

In the fourth subsystem, intelligent caching is made available on the centralised side of the distribution network. The intelligent caches are based on the node of a CDN, where the most used content is stored locally in the access network for quick access by the users. The intelligent cache is connected directly to the access network's centralised SDN switches, enabling the BBU pool, fixed wireless access network and SAT>IP server to access the intelligent cache. This means the mobile and WiFi/femtocell operators can have access to the cache, and the SAT>IP server can offer time-shifted viewing to the user. The intelligent cache also uses SDN network applications running on the controller to best allocate bandwidth and priority to the services on the network.

Figure 17.10 shows the structure of the proposed intelligent caching subsystem utilising SDN.

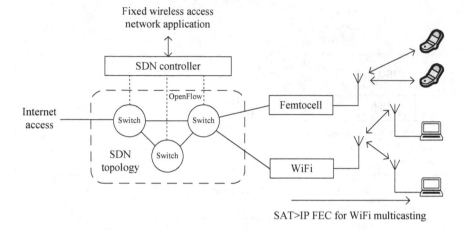

Figure 17.9   Fixed wireless access network using SDN

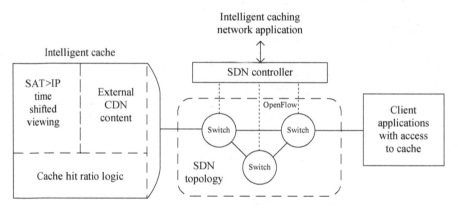

Figure 17.10   Intelligent caching utilising SDN

The final subsystem is the TWDM-PON transportation plane that brings together all of the previous subsystems into a cohesive heterogeneous access network. SDN technology is used to produce an intelligently governed network that is capable of supporting network slices for different technologies, applications, and vendors. The TWDM-PON uses intelligently governed tuneable ONUs and OLTs so wavelengths being used in the PON can be selected by the network controller. The TWDM-PON can also support legacy systems that can't support variable or dynamic wavelength allocations such as native CPRI or support for legacy xPONs. These legacy services can run on their own dedicated wavelengths using their standard fixed ONUs and OLTs. The intelligent controller would be informed by the SDN compliant central-side OLT using an extension to OpenFlow for feedback, but not control. This allows

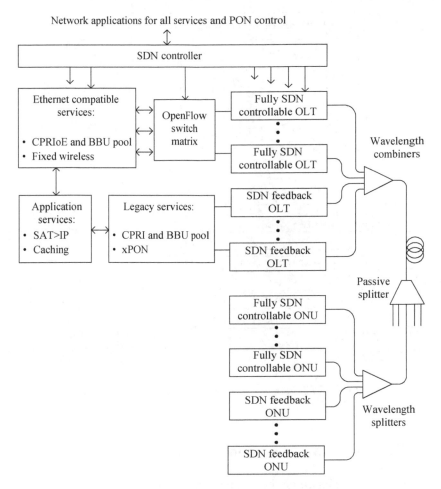

Figure 17.11   SDN-enabled TWDM-PON

the legacy services to work in their native ways, meaning existing equipment can be passed through the new PON without any compromises. Figure 17.11 shows how legacy services can be supported on the SDN-enabled TWDM-PON using OpenFlow feedback.

In addition, Figure 17.12 shows a representative experimental setup of the network under investigation [38]. When new services such as CPRIoE and the fixed wireless access network are being transported over the TWDM-PON, wavelengths can be selected by the centralised controller because the communication for both data and control is performed using standard Ethernet packets. This means additional control layers can be introduced to the ONUs and OLTs compared to current systems by only introducing small changes to the control systems.

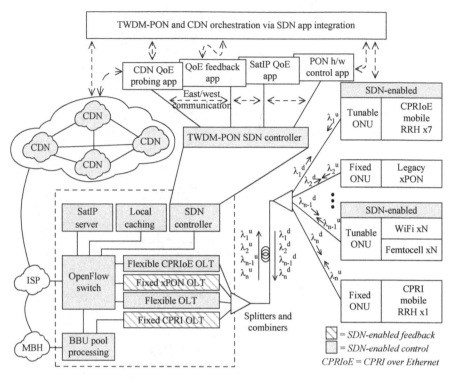

Figure 17.12   Example SDN-enabled TWDM-PON [38]

17.7   Network implementation

To produce a testbed based on the architecture described above, a subset of features have been considered for initial implementation.

The first subsystem implemented is SAT>IP distribution over SDN. The setup for this subsystem is provided in the following subsection along with its validation and an example analysis of the video delivery. This subsystem is going to be extended to include QoE provisioning using the adaptive mechanism explained in the previous section. The heart of this testbed is the Mininet network emulator. Mininet has been chosen because it can create a network of virtual hosts, switches, controllers, and links. In addition, it works using real, unmodified code, meaning applications and SDN features developed using Mininet can be directly used on real SDN hardware platforms without the need for changes to be made to the source code at all.

The second subsystem implemented is the SDN enabled mobile fronthauling. A virtualised LTE core and real hardware small cell is currently being planned for real implementation using a distributed antenna system (DAS). This is currently in an early stage of planning and implementation, and is explained in the following subsections. The combination of these two subsystems is expected to produce a comprehensive

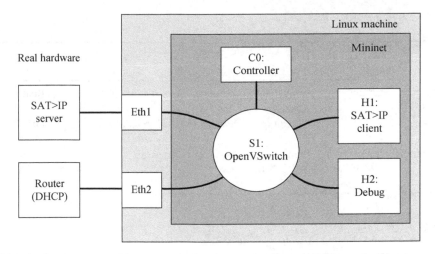

Figure 17.13   SDN-enabled SAT>IP delivery

software/hardware platform able to form the foundation for key experiments expecting to produce high quality high impact outputs.

17.7.1   SDN-enabled SAT>IP delivery

To create an SDN-enabled SAT>IP delivery network, Mininet is initially set up with a simple single switch topology with an SDN controller attached. The virtual Mininet switch includes four Ethernet ports: two are internally connected to respective virtual hosts (for debugging and video delivery), and two are exposed to external Ethernet ports. The external Ethernet ports are then directly connected to real hardware – a Router, for providing dynamic host configuration protocol (DHCP) IP address management; and the SAT>IP server, for providing the video content to the network. Figure 17.13 shows the network setup, including external equipment and internal Mininet hosts.

Ubuntu 16.04 is chosen as the base operating system (OS) [39]. This is mainly due to the SAT>IP client compatibility since older versions of the Ubuntu repositories for prerequisite program installations are not maintained for the latest SAT>IP client developer releases. The SAT>IP client used in this setup is the developer version of VLC Media Player [40] compiled directly from source code. This is because the VLC developer version is the only video client current available that supports SAT>IP streams, is available in Linux, and is open source. Therefore it is suitable for QoE feedback development as described in previous sections.

In Mininet, the standard OpenVSwitch [41] network controller is used for simplicity. Tables 17.1 and 17.2 provide the parameters that were set using the Mininet Python API for emulation.

Table 17.1   Mininet node setup

	H1	H2	S1	C0	SAT>IP server	Router
Real or emulated	Emulated	Emulated	Emulated	Emulated	Real	Real
Purpose	Video client	Debugging	Switch	Controller	Video server	DHCP service
IP address	Static 192.168.1.2	Static 192.168.1.3	N/A	Loopback interface 127.0.0.1	DHCP 192.168.1.4	Static 192.168.1.1

Table 17.2   Mininet link setup

	H1-S1 Ethernet	H2-S1 Ethernet	Eth1-S1 Ethernet	Eth2-S1 Ethernet
Purpose	H1 connection to switch	H2 connection to switch	SAT>IP connection to switch	DHCP connection to switch
Line rate (bps)	Unlimited	Unlimited	Unlimited	Unlimited
Delay (ms)	0	0	0	0

All Mininet virtual Hosts running in the Linux environment share the same user workspace, therefore allowing them to run instances of the same programs at the same time. The switch S1 is emulated in the same way as virtual Hosts, meaning it can also run programs like a traditional Linux user. Wireshark [42] can therefore run on S1 with access to all of the Ethernet ports attached to S1 as well. In addition the SAT>IP and OpenFlow dissector plugins are installed so their respective Ethernet packets could be analysed.

To validate the setup, Wireshark capturing is started on S1 before the hosts in Mininet are activated and while the SAT>IP server and DHCP server are physically disconnected from the system. Two baseline tests were then performed; a latency test and a throughput test. To establish the initial setup time of a link in Mininet due to the controller processing time, a Ping command was used to measure the latency between H1 and H2. This was repeated 10 times to see the different in latency due to the OpenFlow setup time. A graphical display of these Pings can be seen in Figure 17.14.

Figure 17.15 shows the networking timing diagram for this scenario.

From Figure 17.15, there is no reply for the first Ping request because there are no flow entries in S1 to allow the packet to be forwarded from H1 to H2. An OpenFlow flow table miss packet is then sent from H1 to H2, this however is reported incorrectly

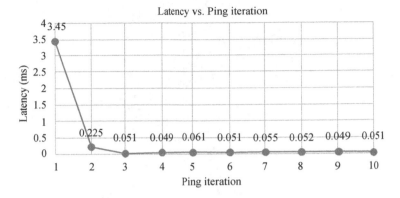

Figure 17.14 Ping H1 to H2 results

and actually is being sent on the LoopBack interface from S1 to C0. An OpenFlow flow table modification packet is subsequently sent from C0 to S1 also using the LoopBack interface. The original Ping packet is then resent by S1 to H2, followed by a Ping reply sent from H2 to H1. Again, there is no flow table entry in S1 for data being sent from H2 to H1 and as a result C0 is notified on the LoopBack interface by S1 about the flow table miss. Subsequently a flow table modification is sent over the LoopBack interface from C0 to S1, and again the original Ping reply packet is resent from S1 to H1, completing the Ping. On the second Ping from H1 to H2, only a Ping request from H1 to H2, and a Ping reply immediately after from H2 to H1, can be seen. There is no OpenFlow interaction due to the flow tables in S1 already being set up.

When comparing the Wireshark and terminal results, it can be seen in the terminal that the first Ping takes 3.45 ms to complete, whereas in Wireshark, the difference in the Ping request and reply time stamps is 1.31 ms. These results for all 10 Pings are shown in Table 17.3.

As can be seen in Table 17.3, the round trip time (RTT) for the Pings reduce to an average of 0.052 ms after initial OpenFlow setup according to the Ping command in terminal, and reduce to 0.022 ms according to Wireshark. The jitter after OpenFlow setup can be seen to be 0.0051 ms in the terminal and 0.0029 ms in Wireshark. This difference in results when comparing the Ping command in terminal and the packet analysis in Wireshark could be put down the processing time within each virtual host, since Wireshark simply records the exact time a packet is sent and received from S1 without including the time for the virtual Ethernet links, host Ethernet port processing, and host application processing. This is because every process is sharing the same computing resources. The difference in jitter between the Ping command in the terminal, and the calculated jitter using Wireshark highlights the variability in processing speed in the host applications.

To establish the maximum possible bandwidth from H1 to H2 based on the current minimal setup, Iperf is used to create a sender and receiver on Mininet hosts. Everything running in Mininet is directly CPU based, and therefore, as more is added

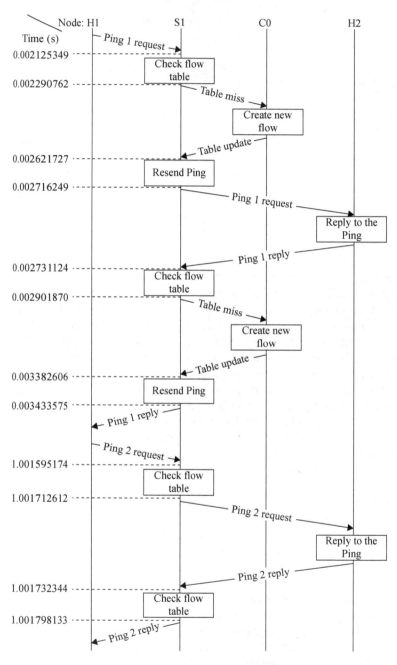

Figure 17.15   Network timing diagram

Table 17.3   Terminal Ping and Wireshark RTT comparison

Ping number	Terminal Ping reported RTT (ms)	Wireshark calculated RTT (ms)
1	3.45	1.308
2	0.225	0.203
3	0.051	0.032
4	0.049	0.029
5	0.061	0.033
6	0.051	0.029
7	0.055	0.032
8	0.052	0.029
9	0.049	0.028
10	0.051	0.030

to the system, the maximum bandwidth is reduced due to the less available CPU time for each host process. A maximum bandwidth of 11.7 Gbps is recorded for both upstream and downstream connections from H1 to H2 via S1. This bandwidth is not affected by the OpenFlow setup procedure because the Iperf test is started after the Ping test demonstrated previously, so flow routes are already set up in S1 for communication between H1 and H2.

To see of the effect of different topologies on the Ping and bandwidth, various topologies were tested. Figure 17.16 shows the different topologies.

The Ping and Iperf tests previously explained were run on each topology. The results are recorded in Table 17.4.

In Table 17.4, the third column displays the initial Ping result in milliseconds. This is the Ping result that also includes the OpenFlow setup time. Each topology was run for the first time with an empty flow table in S1. The fourth column displays the average Ping in milliseconds not including the first 2 Ping results; Figure 17.14 shows how the Ping reduces to a stable level after the first 2 iterations. The fifth column displays the average jitter in milliseconds not including the first 2 Ping results. Finally, the sixth column displays the average Iperf bandwidth for upstream and downstream.

To draw the results for Table 17.4, the system was running with no maximum bandwidth enforced on the system, and the delay on the links was set to 0 ms. To see the effects of a more real network with bandwidth and latency limits, the following tests had respective caps applied to links using the same topologies, shown in Figure 17.16. The bandwidth caps were set at ×10 intervals ranging from 0.1 Mbps up to 1,000 Mbps, concluding in an uncapped scenario. The hosts used for tests in each topology are the same as in Table 17.4. Table 17.5 shows the different scenarios and their respective results.

The results from Table 17.5 are represented in Figure 17.17. Figure 17.17(a) depicts the Initial Ping vs. the Bandwidth Limit. Figure 17.17(b) depicts the Average Ping after OpenFlow setup vs. the Bandwidth Limit. Figure 17.17(c) depicts the average jitter after OpenFlow setup between the test hosts vs. the Bandwidth Limit.

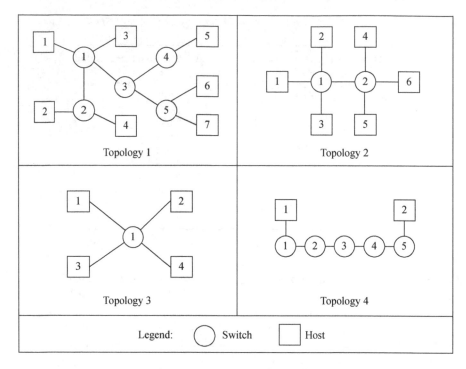

Figure 17.16   Mininet topologies for testing

Table 17.4   Ping and Iperf tests for topologies 1–4

Topology number	Hosts used	Initial Ping OpenFlow setup time (ms)	Average Ping after OpenFlow setup complete (ms)	Average jitter after OpenFlow setup complete (ms)	Iperf bandwidth (Gbps)
1	H1 → H7	8.95	0.036	0.0043	15.4
2	H1 → H6	9.35	0.046	0.0049	15.3
3	H1 → H4	4.08	0.043	0.0059	21.1
4	H1 → H2	13.0	0.054	0.0030	20.3

Figure 17.17(d) depicts the Iperf bandwidth between the test hosts vs. the Bandwidth Limit.

These results show that the OpenFlow setup time, average Ping, and average jitter are generally unaffected by the bandwidth limits applied to the network links. However, the Iperf Bandwidth behaves as expected by closely following the bandwidth limits set in each topology. A similar test to determine the effects of varying latency was also performed with latencies of 0, 1, 5, 10, 50, and 100 ms for all respective

Table 17.5   Varying bandwidth caps

Topology number	Link bandwidth limit (Mbps)	Latency limit (ms)	Initial Ping OpenFlow setup time (ms)	Average Ping after OpenFlow setup complete (ms)	Average jitter after OpenFlow setup complete (ms)	Iperf bandwidth (Mbps)
1	0.1	0	11.90	0.059	0.0140	0.14505
1	1	0	8.07	0.055	0.0077	1.0345
1	10	0	9.82	0.066	0.0084	9.62
1	100	0	8.01	0.061	0.0086	97.65
1	1,000	0	8.2	0.051	0.0031	959
1	Unlimited	0	8.0	0.049	0.0094	3,095
2	0.1	0	4.27	0.055	0.0089	0.1175
2	1	0	3.29	0.044	0.0061	1.0745
2	10	0	2.82	0.044	0.0057	9.775
2	100	0	7.21	0.045	0.0067	97.05
2	1,000	0	4.56	0.046	0.0023	957.5
2	Unlimited	0	6.62	0.042	0.0077	3,900
3	0.1	0	3.77	0.052	0.0064	0.14375
3	1	0	2.87	0.041	0.0060	1.077
3	10	0	4.93	0.058	0.0050	9.765
3	100	0	2.17	0.056	0.0079	96.2
3	1,000	0	4.26	0.041	0.0070	956
3	Unlimited	0	2.91	0.037	0.0060	3,955
4	0.1	0	11.40	0.067	0.0076	0.1181
4	1	0	15.7	0.062	0.0060	1.0345
4	10	0	9.53	0.062	0.0119	9.68
4	100	0	10.2	0.069	0.0087	97.05
4	1,000	0	13.3	0.062	0.0090	943
4	Unlimited	0	10.1	0.054	0.0057	12,400

topologies and with all bandwidths unrestricted. The results for this test can be seen in Table 17.6.

The results from Table 17.6 are represented in Figure 17.18. Figure 17.18(a) depicts the Initial Ping vs. the applied latency. Figure 17.18(b) depicts the Average Ping after OpenFlow setup vs. the applied latency. Figure 17.18(c) depicts the average jitter after OpenFlow setup between the test hosts vs. the applied latency. Figure 17.18(d) depicts the Iperf bandwidth between the test hosts vs. the applied latency.

As can be seen from Figure 17.18, the initial Ping and average Ping scale according to the number of hops between the hosts used in the topology. For example, in topology 4, there are 5 switches between the 2 hosts used for testing, and we can see the average latency after OpenFlow setup is 1,000 ms when 100 ms latency is applied. This is because the Ping packet has to transverse 5 switches in both directions, meaning 10 hops overall. These 10 hops all have 100 ms latency, and therefore the

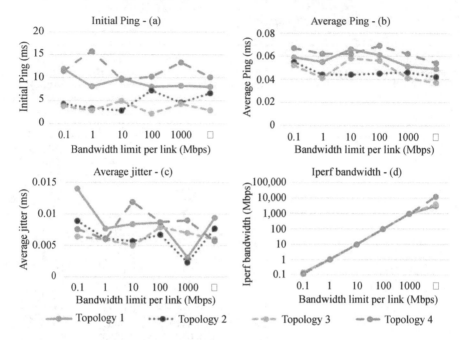

Figure 17.17   Varying bandwidth limit results. (a) Initial Ping, (b) average Ping, (c) average jitter, (d) Iperf bandwidth

overall latency is 10×100 ms $= 1,000$ ms. In addition, the Iperf bandwidth between the 2 hosts can be seen to diminish as the latency is increased. This is due to the TDP throttling in high latency links.

The next important statistic in the Mininet network is the CPU time given to each virtual host. The CPU time is the percentage of overall processing that a host has access to. If 10% is selected for H1, then H1 will only be provisioned 10% of the total CPU time by the OS. This is useful for making sure that virtual hosts do not 'hog' the CPU time, and therefore decrease the CPU time for other applications in the OS. To check the effect of CPU limiting on Mininet hosts, the CPU percentage allocation was changed from 1% to 99% in 1% increments. Topology number 1 was chosen from the previous tests.

Figure 17.19 provides the Iperf bandwidth recorded for tests from 1% to 60% using a set latency of 0 ms, and no bandwidth restriction per link.

As can be seen by Figure 17.19, the Iperf bandwidth between the hosts increases until the 50% mark is reached. At this point, each host is provisioned for example 55%, but cannot realistically exceed 50%, because more than 100% total CPU usage is not possible.

17.7.2   Real video transmission

After the Mininet network has been characterised, a real SAT>IP stream was set up using a Mininet host. To do this, 2 virtual Ethernet ports from a virtual switch were

Table 17.6 *Varying latency limits*

Topology number	Link bandwidth limit (Mbps)	Latency limit (ms)	Initial Ping OpenFlow setup time (ms)	Average Ping after OpenFlow setup complete (ms)	Average jitter after OpenFlow setup complete (ms)	Iperf bandwidth (Mbps)
1	Unlimited	0	6.96	0.049	0.01	3,090
1	Unlimited	1	21.1	8.058	0.01	131
1	Unlimited	5	85.3	40	0	1,570
1	Unlimited	10	165	80	0	34.3
1	Unlimited	50	805	400	0	73.6
1	Unlimited	100	1,604	800	0	5.79
2	Unlimited	0	8.92	0.043	0	4,050
2	Unlimited	1	15.1	6.050	0.01	49.2
2	Unlimited	5	64.2	30	0	2,120
2	Unlimited	10	123	60	0	957
2	Unlimited	50	604	300	0	127
2	Unlimited	100	1,202	600	0	21.5
3	Unlimited	0	4.26	0.038	0.0034	6,690
3	Unlimited	1	10.2	4.037	0.0043	430
3	Unlimited	5	41.5	20	0	3,230
3	Unlimited	10	81.4	40	0	1,260
3	Unlimited	50	401	200	0	235
3	Unlimited	100	803	400	0	73.6
4	Unlimited	0	9.64	0.047	0.0064	7,420
4	Unlimited	1	26.3	10	0	6,320
4	Unlimited	5	105	50	0	1,240
4	Unlimited	10	205	100	0	568
4	Unlimited	50	1,007	500	0	36.2
4	Unlimited	100	2,007	1,000	0	1.73

exposed to the real world. One port was connected to the SAT>IP server, and the other to a DHCP server for IP address provisioning. The video client in this scenario was a version of VLC with SAT>IP capability, running on virtual Mininet Host 2. Figure 17.7 describes the setup used for these tests.

Three tests were performed. First, Wireshark was used to validate the SAT>IP stream flowing through Mininet. Figure 17.20 shows the networking timing diagram for this scenario.

Second, the latency requirements for SAT>IP streaming were determined. To do this, the latency of the link from S1 to H1 was increased from 0 to 2,000 ms in 100-ms steps. The SAT>IP video was requested in each case and left to play for 30 s, the result of the video playback was then determined by looking at the VLC debugging output in the terminal window. One of the debugging output features in VLC informs the user of dropped frames. When dropped frames were indicated by the debugger, and when the video stream was visibly distorted, the result was marked as 'Break Up'.

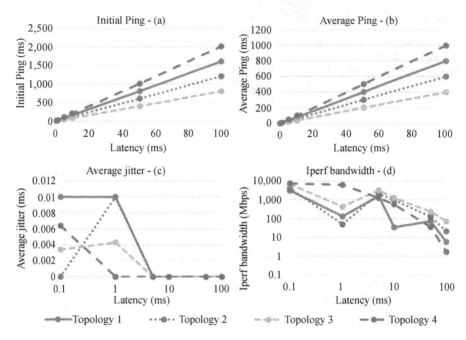

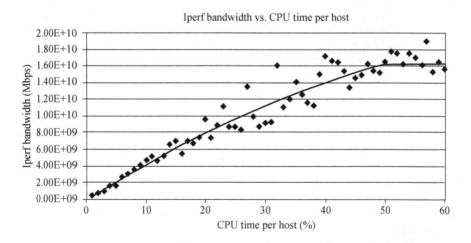

Figure 17.18   Varying latency results. (a) Initial Ping, (b) average Ping, (c) average jitter, (d) Iperf bandwidth

Figure 17.19   Iperf bandwidth vs. CPU time per host

Once video distortion was seen at 1,200 ms latency, further testing concluded that approximately 1,150 ms was the tipping point for video break up and no video break up. Figure 17.21 shows the difference between a good video signal, and a poor video signal.

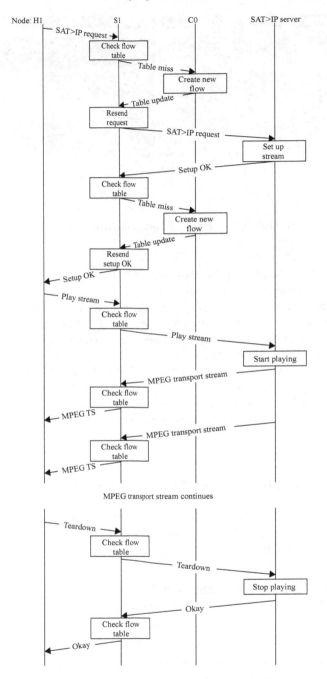

Figure 17.20   SAT>IP stream network timing diagram

Figure 17.21   Poor and good quality video stream

Third, the bandwidth requirements for SAT>IP streaming were determined. To do this, the maximum bandwidth on the link from S1 to H1 was increased from 1 Mbps to 10 Mbps in 1 Mbps steps. The SAT>IP video was requested in each case, the result of the 'Video Working?' test was then recorded. Results indicated that a bandwidth of 6.1 Mbps is required to stream high definition (HD) video from the SAT>IP server to the Mininet host's VLC client.

Finally, the CPU percentage required for SAT>IP streaming by a Mininet host was then determined. To do this, the CPU percentage per host was changed from 1% to 100%. The SAT>IP video was requested in each case, the result of the 'Video Working?' test was then recorded. Measured results confirmed a VLC client requires approximately 60% of CPU time on the particular hardware being used for this test.

17.8   Network optimisation for SDN-enabled video delivery

The next development phase includes the SDN network optimisation for the delivery of video services, with SAT>IP continuing representing the test case. Innovations include the provision of:

1.   Video QoE feedback
2.   Video QoE feedback reception by the SDN application
3.   Network optimisation using an SDN application

17.8.1   Video QoE feedback transmitter

One of the reasons a VLC player has been chosen is because it is open source, with all of the source code readily available to developers. VLC is primarily written in C and C++, and already has many of the features needed to enable QoE feedback. However, to aid quick development within the SDN field, amending VLC is not expected to be the first option for providing autonomous QoE feedback. First, an alternative means of feedback is planned using a dedicated feedback program. Python can be used to send user made QoE feedback to the SDN controller in a standalone application. Python has been chosen because the Mininet control API is also written in Python, therefore allowing the possibility of running the feedback program within the Mininet launching script. The QoE feedback will be inputted by the user when the video quality deteriorates, and this information will then be sent over the network to a receiving application. This will then progress into an automated process whereby the Python QoE transmitter 'looks' at the terminal output of the VLC client. The VLC terminal output provides error messages whenever frames are dropped, with an accompanying reason. By adapting the Python program to 'look' at these Terminal outputs, automatic QoE feedback can be programmed based on predefined QoE algorithms that are based on the VLC error messages. Once the end–end network design has been fully optimised, the QoE feedback mechanism can be written within the VLC player to create a complete SAT>IP player with QoE SDN feedback.

17.8.2 Video QoE reception by the SDN application

The QoE feedback reception is also expected firstly to be created in Python, so the QoE feedback transmitter can be debugged. The feedback reception can initially update a simple text file with the QoE data. The network optimisation application can then use this text file to adapt the network as shown in the following subsection. After the QoE feedback transmitter has been debugged and the network optimisation application is working, the QoE feedback receiver can be re-written within a Java application to create a single program that both receives the QoE information and updates the network.

17.8.3 Network optimisation using an SDN application

The network optimisation program is expected to be built on top of the OpenDaylight [43] network OS. OpenDaylight has a well-defined northbound API that runs within Java.

First, the Mininet network is thought to be set up using the user space switch (USS) instead of the kernel space switch. The USS is slower, but provides more functionality including allowing the use of QoS tags. Within the OpenVSwitch USS, different bandwidth limits can be applied to different QoS classes. The virtual switches within Mininet can be set up with a high QoS class granularity, so there are many different QoS classes to choose between, all with different bandwidth limits. Defining the different QoS limits will be required before the system can be further optimised. The network application can then be created to communicate with OpenDaylight using the northbound API. The application will look at the QoE feedback from the user and then assign the most appropriate QoS tag to the relevant flows within the Mininet network. To do this, it is proposed that RESTCONF and Postman can be used to update the network using OpenDaylight. At this point, the network optimisation needs to be analysed, and the algorithms used to choose the QoS tags for each flow within each switch will need to be studied.

17.8.4   *FlowVisor network slicing*

The next significant step in demonstrating novelty is expected to be produced by introducing FlowVisors [31] to the workflow. The FlowVisor is thought to be able to send different service types to different controllers. Considering the following subsection introduces a real LTE interface for content delivery, FlowVisors will be used to send different services through their own network slices.

17.9   LTE open-air interface

As well as using standard hardwired Ethernet to send SAT>IP over a network to wired clients, other distribution mediums can be investigated. The LTE open-air-interface (OAI) [44] software package allows all functions of a real small cell to be virtualised. The real physical wireless outputs of the small cell can then be sent to a software-defined radio (SDR) in real time and the wireless signals be broadcast on the air. At the same time, the SDR captures wireless signals from the air, analogue to digital converts them and sends them back to the virtualised small cell for uplink processing. By connecting the virtual small cell to the Mininet network, SAT>IP can be streamed through the SDN controlled network and over a real LTE link to the user. The user side can utilise a real LTE SIM card and LTE wireless dongle connected to a Linux machine running the VLC SAT>IP client with QoE feedback mechanism. In addition to a real world LTE network being provided, the small cell can be also connected to a DAS system. The OAI small cell is used to provide real 4G/5G content to the DAS, with the Linux SAT>IP client located within a Faraday cage along with the remote unit of the DAS. Figure 17.22 provides a representation of such network. The LTE OAI software package needs running in a very particular hardware and software environment due to the low latency and consistent processing that is required for real-time communication with real-wireless hardware.

17.10   Conclusions

The mobile communications demand from users over the next 5 years is expected to increase dramatically due to new technologies such as M2M and HD Video becoming more readily available and commercially deployed. M2M and HD Video have drastically different requirements compared to previously used services. M2M communication requires very low latency connections but doesn't require large data bandwidths. Conversely, HD Video requires large data bandwidths but doesn't require low latency connections. This change in services to M2M and HD Video is import because the current generations of mobile communications were developed to handle services with far less stringent latency and bandwidth requirements. As a result, research into the fifth generation of mobile communications is being prioritised in the telecommunications community so these requirements can be supported.

It has also been documented by industry that the fixed wireless access networks comprising of WiFi and femtocells are going to cover a large portion of the bandwidth

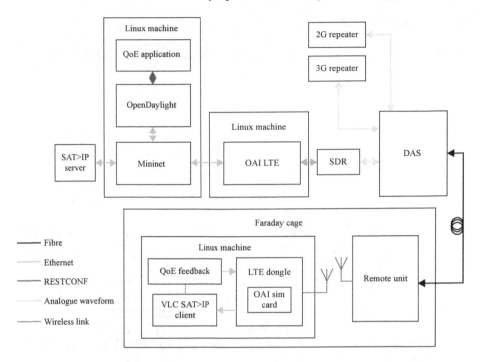

Figure 17.22 LTE and SDN SAT>IP setup

needed by these new services. Therefore, an access network that can support both the next generation of cellular networks and the increasing number of WiFi and femto-cells needs to be created. Based on these requirements, this chapter has discussed a novel SDN enabled Heterogeneous Access network. The proposed architecture uses an SDN-enabled TWDM-PON to allow the fixed wireless access network, cellular network, and legacy PON to coexist in the same infrastructure. In addition, using the SDN platform, on the central side of the access network SAT>IP and intelligent caching video offload is proposed to remove load from the backhaul and CDN network, therefore increasing the QoE to video users and other high bandwidth applications. In addition, the use of SAT>IP QoE feedback is proposed to allow intelligent changes to be made to the network using SDN to help improve the QoE for users.

The SAT>IP subset of this proposed architecture has been demonstrated in the laboratory using Mininet as a foundation, VLC Media Player as a SAT>IP video client, and a real-hardware SAT>IP server to serve the real video data. Plans have been made to create the QoE client feedback, QoE reception server, and QoE-based intelligent network updater. In addition, plans have been made to extend the system to include real LTE distribution using the Open-Air Interface open-source software package and an SDR connected to a DAS for fronthauling. This would allow the SAT>IP content to be streamed over the SDN xhaul.

Acknowledgement

I would like to acknowledge Gary Stafford, Malcolm Burrell, and Jan Treiber of Global Invacom Ltd for co-funding the work.

References

[1]   CISCO. "Cisco visual networking index: global mobile data traffic forecast update, 2015–2020," *White Paper*, 2016.

[2]   I. T. Union. "IMT vision – framework and overall objectives of the future development of IMT for 2020 and beyond," *Recommendation ITU-R M.2083-0*, September 2015.

[3]   Common Public Radio Interface. "CPRI Specification V6.1," 2014. [Online]. Available at: http://www.cpri.info/downloads/CPRI_v_6_1_2014-07-01.pdf. Accessed 21 June 2017.

[4]   A. da le Oliva, J. A. Hernandez, D. Larrabeiti, and A. Azcorra. "An overview of the CPRI specification and its application to C-RAN-based LTE scenarios," *IEEE Communications Magazine*, vol. 54, pp. 152–159, 2016.

[5]   3rd Generation Partnership Project. "LTE," 2017. [Online]. Available at: http://www.3gpp.org/technologies/keywords-acronyms/98-lte. Accessed 21 June 2017.

[6]   IEEE802.1. "Time-Sensitive Networking Task Group," 2017. [Online]. Available at: http://www.ieee802.org/1/pages/tsn.html. Accessed 21 June 2017.

[7]   T. Wan and P. Ashwood-Smith. "A performance study of CPRI over Ethernet with IEEE 802.1Qbu and 802.1Qbv enhancements," in *2015 IEEE Global Communications Conference (GLOBECOM)*, 2015, pp. 1–6.

[8]   N. Shibata, T. Tashiro, S. Kuwano, *et al.* Performance evaluation of mobile front-haul employing Ethernet-based TDM-PON with IQ data compression [invited]," *IEEE/OSA Journal of Optical Communications and Networking*, vol. 7, pp. B16–B22, 2015.

[9]   N. J. Gomes, P. Chanclou, P. Turnbull, A. Magee, and V. Jungnickel. "Fronthaul evolution: from CPRI to Ethernet," *Optical Fibre Technology*, vol. 26, Part A, pp. 50–58, Dec. 2015.

[10]   C. f. Lai, R. h. Hwang, H. c. Chao, M. M. Hassan, and A. Alamri. "A buffer-aware HTTP live streaming approach for SDN-enabled 5G wireless networks," *IEEE Network*, vol. 29, pp. 49–55, 2015.

[11]   H. Nam, K. H. Kim, J. Y. Kim, and H. Schulzrinne. "Towards QoE-aware video streaming using SDN," in *2014 IEEE Global Communications Conference*, 2014, pp. 1317–1322.

[12]   D. Butler. "SDN and NFV for broadcasters and media," in *Optical Communication (ECOC), 2015 European Conference on*, 2015, pp. 1–3.

[13]   L. Hui, H. Yihong, S. Guochu, and G. Zhigang. "Software defined networking for HTTP video quality optimization," in *Communication Technology (ICCT), 2013 15th IEEE International Conference on*, 2013, pp. 413–417.

[14] K. Li, W. Guo, W. Zhang, Y. Wen, C. Li, and W. Hu. "QoE-based bandwidth allocation with SDN in FTTH networks," in *2014 IEEE Network Operations and Management Symposium (NOMS)*, 2014, pp. 1–8.

[15] J. Chandrakanth, P. Chollangi, and C. H. Lung. "Content distribution networks using software defined networks," in *Trustworthy Systems and Their Applications (TSA), 2015 Second International Conference on*, 2015, pp. 44–50.

[16] H. Y. Seo, B. Bae, and J. D. Kim. "Transmission model for next-generation digital broadcasting systems," in *2015 International Conference on Information Networking (ICOIN)*, 2015, pp. 379–380.

[17] Open Networking Foundation. "OpenFlow," 2017. [Online]. Available at: https://www.opennetworking.org/. Accessed 21 June 2017.

[18] N. Cvijetic, A. Tanaka, P. N. Ji, K. Sethuraman, S. Murakami, and W. Ting. "SDN and OpenFlow for dynamic flex-grid optical access and aggregation networks," *Journal of Lightwave Technology*, vol. 32, pp. 864–870, 2014.

[19] K. Kondepu, A. Sgambelluri, L. Valcarenghi, F. Cugini, and P. Castoldi. "An SDN-based integration of green TWDM-PONs and metro networks preserving end-to-end delay," *Optical Fiber Communications Conference and Exhibition (OFC), 2015*, 2015, pp. 1–3.

[20] L. Valcarenghi, K. Kondepu, A. Sgambelluri, *et al.* "Experimenting the integration of green optical access and metro networks based on SDN," in *2015 17th International Conference on Transparent Optical Networks (ICTON)*, 2015, pp. 1–4.

[21] M. Bolea, X. Duan, R. P. Giddings, and J. M. Tang. "Software reconfigurable PONs utilizing digital filter multiple access," in *Networks and Communications (EuCNC), 2015 European Conference on*, 2015, pp. 335–339.

[22] C. H. Yeh, C. W. Chow, M. H. Yang, and D. Z. Hsu. "A flexible and reliable 40-Gb/s OFDM downstream TWDM-PON architecture," *IEEE Photonics Journal*, vol. 7, pp. 1–9, 2015.

[23] W. Jing, N. Cvijetic, K. Kanonakis, W. Ting, and C. Gee-Kung. "Novel optical access network virtualization and dynamic resource allocation algorithms for the Internet of Things," in *Optical Fibre Communications Conference and Exhibition (OFC), 2015*, 2015, pp. 1–3.

[24] M. Fiorani, A. Rostami, L. Wosinska, and P. Monti. "Transport abstraction models for an SDN-controlled centralized RAN," *IEEE Communications Letters*, vol. 19, pp. 1406–1409, 2015.

[25] S. S. W. Lee, K. Y. Li, and M. S. Wu. "Design and implementation of a GPON-based virtual OpenFlow-enabled SDN switch," *Journal of Lightwave Technology*, vol. 34, pp. 2552–2561, 2016.

[26] M. Wichtlhuber, R. Reinecke, and D. Hausheer. "An SDN-based CDN/ISP collaboration architecture for managing high-volume flows," *IEEE Transactions on Network and Service Management*, vol. 12, pp. 48–60, 2015.

[27] C. Yang, Z. Chen, Y. Yao, and B. Xia. "Performance analysis of wireless heterogeneous networks with pushing and caching," in *2015 IEEE International Conference on Communications (ICC)*, 2015, pp. 2190–2195.

[28]   H. Li, Z. Wang, and D. Hu. "Joint wireless and backhaul load balancing in cooperative caches enabled small-cell networks," in *Personal, Indoor, and Mobile Radio Communications (PIMRC), 2015 IEEE 26th Annual International Symposium on*, 2015, pp. 1889–1894.

[29]   N. Golrezaei, K. Shanmugam, A. G. Dimakis, A. F. Molisch, and G. Caire. "FemtoCaching: wireless video content delivery through distributed caching helpers," in *INFOCOM, 2012 Proceedings IEEE*, 2012, pp. 1107–1115.

[30]   M. Broadbent, D. King, S. Baildon, N. Georgalas, and N. Race. "OpenCache: a software-defined content caching platform," in *Network Softwarization (NetSoft), 2015 First IEEE Conference on*, 2015, pp. 1–5.

[31]   GitHub. "Open Networking Laboratory," 2017. [Online]. Available at: https://github.com/opennetworkinglab/. Accessed 21 June 2017.

[32]   Sat>IP. "Technology," 2015. [Online]. Available at: http://www.satip.info/technology-0. Accessed 21 June 2017.

[33]   NS-3. "What is NS-3?," 2015. [Online]. Available at: https://www.nsnam.org/overview/what-is-ns-3/. Accessed 21 June 2017.

[34]   Project Floodlight. "Floodlight," 2017. [Online]. Available at: http://www.projectfloodlight.org/floodlight/. Accessed 21 June 2017.

[35]   Mininet. "Mininet: An Instant Virtual Network on your Laptop (or other PC)," 2017. [Online]. Available at: http://www.mininet.org/. Accessed 21 June 2017.

[36]   GitHub. "OpenCache," 2017. [Online]. Available at: https://github.com/broadbent/opencache. Accessed 21 June 2017.

[37]   Fed4Fire. "Federation for future internet research and experimentation," 2016. [Online]. Available at: https://www.fed4fire.eu/. Accessed 21 June 2017.

[38]   M. Robinson, M. Milosavljevic, P. Kourtessis, G. P. Stafford, M. J. Burrell, and J. M. Senior. "Software defined networking for heterogeneous access networks," in *2016 18th International Conference on Transparent Optical Networks (ICTON)*, 2016, pp. 1–4.

[39]   Ubuntu. "The leading operating system for PCs, IoT devices, servers and the cloud," 2017. [Online]. Available at: https://www.ubuntu.com/. Accessed 21 June 2017.

[40]   VideoLAN. "Official download of VLC media player, the best Open Source player," 2017. [Online]. Available at: http://www.videolan.org/vlc/. Accessed 21 June 2017.

[41]   Open vSwitch. "Production quality, multilayer open virtual switch," 2016. [Online]. Available at: http://www.openvswitch.org/. Accessed 21 June 2017.

[42]   Wireshark. "About Wireshark," 2017. [Online]. Available at: https://www.wireshark.org/. Accessed 21 June 2017.

[43]   OpenDaylight. "OpenDaylight: Open Source SDN Platform," 2017. [Online]. Available at: https://www.opendaylight.org/. Accessed 21 June 2017.

[44]   OpenAirInterface. "5G software alliance for democratising wireless innovation," 2017. [Online]. Available at: http://www.openairinterface.org/. Accessed 21 June 2017.

Chapter 18

Fronthaul and backhaul integration (Crosshaul) for 5G mobile transport networks

Xavier Costa-Perez[1], Antonia Paolicelli[2],
Antonio de la Oliva[3], Fabio Cavaliere[4], Thomas Deiß[5],
Xi Li[1] and Alain Mourad[6]

18.1 Motivation and use cases for fronthaul and backhaul integration

18.1.1 Motivation

According to recent predictions [1], mobile data traffic will increase eight-fold between 2015 and 2020. Fifth generation (5G) radio access network (RAN) technologies serving this mobile data tsunami will require fronthaul and backhaul solutions between the RAN and packet core to deal with this increased traffic load. Furthermore, there will be a sizeable growth in the capillarity of the network as traffic load increase in the 5G RAN is expected to stem from an increased number of base stations with reduced coverage (i.e. mobile network densification).

A promising approach to address this challenge is virtualization, as can be observed in major standardization bodies [e.g. European Telecommunications Standards Institute (ETSI) Network Function Virtualization (NFV)], which exploits the multiplexing gain of softwarized network functions on top of commoditized hardware. This has led to the Cloud RAN (C-RAN) concept where cellular base station functions are hosted in cloud computing centres. Once virtualized, base station functions can be flexibly distributed and moved across data centres, providing another degree of freedom for load balancing. Moreover, base station functions can be decomposed in many different ways, giving rise to the so-called *flexible functional split*, where the radio protocol split between centralized and distributed base station functions can be adjusted on a case-by-case basis.

[1]NEC Labs Europe, Germany
[2]Telecom Italia, Italy
[3]Telematics Department, University Carlos III, Spain
[4]Ericsson, Sweden
[5]Nokia, Germany
[6]InterDigital Europe Ltd, UK

In this context, *the distinction between fronthaul and backhaul transport networks blurs* as varying portions of functionality of 5G points of attachment (5GPoAs) might be moved towards the network as required for cost efficiency reasons. The traditional capacity overprovisioning approach on the transport infrastructure will no longer be possible with 5G. Hence, a new generation of integrated fronthaul and backhaul technologies will be needed to bring capital expenditure (CAPEX) and operational expenditure (OPEX) to a reasonable return on investment (ROI) range. Also, for cost reasons, the heterogeneity of transport network equipment must be tackled by unifying data, control and management planes across all technologies as much as possible. A redesign of the fronthaul/backhaul network segment is a key point for 5G networks as current transport networks cannot cope with the amount of bandwidth required for 5G. Next-generation radio interfaces, using 100 MHz channels and squeezing the bit-per-megahertz ratio through massive multiple-input and multiple-output (MIMO) or even full-duplex radios, require a ten-fold increase in capacity, which cannot be achieved just through the evolution of current technologies.

Considering the new challenges brought by 5G, this chapter presents a novel architecture aimed at integrating fronthaul and backhaul in a common packet-based network defined as *5G-Crosshaul* [2]. This work addresses the increased interest in architectures integrating both the fronthaul and backhaul network segments, currently seen as one of the major bottlenecks of future mobile networks.

18.1.2 Use cases

The 5G-Crosshaul general idea is to integrate both backhaul and fronthaul segments into a single, flexible and software-defined networking (SDN)-controlled transport network in order to transport data between the RAN and its respective operator's metro/core network in an adaptive, cost-efficient and sharable manner.

Multiplexing backhaul and fronthaul traffic is highly beneficial as it enables the use of the same infrastructure and a shared control for multiple purposes, with a subsequent decrease of the total cost of ownership of these network segments. This holds even more in 5G, where new functional split schemes of the radio interface add a plethora of possible intermediate cases between the pure fronthaul and backhaul scenarios. This large variety of cases is impossible to be managed by dedicated infrastructures.

The integration of the fronthaul and backhaul technologies also enables the use of heterogeneous transport technologies leveraging both novel and traditional technologies. This increases the capacity and/or coverage of the 5G technology and removes the need of using specialized fronthaul interfaces. It also optimizes the location of network functions, in order to provide the required services in the most efficient way and dynamically, it accompanies the variation on demand intrinsic to mobile systems.

Different actors will play a significant role in the definition of final services and will take advantage from this new network paradigm: end users who consume the services offered on top of the infrastructure, network operator being in charge of connecting the end user by providing access to telecommunication services and external networks and service providers delivering telecommunication services and application or data providers.

A large number of use cases can be identified from projects, researchers, organizations and standardization bodies. In the following, five of those will be described illustrating flexible and programmable capabilities, dynamic and cost-effective service provision, reduced network deployment costs and efficient energy consumption.

18.1.2.1 Service-oriented use cases

Vehicle mobility
In this use case, the challenge is to provide in a cost-effective way a satisfying quality of experience (QoE) to a crowd of passengers, e.g. more than 500 people, on a high-speed train with top speeds greater than 500 km/h. In this case, due to the obstacles like Doppler effect caused by high mobility and the penetration loss resulted from the metallic carriage, the direct access from on-board terminals to the base stations will likely be hampered. Therefore, one or several points of access (PoA), such as small cells or Wi-Fi APs, will be installed on high-speed trains to provide broadband access to the passengers inside each car of the train. The mobile backhaul for the PoA is provided via outbound gateway(s) which connect to the land base stations and the transport network (see Figure 18.1).

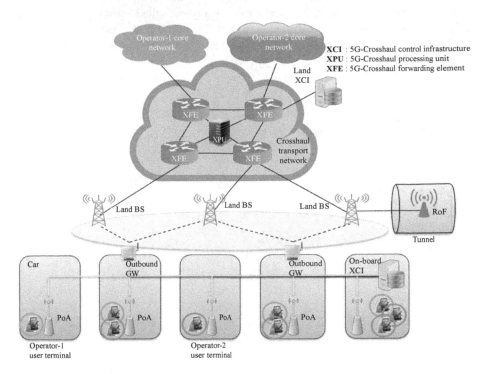

Figure 18.1 High-speed train over Crosshaul transport network

Media distribution: CDNs, TV broadcasting and multicasting

This use case is related to the distribution over 5G networks of media contents, especially video traffic, that is expected to be the prevailing contributor to the mobile data traffic demand.

Relevant situation for content delivery are massive events such as international music festivals or big sport events, in which high volume of traffic is needed maintaining acceptable quality for the end users.

Content delivery networks (CDNs) have evolved to overcome the limitations of the Internet in terms of user perceived quality of service (QoS) when accessing content: CDN replicates content from the origin server to cache servers, in order to deliver content to end users in a reliable and timely manner. In order to save transport resources and reduce transmission delay, CDN end points can be distributed across the network and encoded data and metadata required in CDN networks can be transmitted more efficiently through a unified frame format.

Another important scenario where the media distribution is relevant is the case of media/TV broadcasting and multicasting where the next 5G networks will become a good alternative to classical broadcasting networks, with an additional ability to mix with other media content not coming from the broadcasted TV, using the same network and based on a controlled quality offered as a Broadcast-as-a-service, as shown in Figure 18.2.

The network is configured by selecting the nodes of the tree to optimize the content delivery and make sure a real-time delivery with the lowest possible delay is offered to the users. The use of SDN can optimize in the data plane such a process, besides controlling network parameters and performing QoE monitoring which can be lately needed for reconfiguration purposes.

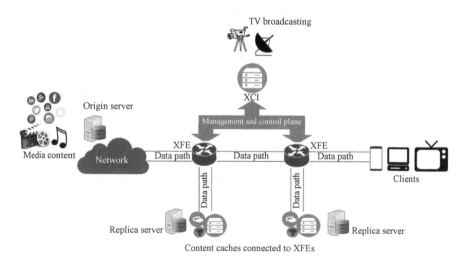

Figure 18.2   Content delivery network architecture

Dense urban information society

This use case takes into account the connectivity required at any place and at any time by humans in dense urban environments, considering both the traffic between humans and the cloud, and direct information exchange between humans or with their environment. Besides a massive increase in the data volumes connected to the usage of public cloud services, a key challenge in communication systems beyond 2020 will lie in the fact that users will expect the same reliable connectivity to the cloud anytime and anywhere and the same QoE at their workplace, in enjoying leisure activities such as shopping, or being or moving on foot or in a vehicle. The particular challenge lies in the fact that users tend to gather and move in 'dynamic crowds', leading to sudden high peaks of local mobile broadband demand.

Similar cases might arise as well in indoor environments such as homes, offices, hotels, shopping malls, hospitals, airports etc. where the required per-user throughput is expected to be very high, as the IT tools for daily use are running on the network side. A big issue due to high penetration loss from the outside-in approach is the capacity and the indoor coverage or the radio network that requires the deployment of appropriate indoor radio systems. The right solution to deliver high capacity to indoor users is the deployment of small cells as they can be directly connected to existing superfast fibre networks, allowing huge access bandwidth.

The problem for this scenario is to integrate several technologies, as a realistic network deployment will include macro sites, C-RAN, virtual RAN, all-in-one small cells (microcells, picocells and femtocells) and distributed RRHs (a simplified scenario is depicted in Figure 18.3). In such an environment, a large plethora of Backhaul/Fronthaul media will be used, including fibre, copper, μWave, mmWave, Free Space Optics, etc. Besides, different protocols (CPRI, OBSAI, Ethernet), multiple RAT (LTE, WiFi, mmWave, ...) and multiple access (3G, 4G, 5G) will be considered and coordinated to provide better services to the end users and ease communication among users and other smart devices.

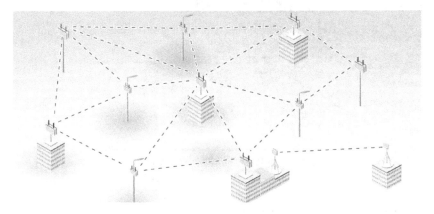

Figure 18.3   Simplified scenario for the dense urban society use case

18.1.2.2 Functional-oriented use cases

Multi-tenancy

This use case takes into account a flexible sharing of backhaul/fronthaul physical resources across multiple network operators or tenants.

Multi-tenancy is a key enabler to maximize the degree of utilization of 5G network deployment and minimize the overhead due to the costs of roll-out and maintenance, e.g. by allowing the exploitation of networking services by a virtual operator in a location where the original (physical) operator has lower load requirements. In this way, the overhead of physical equipment maintenance is minimized, and its utilization is maximized (see Figure 18.4).

A virtual network provider (VNP) often leases resources from one or multiple physical infrastructure providers who own the physical resources and assemble these physical resources into a virtual network (VN) infrastructure. The VNP provides virtual resources over its substrate network to multiple network operators (virtual network operators, VNOs), where each of them owns a slice of the physical resources (computing resources and networking resources). Network virtualization is applied to provision multiple VNs (i.e. multiple tenants) over a substrate network for sharing the computing and network resources.

A VN is a logical topology composed of a set of virtual nodes, e.g. virtual routers and switches, inter-connected by virtual links over the substrate network.

The multi-tenancy application shall provide concurrently and seamlessly resources to each tenant and shall guarantee isolated virtual domains per tenant and dynamic allocation of virtual slices of resources allowing the network to scale to multiple tenants.

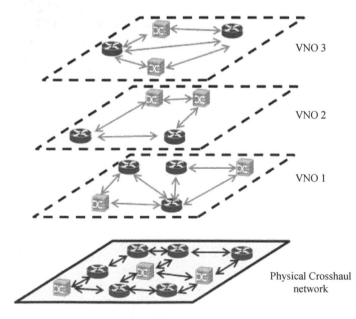

VNO 3

VNO 2

VNO 1

Physical Crosshaul
network

Figure 18.4 Multi-tenancy slicing

This is the main challenge of this use case: to ensure clean isolation across tenants and manage (create, update, delete) the tenants on demand and thus allocate the requested virtual resources dynamically and seamlessly without invoking service disruptions.

Mobile-edge computing

Dense network environments are characterized by a high number of users which simultaneously demand connectivity resources. This increment on the population to be served impacts on the backhaul network deployed and on the radio access part that requires the deployment of small cells in order to increase the radio capacity in that area.

Mobile-edge computing (MEC) proposes the deployment of IT and cloud-computing capabilities within the RAN, in close proximity to mobile subscribers. Content, service and application providers can leverage on such distributed computing capabilities to serve the high-volume, latency-sensitive traffic on dense areas concentrating high number of users. In consequence, the introduction of computing capabilities at the edge of the mobile network can improve the service delivery to the end user on one hand and allow to efficiently deliver the traffic minimizing transport resource consumption on the other introducing savings in the backhaul capacity (see Figure 18.5).

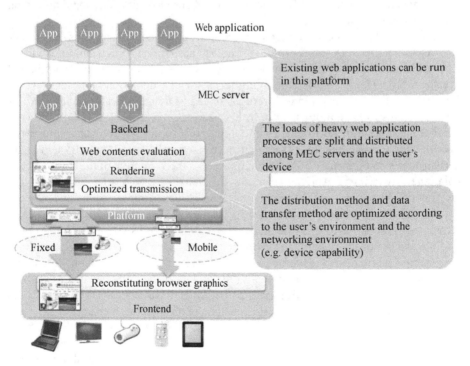

Figure 18.5   Example of mobile-edge computing

Three categories of services benefiting of MEC are as follows:

- Consumer-oriented services, e.g. end users to offload applications or data to the edge network nodes for battery-saving purposes, or for accessing computing capacity beyond their resource-limited handset.
- Operator and third-party services, e.g. services that take advantage of computing and storage facilities close to the edge of the operator's network.
- Network performance and QoE improvements, e.g. services aimed at improving network performance via either application-specific or generic improvements.

Enhancing the introduction and provisioning of novel applications and vertical services at the network edge closer to the end users, MEC can enable innovative services such as e-Health, connected vehicles, industry automation, augmented reality, gaming and Internet of Things services.

18.2   Architectural solutions and components

The goal of the fronthaul and backhaul integration is to build an adaptive, flexible and software-defined architecture for future 5G transport networks. The 5G-Crosshaul architecture presented here thus aims to enable a flexible and software-defined reconfiguration of all networking elements through a unified data plane and control plane inter-connecting distributed 5G radio access and core network functions, hosted on in-network cloud infrastructure.

The control plane needs to include a group of key functional elements (e.g. topology discovery, network monitoring, technology abstraction, provisioning of virtual infrastructure etc.) and their main interfaces towards the applications [northbound interface (NBI)] and towards underlying technologies [southbound interface (SBI)]. The design of the control plane leverages on the SDN (software-defined networking) principles to have a unified control, management and configuration of the 5G multi-technology transport network and apply NFV to the 5G-Crosshaul infrastructure enabling flexible function placement and cost-effective usage of the 5G-Crosshaul infrastructure resources. The SDN principle allows the separation of the data and control planes, fostering network and device programmability. NFV allows infrastructure and function virtualization, where the underlying physical infrastructure and network functions can be virtualized in such a way that they will be appropriately instantiated, connected and combined over the underlying 5G-Crosshaul substrate.

The design of the data plane architecture needs to reflect the integration of heterogeneous technologies for the fronthaul and backhaul links into a single SDN-based controlled network. The main challenge of the data plane is the need for extended flexibility to adapt to the new fronthaul and backhaul technologies arising with 5G as well as to incorporate legacy technologies through abstraction interfaces.

To achieve such a design, our approach is to leverage the state-of-the-art SDN and NFV architectures so as to avoid re-inventing the wheel and maximizing the compatibility and integration of the system design with the existing standard frameworks and reference specifications.

So far, the most well-developed open source SDN controllers which provide carrier grade features and can be used for 5G networks are Open Daylight (ODL) and Open Network Operating System (ONOS). In the NFV case, ETSI NFV ISG is currently studying the ability to deploy instances of network functions running in VMs providing network operators with the ability to dynamically instantiate, activate and re-allocate resources and functions. Based on these open source initiatives and standards, our 5G-Crosshaul architecture keeps the architecture compatibility with the existing ODL/ONOS and ETSI NFV architecture frameworks. For the overall architecture design, we take a bottom-up approach to evolve from current management systems towards the integration of MANO concepts.

18.3   5G-Crosshaul architecture

18.3.1   Overview

The 5G-Crosshaul architecture is depicted in Figure 18.6 and follows the SDN architecture detailed by the Open Networking Foundation (ONF) in [3]:

- Decoupled data plane and control plane;
- Control plane logically centralized;
- Exposure of state and abstract resources to applications.

1. Control plane: As shown in Figure 18.6, the control plane is divided in two different layers: a bottom layer called 5G-Crosshaul Control Infrastructure (XCI) and a top layer for external applications; this top layer exploits 5G-Crosshaul resource orchestration functions to support functionalities like planning, network and service monitoring/prediction, multi-tenancy, CDNs, TV Broadcasting, optimization of resources, energy management, etc.

Successively, the XCI is our 5G Transport Management and Orchestration (MANO) platform, is based on SDN/NFV principles and operates all the available resources (networking and cloud). The top layer applications can use the NBI offered by the XCI, typically based on REST, NETCONF or RESTCONF APIs, to programme and monitor the data plane. The SBI, based on, e.g. OpenFlow, OF-Config, OVSDB, SNMP and/or an ecosystem comprising several of them, will be used by the XCI to interact also with the data plane to:

- Control and manage the packet forwarding behaviour performed by 5G-Crosshaul Forwarding Elements (XFEs) across the 5G-Crosshaul network;
- Control and manage the PHY configuration of the different link technologies (e.g. transmission power on wireless links);
- Control and manage the 5G-Crosshaul Processing Units (XPU) computing operations [e.g. instantiation and management of virtual network functions (VNFs) via NFV].

The scope of operation of the XCI is limited to (physical/virtual networking/ storage/computing) resources within the 5G-Crosshaul transport domain; however,

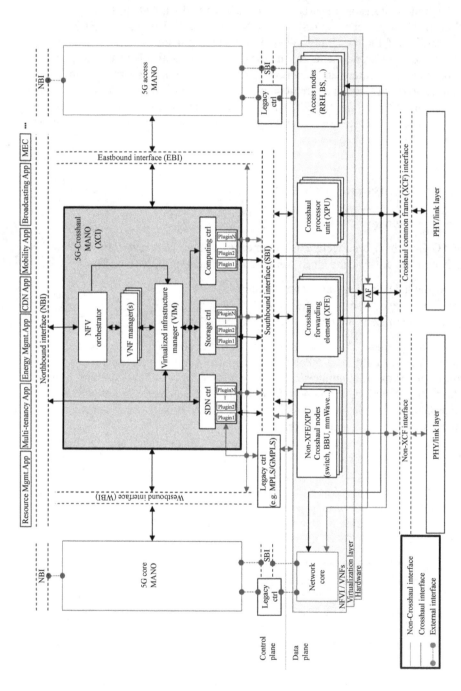

Figure 18.6 5G-Crosshaul architecture illustration [4]

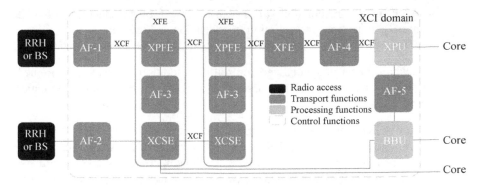

Figure 18.7   5G-Crosshaul data plane architecture

the XCI will require knowledge of the configuration and or other information from the Core network and/or the RAN domains. The communication with the 5G Core MANO will be done through the westbound interface (WBI) and the interaction with the 5G Access MANO through the eastbound interface (EBI).

2. Data plane: 5G-Crosshaul is based on a unified, programmable, multi-tenant enabled and heterogeneous packet-based transport network. This transport network is formed by XEFs (see Figure 18.7), switching units that inter-connect a broad set of links and PHY technologies under a unified frame format called 5G-Crosshaul common frame (XCF). The XCF is designed to handle simultaneously backhaul and fronthaul traffic, which have very different requirements thanks to prioritization and timing.

The XPUs carry out the bulk of the computing operations in the 5G-Crosshaul. The XPU shall support C-RAN, 5GPoA, virtualized functionalities (VNFs) and other services (e.g. CDN-based services).

XCI can also communicate with non-5G-Crosshaul entities like legacy switches or BBUs, using proper plugins. The 5G-Crosshaul data plane elements can communicate also with non-XCF compatible elements thanks to a set of adaptation function entities.

18.3.2   Main components

In the following, we describe the 5G-Crosshaul main components introduced in Section 18.2:

1. XCI: The XCI is the brain controlling the overall operation of the 5G-Crosshaul. The XCI follows the ETSI NFV architecture [2] and is formed by three main functional blocks: (i) NFV Orchestrator (NFVO), (ii) VNF Manager(s) (VNFMs) and (iii) Virtualized Infrastructure Manager (VIM):

- The NFVO performs the orchestration of compute, storage and network resources available to ensure an optimized allocation and provides a network service.

- The VNFMs are responsible of the management of management of VNF instances (e.g. instance instantiation, modification and termination).
- The VIM controls and manages the NFVI compute, storage and network resources.

The XCI includes a set of specialized controllers in addition to the aforementioned; their purpose is the control of the underlying network, storage and computation resources:

- SDN controller is in charge of controlling the networks elements following the SDN principles. 5G-Crosshaul aims at extending current SDN support of multiple technologies used in transport networks (such as micro-wave links) in order to have a common SDN controlled network substrate which can be reconfigured based on the needs of the network tenants.
- Computing/Storage controllers: Storage and computing controllers are included in what we call a cloud controller. A prominent example of this kind of software framework is OpenStack. Note that the SDN/computing/storage controllers are functional blocks with one or multiple actual controllers (hierarchical or peer-to-peer structure) that centralize some or all of the control functionality of one or multiple network domains. We consider the utilization of legacy network controllers (e.g. MPLS/GMPLS) to ensure backward compatibility for legacy equipment. In order to ensure the feasibility of the architecture, the XCI design is based on the ETSI NFV architecture [5]. We have also considered the design and features of two well-known NFV and cloud infrastructures: Open Source Mano and OpenStack, to guide the XCI design.

2. XFE: XFEs are the switching units that integrate different technologies (e.g. mmWave, Ethernet, Optical Fibre) used in the data transmission; they enable a unified and harmonized transport traffic management. XFE supports the XCF format across the various traffic flows (of fronthaul and backhaul) and the various link technologies in the forwarding network. The XFEs are controlled by the XCI which is foreseen to have a detailed (as per the abstraction level defined) view of the fronthaul and backhaul traffic and resources, and to expose this information for intelligent resource, network functions and topology management across the two domains.

As shown in Figure 18.7, XFEs are formed by packet-switching elements [5G-Crosshaul packet forwarding element (XPFE)] and circuit-switching elements. Hence, two paths are defined: (i) a packet-switching path and (ii) an optical circuit-switching path. The packet-switched network is the primary path for the transport of most delay-tolerant traffic; the circuit-switching path complements the packet-switched path for those traffic that is not suited for packet-based transport or has an extremely low delay tolerance (e.g. legacy CPRI); the circuit-switching path can also be used to reduce the offload of the XPFE. This two-path switching architecture is designed to combine bandwidth efficiency, through statistical multiplexing in the packet switch, with deterministic latency ensured by the circuit switch. The modular structure of the 5G-Crosshaul switch, where layers may be added and removed, enables various deployment scenarios with traffic segregation at multiple levels, from dedicated wavelengths to VPN, which is particularly desirable for multi-tenancy support.

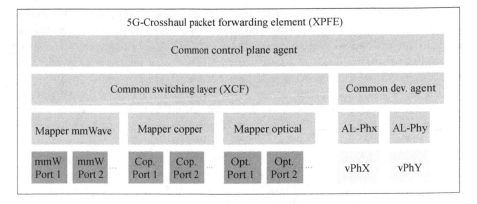

Figure 18.8   5G-Crosshaul packet forwarding element

Figure 18.8 depicts the initial architecture for the XPFE. It includes the following functions:

- A common control plane agent to talk to the XCI.
- A common switching layer based on the XCF to forward packets between inter-faces with different technologies. The switching engine is technology agnostic and relies on (i) an abstract resource model (i.e. bandwidth, latency, BER, jitter, latency, etc.) of the underlying interfaces (i.e. mmWave, copper, etc.) and on (ii) traffic requirements (i.e. jitter, packet loss, etc.) that can be carried in the XCF.
- A common device agent to talk with system peripheral and expose to XCI device-related information (e.g. CPU usage, RAM occupancy, battery status, GPS position etc.).
- Mappers for each physical interface.
- Physical interfaces with different technologies to transmit the data on the link. The common control plane and device agents are relevant for both packet- and circuit-switched forwarding elements of the XFE. It enables the creation of a tech-nology independent and heterogeneous data plane with dynamic reconfiguration of resources, allowing the inter-working with legacy technologies. The XFE is a critical part of the system and as such we have already started its prototyping. The multi-layer switch design of the XFE is based on the evolution of the Ericsson Multi-layer Switch and a prototype version is already under development.

3. XCF: The main goal of the 5G-Crosshaul is the integration between fronthaul and backhaul segments; the first premise needed is a common frame that can be used to transport both traffic types through the same network. The XCF is the frame format used by the XPFE and is supported by all physical interfaces where packets are transported. The circuit-switched forwarding is independent of the XCF. Where necessary, the frame format is mapped to the XCF format for forwarding by the XPFEs (e.g. CPRI over Ethernet). The XCF is based on Ethernet, taking advantage of MAC-in-MAC [6] (or Provider Backbone Bridged Network). MAC-in-MAC frames

allow multi-tenancy, keeping the traffic of different tenants separated via the outer MAC header and allowing different forwarding behaviours per tenant. The priority of different traffic flows is enabled through the priority bits of the Ethernet header. Basing the XCF on Ethernet eases reuse of legacy switches and increases synergies with the development of more generic switches. The use of MAC-in-MAC encapsulation as XCF ensures the feasibility of the solution, although its development requires some extensions over the OpenFlow Protocol. Initial features in order to support MAC-in-MAC in OpenFlow have been already included in the latest versions of the protocol [7].

4. XPU: Although the SDN control platform is responsible for the configuration of the network elements (i.e. the XFEs), the XPUs carry out the bulk of the operations in the 5G-Crosshaul supporting C-RAN, 5GPoAs, VNFs and a heterogeneous set of other services (e.g. CDN-based services). Each virtual infrastructure is instantiated, configured and operated by the XCI in the XPUs. The different functional distributions between the 5GPoA and XPU and the different services that can be hosted in the XPUs are one of the pillars of the flexibility provided by the 5G-Crosshaul architecture.

18.4   Common framing and switching elements

A RAN includes different levels of inter-connection between base stations (BSs), digital units (DUs) and radio units (RUs). The Metro Ethernet Forum (MEF, [8]) distinguishes three types of inter-connection (Figure 18.9):

- Backhaul: network between BS, or DU, and network controller site;
- Midhaul: carrier Ethernet network between BS, or DU, when one site is a small-cell site;
- Fronthaul: connection from a BS, or DU, to a remote RU (RRU).

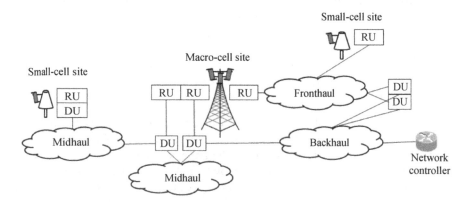

Figure 18.9   Mobile backhaul, midhaul and fronthaul definitions [8]

Usually, backhaul and midhaul networks are Ethernet packet networks, whereas, in the fronthaul link, constant bit rate (CBR) data flows are obtained by sampling and digitizing the antenna signal, as defined by CPRI [9]. Due to the analogue-to-digital conversion, the fronthaul interface is very bandwidth consuming: a 20 GHz LTE signal on air could require up to 2.5 Gbit/s in fibre. Moreover, the tolerated latency is low (about 100 μs), as required by the retransmission protocol (Hybrid Automatic Repeat request, HARQ) between DU and RRU.

In the 5G mobile systems generation, where the user capacity can be as high as 10 Gbit/s, transmitting at the bit rate required by a single fronthaul connection could become extremely expensive. This consideration is leading to rethink the repartition of functionalities between DU and RRU, so that the bit rate is proportional to the information rate rather than the radio bandwidth, as today happens with CPRI, where the sampling process occurs even in absence of transmitted data.

Depending on the split point of the radio protocols stack between RRU and DU, also the latency requirement may be relaxed, at least for those options where the HARQ is performed locally at the RRU.

It is likely that the 5G transport network will have to support different, vendor-specific, options of protocols split or even situations where the split is configurable. As a consequence, the boundaries between backhaul, midhaul and fronthaul will fade to be replaced by a common network infrastructure.

A multi-layer network architecture is probably the most suitable to transport and switch radio client signals with heterogeneous characteristics and requirements.

We will take the optical network as a relevant example, due to its high aggregate capacity and link distance capabilities, both key aspects in centralized 5G deployments (C-RAN).

In the optical network, signals transport and switching can rely on both circuit and packet multiplexing techniques. Circuit multiplexing can be performed at wavelength or time-slot level, or both. Wavelength-division multiplexing (WDM) and OTN [10] are two examples of multiplexing techniques at wavelength and time-slot level, respectively. As it will be discussed later, OTN might be replaced by simpler and more cost-effective time-division multiplexing (TDM) schemes, designed on the requirements of the 5G-Crosshaul network.

In real networks, not all the layers might be present. A mesh of Ethernet switches connected by point-to-point fibre links is an example where only the packet layer is present. In another example, the time-slot level may be missed, interfacing the packet switch with wavelength switches or reconfigurable add-drop multiplexers (ROADMs, [11]) by means of WDM transponders.

Figure 18.10 outlines the multiplexing hierarchy and the pros and cons of the three layers.

WDM networks guarantee the segregation of heterogeneous traffic data from different operators, radio vendors or tenants, simply assigning different data to different wavelengths. Moreover, the absence of digital processing in all-optical switches (e.g. wavelength selective switches [11]) is helpful when the latency and jitter specifications are tight.

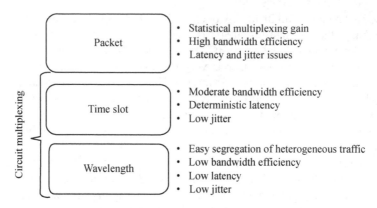

Figure 18.10   Layered 5G-Crosshaul network architecture

WDM can achieve a very high capacity: with the current technology, it is possible to transmit 96 wavelengths, 100 Gbit/s each, on the same fibre. This allows significant cables saving when optical fibres are not available or expensive to lease, a frequent situation in some countries or for non-incumbent operators.

However, the WDM transceivers are expensive: to lower the cost per bit/s, it is important to guarantee that the wavelength channels are always 'filled' with real traffic data and never sent empty. This is the reason why the second layer, time-division multiplexing, is introduced.

An example of TDM scheme for fronthaul is described by CPRI [9] but the multiplexing can also be performed in the transport network, especially when it is shared by different radio systems vendors.

OTN [10] is the most common standard to enable multivendor inter-operability, switching and grooming. It provides features such as forward error correction (FEC), multiple levels of tandem connection monitoring, transparent multiplexing of different client signals and fine granularity switching capability.

CPRI mapping over OTN has been recently defined [12].

However, the transport of latency sensitive signals, like CPRI, over OTN still presents challenges. One is the dimensioning of the so-called de-synchronizer bandwidth, to mitigate jitter and wander introduced when mapping CPRI clients over OTN frames.

An analysis of root mean square value of the frequency offset ('jitter') and mean time interval error ('wander') is performed in [12]. These results show that to comply with the CPRI jitter specification (2 parts per billion), a de-synchronizer bandwidth lower than 300 Hz (the value normally used in OTN) is required. In some cases, sub-Hertz bandwidths are needed, requiring expensive stable oscillators and sharp filters.

Another issue is that the fronthaul connections require an equal delay in uplink and downlink. There are different causes of delay asymmetry in optical networks: different fibre lengths; use of different wavelengths and, especially, different processing times at OTN framers and de-framers.

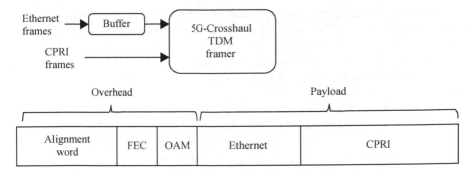

Figure 18.11   CPRI and Ethernet over the same frame

Today, the practical use of CPRI over OTN is limited to the case of synchronous mapping of CPRI signals belonging to a single synchronization island, quite in contrast with the concept of a common 5G-Crosshaul network.

To overcome these issues, alternative framing structures are possible, designed for fronthaul networks characterized by short distances, limited number of hops between add/drop nodes, demanding latency requirements and symmetric delay in downlink and uplink, but open to support other types of clients, as Ethernet signals used for backhauling. The following design guidelines can be identified:

– The frame should be synchronous to the client signal having the most accurate jitter specifications (e.g. CPRI), to avoid any degradation of the synchronization accuracy;
– Optional FEC can be present, with limited gain and complexity to keep low the additional latency;
– When spectrally inefficient codes are used, like 8B10B in CPRI, they could be replaced by more bandwidth efficient scramblers;
– Space should be left for in band signalling channels.
– Means to monitor possible uplink/downlink delay asymmetries and dimension the buffers to compensate for it, should be included.

A principle scheme for multiplexing fronthaul and backhaul signals on the same frame is shown in Figure 18.11, consisting in allocating two separate portions of the frame to two different types of signals, for example Ethernet and CPRI. Size and position of CPRI and Ethernet sub-frames are known but can be configured depending on network configuration and traffic load. The frame is synchronous to the clock of the most jitter-sensitive client (CPRI, in this example), to minimize the impact on its performance. The other client, Ethernet, is buffered to absorb differences in clock value and accuracy.

So far, we discussed circuit multiplexing, at wavelength or time-slot level, which is especially suitable for CBR signals, like CPRI. Packet multiplexing makes sense instead when the data rate is load dependent, so that statistical multiplexing gain can be exploited.

It is likely that new generation fronthaul interfaces, thanks to a different functional split between RRU and DU, will be packetized to absorb much less bandwidth than CPRI.

However, packet interfaces could not always fit latency critical services. A first solution to this issue is to adopt a layered architecture (Figure 18.10), where circuit switches could be used to limit the occurrence of overload situations or long queueing times in the packet switch, for example aggregating packets that have the same destination in big CBR pipes that can managed by the circuit switch.

A second solution is to introduce a more deterministic timing behaviour in packet networks, through the definition of new link-level features.

It is likely that the frame format for new 5G-Crosshaul packet interfaces will be an evolution of the standard Ethernet frame. This will also help the back compatibility with legacy systems.

At this purpose, the IEEE 802.1 Time Sensitive Networking (TSN, [13]) task group defines a set of standards providing means to reduce jitter by pre-emption and scheduling, and to guarantee bandwidth by reserving resources along the source-destination path, with important hardware modifications required in the switches, mainly affecting buffers, shapers and schedulers.

Some of the requirements of TSN are: time synchronization of relay and end systems with an accuracy better than 1 μs (IEEE 802.1AS); QoS in terms of low latency and high probability of packet delivery; protection against behaviour of systems that can disrupt QoS or time synchronization (IEEE 802.1Qci); ability to interrupt the transmission of non-latency-critical frames in order to transmit other frames with more stringent latency requirements (IEEE 802.1Qbu).

TSN describes three configuration models. In the Fully Distributed Model, the end station that contains TSN stream users communicates the user requirements directly over the TSN User Network Interface (UNI). The network is configured in a distributed way without a centralized network configuration entity.

The Centralized Network/Distributed Model is similar to the fully distributed one: the end stations communicate their requirements directly over the TSN UNI but the configuration information is directed to a Centralized Network Configuration entity (CNC).

Finally, the Fully Centralized Model enables a Centralized User Configuration entity (CUC) to discover end stations.

A list of features for a new generic 5G-Crosshaul packet interface is reported in Table 18.1.

MAC-in-MAC and MPLS-TP are examples of Ethernet frame format compatible with these requirements.

MAC-in-MAC allows to hide the tenant MAC address from the provider network, so that changes to the tenant address do not affect the packet forwarding behaviour in the provider network, and only the edge of the provider network needs to be reconfigured.

The MAC frames of each tenant are encapsulated by two additional headers and transported unchanged across the provided network.

Table 18.1 Features of a 5G-Crosshaul packet interface

Functional split

Support multiple functional splits — Support different functional splits of the radio protocol stack, ranging from CPRI-like constant bit rate to backhaul packet traffic

Multi-tenancy

Traffic isolation	QoS policy and SLA of one tenant should not affect other tenants
Traffic separation	One tenant should not be able to listen the traffic of other tenants
Multiplexing gain	Support statistical multiplexing gains across different tenants

Standardization

Ethernet	Compatibility with conventional Ethernet switches
Security support	Support of standard encryption and authentication methods
Synchronization	Compatibility with IEEE 1588v2 or IEEE 802.1AS

Transport efficiency

Short overhead	The additional headers should be short
Multi-path	Carry traffic towards the same destination through different paths
Flow differentiation	QoS should be set for individual flows in addition to traffic classes
Class of Service	Support different classes of service for different types of traffic

Management

In-band signalling — Carry management and signalling data on the same link of the traffic data

Multi-media support

Multi-media support — Compatibility with 802.3, 802.11ad and their evolution

Energy efficiency

Energy consumption proportional to traffic — Support of features such as sleep modes, reduced line rates etc.

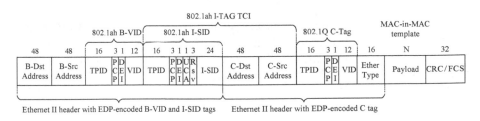

Figure 18.12   MAC-in-MAC header [6]

The header has an optional F-Tag field to support equal cost multi-path (ECMP) and contains a new Ethernet header with MAC addresses, a B-Tag field to support VLANs and an I-TAG field to support further service differentiation.

The MAC-in-MAC frame format is shown in Figure 18.12.

The labels of each packet must contain fields to support the mentioned requirements.

The label used to identify a tenant is the B-VID tag contained in the field VID that has 12 bits of length and allows to identify 4096 different tenants.

The I-SID tag in the I-SID field is used as service ID, to differentiate the type of service. The I-SID value can be shared among all the tenants and defined by the infrastructure owner; alternatively, each tenant can define its own value.

The frame provides three bits that allows to distinguish eight different priority classes, where the priority is established in terms of latency and jitter requirements. For example, the class with the highest priority might even pre-empt the transmission of other frames, if this is deemed necessary to reduce the jitter.

18.5   Control infrastructure

The XCI provides control and management functions to operate the different types of resources composing the Crosshaul physical infrastructure, including both network nodes (XFEs) and processing units (XPUs).

The SDN-based approach enables the on-demand and automated provisioning of VN slices and chains of VNFs to compose dynamic and converging 5G fronthaul and backhaul virtual infrastructures and services.

The XCI is the central part of the overall 5G-Crosshaul system architecture as shown in Figure 18.6. The XCI has interfaces to upper and lower as well as to neighbouring layers, in particular:

- the east–west interfaces (EBI, WBI) towards core network and RAN domains;
- the southbound interface (SBI) towards the data plane;
- the northbound interface (NBI) with applications on top.

For the sake of simplicity, the core and access domains are shown here with a dedicated MANO each, but it is also possible to operate them with a common MANO. The XCI is limited to the operation of the resources in the Crosshaul domain. It has to interact through the EBI/WBI with the core and access MANO e.g. for monitoring purposes.

18.5.1   XCI high-level architecture

The XCI part dealing with NFV comprises three main functional blocks, namely: NFV orchestrator, VNFM(s) and VIM (following the ETSI NFV architecture, [14]):

- The **NFVO** manages a Network Service (NS) lifecycle. It coordinates the VNF lifecycle (supported by the VNFM) and the resources available at the NFV Infrastructure (supported by the VIM) to ensure an optimized allocation of the necessary resources and connectivity to provide the requested VN functionality.
- The **VNFMs** are responsible for the lifecycle management of VNF instances (e.g. instance instantiation, modification and termination).

- The **VIM** is responsible for controlling and managing the NFVI computing (via *Computing ctrl*), storage (via *Storage ctrl*) and network resources (via *SDN ctrl*). The VIM can be extended with planning algorithms, which take efficient decisions about VM placement and network configuration. Such an extended VIM is called an integrated Virtual Infrastructure Management and Planning (VIMaP) function.

In addition to these modules, which are in charge of managing the different VNFs running on top of the 5G-Crosshaul, the XCI includes specialized controllers to deal with the control of the underlying network, storage and computation resources:

- SDN Controller: This module controls the underlying network elements following the conventional SDN paradigm. The SDN controllers may have to deal multiple technologies used in transport networks (such as micro-wave links) to have a common SDN controlled network substrate, which can be reconfigured based on the needs of the network tenants.
- Computing/Storage Controllers: Storage and computing controllers are included in what we call a Cloud Controller. A prominent example of this kind of software framework is OpenStack.

Note that the **SDN/Computing/Storage controllers** are functional blocks with one or multiple actual controllers (hierarchical or peer-to-peer structure) that centralize some or all of the control functionality of one or multiple network domains. One such network domain could be a domain with legacy MPLS or GMPLS equipment, a dedicated network controller ensures backward compatibility for such legacy equipment and at the same time control via a central controller.

18.5.2 XCI SDN controller

The SDN controller consists of three internal layers, network applications, network core services and southbound plugins. Figure 18.13 shows in addition the applications using the services of an SDN controller and physical infrastructure controlled at the southbound interface.

The southbound plugins establish an abstraction layer. This layer provides a common, vendor- and protocol-independent set of commands and notifications to/from the data plane devices. This layer allows the network services implemented in the controller to interact with the variety of data plane technologies in the physical infrastructure without taking into account the specific protocol supported by each network node at the XCI southbound interface.

The network services provide an infrastructure control layer. These services interact with the different devices making use of the unified APIs provided by the drivers.

The network applications create a network service control layer, which implements the highest level of network services. It introduces a certain level of automation and intelligent control in the physical infrastructure. The applications in this layer operate on the network only, they do not consider the compute and storage part of the physical infrastructure.

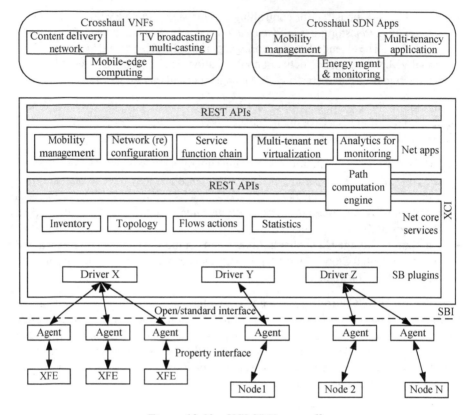

Figure 18.13   XCI SDN controller

Typical network applications embedded in the SDN controller are as follows:

- Topology and inventory: maintains an updated view of the resources (Inventory) and their connection (Topology).
- Flows actions: configures the flow entries in the different devices, guaranteeing their internal consistency and maintenance (e.g. for periodical refresh).
- Statistics: collects monitoring data about status and performances of devices (e.g. alarms, failures, number of dropped packets, congestion status etc.).
- Path computation engine: provides basic routing functions. Can be extended through dedicated computation algorithms for end-to-end paths or VN slice composition with more complex objective functions.
- Analytics for monitoring: used to correlate raw network monitoring data originally provided by the statistics service. It can be used, for example, to elaborate monitoring information related to a VN based on statistics data on flows and physical ports.
- Network (re-)configuration: used to re-configure specific network elements, mainly for management purposes.

- Path provisioning: establishes a network path between a source and one (or more) destination endpoint(s). The internal path computation engine core service can consider constraints on the path.
- Multi-tenant network virtualization: builds isolated and virtualized networks on the shared physical infrastructure. The multi-tenancy logic, i.e. the mapping between a virtual infrastructure and a tenant, can be kept at the SDN controller level but may be also implemented at the upper layers only, e.g. at the VIMaP, the NFVO or the application level.

18.5.3 Deployment models of XCI

For a large or complex network, a single SDN controller may not be sufficient. The network size, in terms of controllable elements, has a direct impact on the controller requirements on aspects such as the number of, e.g. active and persistent TCP connections for control sessions or memory requirements to store the network graphs. Also, the network complexity in terms of multiple deployed technologies (such as a packet-switched layer for Layer2/Layer3 transport over a circuit-switched optical layer) has an impact on functionalities and protocols to be implemented by the controller. In such a setup, an inter-layer coordination function is needed to deal with end-to-end connections and inter-layer technology adaptation.

To address such shortcomings, an SDN control plane may be deployed as multiple controllers. Two simple models to arrange the controllers are a flat or a hierarchical model. In the flat or peer model, a set of controllers is connected in a mesh topology and cooperates, to provision end-to-end services. Each controller may be responsible for a network sub-domain. The controllers hide the internal control technology and synchronize state using, e.g. East/West interfaces. In the hierarchical model, controllers are arranged in a tree-like topology. For a given hierarchy level, a parent controller handles the automation and has a certain number of high-level functions, while low-level child controllers cover detailed functions and operations.

These simple controller arrangements can be combined to cover relevant aspects of more complex networks such as network segment splitting, constraints among controllers of different vendors, redundancy and high availability, administrative domains and provisioning workflows.

18.6 Enabled applications

The 5G-Crosshaul requires a fully integrated and unified management of fronthaul and backhaul resources in a sharable, scalable and flexible manner. The key to offer context-aware resource management and to enable system-wide optimization lies on developing an ecosystem of SDN/NFV applications that exploits 5G-Crosshaul resource orchestration functions to support diverse functionalities such as planning, resource management and energy management, etc. These applications serve as the logical decision engines to programme the underlying transport network and packet forwarding behaviour. This section introduces a number of SDN/NFV applications designed for managing the transport network, offering context-aware resource orchestration to manage different types of resources (including networking, computing and storage resources) of physical and/or virtual infrastructures, so as to ensure top-notch service delivery and optimal resource utilization both cost-wise and performance-wise.

18.6.1   Resource management application (RMA)

Resource management is crucial to ensure that all users, including high-mobility users, receive the requested services with adequate level of QoS (enjoying a satis-factory experience), whereas on the other hand, aiming for an effective utilization of available resources so as to reduce the overall CAPEX and OPEX, as well as to provide resilience in case of network failures. Over-provisioning of transport resources is no more a feasible and economic solution for the network operators to meet the stringent fronthaul and backhaul traffic requirements. Hence, it is necessary to develop a new resource management solution to cope with the new requirements and challenges of 5G transport.

RMA aims to offer logically centralized and automated management of 5G-Crosshaul resources to promptly provision transport services according to their SLAs and specific requirements of different client applications, while ensuring the infrastructure resources are effectively utilized. The RMA relies on the XCI controllers for the actual provision and allocation of resources. The RMA can operate over phys-ical or VN resources, on a per-network or a per-tenant basis, respectively. Essentially, the RMA has two main functional pillars: (i) dynamic resource allocation and (re-) configuration (e.g. new or updated routes, load balancing, adaptation of physical parameters etc.) as the demand and network state changes and (ii) dynamic NFV placement, e.g. enabling multiple Cloud-RAN functional splits flexibly allocated across the transport network, or allocating and managing virtual services like CDN (Figure 18.14).

18.6.2   Multi-tenancy application

Multi-tenancy is a desired feature of 5G-Crosshaul to enable a generalized, flexible sharing of infrastructures among multiple VNOs or service providers (i.e. multi-ple tenants), as depicted in Section 18.1.2.2. The end target is to significantly reduce the CAPEX and OPEX by sharing the infrastructure resources and maximize their utilization in a cost-efficient manner.

We consider two main multi-tenancy services that enable different degrees of explicit control and are characterized by different levels of automation of the network slices management, which are as follows:

- The provisioning of virtual infrastructures under the control and operation of different tenants – in line with an Infrastructure-as-a-Service (IaaS) model;
- The provisioning of tenant's owned Network Services as defined by the ETSI NFV architecture (Figure 18.15).

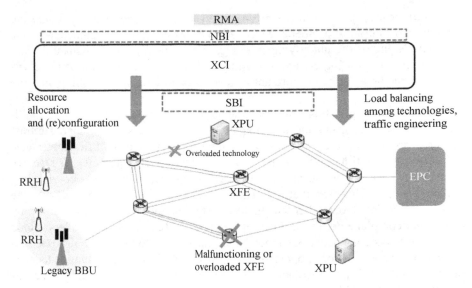

Figure 18.14 Resource management application

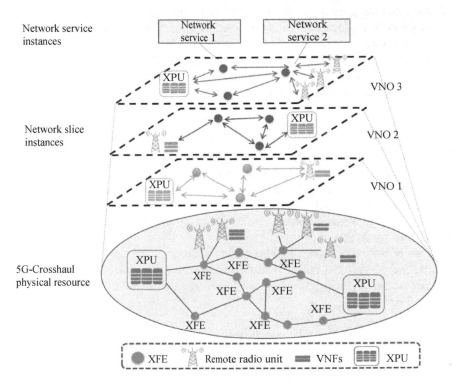

Figure 18.15 Multi-tenancy application

Multi-tenancy is an orthogonal characteristic of both services, guaranteeing separation, isolation and independence between different slices coupled with the efficient sharing of the underlying resources. Consequently, 5G-Crosshaul defines the term Tenant as a logical entity owning and operating either one or more virtual infrastructures or one or more network services, ultimately controlling their life cycles.

MTA is the application that implements the 5G-Crosshaul support for multi-tenancy [15], by coordinating and managing tenants access to the shared infrastructure, driving resource allocation for instances assigned to different tenants, and delivering multi-tenancy-related services. A high-level requirement is resource isolation, understood as the function of partitioning, separating and book-keeping of resources in such a way that a given tenant has no visibility of or access to the resources associated to another tenant. To perform this function, the MTA uniformly wraps and complements the infrastructure elements (e.g. SDN controllers, cloud management systems, network elements etc.) capabilities to provide multi-user support and resource isolation support, offering uniform and abstracted views to tenants. Regarding mechanisms for isolation, our approach is to rely on and leverage on existing ones, with the MTA acting as middleware and hypervisor. Full resource isolation requires system and infrastructure support, and it is not straightforward or cannot even be achieved, e.g. without hardware redundancy. 5G-Crosshaul provides soft-resource isolation including, notably, driving the SDN controllers capabilities to create per-tenant networks, allocating software switches within XPUs dedicated to per-tenant traffic, defining security groups and per-tenant addressing, switching and routing within XPUs and logically separating traffic within XPEs. Similarly, from the ETSI NFV/MANO perspective, the MTA manages state regarding to allocation of Network Services mapping tenants to actual instances and relying on implementations support.

18.6.3 Energy management and monitoring application

The goal of EMMA is to monitor energy parameters of RAN, fronthaul and backhaul elements, estimate energy consumption and trigger reactions to optimize and minimize the energy footprint of the virtual Crosshaul network while maintaining QoS for each VNO or end user. The main driver is to reduce the grid energy consumption across all Crosshaul components, which has a major environmental impact and a sizable cost for operators.

One of the main tasks of EMMA is to monitor the system power consumption and the energy status of physical and virtual nodes, both network and computing, to ensure that a complete view of the system energy consumption is available. This is achieved by:

- periodically collecting the values of power consumption and the energy status of the physical network nodes (e.g. XPEs, base stations) and their communication drivers;
- periodically collecting information on the power consumption of the servers (XPUs) due to the currently deployed virtual machines;

- computing and providing energy-related metrics of interest to the network operator.

Together with energy-specific parameters like power consumption and CPU loads, EMMA will also collect information about several network aspects such as traffic routing paths, traffic load levels, user throughput and number of sessions, radio coverage, interference of radio resources and equipment activation intervals. All these data can be used to compute a virtual infrastructure energy budget to be used for subsequent analyses and reactions/optimizations.

The application will optimally schedule the power operational states and the levels of power consumption of network nodes, jointly performing energy-efficient routing, load balancing and frequency bandwidth assignment, in a highly heterogeneous environment. Also the re-allocation of virtual functions across Crosshaul will be done as part of the optimization actions. This will allow moving fronthaul or backhaul VNFs to less power-consuming or less loaded servers, thus reducing overall energy demand of the network.

By leveraging the knowledge of physical and VN topology, data paths and traffic load over the links, EMMA:

- identifies which physical network nodes are idle (i.e. nodes whose communication cards are all inactive) and hence can be switched off (BASIC FUNCTION);
- computes alternate paths for traffic flows on the physical network topology, or it selects a path among the available ones for a new traffic flow, triggering path (re-)configuration with the objective of minimizing the energy consumption (NETWORK-RELATED FUNCTION);
- computes the optimal placement of the requested virtual infrastructures over the available physical infrastructure, triggering their re-planning to maintain the energy consumption under configurable thresholds (COMPUTING-RELATED FUNCTION) (Figure 18.16).

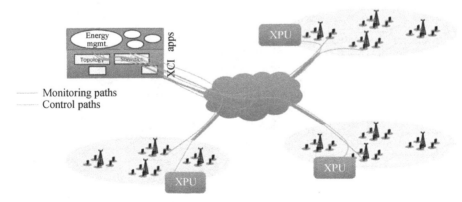

Figure 18.16   Energy management and monitoring application

18.6.4   Mobility management application

MMA is *Distributed Mobility Management* (DMM) based, and its main goal is to (i) provide mobility management for mobility scenarios such as vehicle mobility use cases like high-speed train scenarios and (ii) also to optimize traffic offloading for media distribution like CDN and TV Broadcasting.

The challenge in the high-speed train scenario (see Section 18.1.2.1) is to provide high-mobility support (e.g. up to or greater than 500 km/h) to a crowd of passengers (e.g. more than 500 people) using 5G services (video in particular) and satisfying the user QoE in a cost-effective way. Handover (HO) is a critical operation in mobility management, especially in such environment that suffers from frequent HOs and high sensitivity to link degradation. The focus of the MMA is to exploit the context information, including train location, speed and direction, etc. as well as the loading of some candidate target BSs, in determining the target BS and the corresponding resource allocation. The MMA will exploit the deterministic trajectory of the node for the proactive creation of paths in advance, placing cache nodes and even core nodes (e.g. authentication servers) on the path of movement of the train.

On the other hand, the task for traffic offloading is to optimize the location and relocation procedures for services such as CDN in combination with resource management decisions. The MMA will focus on providing traffic offload to the Internet and/or mobility of applications located at the edge as near as possible to the user, as shown in Figure 18.17. Its main goal is to optimize the route or path followed by users' traffic towards the Internet or to an application located in the XPUs. This

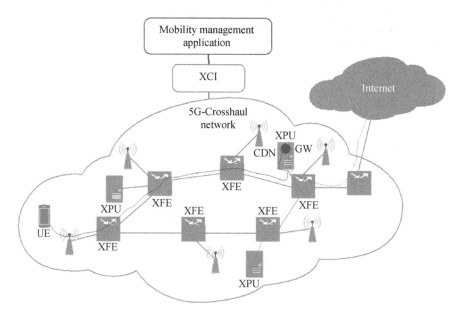

Figure 18.17   Mobility management application

will be done by computing the best allocation of services to the user, such as Points of Connection (PoC) to the Internet, CDN node and BBU allocations, based on the user location. After computing the best set of elements to provide a service to the user, it will request the RMA (see Section 18.5.1) to find the best path connecting these points based on the status of the network. Regarding the mobility management mechanism itself, the mobility will be based on a flat IP network, on which traffic is forwarded to the nearer point of connection to the Internet. The forwarding will be based on direct modification of flow tables at the XFE, using standard OpenFlow protocol or similar.

18.6.5 Content delivery network management application

The CDNMA is an OTT (Over-The-Top) application for managing the distribution of media content over 5G networks. Content distribution, especially video traffic, is expected to be the dominant contributor to the mobile data traffic demand. Hence, providing efficient ways of delivering content to the end users is a must. A CDN is a combination of a content-delivery infrastructure (in charge of delivering copies of content to end-users), a request routing infrastructure (which directs client requests to appropriate replica servers) and a distribution infrastructure (responsible for keeping an up-to-date view of the content stored in the CDN replica servers).

In 5G-Crosshaul, the content delivery infrastructure is deployed with networks of content caches – replica servers – which are deployed at the XPUs that are connected via the XFEs across the Crosshaul network. The origin server will be located in the Crosshaul network. The request on routing and distribution infrastructure functions will be enforced through optimal content routing and delivery based on the information from the network.

This application is designed to manage the transport resources to deploy and manage a CDN infrastructure, controlling load balancing over several replica servers strategically placed at various locations to deal with massive content requests while improving content delivery based on efficient content routing across the 5G-Crosshaul fronthaul and backhaul network segments and the corresponding user demands. The aim is to improve the content delivery maximizing bandwidth and improving accessibility. In summary, the CDNMA offers three functionalities:

- Deployment and management of a virtual CDN infrastructure;
- Orchestration of CDN nodes deployed on virtual instances in different XPUs along the Crosshaul network;
- Management the content delivery rules based on the monitoring information from the network and users.

Figure 18.18 depicts the use of CDMNA in 5G-Crosshaul. A content provider accesses the 5G-Crosshaul network and requests a CDN infrastructure through the CDNMA. Over the system runtime, the CDNMA would receive information from the XCI about the status of the network, information from the CDN infrastructure (computing, storage and networking resources) and information from the user about his location. Taking this information as input and applying optimization logic defined

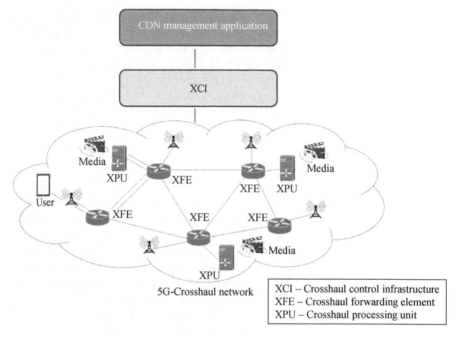

Figure 18.18   CDN management application

by the CDN operator based on the CDN and network metrics, the application strategically deploys and optimizes the replica servers in order to optimize resource utilization
and deliver the content to the end users/content consumers with maximum possible
QoS/QoE.

18.7   Standardization for the 5G-integrated fronthaul and backhaul

The integration of fronthaul and backhaul implies the definition of a number of
interfaces in the data, control and application planes. These interfaces ought to be
standardized in order to ensure inter-operability across different players (vendors
and operators alike) in the value chain and hence lower the deployment costs and
guarantee economies of scale. The standardization landscape for integrated fronthaul and backhaul in 5G is particularly complex in view of the various components
and heterogeneous technologies which need to be integrated. This section presents
the standardization activities underway for defining the 5G integrated fronthaul and
backhaul. These standardization activities are reviewed and summarized in a 2020
standardization roadmap. Table 18.2 presents the classification of the standardization
activities for the 5G integrated fronthaul and backhaul.

A standardization roadmap by 2020 is then depicted in Figure 18.19 with an
indicative timeline towards the year 2020.

Table 18.2 Classification of standardization activities for 5G fronthaul and backhaul

No.	Topic	Standardization groups
1	Use cases, gaps, requirements, architectures	NGMN, ITU-T 2020 FG, ITU-R WP5D, 3GPP, BBF, SCF
2	Gbps transmission technology (wired/wireless)	Wired: ITU-T SG15, NG-PON2, 100GE, CPRI Wireless: ETSI mWT, IEEE 802.11ay
3	Wireless access protocol functional splits	3GPP, IEEE 802.11, IEEE 1914.1, SCF
4	Fronthaul and backhaul traffic packetization (formatting)	Fronthaul: CPRI, NGFI (IEEE 1914.1/1904.3) Backhaul: VLAN (IEEE 802.1Q), MPLS
5	Fronthaul and backhaul traffic forwarding (switching protocols)	IEEE 802.1CM (Time Sensitive Networking), IETF DETNET (Deterministic Networking)
6	SDN control	ONF (OpenFlow), OpenDayLight, ONOS, IRTF SDNRG, ITU-T SG13, IEEE 802.1CF
7	NFV-based management and orchestration	ETSI NFV, IRTF NFVRG, OPNFV, OpenMANO, OpenStack, ETSI MEC

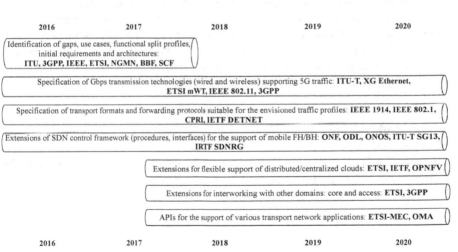

Figure 18.19 Initial standardization activity roadmap towards 2020

As shown, the 2016 and 2017 time period will continue to see the identification of gaps, definition of functional split profiles for the fronthaul, as well as requirements and architectures. The key SDOs involved here are: ITU (T/R), ETSI, 3GPP, IEEE, NGMN, BBF and SCF.

In parallel, the specification of Gbps transmission technologies both wired and wireless for fronthaul and backhaul is anticipated to continue until 2020 and beyond. These are taking place in ITU-T, 100G-Ethernet, ETSI mWT, IEEE 802.11 (ax, ay) and 3GPP.

The transport formats and forwarding protocols are also underway in several groups such as IEEE 1914, (e)CPRI, IEEE 802.1CM, IETF DetNet, and these are anticipated to continue until 2020.

Now on the SDN-based control plane, several activities are ongoing in forums such as ONF, ONOS, ODL, ITU-T, IETF/IRTF, including extensions for support of wireless and wired transport. These are anticipated to continue until 2020.

Further ahead, standardization related to areas for extensions of the integrated fronthaul and backhaul framework to cover emerging aspects such as edge cloud, fog computing, cross-domain orchestration (across core, transport and access), as well network applications for running on top, is envisioned to start in the timeframe of 2017/2018 and continues until 2020 and after.

The ITU-T 2020 Focus Group is developing recommendations for identified standardization gaps in the area of mobile fronthaul and backhaul slicing. These recommendations are due end of 2016 and are expected to feed into the work plan of specification groups inside ITU-T but also external organizations such as ETSI, 3GPP, IEEE and IETF.

As shown, during 2017 we will continue to see the identification of gaps, definition of functional split profiles for the fronthaul, as well as requirements and architectures. In parallel, the specification of Gbps transmission technologies both wired and wireless for fronthaul and backhaul is anticipated to continue until 2020 and beyond. The transport formats and forwarding protocols are also anticipated to be developed and continuously improved. Extensions of the SDN control framework are anticipated to continue until 2020.

Acknowledgement

This work has been partially funded by the EU H2020 project '5G-Crosshaul: The 5G Integrated Fronthaul/Backhaul' (Grant no. 671598).

References

[1]   Cisco, "Cisco visual networking index: forecast and methodology, 2015–2020," *White Paper*, 2016.
[2]   Horizon 2020, 5GPPP: 5G-Crosshaul Project. http://5g-crosshaul.eu/.
[3]   Open Networking Foundation (ONF), "SDN architecture, issue 1.1," 2016. [Online]. Available: https://www.opennetworking.org/images/stories/down loads/sdn-resources/technical-reports/TR-521_SDN_Architecture_issue_1.1. pdf.

[4] X. Costa-Perez, A. Garcia-Saavedra, X. Li, *et al.*, "5G-Crosshaul: an SDN/NFV integrated fronthaul/backhaul transport network architecture," *IEEE Wireless Communications Magazine*, February 2017.

[5] ETSI, Network Functions Virtualisation, "Network functions virtualisation (NFV): management and orchestration," December 2014. [Online]. Available: http://www.etsi.org/deliver/etsi_gs/NFV-MAN/001_099/001/01.01.01_60/gs_nfv-man001v010101p.pdf.

[6] IEEE 802.1 Task Group, "IEEE 802.1ah-2008 – IEEE standard for local and metropolitan area networks – Virtual Bridged Local Area Networks Amendment 7: provider backbone bridges."

[7] Open Networking Foundation (ONF), "OpenFlow switch specification 1.5.1," [Online]. Available: https://www.opennetworking.org/images/stories/downloads/sdn-resources/onf-specifications/openflow/openflow-switch-v1.5.1.pdf.

[8] Metro Ethernet Forum (MEF), "Implementation agreement MEF 22.1.1 Mobile Backhaul Phase 2 Amendment 1 – small cells" July 2014 [Online]. Available: https://www.mef.net/resources/technical-specifications/download?id=69&fileid=file1

[9] Common Public Radio Interface (CPRI), "Interface specification 7.0" [Online]. Available: http://www.cpri.info/spec.html.

[10] ITU-T G.709/Y.1331 Recommendation: "Interfaces for the optical transport network" [Online]. Available: https://www.itu.int/rec/T-REC-G.709/en.

[11] Sterling Perrin: "Building a fully flexible optical network with next-generation ROADMs, heavy reading white paper" [Online]. Available: https://www.lumentum.com/sites/default/files/technical-library-files/HR-ROADM-2011-Final.pdf.

[12] ITU-T G Suppl. 56 (02/2016) "Supplement: OTN transport of CPRI signals" [Online]. Available: https://www.itu.int/rec/T-REC-G.Sup56/en.

[13] IEEE 802.1 Task Group, "Time-sensitive networking task group" [Online]. Available: http://www.ieee802.org/1/pages/tsn.html.

[14] ETSI, "Network functions virtualisation (NFV): virtual network functions architecture," December 2014. [Online]. Available: http://www.etsi.org/deliver/etsi_gs/NFV-SWA/001_099/001/01.01.01_60/gs_nfv-swa001v010101p.pdf.

[15] X. Li, R. Casellas, G. Landi *et al.*, "5G-Crosshaul network slicing: enabling multi-tenancy in mobile transport networks," *IEEE Communications Magazine*, 2017.

Chapter 19

Device-to-device communication for 5G

Rafay Iqbal Ansari[1], Syed Ali Hassan[2],
and Chrysostomos Chrysostomou[1]

Abstract

Fourth generation (4G) communication technology was introduced to accommodate the rising demands of mobile user equipments (UEs) in terms of data rates and network reliability. However, a sharp rise in the number of mobile users has resulted in recent years and is expected to grow exponentially in the years to come, which prompted the need for a futuristic standard which could support a rather complex infrastructure of communication. Fifth generation (5G) communication technology is labelled as a platform for providing higher data rates with efficient utilisation of resources. The techniques which would undergo development under 5G communication standard include heterogeneous networks (HetNets), machine-to-machine (M2M) communications, device-to-device (D2D) networks and Internet of things (IoT) among others. Spectrum constraints have given impetus to the concept of offloading traffic from cellular network. D2D networks are viewed as an apt technology for providing direct peer-to-peer (P2P) links for data transfer, thus minimising their dependence on the base station (BS). Direct communication between devices will open up a window of opportunity for realising proximity services such as public safety networks, health monitoring, disaster area networks and numerous multimedia services. However, there are several challenges associated with D2D communication that include interference and resource management, network discovery, context aware services and network security. Viewing the resource constraints such as limited battery of the hand-held devices, it is important that a D2D network is able to cope up with the energy requirements to ensure network sustainability. In this chapter, we provide an overview of the techniques reported in the literature regarding the aforementioned challenges. Moreover, we also discuss some emerging trends and new aspects related to D2D networks such as energy harvesting and simultaneous wireless information and power transfer (SWIPT), pricing and incentive mechanisms and millimetre wave

[1]Department of Computer Science and Engineering, Frederick University, Cyprus
[2]School of Electrical Engineering and Computer Science (SEECS), National University of Sciences and Technology (NUST), Pakistan

(mmWaves) spectrum. We conclude by providing emerging aspects with regards to D2D communication and delineate issues which require further research and analysis.

19.1 Introduction

D2D communication has been envisioned as a futuristic technology for enhancing the performance of proximity P2P networks. D2D networks exploit the direct connection between devices in proximity instead of relying on the BS for communication. The mushroom growth of mobile UEs signified the need for a technology, which can cater the demands of higher data rates and reliable connectivity. 5G communication standard promises higher data rates, enhanced spectral efficiency and an interconnection between thousands of devices. D2D networks are considered as one of the technologies which would be developed under the purview of 5G network standardisation. Devices in proximity would be able to establish high data rate links for applications such as video transmission and other multimedia services. Moreover, lower transmit powers and capacity gains achieved through D2D networks makes them an apt technology for future 5G networks. Recently, researchers have also drawn attention towards utilisation of multi-hop D2D networks in public safety aspects. In the next subsection, we present the classification of D2D networks according to the frequency bands employed for transmission.

19.1.1 Classification of D2D networks

D2D networks are classified into In-band and Out-band, networks, which are explained below.

- *In-band*: In-band D2D networks involve utilisation of cellular spectrum for D2D communication.
- *Out-band*: Out-band D2D networks utilise other unlicensed spectrum bands such as the industrial, scientific and medical (ISM) or WiFi band for communication.

The In-band D2D network is further classified into *underlay* and *overlay* networks. In underlay D2D, the D2D and cellular users share the same spectrum resources. On the other hand, in overlay D2D networks, separate frequency bands are assigned to D2D and cellular users.

D2D networks open up a number of opportunities for new applications but there are several issues which need to be addressed to make D2D networks a viable option for future 5G networks. Cellular offloading will allow the BS to utilise the vacant spectrum for other applications but co-existence of D2D and cellular users would lead to several issues such as interference. Interference management in D2D networks has attracted significant attention of the researchers and several interference cancellation techniques have been proposed for D2D network environments. Interference free communication between a pair of D2D devices can be ensured by exploiting spatial degrees of freedom (DOF) [1] for cross tier interference cancellation. High density of nodes require efficient resource allocation techniques for judicious utilisation of resources. Dynamic spectrum allocation provides significant performance gains when compared to static spectrum allocation schemes [2]. D2D network discovery allows the devices in proximity to build clusters and retain information about the neighbouring nodes. The network discovery mechanism entails considerable interference issues which adversely impact the network performance [3]. The signalling overhead accumulated over time due to network discovery can lead to extra processing which can affect performance. In the following subsection, we delineate the D2D network topologies and explain the transmission flow in each topology.

19.1.2 D2D network topologies

Researchers have identified several network topologies which are feasible for establishing D2D networks. The BS controls the signalling in most of the scenarios but 3GPP has envisioned cluster-based multi-hop D2D networks for public safety networks where the BS is partially or fully damaged. The D2D network topologies can be categorised into the following four major types:

1. *BS assisted D2D networks (In-coverage)*: D2D devices can form pairs to establish direct communication with the help of the BS. The BS is responsible for network setup and other signalling activities. All the D2D users are assumed to be present in the coverage area of the BS.
2. *BS assisted D2D networks (Out of coverage)*: In this topology, one of the devices is assumed to be far enough that it is unable to access the cellular network. In such cases, the device in proximity can cooperate with such a node to facilitate access to cellular network.
3. *D2D cluster network (Out of coverage and In-coverage)*: D2D cluster networks with a Cluster Head (CH) node for managing signalling activities have been explored for scenarios with no cellular coverage. Such networks have been envisioned for public safety and security networks. In-coverage cluster networks provide cellular offloading with the assistance of the BS.
4. *D2D in heterogeneous networks*: HetNets involving pico, micro and macro cells provide numerous opportunities for future 5G D2D scenarios. However, such networks require robust interference management techniques for ensuring the quality-of-service (QoS) requirements.

Figure 19.1 depicts the aforementioned network topologies, highlighting the signalling and data transfer between D2D devices and BS.

Recently, another deployment scenario involving M2M traffic aggregation with D2D networks has been proposed. The dense deployment of M2M devices generates huge amounts of data which requires instant processing. The D2D user can collect the packets generated by the M2M devices in vicinity and utilise the cellular link to transmit the aggregated packets to the BS [4]. This technique allows to exploit proximity gain for low latency and energy-efficient transmissions [5].

The remainder of this chapter is organised as follows. We enlist different techniques associated with D2D communication that have been proposed for resource allocation, interference management, network discovery and proximity services,

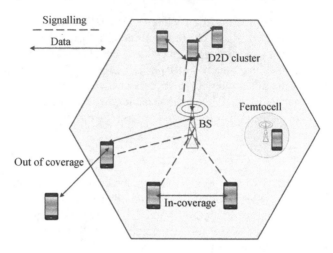

Figure 19.1   D2D network topologies

network security and network coding. Next, we discuss some emerging trends in D2D communications and identify issues which require further deliberation.

19.2   Resource management

Resource constraints in terms of available spectrum highlight the need for an effective resource management mechanism. The resource management techniques also differ according to the aforementioned network topologies. Cellular network environment with multiple cellular and D2D users poses several challenges such as power management to control interference. Moreover, the allocation of resources depends on other factors which include fairness and quality-of-experience (QoE) [6,7]. Knowledge of channel state information (CSI) is pivotal for ascertaining the apt technique of resource management. The resource allocation scheme also depends on the number of D2D and cellular users supported by the network.

Network-assisted 5G D2D involves mode selection between D2D and device-to-infrastructure networks. Dynamic uplink and time division duplexing (TDD) in the presence of instantaneous CSI is suitable for 5G D2D network deployment. The desired D2D communication range is considered as one of the major factors while choosing the D2D mode of operation [8]. Network assists the devices in the reuse of resource blocks. Resource allocation in heterogeneous networks becomes tedious due to the presence of macro, femto and pico cells. The multi-tier structure provides several benefits but leads to issues such as interference. The BS can allow the devices to operate simultaneously by intimating the D2D users with tolerable transmission power levels. Several factors such as path loss, shadowing, frequency selective multipath fading and wall penetration losses affect the network performance [9] by having an impact on the reliability of the links.

Previously, the analysis of D2D networks was limited to single D2D pairs existing in a cellular environment, but 5G networks promise dense deployment of nodes. New techniques have evolved which allow multiple D2D users to co-exist in underlay network scenarios. Jointly addressing spectrum allocation to cellular and multiple D2D links allows enhanced efficiency of the system but at the cost of increase in complexity [10]. 5G D2D communication can also support vehicular networks (VANETs). The major issue faced by VANETs is the latency which could be handled through efficient resource allocation. The collision avoidance mechanism such as CSMA/CA can lead to delays due to retransmission of signals [11]. Secondly, dynamic relay selection also causes delay due to mobility of vehicles and the limited range of D2D networks. BS could assist the D2D network setup for VANETs allowing optimal receiver selection and channel allocation to avoid delays.

The resource allocation problem can be divided into *centralised* and *decentralised*, depending on the decision-making mechanism. In a decentralised scheme the decisions related to resource allocation are taken locally, e.g., at the cluster head. In a centralised scheme, the central BS is responsible for taking decisions regarding the resource allocation. The centralised scheme, which involves multi-cell resource allocation allows significant performance gains [12]. The utilisation of centralised and decentralised schemes also depends on the network setup and the QoS requirements. In scenarios which involve both indoor and outdoor communications, it is pivotal to ascertain the scheme that performs best under the given conditions.

The resource allocation schemes for 5G D2D networks should involve less complexity and should be dynamic in nature. The resource allocation could be managed by the BS or could be handled locally by a CH. The performance of each technique depends on multiple factors. Scheduling in allocation of resources could be viable in some scenarios, but is unsuitable for 5G networks involving sharing of sensitive information. Applications involving real-time decisions regarding resource management require techniques, which involve less complexity and signalling overhead. The resource allocation also depends on the mode of transmission. In only D2D mode the resources may be shared by the D2D users. In scenarios involving co-existence of CU and D2D users, resource allocation is done in a manner which ensures QoS for both cellular and D2D transmissions. Moreover, the resource allocation restriction is one of the major factors in resource management algorithm design. Distributed resource allocation schemes seem feasible with regards to future 5G D2D networks. The optimisation problems related to resource allocation are aimed at maximising network capacity. The resource allocation can be segregated into device discovery and data transmission stages. The spatial distribution of the nodes and the channel state information can provide the decision matrix for designing an efficient resource allocation mechanism.

19.3   Interference management

Proximity D2D networks suffer from interference from cellular users and neighbouring D2D users. Several interference management techniques have been proposed to mitigate the impact of interference. Power management schemes allow to control the transmit power to limit interference. The channels can be assigned in a manner which avoids interference. Interference management techniques can be grouped into: *Interference Cancellation* and *Interference Avoidance*.

19.3.1 Interference management (underlay D2D)

Interference avoidance for underlay D2D networks allows the devices to utilise the frequency of cellular users without compromising the integrity of the data due to interference. Dynamic resource allocation technique helps the D2D users in avoiding the interference due to transmissions from neighbouring users. The D2D users monitor the tolerable transmit power with the help of the BS to avoid interference with the cellular users [13]. It is assumed that the devices remain static during one transmission cycle. The role of the BS is limited in a sense that it only conveys the required QoS to the D2D devices. The devices process the information received from the BS to determine the optimal transmit power to avoid interference.

Cross cell D2D communications in underlay network environment provide an interesting analysis of situations in which D2D users exist in separate cells. Graph theory can help in modelling such networks. A hyper-graph-based resource allocation scheme can help in modelling the interference in cross cell D2D networks [14]. The hyper-graph model helps in ascertaining the impact of interference in a particular network deployment. The cross cell CU to D2D and intra-cell D2D to D2D interference is modelled using the proposed technique.

Interference in a multi-tier environment can be avoided by using guard zones. The D2D users are able to avoid interference from nearby BSs because of guard zones. Stochastic geometry tools can be employed to model the distribution of nodes. No active D2D users are able to operate in the guard zone. The concept of interference limited area (ILA) [15] works on a similar principle. The cellular users in the vicinity of ILA and the D2D users are not allowed to transmit simultaneously. The geographical ILA is identified by the SINR. ILA approach allows rate maximisation for the D2D network operation. However, knowledge of CSI is required for executing this scheme which makes it unsuitable for scenarios where channel states change due to node mobility.

Power control schemes combined with scheduling techniques allow interference avoidance at D2D pairs. The D2D users can share the spectrum resources with cellular users while keeping the interference levels at tolerable limits. These schemes could be modified for use in scenarios where perfect CSI is not available. Successive interference cancellation (SIC) [16] allows the D2D receiver to cancel the interference. A stochastic model can be used to combine the multi-tier interference into single interference expression for ease of analysis. The SIC technique is easily implementable on existing systems due to low hardware complexity as it uses simple decoding mechanisms. SIC is considered as an apt technology for spectrum sharing scenarios where nodes suffer from interference issues. Multi-level SIC techniques can be employed for better results with regards to interference cancellation.

19.3.2   Interference management (overlay D2D)

Most of the works in the literature assume perfect time and frequency synchronisation which is not always the case if we refer to future 5G D2D networks. New techniques have been introduced to counter the time and frequency misalignment in D2D networks. The expected interference due to each D2D user is evaluated to identify the suitable transmit powers. A comparison with the OFDM technique for D2D reveals that the impact of D2D transmissions not synchronised in time is marginal. Different waveforms are assigned to D2D users for throughput maximisation. The type of waveform depends on the available time and frequency resources [17]. The analysis is expanded by ascertaining the impact of D2D transmission window on the interference. Techniques such as OFDM which have already been reported in literature are evaluated on the basis of their performance in large and small transmission window scenarios.

The interference from cellular users can be avoided by using overlay networks, but the interference from neighbouring D2D transmitters can cause interference. Most of the works address the interference to and from cellular users, but recently the intra D2D interference has drawn attention. Power control mechanism combined with channel allocation can mitigate the impact of interference [18]. Energy efficiency is enhanced due to the power control scheme, while efficient resource allocation helps in conserving the spectrum.

Several interference mitigation techniques have been proposed for intra-cell and inter-cell communication. In a multi-tier network scenario, such as heterogeneous networks, interference coordination is of prime importance for maintaining the network reliability. Future 5G networks with dense deployment of nodes will pose great challenges with regards to interference. The interference mitigation techniques for both in-band and out-band communications will be essential for proper handling of the network. The number of proximity services envisioned in future 5G networks will not be able to operate to their full potential if effective interference management systems are not designed. The research should also be focused on finding interference management techniques for mobile nodes where perfect CSI might not be available.

19.4   Network discovery and proximity services

19.4.1   Network discovery

D2D connection setup involves discovery of devices in proximity for establishing communication links. The devices in proximity can be discovered through network-assisted protocol [19]. Several discovery protocols have been discussed in the literature and 3GPP has categorised model A and model B as major protocols for device discovery. In model A discovery protocol, the cellular users can be segregated into potential *announcers* or *monitors*. The *announcers* broadcast a message to the nodes in proximity while the *monitors* (in the listening mode) are able to identify the message sent by the announcer. This mechanism allows the announcers to disclose their spatio-temporal location to other nodes in proximity. In model B discovery protocol, the nodes are segregated into *discoverer* or *discoveree*. The message sent by the *discoveree* is captured by the *discoverer* to identify the nodes in proximity. This protocol is suitable for scenarios where a priori information about the nodes in proximity may or may not be known [20].

Direct D2D discovery involves signalling and processing which can lead to energy issues in the battery powered devices. Network-assisted discovery could exploit the available information such as device location, channel conditions and interference issues. Stochastic geometry tools could help in finding the analytical expressions for distance distribution between the devices in proximity. The discovery mechanism depends on the network density and the transmit power requirements. A thorough analysis of such networks helps in ascertaining the conditions which ensure quick and reliable network discovery [21].

Quick network discovery is particularly important for scenarios such as disasters where the destruction of infrastructure limits the communication options to local D2D networks. Signature-based discovery has also been proposed as a quick network discovery technique. In this technique a discovery channel is assigned to the device which announces its presence to other nodes in proximity [22]. Although, signature-based discovery is quick as compared to packet-based discovery, it is not secure as physical layer waveforms are used to identify the nodes. A hybrid scheme based on a combination of both schemes is quite suitable in the context of 5G networks.

The discovery mechanism should be scalable so that all the devices in proximity are identified and any new nodes entering the proximity area are able to join the D2D network. Moreover, the discovery signals from the devices which are close to each other are very strong which causes problems at the receiver as it is unable to detect discovery signals from devices which are relatively farther. This issue has been labelled as near-far problem in the literature [23]. It is important to address the near-far issue for quick network discovery. 5G networks will involve dense deployment of nodes along with high node mobility. Quick and self adjusting network discovery mechanisms are of prime importance for providing seamless connectivity. Device discovery mechanism is also beset with security issues which need to be addressed [24]. The geographical environment can also impact network discovery. If we consider a disaster environment the destruction of infrastructure can lead to different path losses. The difference between indoor and outdoor environment signifies the need for network discovery protocols which can work in both environments.

19.4.2 *Proximity services*

Proximity multimedia services are widely considered as an important part of future 5G D2D networks [25]. The BS can assist proximity services (ProSe) by gathering the requirements of the D2D users. The D2D pairs can then be activated to support ProSe. The BS assists in network setup and channel assignment tasks. Some ProSe involve transmission of data that is sensitive and delays can prove to be critical. BS can evolve a prioritisation mechanism depending on the sensitivity of the data.

In multi-hop ProSe scenarios the relay selection problem is of prime importance as the relay nodes impact the network rate and reliability of the transmissions.

Applications involving ProSe in a cluster environment include E-commerce, advertisement, common content sharing, health monitoring, neighbourhood social networks and public safety networks. D2D neighbourhood social networks will open up a numerous avenues for building new applications. ProSe allows traffic offloading, which enhances the spectrum efficiency. Recently, several proximity services, such as live gaming have been explored as new applications of D2D ProSe. The fundamental issues, which might hamper the network performance, are energy constraints and proximity device discovery. Cooperation among devices requiring same services could ease the pressure on BS in terms of spectrum. Handover mechanisms for ProSe have drawn attention of researchers. The mobility of devices highlights the need for a handover mechanism for nodes moving from one cell to another. The impact of handover might become more pronounced in urban environments, where frequent handovers between small cells are required. Emulating the concept of handover of purely cellular environment, similar concepts of hard and soft handover can be employed in ProSe D2D networks. Handover failure can lead to disruption in transmissions and could prove to be critical for overall network performance. Interference mitigation is also an important aspect of ProSe. Efficient synchronisation between D2D users can help in avoiding the interference [26]. Several commercial applications have been envisioned for ProSe due to the increasing usage of smart phones.

D2D caching techniques have been proposed to realise ProSe for context aware common content sharing. Caching allows the backhaul network to offload traffic to D2D network. The caching techniques can be further classified into centralised and decentralised caching. The BS can coordinate content dissemination among D2D users by helping to establish a D2D link. This technique comes under the purview of centralised scheme, where the BS maintains the record of the cached content at the individual users and facilitates content sharing accordingly. Decentralised caching schemes have also been proposed for infrastructureless scenarios. The subpackets of a file can be shared by the users in proximity, which act cooperatively to satisfy the demands of each other. A decentralised caching mechanism allows D2D users to share content without the help of a central authority, e.g., the BS [27]. The cache storage capacity determines the performance bounds for the network rate. Proactive caching techniques have been proposed to allow the content distribution among users in vicinity, based on the history of interactions between users and the popularity of the content. The interaction between users can be modelled by analysing the social network relationships between users. Prediction-based proactive caching mechanisms can help in enhancing the efficiency of future 5G D2D networks [28].

ProSe for static networks have been analysed by researchers, but mobility models for ProSe require further research as high mobility could lead to significant signalling overhead. Location based ProSe could lead to computational overheads, which in turn could lead to extra energy consumption. Delays can also overload the buffers of D2D devices, which can lead to data overflow at the devices. Recent works on ProSe suggest its commercial benefits due to numerous applications that could facilitate the users as well as commercial enterprises.

19.5   Network security and trust

Increase in the usage of social networks, such as Facebook and Twitter, prompted the research towards exploiting social ties to build trusted networks. Human relationships play an important role in how they communicate with each other. Social aware D2D networks have been explored as a possible solution for building reliable networks. Concepts of 'social trust' and 'social reciprocity' have been discussed as possible solutions for allowing devices in proximity to exploit the trust metric. Social trust is the direct trust relationships that exists between members of the same family, group or organisation. Social reciprocity is a dynamic relationship that may be developed on the basis of mutual help between devices. The network is observed from two domains: physical domain and the social domain. The devices in proximity may be able to communicate in the physical domain, but it is the social domain that helps in ascertaining a trusted connection between devices. A social trust-based network can help the nodes in proximity to cooperate and share common content among themselves. This approach relieves the BS from sending the same information to the devices and hence helps in conserving spectrum [29]. The decision to build a social trust or social reciprocity-based network depends on the network topology and the type of application. Figure 19.2 depicts a social trust-based D2D network, where the links between devices are evaluated in two domains: physical domain and the social domain. As shown in Figure 19.2, the nodes *B–C*, *C–E* and *D–E* can establish links in the physical domain for common content sharing. However, the mapping of nodes onto the social domain suggests that a trusted connection cannot be established between nodes *C* and *E*. The pair of nodes having links in both physical and social domain are able to establish trusted links and cooperate with each other for common content sharing.

Social trust also leverages the multi-hop D2D network by allowing the devices to select trusted relays. Social relationship traces can be used to build a statistical model

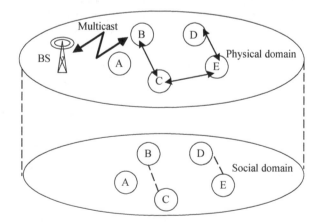

Figure 19.2   Social aware multicast (SoCast) system [29]

of the spatio-temporal behaviour of the D2D users. Proximity-based trust is one of the techniques, where the distance between devices is considered to determine a factor of trust. Experience-based trust networks determine the trust metric by analysing the history of interaction between devices [30]. The experience is observed over a period of time and is averaged out so that any occasional unusual activity does not impact the trust relationship between D2D users. Building a proximity-based trust network is a simpler and quicker approach, but it can allow malicious nodes to access the network. On the other hand, building an experience based network is a cumbersome task, but it limits the access of malicious nodes.

Trust-based networks can also help in identifying and isolating malicious nodes. The trusted network protocol may also vary with the type of network deployment, i.e., network controlled or device controlled. The experience of neighbouring devices can also be exploited to build trusted networks. The untrusted or malicious nodes can impact the network in several aspects. Some of the ways in which the malicious nodes can impact the network are given as follows:

- *Fake identity:* The malicious nodes can create fake identities to join a D2D cluster. The nodes with a history of malicious activity can be blocked by the nodes forming a trusted group, but if the nodes are able to create and exploit fake identities then it becomes difficult to identify such nodes.
- *False trust reports:* The malicious nodes can also produce false trust reports about other nodes in order to sideline them. The nodes in a cluster might be misled to believe that a particular node is involved in malicious activity, which might not be the case in reality.
- *Unwarranted coalitions:* Malicious nodes can form coalitions of their own which can compromise the integrity of the network. Such nodes can greatly influence the reliability of the network by disseminating false information and blocking information from neighbouring nodes.
- *Network sinks:* In multi-hop D2D networks the malicious relay nodes can also act as information sinks by not relaying the information to the destination.

It is also important that the social network trace that is utilised for determining the trusted relationships is reliable. If the social network trace itself is compromised then it is unsuitable to rely on it for designing the transmission flow strategy. The security mechanisms designed for future 5G D2D networks should be able to cater to both the in-coverage and the out-of-coverage scenarios. In the in-coverage scenario the BS might help the D2D users in building a trusted network and can support in signalling. However, in the out-of-coverage scenarios the cluster head needs to handle the security mechanism and aid in all the signalling activities. The mobility of users presents a challenging task for researchers to ensure security of the network. Dynamic adjustments on such a large scale would not only increase the signalling cost, but also lead to network delays. Moreover, the transmission of private profiles of nodes can pose threats to the privacy of nodes as the private profiles can be misused by malicious nodes.

19.6 Network coding

Network coding in cooperative D2D networks helps in exploiting the network diversity, which enhances the transmission success. Two major approaches of network coding have been reported in the literature [31].

- *Random network coding (RNC):* The packet combination is based on independent and random coefficients. Feedback information is not required to recover the message. However, in situations where real-time data recovery is required the performance of RNC is not up to the mark due to high computational cost.
- *Opportunistic network coding (ONC):* In this technique only the packets which are lost and not received at the receiver undergo network coding.

Several network coding techniques for D2D networks have been discussed in the literature. A summary of some of the techniques is as follows:

- *Exclusive Or (XOR):* In relay aided D2D communication the relay performs the XOR operation on the packets sent by two nodes, and then transmits them to both the nodes. The nodes are then able to receive multiple copies of the required packet, thereby enhancing the network reliability.
- *Instantly decodable network coding (IDNC)*: In this technique, the encoding mechanism is based on the information that is already stored in the receiver cache. This technique also uses the XOR function to encode the packets, and it is considered very suitable for applications requiring quick coding for real-time applications. The technique yields delay reduction in D2D networks which makes it suitable for future 5G networks.
- *Physical layer network coding (PNC)*: This technique superimposes electromagnetic waves for code formulation. PNC is an example of finite field network coding. The relay applies PNC to the signals sent by two nodes in one time slot. In the next time slot it transmits the coded message to the nodes. This technique is considered to be more efficient with regards to the number of time slots required to code and transmit the message. Traditionally, three time slots are required to complete the transmission of the coded message. However, PNC is able to complete the task in two time slots.
- *Analog network coding*: Analog network coding works on the same concept as PNC with the difference that the codewords are chosen from the infinite field.

Joint network coding and resource allocation techniques are very suitable for 5G D2D networks. An energy efficient cooperative network protocol, namely, the adaptive cooperative network coding-based MAC (ACNC-MAC) [32] has been presented as a suitable technique for relay aided D2D networks. The MAC protocol allows avoiding any collisions at the relays, while the network coding provides the coding gain.

In our work on network coding in relay-aided D2D networks [33], we presented the idea of assigning priorities to the D2D pairs. The node suffering from the work channel conditions is assigned the first priority. Similarly, other nodes are assigned priorities according to their channel conditions. The transmission cycle consists of three time slots in which the Sources S_1, S_2 and S_3 transmit the data in succession. The relay R transmits the coded message in the fourth time slot. Linearly independent codewords are selected from the Galois field. The receiver is able to decode the desired information by using the Gaussian elimination method. The results highlight enhanced performance of the proposed scheme as compared to the simple relay-aided network.

19.7 Emerging aspects in D2D

19.7.1 Millimetre wave (mmWave)

The management of existing spectrum bands for providing resources to new technologies is becoming extremely difficult due to the spectrum crunch. Recently, unlicensed mmWave technology has been introduced for its usage in 5G D2D networks. Providing low interference transmissions, mmWave in D2D opens up a host of opportunities for ProSe. Multimedia services and other proximity common content sharing applications could exploit mmWave technology for high data rate transmissions.

Interference issues in a multi-tier network could be avoided by using mmWave technology along with higher network capacity [34]. mmWave techniques are highly directional in nature, which opens up the possibility of having multiple links [35]. The analysis of mmWave in a cellular environment signifies its suitability for supporting D2D communication for heterogeneous networks. Applications involving real-time video sharing between hand-held devices could utilise the mmWave technology to set up high data rate links. Research has also been directed towards the probable hardware changes that might ensue the introduction of mmWave in D2D.

mmWave technology also have some limitations, which make it unsuitable for some network scenarios. Limited range is one of the drawbacks of mmWave technology. Physical obstacles, such as buildings, hinder the transmissions, compromising the range even further. Multi-hop D2D networks could also exploit the mmWave to increase network coverage. Transmission paths involving less attenuation can be identified and the transmissions can be routed through such paths. Relay mechanisms can also help in avoiding obstacles, but the transmission delays need to be taken into consideration. mmWave in D2D has been envisioned for line of sight (LOS) transmissions. In common content sharing, the D2D users could be paired to sustain the downloading mechanism. However, the pairing mechanism should be designed as such to reach all the users who wish to download the content. Proximity node discovery could help in the selection of a suitable relay from a choice of potential relays, but it could lead to signalling overhead. There might be situations in which one relay could suffice instead of using multiple relays for communication. The access point (AP) location could greatly impact the performance of the network as the channel conditions from AP to the D2D user can vary due to the location. The impact of delays need to be quantised to ascertain the workability of the relay mechanism [36].

mmWave technology could also help in setting up public safety and disaster area networks. Unlike the conventional wireless sensor networks, which suffer from interference, mmWave technology could greatly improve transmission between nodes in proximity by avoiding interference. Though, some hardware changes for operating mmWave technology on hand-held devices would become inevitable. Directional antennas would be required to communicate through mmWave technology. Previously, most of the research was focussed on determining interference mitigation techniques, but with the advent of mmWave technology for 5G, the research could be directed towards other aspects. Due to limited range of mmWave technology, a network deployment for supporting high data rate D2D communication could require several mmWave technology enabled BSs in close vicinity. As the number of BSs would increase, issues such as quick handover and network discovery need to be taken into consideration. Smaller antenna dimensions is one of the key advantages of mmWave technology which could help in reducing the size of communication equipment. Mode switching between mmWave and the conventional cellular network could take place to avoid transmission blockage in the case of physical obstacles.

19.7.2   Pricing and incentives

The D2D users can exhibit selfish behaviour due to the battery and storage constraints along with processing limitations of the D2D equipment. It also seems unfair to seek cooperation of a user without providing any incentive in return. Several incentives for D2D user cooperation have been proposed to encourage the devices to work on the concept of mutual help. The device might be reluctant to switch to D2D mode from cellular mode and it could make a decision to choose a particular mode depending upon the QoS required by the devices and the availability of resources. The type of incentive may also vary depending upon the in-band or out-band communication. The cellular network could also provide incentives for users who are willing to wait in the queue for downloading the data. Moreover, there is a need to develop a secure incentive mechanism so that any unwarranted users are not able to enjoy incentives intended for other users. As an incentive, energy harvesting could be allowed for idle relays. The devices that have allowed the usage of their cache storage for helping other users can be provided with a priority for D2D network setup in future. Recently, an interesting concept of incentives has been introduced for mobile D2D users. In order to support the cooperative mechanism in a geographical region with higher data demands the potential D2D relay nodes can be persuaded to move to such regions and support the communication [37]. In return they could be provided incentives for their cooperation. Figure 19.3 depicts a BS-assisted relay selection scenario, where the potential relay nodes share their bids with the BS, which in turn selects the relay that best suits the transmission flow. For example, the nodes in the cluster require packet $D1$, which is available with two relay nodes R_1 and R_2. The BS will initiate a bidding process and collect bids from the relays. As shown in Figure 19.1, the relay node R_1 possesses the desired packet and is close to the cluster, making it the most suitable candidate for relaying the packet. The BS selects the best relay and provides incentive to the relay according to the distance travelled by the relay to reach closer to the destination. This concept is in consonance with the idea of future 5G D2D networks,

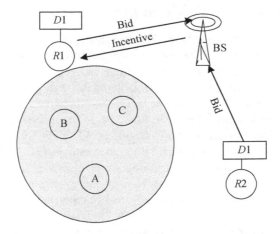

Figure 19.3 Incentive mechanism for relay nodes [37]

where D2D users are able to change their locations according to their preferences. Optimal traffic offloading can also be achieved in this manner. The decision metrics which could be analysed for determining the incentives include: connection period, amount of data transmitted, number of users served and battery usage for cooperation. The incentives provided to the devices could include priority in resource allocation and free data services, which are commensurate with the role of the D2D user in cooperative network.

The pricing mechanism depends on the type of services demanded by the D2D user. The D2D user can demand for dedicated connection for high data rate services or it could demand a shared connection for common content downloading. The price charged by the cellular network operator would depend on the type of service demanded and the data rate requirements of the device. The pricing mechanism can also be designed on the following principles:

- *Prepaid:* The D2D users have already paid for a particular services and are able to access the services until all the prepaid amount is exhausted.
- *Postpaid:* The D2D users are able to use unlimited services and are charged periodically.

For the pricing mechanism, the D2D devices can be segregated into potential data sellers and data purchasers. The nodes in a cluster can sell the data demanded by neighbouring users for a cost. The buyer can also identify several data sellers to gather the required data, and the prices can differ depending upon the seller. For example, the video streams can be bought from several sellers and combined at the receiver [38]. The selection of the seller could also depend on the channel conditions, and the sellers with the best channel conditions can be selected. However, such a scheme could only work in scenarios where perfect CSI is available with the buyer.

19.7.3   Energy harvesting and SWIPT

Recently, energy harvesting (EH) in D2D communication has drawn considerable attention of the researchers. The battery constraints of the D2D communication devices highlights the need for an EH capability, which replenishes the energy for network longevity. EH-based devices can be deployed in scenarios, where conventional power sources are not available. The idle D2D users can harvest ambient electromagnetic energy and store it for future use. Some hardware changes seem inevitable to make the D2D devices compatible with the concept of EH. The D2D users that are able to harvest enough energy to transmit are allowed to establish D2D links and transmit the messages [39]. Stochastic geometry tools can be employed to determine the outage probability of D2D users, keeping in view the energy requirements for successful transmission. The decision regarding the channel assignment for D2D operation, i.e., uplink or downlink, depends on the BS density.

Research has also been conducted on using renewable energy sources, such as solar energy for harvesting. The routing mechanism for EH-based multi-hop D2D network is important as the devices suffering from low energy can refuse to cooperate, leading to information blockages. Keeping the energy level trace of the devices in proximity or a cluster would involve extra signalling and storage resources. The knowledge of CSI impacts the type of routing mechanism that is adopted for ensuring reliable connectivity in EH-based D2D networks. EH-based relay network is a suitable solution for disaster scenarios, where the devices are unable to access the cellular infrastructure. The EH-based relays can sustain the information flow by serving the nodes in proximity. The magnitude of energy harvested by the relay depends on the distance of relay from the AP and the path loss parameters.

EH in heterogeneous network environment depends on the interoperability between AP and the D2D users [40]. Several approaches have been analysed to direct the transmissions from AP in a manner which allows energy harvesting. The distribution and density of APs within a geographical area could greatly impact the magnitude of EH by D2D users. High data rates provided by the heterogeneous networks compounded with an additional feature of EH could greatly help in designing sustainable future 5G D2D networks. Recently, the concept of simultaneous wireless and information power transfer (SWIPT) has gained ground for wireless networks. Figure 19.4 shows the major techniques, which have been proposed for realising SWIPT in wireless networks. The wireless power transfer model impacts the network performance in terms of power outage probability. Moreover, it is also important to ensure secure exchange of information between a D2D pair [41]. Analytical models for power outage probability and secrecy outage probability could help in ascertaining the transmission flow mechanism which fulfils the security requirements of the network and allows network longevity.

Research on SWIPT in D2D is still in its nascent stages as we cannot find many works in the literature discussing the technology for D2D networks. SWIPT will help in providing network longevity along with sustaining transmissions. D2D devices must be equipped with a dual antenna interface for the antenna switching technique of SWIPT. However, for mode selection technique, a single antenna would suffice as

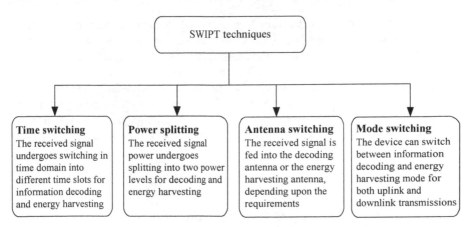

Figure 19.4 SWIPT techniques

the EH or the information decoding (ID) would depend on the mode of the D2D user. Identifying the EH and the ID nodes in a cluster and avoiding interference is one of the major challenges for adopting SWIPT in D2D networks. Moreover, the SWIPT techniques might differ for different network topologies of D2D networks, and it is important to identify the technique that provides optimal performance for a particular network topology.

19.8 Conclusion

In this chapter, we presented an overview of D2D communication in the context of future 5G network deployment. Several D2D network deployment topologies are identified, depending upon the coverage scenarios and spatial distribution of nodes. D2D networks allow cellular offloading, which leads to conservation of spectrum resources at the BS. Moreover, devices in proximity operate on low transmit powers, which helps in saving energy. Resource allocation techniques are highlighted for optimal allocation of resources for both D2D and cellular users. The need for dynamic resource allocation scheme is highlighted, viewing the dense deployment and mobility of nodes envisioned in 5G networks. This chapter explains different interference cancellation and interference avoidance techniques reported in the literature for ensuring coexistence of D2D and cellular users. Interference issues become more pronounced in a multi-tier environment such as HetNets. D2D networks can allow numerous proximity services between neighbouring devices. However, quick network discovery is important to identify the devices in proximity and the discovery mechanism should possess the capability to dynamically adjust the network according to node. The communication between devices in proximity also poses security challenges as sensitive information about devices could be exploited by the eavesdropping nodes. In this context, the concept of social trust-based networks has been introduced as a possible solution for establishing trusted links. Moreover, we also ascertain some emerging aspects with regards to D2D networks, such as mmWave communication, SWIPT-based networks and pricing/incentive mechanism for D2D communication. Energy harvesting techniques for D2D networks have drawn considerable attention of the researchers due to the battery constraints of the hand-held devices. Energy harvesting-based transmission protocols could ensure network longevity, while providing the desired QoS. We also highlight future challenges with regards to these emerging aspects, especially in the context of future 5G D2D networks.

References

[1]   Chiu SL, Lin KCJ, Lin GX, and Wei HY. "Empowering device-to-device networks with cross-link interference management," *IEEE Transactions on Mobile Computing*. 2017;16(4):950–963.

[2]   Ding J, Jiang L, and He C. "Dynamic spectrum allocation for energy harvesting-based underlaying D2D communication," in: *2016 IEEE 83rd Vehicular Technology Conference (VTC Spring)*. IEEE; 2016; Nanjing, China. p. 1–5.

[3]   Albasry H and Ahmed QZ. "Network-assisted D2D discovery method by using efficient power control strategy," in: *2016 IEEE 83rd Vehicular Technology Conference (VTC Spring)*. IEEE; 2016; Nanjing, China. p. 1–5.

[4]   Rigazzi G, Pratas NK, Popovski P, and Fantacci R. "Aggregation and trunking of M2M traffic via D2D connections," in: *2015 IEEE International Conference on IEEE Communications (ICC)*, 2015. p. 2973–2978.

[5]   Tsai SY, Sou SI and Tsai MH. "Reducing energy consumption by data aggregation in M2M networks," *Wireless Personal Communications*. 2014;74(4): 1231–1244.

[6]   Ciou SA, Kao JC, Lee CY and Chen KY. "Multi-sharing resource allocation for device-to-device communication underlaying 5G mobile networks," in: *2015 IEEE 26th Annual International Symposium on Personal, Indoor, and Mobile Radio Communications (PIMRC)*, IEEE; 2015; Hong Kong, China. p. 1509–1514.

[7]   Maghsudi S and Stańczak S. "Hybrid centralized-distributed resource allocation for device-to-device communication underlaying cellular networks," *IEEE Transactions on Vehicular Technology*. 2016;65(4):2481–2495.

[8]   Militano L, Orsino A, Araniti G, Molinaro A, Iera A, and Wang L. "Efficient spectrum management exploiting D2D communication in 5G systems," in: *2015 IEEE International Symposium on Broadband Multimedia Systems and Broadcasting*. IEEE; 2015; Ghent, Belgium. p. 1–5.

[9]   Tsai AH, Wang LC, Huang JH, and Lin TM. "Intelligent resource management for device-to-device (D2D) communications in heterogeneous networks," in: *2012 15th International Symposium on Wireless Personal Multimedia Communications (WPMC)*. IEEE; 2012; Taipei, Taiwan. p. 75–79.

[10] Wang R, Zhang J, Song SH, and Letaief KB. "QoS-aware joint mode selection and channel assignment for D2D communications," in: *2016 IEEE*

International Conference on Communications (ICC). IEEE; 2016; Kuala Lumpur, Malaysia. p. 1–6.

[11] Cao X, Liu L, Cheng Y, Cai LXX, and Sun C. "On optimal device-to-device resource allocation for minimizing end-to-end delay in VANETs," *IEEE Transactions on Vehicular Technology*. 2016;65(10):7905–7916.

[12] Venkatasubramanian V, Moya FS, and Pawlak K. "Centralized and decentralized multi-cell D2D resource allocation using flexible UL/DL TDD," in: *Wireless Communications and Networking Conference Workshops (WCNCW), 2015 IEEE*. IEEE; 2015; New Orleans, LA, USA. p. 305–310.

[13] Chiti F, Di Giacomo D, Fantacci R, and Pierucci L. "Interference aware approach for D2D communications," in: *2016 IEEE International Conference on Communications (ICC)*. IEEE; 2016; Kuala Lumpur, Malaysia. p. 1–6.

[14] Zhang H, Ji Y, Song L, and Zhu H. "Hypergraph based resource allocation for cross-cell device-to-device communications," in: *2016 IEEE International Conference on Communications (ICC)*. IEEE; 2016; Kuala Lumpur, Malaysia. p. 1–6.

[15] Min H, Lee J, Park S, and Hong D. "Capacity enhancement using an interference limited area for device-to-device uplink underlaying cellular networks," *IEEE Transactions on Wireless Communications*. 2011;10(12):3995–4000.

[16] Ma C, Wu W, Cui Y, and Wang X. "On the performance of successive interference cancellation in D2D-enabled cellular networks," in: *2015 IEEE Conference on Computer Communications (INFOCOM)*. IEEE; 2015; Kowloon, Hong Kong. p. 37–45.

[17] Bodinier Q, Farhang A, Bader F, Ahmadi H, Palicot J, and DaSilva LA. "5G waveforms for overlay D2D communications: Effects of time-frequency misalignment," in: *2016 IEEE International Conference on Communications (ICC)*. IEEE; 2016; Kuala Lumpur, Malaysia. p. 1–7.

[18] Swain SN, Mishra S, and Murthy CSR. "A novel spectrum reuse scheme for interference mitigation in a dense overlay D2D network," in: *2015 IEEE 26th Annual International Symposium on Personal, Indoor, and Mobile Radio Communications (PIMRC)*. IEEE; 2015; Hong Kong, China. p. 1201–1205.

[19] Panigrahi B, Rath HK, and Simha A. "Interference-aware discovery and optimal uplink scheduling for D2D communication in LTE networks," in: *2015 Twenty First National Conference on Communications (NCC)*. IEEE; 2015; Mumbai, India. p. 1–6.

[20] Pratas NK and Popovski P. "Network-assisted device-to-device (D2D) direct proximity discovery with underlay communication," in: *2015 IEEE Global Communications Conference (GLOBECOM)*. IEEE; 2015; San Diego, CA, USA. p. 1–6.

[21] Xenakis D, Kountouris M, Merakos L, Passas N, and Verikoukis C. "Performance analysis of network-assisted D2D discovery in random spatial networks," *IEEE Transactions on Wireless Communications*. 2016 Aug;15(8):5695–5707.

[22] Seo SB, Kim JY, and Jeon WS. "Robust and fast device discovery in OFDMA-based cellular networks for disaster environment," in: *2016 18th International*

Conference on Advanced Communication Technology (ICACT). IEEE; 2016; Pyeongchang, South Korea. p. 498–502.

[23]   Bagheri H, Sartori P, Desai V, Classon B, Al-Shalash M, and Soong A. "Device-to-device proximity discovery for LTE systems," in: *2015 IEEE International Conference on Communication Workshop (ICCW)*. IEEE; 2015; London, UK. p. 591–595.

[24]   Abd-Elrahman E, Ibn-khedher H, Afifi H, and Toukabri T. "Fast group discovery and non-repudiation in D2D communications using IBE," in: *2015 International Wireless Communications and Mobile Computing Conference (IWCMC)*. IEEE; 2015; Dubrovnik, Croatia. p. 616–621.

[25]   Shen Y, Jiang C, Quek TQ, and Ren Y. "Device-to-device-assisted communications in cellular networks: an energy efficient approach in downlink video sharing scenario," *IEEE Transactions on Wireless Communications*. 2016;15(2): 1575–1587.

[26]   Lee D, Lee D, Hwang W, and Choi HJ. "A timing synchronization method for D2D communication in asynchronous cellular system," in: *The 20th Asia-Pacific Conference on Communication (APCC2014)*. IEEE; 2014; Pattaya, Thailand. p. 366–371.

[27]   Ji M, Caire G, and Molisch AF. "Fundamental limits of caching in wireless D2D networks," *IEEE Transactions on Information Theory*. 2016;62(2):849–869.

[28]   Bastug E, Bennis M, and Debbah M. "Living on the edge: the role of proactive caching in 5G wireless networks," *IEEE Communications Magazine*. 2014;52(8):82–89.

[29]   Cao Y, Jiang T, Chen X, and Zhang J. "Social-aware video multicast based on device-to-device communications," *IEEE Transactions on Mobile Computing*. 2016;15(6):1528–1539.

[30]   Coon JP. "Modelling trust in random wireless networks," in: *2014 11th International Symposium on Wireless Communications Systems (ISWCS)*. IEEE; 2014; Barcelona, Spain. p. 976–981.

[31]   Douik A, Sorour S, Tembine H, Al-Naffouri TY, and Alouini MS. "A game-theoretic framework for network coding based device-to-device communications," *IEEE Transactions on Mobile Computing*. 2017;16(4):901–917.

[32]   Datsika E, Antonopoulos A, Zorba N, and Verikoukis C. "Adaptive Cooperative Network Coding based MAC protocol for device-to-device communication," in: *2015 IEEE International Conference on Communications (ICC)*. IEEE; 2015; London, UK. p. 6996–7001.

[33]   Ansari RI, Hassan SA, and Chrysostomou C. "RANC: relay-aided network-coded D2D network," in: *2015 10th International Conference on Information, Communications and Signal Processing (ICICS)*. IEEE; 2015; Singapore. p. 1–5.

[34]   Omar MS, Anjum MA, Hassan SA, Pervaiz H, and Niv Q. "Performance analysis of hybrid 5G cellular networks exploiting mmWave capabilities in suburban areas," in: *2016 IEEE International Conference on Communications (ICC)*. IEEE; 2016; Kuala Lumpur, Malaysia. p. 1–6.

[35] Niu Y, Su L, Gao C, Li Y, Jin D, and Han Z. "Exploiting device-to-device communications to enhance spatial reuse for popular content downloading in directional mmWave small cells," *IEEE Transactions on Vehicular Technology*. 2016 July;65(7):5538–5550.

[36] Wei N, Lin X, and Zhang Z. "Optimal relay probing in millimetre wave cellular systems with device-to-device relaying," *IEEE Transactions on Vehicular Technology*. 2016;65(12):10218–10222.

[37] Tian F, Liu B, Xiong J, and Gui L. "Movement-based incentive for cellular traffic offloading through D2D communications," in: *2016 IEEE International Symposium on Broadband Multimedia Systems and Broadcasting (BMSB)*. IEEE; 2016; Nara, Japan. p. 1–5.

[38] Wang J, Jiang C, Bie Z, Quek TQS, and Ren Y. "Mobile data transactions in device-to-device communication networks: pricing and auction," *IEEE Wireless Communications Letters*. 2016 June;5(3):300–303.

[39] Sakr AH and Hossain E. "Cognitive and energy harvesting-based D2D communication in cellular networks: stochastic geometry modelling and analysis," *IEEE Transactions on Communications*. 2015;63(5):1867–1880.

[40] Yang HH, Lee J, and Quek TQ. "Heterogeneous cellular network with energy harvesting-based D2D communication," *IEEE Transactions on Wireless Communications*. 2016;15(2):1406–1419.

[41] Liu Y, Wang L, Zaidi SAR, Elkashlan M, and Duong TQ. "Secure D2D communication in large-scale cognitive cellular networks: a wireless power transfer model," *IEEE Transactions on Communications*. 2016;64(1):329–342.

Chapter 20

Coordinated multi-point for future networks: field trial results

Selcuk Bassoy[1], Mohamed Aziz[1],
and Muhammad A. Imran[2]

Abstract

Wireless networks face big capacity challenges, struggling to meet ever increasing user data demands. Global mobile data traffic grew by 74% in 2015 and it is expected to grow 8-fold by 2020. Future wireless network will need to deploy massive number of small cells and improve spectral efficiency to cope with this increasing demand. Dense deployment of small cells will require advanced interference mitigation techniques to improve spectral efficiency and enhance much needed capacity. Coordinated multi-point (CoMP) is a key feature for mitigating inter-cell interference and to improve throughput and cell edge performance.

In this chapter, we first provide the motivation for CoMP deployment from an operator perspective. Then, we discuss different types of CoMP schemes, their associated challenges and the third generation partnership project (3GPP) standardisation roadmap. Next, we provide insights into operational requirements for CoMP implementation and discuss potential solutions to enable cost-effective CoMP deployment. We then provide results for an intra-site uplink (UL) joint reception (JR) CoMP trial in a large Long-Term Evolution-Advanced (LTE-A) operator in the United Kingdom (UK) for three different deployment scenarios namely, dense, medium density and sparse deployment at macro layer. CoMP sets consist of co-located cells only with joint baseband processing units, and hence no backhaul is required for data exchange. On the other hand, interference between cells from different locations are not mitigated. Only two cells are allowed for coordination and interference rejection combining (IRC) is employed for joint processing. Trial performance is measured based on average network counters. Results show an average increase in signal-to-interference-plus-noise ratio (SINR) on physical uplink shared channel (PUSCH) by 5.56% which

[1]Telefonica UK Ltd, UK
[2]School of Engineering, University of Glasgow, UK

is then reflected in improved usage of higher modulation schemes and better UL user throughput. An average increase of 11.32% is observed on UL user throughput. Additionally, we discuss the limitations of the trialled CoMP scheme and suggest improvements for better CoMP gains. Furthermore, we review the evolution of CoMP into 5G and potential improvements CoMP can provide for some of the key 5G network objectives such as spectral efficiency, energy efficiency, load balancing and backhaul optimisation.

20.1   Introduction

One of the biggest challenges of mobile communication networks is to be able to cope with the ever increasing data demand, primarily driven by growing availability of wide range of high bandwidth applications combined with proliferation of smartphone devices. In 2015, global mobile data traffic increased by 74% and this growth is expected to accelerate to 8-fold increase by 2020 [14]. In the next decade, 1,000 times increase in mobile data traffic is expected [25].

With evolving customer needs, increased demand for higher data speed and improved quality of experience, mobile operators are increasingly coming under pressure to find innovative and cost-efficient solutions to grow the capacity of their networks. This includes range of solutions such as dense deployment of small cells, wireless local area networks (WLANs) offloading, interference mitigation to improve spectral efficiency (SE) and additional licensed/unlicensed spectrum usage [25,29]. Advanced interference mitigation techniques are required to limit the inter-cell interference, especially in dense deployment scenarios where there is high interference. Given that, there is only limited available spectrum, SE will need to be improved to provide the required uplift in capacity. Network multiple-input multiple-output (MIMO) or CoMP has been extensively studied in the literature as an emerging technology to reduce interference and increase throughput especially at the cell edge in densely deployed networks. It has been already identified as a key feature for LTE-A by 3GPP [4] and been discussed as a strong feature for 5G [25].

In this chapter, we aim to present an operator's perspective on deployment of CoMP. Firstly, we present the main motivation and benefits of CoMP from an operator's viewpoint. Next, we provide an introduction to CoMP and the 3GPP standardisation roadmap for LTE-A. We give a short review of available coordination types and discuss associated challenges for each CoMP scheme. We then present the operational requirements for CoMP implementation and discuss practical considerations and challenges of such deployment. Possible solutions for these experienced challenges are reviewed. We then present initial results from an UL CoMP trial and discuss changes in key network performance indicators (KPI) during the trial. Additionally, we propose further improvements to the trialled CoMP scheme for better potential gains. Moreover, we give our perspective on how CoMP will fit into the future 5G networks and finally conclude the chapter with a summary of lessons learnt.

20.2 Motivation and benefits of CoMP – operator perspective

Mobile network operators have been looking for various solutions to improve network capacity as briefly discussed in Section 20.1. Driven from the capacity improvement objective, LTE networks have been rapidly deployed, degree of sectorisation is increased in densely populated areas, micro, pico and Wi-Fi cells are deployed to offload the macro base station (BS) network. Furthermore, multiple antenna solutions (i.e., MIMO) is introduced especially with the relatively new LTE network deployment. Moreover, additional spectrum is utilised where possible and solutions for increasing SE on existing spectrum has been explored extensively. Operators globally have been re-purposing their existing frequency spectrum, usually used for 2G/3G technologies, to LTE/4G in the drive to achieve increased efficiency, grow capacity and offer improved user experience.

Further to all above, network coordination, i.e., CoMP has also been in the interest of operators to further improve SE in densely deployed LTE-A networks. CoMP can eliminate interference from cells within the coordinating cluster and even can exploit this signal as useful signal. In interference-limited, densely deployed networks, CoMP has got the potential to improve SE and increase user throughput especially at the cell edge where there is significantly more interference. Given that some of the CoMP types do not require additional infrastructure and relatively low cost (such as intra-site uplink CoMP), it is in the interest of operators to deploy CoMP for LTE-A to improve much needed capacity. More complex CoMP deployment scenarios introduce challenges such as high backhaul bandwidth requirement and very low latency, UE capability, clustering challenge, complex precoding and precise synchronisation requirement, etc. These challenges and possible solutions are briefly discussed in the next section.

20.3 What is CoMP and standardisation roadmap

20.3.1 What is CoMP?

CoMP proposes to make use of the shared data between the BSs (i.e., user data, channel state information (CSI) and scheduling information) and performs joint scheduling/precoding function centrally at CoMP control unit (CCU), so that interference between the cells can be minimised or even exploited as useful signal. It reduces interference especially at the cell edge and improves overall cell throughput and network capacity. It is expected to be more effective in interference limited, densely deployed network scenarios. CCU can be located at a central location where all coordinating BSs are connected with fast backhaul links. In a heterogeneous network (HetNet) scenario, this location can be at the macro BS for coordination between all small cells within the coverage area of the macro BS as depicted in Figure 20.1.

CoMP and network MIMO in general has been studied extensively in the literature [17,20] to combat inter-cell interference and improve SE. It has been proposed as one

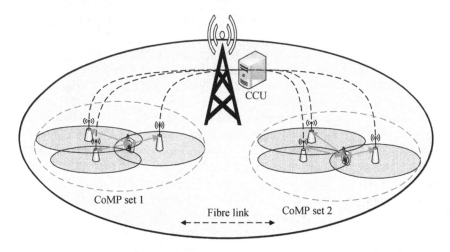

Figure 20.1   An illustration of a typical HetNet with CoMP

of the main features of LTE-A by 3GPP [4] and been discussed as a key technology for 5G [25].

20.3.2   Standardisation roadmap

CoMP has been in the interest of 3GPP for standardisation activities for LTE-A as one of the key features. Initial requirements for LTE-A enhancements were published by 3GPP in 2008 [1]. In response to these requirements, CoMP has been proposed as a key feature to improve performance in [2]. 3GPP published a detailed feasibility study in Release 11 on the physical layer aspects of CoMP in [4] where different categories of CoMP and various deployment scenarios are set and simulation results are published. Published results from several vendors show a significant improvement especially on throughput at the cell edge and also on coverage of high data rates [4]. In Release 12, 3GPP published a further study to understand inter-eNodeB CoMP performance under different backhaul latency scenarios [5] with a view to support standardisation activities on network interface and signalling required for CoMP involving multiple eNodeBs with non-ideal backhaul. Following this study, a new work item is initiated in Release 13 to support signalling for inter-eNodeB CoMP for non-ideal backhaul scenarios [6]. Existing X2 interface is improved with enhanced coordination information exchange between the eNodeBs for CoMP [8]. Further enhancements on CoMP for LTE-A including increase in supported MIMO transmission rank and CSI measurement enhancement especially in dense deployment scenarios are also initiated as a new study item for Release 14 [7]. Followed by this study item, a technical report is issued recently by 3GPP focusing on non-coherent (NC) joint transmission (JT) CoMP and coordinated scheduling/beamforming (CS/CB)CoMP with full dimension (FD) MIMO. Both CoMP improvement proposals show promising improvements on user perceived throughput [9].

20.3.3 Types of CoMP

Coordination between transmission points (TPs) can be at different levels ranging from simple CS schemes to more complex precise coherent joint transmission/
reception (JT/JR) CoMP [37]. TPs may or may not be geographically co-located.
3GPP study mainly focuses on 3 CoMP schemes on downlink (DL) for LTE-A [4]
based on backhaul bandwidth requirements and scheduling/precoding complexity.
DL CoMP schemes are illustrated in Figure 20.2 and briefly discussed below. UL
CoMP is introduced in Section 20.5.1.

- Joint Processing (JP): This type of coordination improves the signal quality and
  throughput for the user by simultaneous data transmission from multiple TPs in
  a time/frequency resource. Data transmission can be possible to a single user
  or multiple users from all or a subset of the TPs in the coordinating set. User
  data need to be available on all TPs serving to the same user, hence this type
  of coordination require high backhaul bandwidth to share user data between the
  TPs. Joint transmission can be implemented coherently or non-coherently. Joint
  precoding is performed in coherent transmission to eliminate inter-cell interference however this scheme require precise synchronisation and CSI exchange
  between all TPs with minimal latency. Multi-user coherent joint precoding CoMP
  is illustrated in Figure 20.3b. Non-coherent transmission does not require joint
  precoding, i.e., user data is transmitted from multiple TPs with independent local
  precoding at each TP in the coordinating set. So non-coherent transmission is
  a simpler scheme which does not require CSI exchange and hence the absence
  of CSI delay problem. Multi-user non-coherent joint processing is depicted in
  Figure 20.3a.
- Dynamic Point Selection (DPS): User is served by only one TP in this CoMP
  type, however the serving TP dynamically changes at each subframe (i.e., 1 ms)
  to the best preferred signal, exploiting the fast fading variations in the wireless
  channel. Similar to JP, user data needs to be available at each TP in the CoMP
  set, hence this scheme also require high backhaul bandwidth. Non-serving TPs in
  the CoMP set can remain silent to improve SINR for the user [22]. Figure 20.2c
  depicts the DPS scheme between two cells where time/frequency resource at one
  cell is muted while the other cell is transmitting to the user.
- Coordinated Scheduling/Beamforming (CS/CB): CSI and scheduling information
  is shared between the TPs in the CoMP set where scheduling and beamforming
  decisions are made centrally to reduce interference between the TPs. User data
  is only available at one TP and it's transmitted from the same TP while reducing
  interference to other TPs in the CoMP set. This scheme is a lighter version of
  JP where user data does not need to be shared between the TPs, hence backhaul
  bandwidth requirement is reduced. CB CoMP between two cells is depicted in
  Figure 20.2b.

A brief summary of each DL CoMP type and its associated challenges and
benefits are given in Table 20.1.

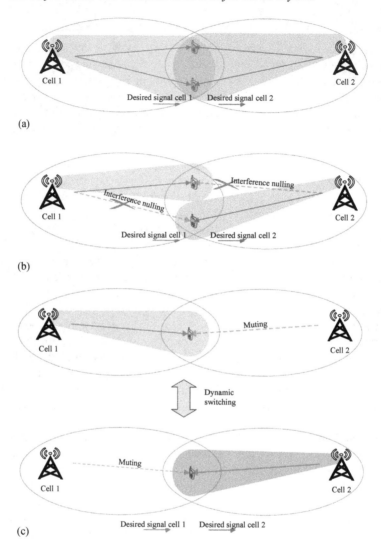

*Figure 20.2   Main downlink CoMP types for LTE-A [4]: (a) joint processing,
(b) coordinated beamforming, (c) dynamic point selection*

20.3.4   Challenges of CoMP

Deployment of CoMP for the whole network is very complex where precise synchronisation between all cells would be required. Additionally, scheduling and processing
complexity increases with the number of TPs in coordination. On the other hand, wide
network coordination is probably not required as capacity gains are only realised from
TPs with high geographical correlation. As discussed earlier, CoMP is also dependant
on high backhaul bandwidth especially for JP-CoMP and this requirement increases

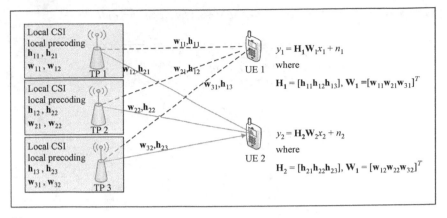

(a)

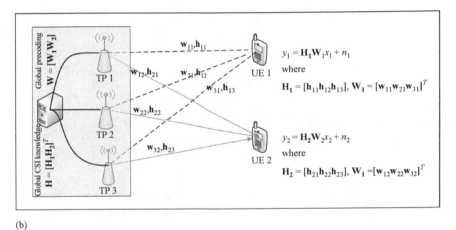

(b)

Figure 20.3 Coherent and non-coherent joint processing: (a) multi-user non-coherent joint processing and (b) multi-user coherent joint processing

Table 20.1 DL CoMP types, associated benefits and challenges

CoMP type	Method	Challenges	Benefits
JP	User data/CSI is shared UE is served by multiple TPs	High backhaul bandwidth requirement	Interference is mitigated as useful signal
DPS	User data is shared UE is served by one TP only	High backhaul bandwidth requirement	Fast fading changes are exploited
CS/CB	Only CSI is shared UE is served by one TP only	Lower backhaul bandwidth requirement	Interference is reduced/eliminated

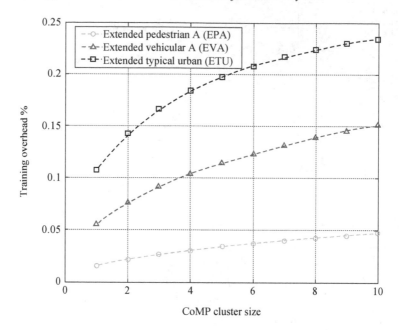

Figure 20.4   Optimum pilot overhead, as a function of number of antennas for Clarke–Jakes spectrum with SNR = 10 dB [19]

with the size of the coordination set. Additional reference/feedback signalling is required for CSI estimation for CoMP operation and this overhead also grows with the CoMP cluster size. A significant amount of time/frequency resources need to be allocated for additional reference signalling to be able to estimate CSI for the CoMP set accurately [19]. Figure 20.4 shows the optimum overhead required for three typical wireless channels widely used by 3GPP [3] for Clarke–Jakes spectrum with SNR = 10 dB based on the work presented in [19]. As an example, for extended vehicular A (EVA) channel, optimum training overhead will increase from about 5% with one antenna port to about 15% with 10 antenna ports. All mentioned challenges imply that, CoMP needs to be implemented in smaller groups of cells to limit complexity and other overheads. Trade-off between additional overheads and CoMP gains will need to be taken into account for effective CoMP cluster design. In [16], it is shown that there is no significant CoMP gain in high SINR region due to additional pilot overhead reducing SE more than the potential SE gain from CoMP. SE gain is maximised when the signal level from cells within the cooperating set are similar to each other, creating severe interference in the absence of CoMP. On the other hand, CoMP gains are less for users located at the cell centre where there is a dominant signal from one cell. Formation of CoMP clusters with the right cells in each cluster is key for maximising CoMP gains. Too large clusters will increase CoMP overheads and can be detrimental as discussed above [30] and it can also reduce energy efficiency [13]. On the other hand, too small clusters will fail to give the potential SE gains.

Initial deployment of CoMP for LTE-A is likely to implement basic static clustering approach, i.e., intra-site clustering only where co-located cells are in the CoMP cluster, hence it will not depend on high backhaul between non-colocated cells. Once network infrastructure is improved with especially new small cell deployment, backhaul availability will also increase and this will enable coordination between BSs from different locations. Furthermore, CoMP cluster sets can be designed to dynamically adapt to spatio-temporal changes in network and user profiles to further improve CoMP gains based on various network objectives like spectral efficiency, energy efficiency and load balancing. We briefly discuss dynamic CoMP clustering for future 5G networks in Section 20.6.

20.4 Operational requirements

As already mentioned in previous sections, CoMP can generally be categorised into three main types of coordination schemes; coordinated scheduling and coordinated beamforming (CS/CB), joint processing (JP) and DPS. Each of these techniques impose varying delay and bandwidth requirements on the backhaul which in turn depend on backhaul technology employed (fibre, ADSL, copper, etc.). Studies conducted by the 3GPP and other literature [4,35,36] show that highest gain in spectral efficiency, and thus consequent capacity requires virtually zero latency and unlimited capacity backhaul. This practically requires cloud radio access network (C-RAN) architecture using with fibre connectivity to connect remote radio heads to central baseband unit. In practice, there will be constraints that will limit achievable gain in distributed RAN architecture [34].

Operators with fibre optics assets face relatively easier route to deployment of CoMP than those with no or limited fibre connectivity. Although varies by market conditions, acquiring fibre optic connectivity is likely to attract certain operational expenditure (OPEX), capital expenditure (CAPEX) and internal organisation resources. Justifying such investment and upgrade programmes solely on the basis of CoMP benefits is likely to be a difficult business decision.

CoMP algorithms and associated functionality are implementation dependent and requires certain processing power. Even simplest deployment on intra-site CoMP requires additional uplifting of the physical baseband resources. In some cases, where there is no baseband capacity headroom left, additional site visits are required to upgrade site physical processing as part of upgrade process. This incurs certain costs and potentially a degree of service disruption if outage is required to carry out physical upgrade activities. Operational and cost impact will scale up, proportionate to network size and the number of sites to be upgraded. Service providers will need to factor that in their upgrade plans. Unnecessary multiple site visits and potential service disruption resulting from upgrade activity could be prevented by including CoMP requirements at the early stage of the design process to ensure site physical resources are optimally dimensioned.

Propagation environment and carried traffic load will influence the achievable gain in spectral efficiency everything being equal. Selection of cells that make up the

CoMP set is very important factor for inter-site deployment. A planning process is required to identify cooperating cells and group them. Such a process will take into account coverage overlap, traffic load, cell type, availability and type of backhaul. After completing an initial planning processing and identifying the cells that make up the CoMP set, it is important to ensure CoMP set is kept optimum by having a feedback loop to cater for changes in the RF environment resulting from various reasons such as new sites, new building development, amongst others. Managing this process manually is likely to be tedious and inaccurate and therefore automation is very important capability to enable large scale deployment of CoMP.

20.5   Uplink CoMP field trial for LTE-A

In this section, we briefly introduce UL CoMP for LTE-A in general and provide intra-site UL CoMP trial results from an operational LTE-A network for three deployment scenarios in two trial areas. Improvements in various KPIs are reviewed and limitations of the trialled CoMP scheme are critically discussed. Further potential enhancements for future networks are highlighted to maximise CoMP gain.

20.5.1   *Uplink CoMP introduction*

Uplink CoMP makes use of the UE signal at different BSs to improve SINR and hence cell throughput especially at the cell edge where severe inter-cell interference is usually experienced. Unstandardized applications of exploiting uplink signal from different receiver ports may already be available in conventional networks [23] but recent standardised efforts by 3GPP introduced UL CoMP for LTE-A networks [4]. Two types of UL CoMP categories are identified by 3GPP, namely joint reception(JR) and coordinated scheduling/beamforming (CS/CB). In JR, UE signal is received by multiple points and processed jointly to eliminate interference. Signal processing techniques like IRC are employed to mitigate interference and improve SINR. A tutorial on IRC in LTE networks can be found in [24]. In CS/CB scheme, UE data is intended for one point only and interference is minimised within the coordinated set by central scheduling and beamforming.

UL CoMP is relatively a less complex option for operators to deploy initially, as it is transparent to the UE, and also channel estimation is done by uplink reference signals, i.e., there is no feedback signalling overhead unlike downlink CoMP. Current LTE-A networks are under increasing pressure for DL capacity than UL as data demand for DL is usually eight times more than UL. However, in special scenarios such as football matches, music festivals, etc. UL data demand increases up to half of the DL data demand. This is evidently explained by consumption of application and services that generate significant data in UL direction like users taking photos/video and sharing with friends and family on social networks like Instagram and Facebook. This, combined with relatively low complexity and operational overhead makes deployment of UL CoMP in such scenarios very attractive proposition to enhance capacity and improve user experience.

As mentioned in previous sections, one of the important factors for maximising CoMP gains is to decide which and how many cells to coordinate for finding the right balance between complexity/overhead costs and CoMP gain. On the other hand, these clusters will need to dynamically change in response to the spatio-temporal changes in user profile/demand distribution and network elements. Although, fully dynamic CoMP clustering solutions are studied extensively in the literature, such solutions are not yet available for current LTE-A networks to our knowledge. Furthermore, available backhaul is mostly limited, preventing such dynamic design for non-co-located BSs in the CoMP cluster. Therefore, realistic CoMP cluster solution for current LTE-A networks for initial deployment is the intra-site CoMP where coordination only takes place within the same site with joint baseband processing unit. Since all cells are geographically collocated and belong to the same site, backhaul delay can be assumed to be virtually zero for intra-site deployment.

20.5.2   UL CoMP in trial area

Intra-site UL JR-CoMP was trialled in two different major cities in the UK where LTE-A BSs are deployed at macro layer with typically three cells at each BS. Trial area-A covers a middle-size city in the United Kingdom (UK) with dense deployment. One frequency layer at 800 MHz is deployed in this area. Trial area-B covers outskirts of another city in the UK where LTE-A carrier at 800 MHz is deployed at medium site density in this area. An additional layer at 1,800 MHz is also deployed in trial area-B as a capacity layer in local areas where it is required for additional capacity. Hence deployment at 1,800 MHz is at lower density than 800 MHz. In the rest of the chapter, we refer to trial area-A as "dense deployment at 800 MHz", and trial area-B as "medium density deployment at 800 MHz" and "sparse deployment at 1,800 MHz". Figure 20.5 depicts the site layout and Table 20.2 shows the approximate areas covered and average inter-site distance for each deployment scenario. We discuss UL CoMP gains achieved separately for each of these three deployment scenarios. CoMP sets are formed from co-located cells on the same frequency band of the same site, where each cell consists of two RX antennas. Co-located cells share the same baseband processing unit and hence there is no backhaul requirement for data/signalling exchange between the cells. Maximum of two cells are allowed (i.e., four RX antennas in total) to coordinate at any time where IRC receiver is employed to extract the main signal and eliminate inter-cell interference. For each UE, serving cell and the strongest neighbour cell from the CoMP set form the coordination set. CoMP is enabled on the uplink data channel only, i.e., PUSCH, it is not deployed on the control channels, i.e., physical random access channel (PRACH) or physical uplink control channel (PUCCH). Uplink CoMP scheme in the trial area is illustrated for 2 cells within the same BS in Figure 20.6.

20.5.3   Trial performance results

Intra-site UL CoMP is enabled in the three LTE-A network layers at two trial areas as described above for two weeks and various KPIs are benchmarked against the identified benchmarking time window of two weeks before and after the trial. Changes

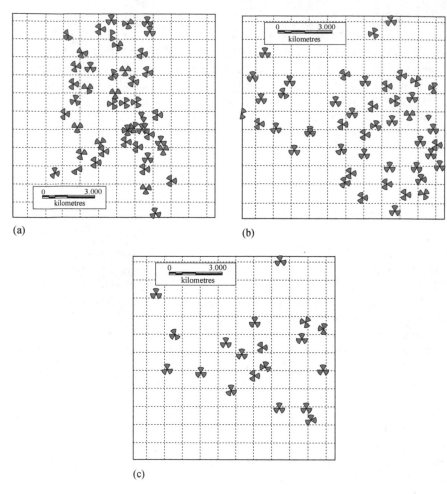

(a)

(b)

(c)

Figure 20.5   *Network topology of the three deployment scenarios in the trial:*
(a) dense deployment topology, (b) medium density deployment
topology and (c) sparse deployment topology

Table 20.2   Site density details for the three deployment scenarios

Deployment scenario	Approx. area (km²)	Aver. inter-site dist. (km)
Trial area-A – dense deployment	63.6	0.733
Trial area-B – medium density deployment	112.4	0.937
Trial area-B – sparse deployment	112.4	1.509

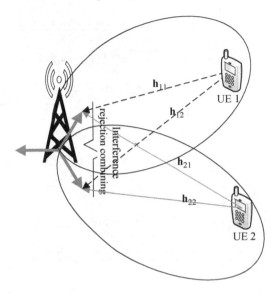

Figure 20.6 An illustration of intra-BS uplink CoMP: joint processing with IRC

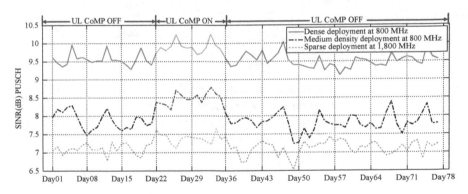

Figure 20.7 PUSCH SINR with/without UL CoMP in the three deployment scenarios

in the KPIs are also presented for a wider time window. Figure 20.7 depicts the average SINR change in logarithmic scale on PUSCH based on network counters for the three deployment scenarios in the trial. As expected, SINR value is higher on dense deployment in general and it reduces in medium density deployment and further reduced in sparse deployment. Average SINR is improved with UL CoMP in all three deployment scenarios. Largest increase in average SINR is observed in medium density deployment scenario with 8.57% (+0.67 dB) SINR improvement, followed by 4.55% (+0.32 dB) and 3.56% (+0.34 dB) increase in sparse and dense deployment scenarios, respectively.

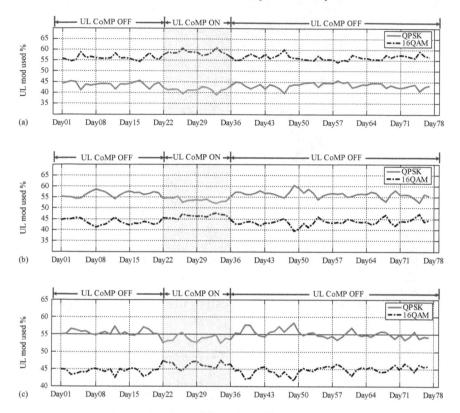

Figure 20.8   Employed uplink modulation types with/without UL CoMP in the three deployment scenarios: (a) employed uplink modulation type in dense deployment at 800 MHz, (b) employed uplink modulation type in medium density deployment at 800 MHz and (c) employed uplink modulation type in sparse deployment at 1,800 MHz

Average SINR gain due to UL CoMP is also reflected in the employed modulation type on UL. Figure 20.8 shows the average percentage usage of the 2 modulation types employed in uplink, i.e., QPSK and 16QAM. Intuitively, higher SINR in dense deployment scenario is reflected in wider use of higher modulation type when compared to lighter deployment scenarios. The average percentage usage of 16QAM against QPSK is increased by 7.46%, 4.68% and 4.00% in medium density, sparse and dense deployment scenarios, respectively, following a similar gain pattern from average SINR increase. UL CoMP can be more effective in dense deployment scenarios, where inter-cell interference is expected to be higher, however intra-site CoMP is employed in this trial where interference from cells located at different BS locations are not mitigated.

Figure 20.9 depicts the average achieved block error rate (BLER) for each modulation type on UL during the trial. BLER improvement is observed in all modulation

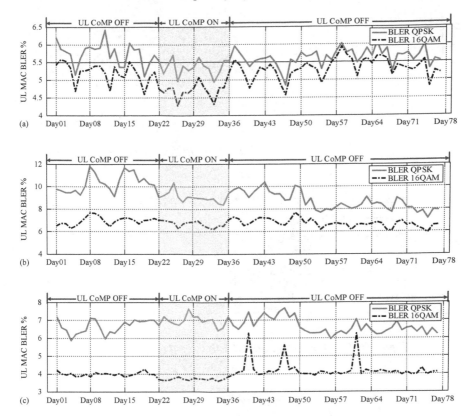

Figure 20.9 Uplink BLER with/without UL CoMP in the three deployment scenarios: (a) uplink BLER in dense deployment at 800 MHz, (b) uplink BLER in medium density deployment at 800 MHz and (c) uplink BLER in sparse deployment at 1,800 MHz

types for all deployment scenarios apart from inconclusive results in BLER for QPSK modulation for sparse deployment scenario. Overall, an increase is observed on the average ratio of higher modulation usage and also average BLER for each modulation is improved by 7.69%, 9.51% and 4.28% for dense, medium density and sparse deployment scenarios, respectively.

Improvement in the employment of higher order modulation type is also reflected in average UL user throughput. Figure 20.10 depicts the increase in average UL user throughput in all three scenarios. Average improvement observed during the trial are 17.86%, 8.68% and 7.42% in medium density, sparse and dense deployment scenarios, respectively.

CoMP is more effective at cell edge, as there is more inter-cell interference expected and UEs use full power to overcome this interference. Figure 20.11 depicts the percentage of UEs which are power limited, where a slight reduction can be seen

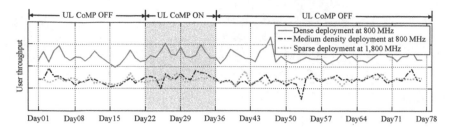

Figure 20.10   Average UL user throughput in three deployment scenarios

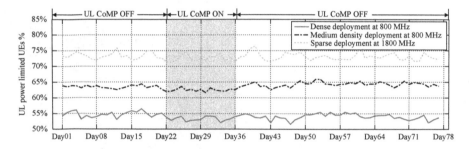

Figure 20.11   Power limited UEs in three deployment scenarios

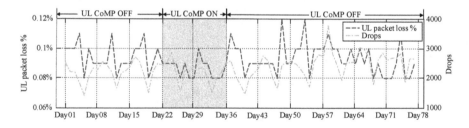

Figure 20.12   Uplink packet loss and drops in dense deployment at 800 MHz

during the trial. An average of 1.92%, 1.85% and 1.17% reduction is observed in the amount of power limited UEs for dense, medium and sparse deployment scenarios, respectively, during the trial.

From customer experience point of view, number of drops and overall UL packet loss ratio is also monitored during the trial. A clear reduction in UL Packet loss rate is observed in all three scenarios, where the average improvement has been 8.20%, 10.48% and 14.68% for dense, medium and sparse deployment scenarios, respectively. Number of drops is also reduced in dense and medium deployment by 1.75% and 7.19%, respectively, and the results for sparse deployment was not conclusive. Figure 20.12 shows the radio drops and UL packet loss rate for dense deployment case where an improvement on both metrics can be observed clearly during the trial.

In summary, a significant performance gain is achieved by enabling intra-site UL CoMP in three different macro network deployment scenarios. A notable increase in average SINR on PUSCH is observed which is also reflected on the usage of higher order modulation in UL. Consequently, average UL user throughput is also increased and overall session drops are reduced. However, a number of factors have limited further CoMP gains achievable. Firstly, current network load on UL is quite low and inter-cell interference is mostly avoided by inter-cell interference cancellation (ICIC) schemes available in the network so it is expected that the observed gain from the trial will increase when UL network load increases. Additionally, trial results are based on daily average network counter values so higher CoMP gains are expected at the busy hour of the day, where UL load is usually higher than daily average. Secondly, CoMP sets are formed from intra-site cells only during the trial, i.e., interference due to cells from other BSs are not mitigated. Inter-BS interference is increased in dense deployment scenarios where CoMP gains can be maximised when inter-BS CoMP is enabled. Furthermore, instead of static clustering schemes, intelligent dynamic clustering algorithms can be deployed to design CoMP sets to dynamically adapt to spatio-temporal network/user profile changes and increase CoMP gain further.

20.6   Evolution into 5G

In this section, we present the potential deployment of CoMP in future 5G networks as one of the key features. We discuss how CoMP can provide solutions for some of the key challenges of 5G like spectral efficiency, energy efficiency, backhaul bandwidth challenge and load balancing.

20.6.1   CoMP for improved spectral efficiency in 5G

As discussed extensively in previous chapters, cellular networks need to increase their capacity extensively to be able to meet the ever increasing mobile data demand. Given the spectrum shortage to meet with this demand, 10 times more spectral efficiency is envisioned for 5G [42] (i.e., from 2–3 b/s/Hz on LTE to 20 b/s/Hz for 5G [20]).

Network densification, i.e., extensive small cell deployment and indoor DAS systems, enhanced use of massive MIMO and network MIMO (i.e., CoMP) are proposed as key development areas to meet the challenging capacity targets of 5G [18,20,25].

Small cell deployment will need a different approach for network coordination. Different frequency allocation for small cell and macro layer will not give the sufficient SE gains. New JT-CoMP schemes are required to achieve better SE gains in multi-layer complex 5G network architectures [20].

JT-CoMP will play a key role on improving SE in such dense small cell deployment scenario underlayed to macro network, where there will be extensive inter-cell interference presence due to overlapping coverage. An integrated approach will be required to adapt massive MIMO and JT-CoMP where required for maximal efficiency gains with minimal effort [20].

To further improve CoMP gains, CoMP clusters need to be designed efficiently and should be dynamically changing to adapt to changing network and user profiles [10]. Dynamic CoMP clustering algorithms for future networks are extensively studied in the literature where cells within each cluster and the cluster size dynamically change, responding to dynamic user demand and network profile changes to maximise spectral efficiency [15,28,38].

20.6.2 CoMP and backhaul bandwidth challenge in 5G

One of the challenges for JT-CoMP for 5G is the high bandwidth, low latency backhaul requirement due to user data and CSI exchange required between the coordinating cells. Improved JT-CoMP schemes will need to take limited backhaul bandwidth availability into account for better gain. In [12], a dynamic serving BS re-assignment algorithm is studied for JT-CoMP scheme to distribute backhaul load evenly to avoid backhaul congestion on the serving cell link. In [43], a JT-CoMP scheme is presented to optimise the number of links for CoMP clusters for min SINR requirement to reduce the backhaul requirement. Qi *et al.* propose to employ distributed compression at the TPs for UL CoMP in cloud radio access network (C-RAN) architecture to reduce fronthaul traffic between the TPs and the central processing unit. Decompression and decoding of the user message is conducted centrally, reducing the required fronthaul rate by more than 50% [39,40].

Caching popular multimedia at the RAN is an increasingly popular concept to reduce user data sharing between the BSs and hence reducing the high backhaul bandwidth requirement for JT-CoMP. In [27], an opportunistic JT-CoMP scheme is proposed aiming for each BS in the CoMP cluster to cache the same user data eliminating the need to exchange this data over the backhaul.

20.6.3 CoMP for energy efficient 5G networks

Alongside with much needed spectral efficiency, energy efficiency is another key challenge for 5G networks for environmental reasons and to reduce energy costs. Enabling CoMP will also improve energy efficiency [41]. CoMP can reduce UE/BS power requirements, however it requires additional energy for additional signal processing and backhaul. In [21], authors studied the trade-off between energy efficiency and throughput gain for JT-CoMP, identifying inter-site distance, minimum SINR requirement and transmit power as the key inputs to this trade-off. CoMP deployment can also improve energy efficiency by maximising the number of sleeping cells when the user demand is low. Small cells are switched on, only when additional capacity is required in the network due to increasing demand. Authors in [26] proposed a dynamic CoMP clustering algorithm which improves energy efficiency by maximising BS sleeping while keeping high achievable throughput. A user-centric CoMP clustering mechanism is presented in [13], which proposes to switch-off lightly loaded cells to maximise energy efficiency.

Motivated by the energy efficiency concerns and heterogeneous deployment, control and data plane separation architecture (CDSA) is envisioned for 5G where most control signalling is handled by the macro layer and user data is provided by the small cell layer [31,32]. Macro BSs are envisioned to be always switched on to provide basic coverage and control signalling whereas small cells are switched on/off based on the service demand to improve energy efficiency. CDSA architecture is a key enabler for CoMP in 5G, where macro BS can be enhanced to include CCU functionality and small cells within its coverage can be connected to the macro BS with high backhaul bandwidth links.

20.6.4 *CoMP for cost effective load-aware 5G network*

CoMP is also envisioned to be deployed in future network to reduce high deployment costs. In [33], it is shown that the number of required BSs to provide a certain quality of service (QoS) in a given area can be reduced by deploying CoMP.

As mentioned in previous sections, ever increasing capacity demand is one of the biggest challenges for 5G. Inevitably, traffic load is usually not evenly distributed, traffic hotspots are formed in some areas at certain times. Intelligent load balancing algorithms are required to distribute load from congested cells to its relatively unloaded neighbours. Deployment of CoMP, especially JT-CoMP MU-MIMO can improve capacity of the congested BS, and intelligent algorithms can be deployed to design load-aware CoMP clusters. In our previous study [11], we presented a load-aware user-centric CoMP clustering algorithm and shown a significant reduction in unsatisfied users due to overload conditions.

In summary, CoMP will have an important role in 5G networks to mitigate inter-cell interference in densely deployed, multi-layer complex networks. An integrated approach to adapt massive MIMO and JT-CoMP is envisioned to maximise spectral efficiency. Additionally, CoMP is key for some of the other challenges for future 5G networks like energy efficiency, load balancing and high deployment costs. CoMP is envisioned to increase capacity for the congested cells, and load-aware CoMP clusters can support load balancing further. Moreover, CoMP will enable maximising sleeping cells for energy efficiency. The challenges for JT-CoMP deployment such as high backhaul bandwidth requirement and the need for dynamic clustering are extensively studied in the literature. Furthermore, wider availability of high bandwidth backhaul such as fibre is expected in future which will enable larger scale deployment of CoMP in 5G.

20.7 Conclusions

In this chapter, we first presented the motivation for CoMP deployment from an operator perspective, then we provided a brief introduction to CoMP and gave an overview of the 3GPP standardisation activities. Next, we discussed different levels of coordination and the associated challenges. We then analysed the practical limitations for CoMP deployment from an operator perspective and provided possible solutions. Moreover, we presented performance results for an intra-site UL CoMP trial for three different deployment scenarios for a commercial LTE-A operator. We briefly introduced the CoMP scheme deployed in the trial and discussed improvement on various KPIs. Trial was conducted for a basic UL CoMP scheme where only intra-site cells were allowed in the same CoMP set and only two cells could cooperate for the same UE at the same time. An average of 5.56% SINR improvement is observed on PUSCH. This improvement is reflected on the employed modulation types, where percentage usage of higher order 16QAM modulation on UL is increased by 5.38% on average. One of the main objectives of UL CoMP deployment is to improve user throughput especially at the cell edge. An average of 11.32% increase in user throughput is observed during this trial. Ratio of power limited UEs is reduced by 1.65% which shows the improvement observed especially at the cell edge where UE power is maximised and limited on extra power. Overall customer experience is improved with reduced radio related drops by 2.68% and UL packet loss rate is reduced by 11.12% on average. We further discussed the limitations of the UL CoMP scheme trialled and presented potential improvements for better CoMP gains for future networks. Effect of network load on CoMP gain is discussed and inter-BS CoMP and further dynamic clustering techniques are proposed as potential improvements. Finally, we presented the evolution of CoMP into 5G from an operator perspective. The need for CoMP to support some of the 5G challenging network objectives like energy efficiency, load balancing and spectral efficiency are presented.

References

[1] 3GPP. Requirements for Further Advancements for Evolved Universal Terrestrial Radio Access (E-UTRA) (LTE-Advanced). TR 36.913 V9.0.0, 12 2009.

[2] 3GPP. Evolved Universal Terrestrial Radio Access (E-UTRA); Further Advancements for E-UTRA Physical Layer Aspects. TR 36.814, 03 2010.

[3] 3GPP. Evolved Universal Terrestrial Radio Access (E-UTRA); Base Station (BS) Radio Transmission and Reception. TS 36.104, 06 2011.

[4] 3GPP. Coordinated Multi-Point Operation for LTE Physical Layer Aspects. TR 36.819 R11 v11.2.0, Sep 2013.

[5] 3GPP. Coordinated Multi-point Operation for LTE with Non-ideal Backhaul. TR 36.874 R12 v12.0.0, Dec 2013.

[6] 3GPP. RP-141032 Work Item on Enhanced Signalling for Inter-eNodeB CoMP. Technical Report, 3rd Generation Partnership Project (3GPP), 2014.

[7] 3GPP. RP-160954 Study Item on Further Enhancements to Coordinated Multi-point Operation. Technical Report, 3rd Generation Partnership Project (3GPP), 2016.

[8] 3GPP. Evolved Universal Terrestrial Radio Access Network (E-UTRAN); X2 Application Protocol (X2AP). TS 36.423, 3rd Generation Partnership Project (3GPP), 01 2017.

[9] 3GPP. Study on Further Enhancements to Coordinated Multi-point (CoMP) Operation for LTE. TR 36.741, 3rd Generation Partnership Project (3GPP), 02 2017.

[10] Selcuk Bassoy, Hasan Farooq, Muhammad A Imran, and Ali Imran. "Coordinated multi-point clustering schemes: a survey," *IEEE Communications Surveys & Tutorials*, 19(2):743–764, 2017.

[11] Selcuk Bassoy, Mona J Jaber, Muhammad A Imran, and Pei Xiao. "Load aware self-organising user-centric dynamic CoMP clustering for 5G networks," *IEEE Access*, 4:2895–2906, 2016.

[12] Thorsten Biermann, Luca Scalia, Changsoon Choi, Holger Karl, and Wolfgang Kellerer. "Improving CoMP cluster feasibility by dynamic serving base station reassignment," in *22nd International Symposium on Personal Indoor and Mobile Radio Communications (PIMRC)*, p. 1325–1330. IEEE, Toronto, ON, Canada, 2011.

[13] Gencer Cili, Halim Yanikomeroglu, and F. Richard Yu. "Cell switch off technique combined with coordinated multi-point (CoMP) transmission for energy efficiency in beyond-LTE cellular networks," in *International Conference on Communications (ICC)*, p. 5931–5935. IEEE, Ottawa, ON, Canada, 2012.

[14] Cisco. Cisco Visual Networking Index: Global Mobile Data Traffic Forecast Update, 2015-2020. *White Paper*, Feb. 2016.

[15] Si Feng, Wei Feng, Hongliang Mao, and Jianhua Lu. "Overlapped clustering for comp transmissions in massively dense wireless networks," in *International Conference on Communication Systems (ICCS)*, p. 308–312. IEEE, Macau, China, 2014.

[16] Ian Dexter Garcia, Naoki Kusashima, Kei Sakaguchi, Kiyomichi Araki, Shoji Kaneko, and Yoji Kishi. "Impact of base station cooperation on cell planning," *EURASIP Journal on Wireless Communications and Networking*, 2010(1):5, 2010.

[17] David Gesbert, Stephen Hanly, Howard Huang, Shlomo Shamai Shitz, Osvaldo Simeone, and Wei Yu. "Multi-cell MIMO cooperative networks: a new look at interference," *IEEE Journal on Selected Areas in Communications*, 28(9):1380–1408, 2010.

[18] Ali Imran, Ahmed Zoha, and Adnan Abu-Dayya. "Challenges in 5G: how to empower SON with big data for enabling 5G," *IEEE Network*, 28(6): 27–33, 2014.

[19] Nihar Jindal and Angel Lozano. "A unified treatment of optimum pilot overhead in multipath fading channels," *IEEE Transactions on Communications*, 58(10):2939–2948, 2010.

[20] Volker Jungnickel, Konstantinos Manolakis, Wolfgang Zirwas, *et al.* "The role of small cells, coordinated multipoint, and massive MIMO in 5G," *IEEE Communications Magazine*, 52(5):44–51, 2014.

[21] Efstathios Katranaras, Muhammad Ali Imran, and Mehrdad Dianati. "Energy-aware clustering for multi-cell joint transmission in LTE networks," in *International Conference on Communications Workshops (ICC)*, p. 419–424. IEEE, Budapest, Hungary, 2013.

[22]   Daewon Lee, Hanbyul Seo, Bruno Clerckx, *et al*. "Coordinated multi-point transmission and reception in LTE-advanced: deployment scenarios and operational challenges," *Communications Magazine, IEEE*, 50(2):148–155, 2012.

[23]   Juho Lee, Younsun Kim, Hyojin Lee, *et al*. "Coordinated multipoint transmission and reception in LTE-advanced systems," *IEEE Communications Magazine*, 50(11):44–50, 2012.

[24]   Yann Léost, Moussa Abdi, Robert Richter, and Michael Jeschke. "Interference rejection combining in LTE networks," *Bell Labs Technical Journal*, 17(1): 25–49, 2012.

[25]   Qian Clara Li, Huaning Niu, Apostolos Tolis Papathanassiou, and Geng Wu. "5G network capacity: key elements and technologies," *IEEE Vehicular Technology Magazine*, 9:71–78, 2014.

[26]   Yun Li, Yafei Ma, Yong Wang, and Weiliang Zhao. "Base station sleeping with dynamical clustering strategy of CoMP in LTE-Advanced," in *Proc. IEEE International Conference on Green Computing and Communications*, p. 157–162, Aug. 2013.

[27]   An Liu and Vincent Lau. "Exploiting base station caching in MIMO cellular networks: opportunistic cooperation for video streaming," *IEEE Transactions on Signal Processing*, 63(1):57–69, 2015.

[28]   Jingxin Liu and Dongming Wang. "An improved dynamic clustering algorithm for multi-user distributed antenna system," in *International Conference on Wireless Communications & Signal Processing, WCSP*, p. 1–5. IEEE, Nanjing, China, 2009.

[29]   David López-Pérez, Ming Ding, Holger Claussen, and Amir H Jafari. "Towards 1 Gbps/UE in cellular systems: understanding ultra-dense small cell deployments," *IEEE Communications Surveys & Tutorials*, 17(4):2078–2101, 2015.

[30]   Patrick Marsch and Gerhard Fettweis. "Static clustering for cooperative multi-point (CoMP) in mobile communications," in *Proceedings of IEEE International Conference on Communications*, p. 1–6. IEEE, Kyoto, Japan, 2011.

[31]   Abdelrahim Mohamed, Oluwakayode Onireti, Muhammad Imran, Ali Imran, and Rahim Tafazolli. "Control-data separation architecture for cellular radio access networks: a survey and outlook," *IEEE Communications Surveys & Tutorials*, 18(1):446–465, 2015.

[32]   Hafiz Atta Ul Mustafa, Muhammad Ali Imran, Muhammad Zeeshan Shakir, Ali Imran, and Rahim Tafazolli. "Separation framework: an enabler for cooperative and D2D communication for future 5G networks," *IEEE Communications Surveys & Tutorials*, 18(1):419–445, 2016.

[33]   Zhisheng Niu, Sheng Zhou, Yao Hua, Qian Zhang, and Dongxu Cao. "Energy-aware network planning for wireless cellular system with inter-cell cooperation," *IEEE Transactions on Wireless Communications*, 11(4):1412–1423, Apr. 2012.

[34] 3GPP TSG-RAN WG1 Meeting no:64. R1-110651, Deployment and Back-haul Constraints for CoMP Evaluations. Technical Report, 3rd Generation Partnership Project (3GPP), 2011.

[35] 3GPP TSG-RAN WG1 Meeting no:64. R1-111174, Backhaul Modelling for CoMP. Technical Report, 3rd Generation Partnership Project (3GPP), 2011.

[36] 3GPP TSG-RAN WG1 Meeting no:65. R1-111628, Phase 1 CoMP Simulation Evaluation Results and Analysis for Full Buffer. Technical Report, 3rd Generation Partnership Project (3GPP), 2011.

[37] P. Marsch and G. Fettweis. *Coordinated Multi-Point in Mobile Communications. From Theory to Practice.* Cambridge University Press, 1st edition, 2011.

[38] Agisilaos Papadogiannis, David Gesbert, and Eric Hardouin. "A dynamic clustering approach in wireless networks with multi-cell cooperative processing," in *International Conference on Communications, 2008. ICC'08*, p. 4033–4037. IEEE, Beijing, China, 2008.

[39] Yinan Qi, Muhammad Z Shakir, Muhammad A Imran, Atta Quddus, and Rahim Tafazolli. "How to solve the fronthaul traffic congestion problem in H-CRAN? in *International Conference on Communications Workshops (ICC)*, p. 240–245. IEEE, Kuala Lumpur, Malaysia, 2016.

[40] Yinan Qi, Muhammad Zeeshan Shakir, Muhammad Ali Imran, Khalid A Qaraqe, Atta Quddus, and Rahim Tafazolli. "Fronthaul data compression for uplink CoMP in cloud radio access network (C-RAN)," *Transactions on Emerging Telecommunications Technologies*, 27(10):1409–1425, 2016.

[41] Jaya B. Rao and Abraham O. Fapojuwo. "A survey of energy efficient resource management techniques for multicell cellular networks," *IEEE Communications Surveys & Tutorials*, 16(1):154–180, 2014.

[42] Cheng-Xiang Wang, Fourat Haider, Xiqi Gao, *et al.* "Cellular architecture and key technologies for 5G wireless communication networks," *IEEE Communications Magazine*, 52(2):122–130, 2014.

[43] Jian Zhao and Zhongding Lei. "Clustering methods for base station cooperation," in *Wireless Communications and Networking Conference (WCNC)*, p. 946–951. IEEE, Shanghai, China, 2012.

Index

Printed in the USA
CPSIA information can be obtained
at www.ICGtesting.com
JSHW011507221024
72173JS00005B/1229